Product and Process Modelling

Product and Process Modelling

A Case Study Approach

Ian Cameron
School of Chemical Engineering
The University of Queensland
Australia

Rafiqul Gani
Department of Chemical & Biochemical Engineering
Technical University of Denmark
Denmark

ELSEVIER
AMSTERDAM • BOSTON • HEIDELBERG • LONDON • NEW YORK • OXFORD
PARIS • SAN DIEGO • SAN FRANCISCO • SINGAPORE • SYDNEY • TOKYO

Elsevier
The Boulevard, Langford Lane, Kidlington, Oxford OX5 1GB, UK
Radarweg 29, PO Box 211, 1000 AE Amsterdam, The Netherlands

First edition 2011

Library of Congress Cataloging-in-Publication Data
A catalog record for this book is available from the Library of Congress

British Library Cataloguing in Publication Data
A catalogue record for this book is available from the British Library

ISBN: 978-0-444-53161-2

For information on all Elsevier publications
visit our web site at www.elsevierdirect.com

Contents

Contributors

In addition to the authors (Cameron and Gani), the following persons contributed to the chapters-sections listed below.

Author	Address	Chapter/ section	Title
Dr. Ricardo Morales-Rodriguez	CAPEC, Department of Chemical and Biochemical Engineering, Building 229, Technical University of Denmark, DK-2800 Lyngby, Denmark	7.2	Complex integrated operation: Direct methanol fuel cell
		12.1	Microcapsule-based controlled release
Martina Heitzig	CAPEC, Department of Chemical and Biochemical Engineering, Building 229, Technical University of Denmark, DK-2800 Lyngby, Denmark	5.3	Evaporation from a droplet
		7.3	Multiscale fluidised bed reactor
		11.6	Parameter regression – dynamic optimisation
Dr. Chiara Piccolo	CAPEC, Department of Chemical and Biochemical Engineering, Technical University of Denmark, Building 229, DK-2800 Lyngby, Denmark	5.2.3	Activity coefficient model -2: Elec-NRTL
Dr. Ravendra Singh	PROCESS, Department of Chemical and Biochemical Engineering, Building 229, Technical University of Denmark, DK-2800 Lyngby, Denmark	12.2	Fermentation process modelling
		12.3	Milk pasteurisation modelling
		12.4	Milling process model
		12.5	Granulation process model
		12.6	Pharmaceutical tablet pressing process modelling
Noor Asma Fazli Abdul Samad	CAPEC, Department of Chemical and Biochemical Engineering, Building 229, Technical University of Denmark, DK-2800 Lyngby, Denmark	10.1	Batch cooling crystallisation modelling (population balance modelling)

(continued)

Author	Address	Chapter/ section	Title
Prof. Maurico Sales-Cruz	Departamento de Procesos y Tecnología, División de Ciencias Naturales e Ingeniería, Universidad Autónoma Metropolitana – Cuajimalpa Artificios No, 40. Segundo Piso, Col. Hidalgo, Del. Álvaro Obregón, 01120, México D.F.	7.4	Dynamic chemical reactor
		7.5	Dynamic polymerisation reactor
		8.3	Short-path evaporation model
		9.1 – 9.2	Tennessee Eastman Challenge Problem
		11.1 – 11.5	Model Identification (Parameter Estimation)
		14.2	ICAS-MoT model library

Preface

This is a case study book on product and process modelling. It has arisen from over 30 years of working in process system applications that required models to be conceptualised, formulated, built and deployed for a range of uses. Modelling is ultimately goal-driven and we are particularly mindful that for modelling "one size does not fit all" applications. We need to be cognizant of the end-goal as we carry out the modelling activity. One thing that has been reinforced over those 30 years is that modelling is a structured activity – we need a modelling methodology to be effective in model building and deployment. Through these case studies you will see that methodology is being exercised across a wide range of modelling applications. Case studies can be a powerful means of understanding how models are formulated, built and used, so we hope this understanding might be enhanced through these examples.

The book is essentially practical in nature, as there are several well-known books that deal with the theory and methodologies that underpin modelling practice. Our intention is not to repeat those ideas but give practical expression to those principles.

To this end, we give a brief coverage of the use and nature of modelling to help set the context of what follows. This includes the practical aspects that are important in modelling and some reflections on the burgeoning availability of modelling tools, many of which can help conceptualise models, provide expert analysis of the properties of the model as well as provide efficient solution of those models. We are now accustomed to seeing large-scale commercial models solved as part of real-time applications in the industry.

The following chapters provide insights into such areas as constitutive models as well as models for steady-state and dynamic applications. We also consider aspects of lumped parameter and distributed parameter modelling via industrial applications drawn from a wide range of industries. There are examples around specific industry sectors which are of current interest. The case studies illustrate the ubiquitous nature of modelling applications and show, in a measured way, just what can be derived from such activities.

In concluding the case studies, we make a number of reflections around the future of product and process modelling – a dangerous activity, given how quickly developments and innovations take place! Not only the benefits from modelling need to be appreciated by practitioners but a real effort needs to be made by practitioners to more effectively communicate the benefits of appropriate modelling and deployment across the product and process life cycle. Much still needs to be done.

Of course this book is the product of many people, not just the principal authors. We are very thankful to the large number of research students who have worked with us over many years in developing methodologies, exercising the principles and applying the resultant modelling to a range of applications. In this matter, we also acknowledge the many industrial supporters of work leading to model development and application. This has been through collaborative research programmes and industrial consulting activities. We thank those who have given permission to present some of this work.

Notably, we would also thank many close colleagues in our academic institutions as well as many in the process systems engineering community worldwide from whom we have benefitted from insights, concepts and practice.

We are hopeful that these case studies will be helpful to many who are in the various stages of the modelling experience – from the novice through experienced practitioner to the expert.

Ian Cameron and Rafiqul Gani

Chapter 1

Modelling: Nature and Use

This book deals with a broad coverage of modelling concepts and applications for the development of products and processes. It presents these ideas through case studies, showing the importance of modelling to a wide spectrum of discovery activities and decision making processes, that ultimately bring new insights and ideas to reality for social, industrial and economic applications.

There are numerous books and papers available that deal with specific issues in the conceptualisation, development, solution and deployment of models, for applications in many domains of research, business, manufacturing and production. Models are ubiquitous—from those embedded as fuzzy model applications in washing machines through medical applications in anaesthetics, to models that attempt to predict climate variability. Models are routinely used in all areas of human endeavour and appear in various types and forms.

This book presents a number of applications areas, model forms and types, giving insights into some of the important concepts behind modelling, and the drivers for using models in various applications.

1.1. MODELLING FUNDAMENTALS

Modelling in process and product engineering has a long history. It is now clear that much of the design and discovery processes in product and process engineering, are facilitated by various forms of modelling. The decision making processes that are interwoven into the whole of the life cycle, are heavily influenced by modelling, simulation and visualisation.

Modelling, in its minimal form, is the representation of a real or virtual physico-chemical, economic, social or human situation, in an alternate mathematical or physical form, for an envisaged purpose.

This simple definition has three important concepts: an identified system (S) which is the subject of interest, an intended purpose (P) for the model in terms of decision making, and a representational form (M) of the model. The key concepts have associated with them a wide range of issues, relating to details of the system, purpose and form. An extended discussion of such issues can be found elsewhere, as it is not the purpose in this book to dwell on those matters (Aris 1999, Hangos & Cameron 2001).

However, the following sections sketch out some of the key ideas that underpin what is presented in the case studies that appear in subsequent chapters.

1.1.1. Systems Perspectives for Model Development

We need systematic ways of tackling modelling problems. Of particular importance is the application of systems perspectives to modelling activities.

Figure 1 shows a typical systems framework used for conceptualisation of a model. All modelling applications can be cast into such a framework, allowing the modeller to formally describe the system under study. It is equally applicable to the modelling of a complex reactor, as to the interaction model of a human operator with technology, or to the model of active ingredient take-up of sprayed agrochemicals onto a leaf surface. As such, the formalism has a significant descriptive power. The challenge for the modeller is in understanding the individual aspects of the system, which require modelling. Modelling requires significant insight into the system under study.

In Figure 1 a general system (S) with its boundary as the box is given. The states of the system plus inputs, outputs and disturbances, are shown.

The following meaning can be given to the components of the system–

- ***Boundary***: This provides the limits of the model consideration. It is vital to restrict the modelling through a clearly defined boundary. Modelling the behaviour of a single component in a process operation will use a different boundary, to that for a complete production unit.
- Boundaries can also be hierarchical in nature, in that boundaries for lower level detail can be agglomerated into higher levels of view. Likewise, the decomposition of higher level views into finer detail can be achieved as the modelling goal changes.
- ***System:*** The principal entities within the boundary, and their interconnections, including the primary mechanisms operating in that system. The entities can be pieces of processing equipment, phases, people, particles etc.

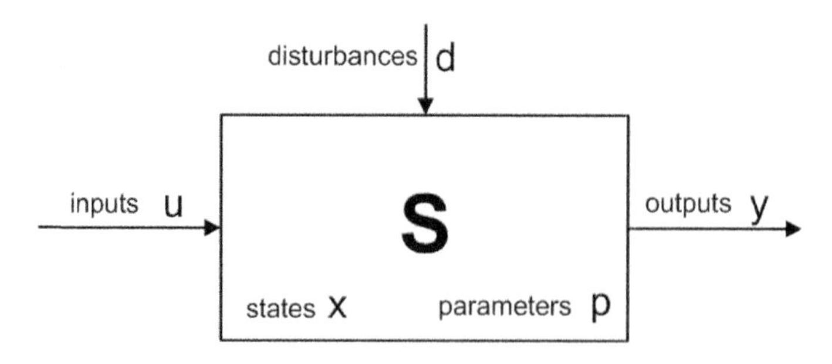

FIGURE 1 **A systems perspective for modelling purposes.**

- ***States***: These are the variables (**x**) which indicate the 'state' of the system, at a point in time and space. They characterise the important properties of the systems. They are often associated with variables that represent the amount of mass, energy or momentum in a system.
- ***Inputs***: They refer to those variables (**u**) which are associated with the properties of the system that can be chosen to affect the behaviour of the system. They are usually known. They can be flows, temperatures, money, training etc.
- ***Outputs***: They refer to the variables (**y**) that reflect internal properties of the system. They are often linked to the states of the system. They could be production rates, quality measures, target temperatures, efficiencies or any other performance measure of interest.
- ***Disturbances***: They refer to the variables (**d**) that reflect those effects on the system that are normally uncontrolled. They can in some circumstances be measured. These variables could be ambient conditions, raw material quality changes or performance shaping factors affecting people.
- ***Parameters***: They refer to the variables (**p**) that are associated with constants, geometric, physical or chemical properties within the system. In some cases, they can be functions of the states.

The immediate challenge for modelling is to take the real or envisaged situation, and decide what aspects of that reality must be associated with the general systems concepts, so as to answer important questions about the real-world application. This is a 'model conceptualisation' stage.

The action of 'modelling' seeks to replace the real system **S** with a model **M** of sufficient fidelity that will help answer questions about the original system. The required fidelity is often difficult to specify *a priori*, hence leading to iterative activity in the model-building life cycle.

Model conceptualisation involves the decision, about what are the boundaries of the system? How much needs to be captured in the model, in terms of the extent of the system? What are the key mechanisms, and what are the associated system states? What variables will be considered as the inputs, and what will be the outputs that will be an indication of the important performance measures? Can the disturbances be identified, and if so, could they be measured?

These are the initial issues, around capturing the important ideas of the real-world system. However, they are also affected by the modelling goal. Why is this modelling being created, and what will the goal mean for the fidelity and complexity of the model that is needed for a particular application, or range of applications?

1.1.2. Modelling Goals in Applications

Models are developed for a purpose and as suggested in Figure 2, achieving the modelling goal requires a system description, clearly defined application area and the model that represents the real-world phenomena, in sufficient

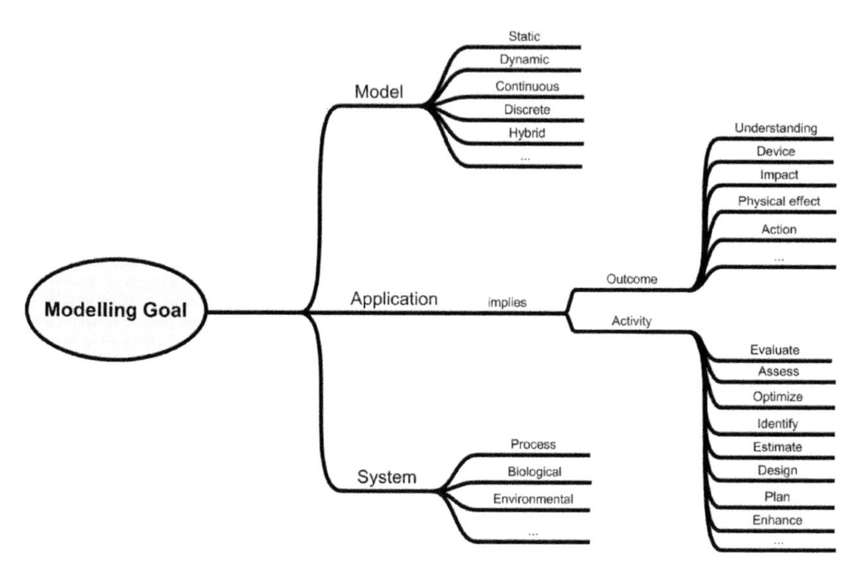

FIGURE 2 Modelling goal components.

fidelity to enable useful questions to be answered. This key focus on the achievable model end-point could be termed 'goal oriented' or 'teleological' modelling.

It was George Box, the famous British statistician and time series modeller who stated that, '. . . all models are wrong, but some are useful . . .'—an excellent truism to bear in mind when building efficient and effective models of real-world systems. There is ultimately an efficiency-thoroughness trade-off that operates when modelling for a purpose (Hollnagel 2009). We need to make decisions about the degree to which we capture system phenomena into the model, and this requires a trade-off between time and ability to provide answers useful in the application of the model. 'Over-kill' in model complexity, is an ever-present challenge for the modeller.

Figure 2 shows that models can have various attributes related to their form and function. This reflects the nature of the underlying phenomena, captured in the model form. The form can reflect the fact that the system is essentially continuous in time and space, or possibly represented by discrete entities such as particles or functions, through discrete actions. The system modelled can be taken from a wide variety of domains in the natural or engineered world, including process, product, economic, biological, human etc.

The model application can be regarded as the combination of a desired *outcome,* and an *activity,* that drives the reason for the modelling in the first instance. Those *outcomes* can relate to *understanding* of the role of complex interactions, in *determining* and *manipulating* behaviour. Hence the activity is clearly an action being carried out, through using the model (evaluate, assess, optimise etc.).

TABLE 1 Model Types and Characteristics

Type of model	Criterion of classification
Mechanistic	based on mechanisms/underlying phenomena
Empirical	based on input-output data, trials or experiments
Stochastic	contains model elements that are probabilistic in nature
Deterministic	based on cause-effect phenomena
Lumped parameter	dependent variables not a function of spatial position
Distributed parameter	dependent variables are functions of spatial position
Linear	superposition principle applies
Nonlinear	superposition principle does not apply
Continuous	dependent variables defined over continuous space-time
Discrete	only defined for discrete values of time and/or space
Hybrid	capturing both continuous and discrete behaviour in the one description

Table 1 gives some further insights into the types and characteristics of models that are often important, in product and process modelling. Section 1.3.2 gives further details of these model types, in the form of a simple classification scheme.

The model types reflect our understanding of the underlying attributes of the system being modelled. All can be observed in the voluminous literature on this topic.

In understanding the modelling goals, we could also classify the major types of generic problems that modelling seeks to answer. The following section sets out a number of such generic problems, commonly addressed by modelling practice.

1.1.3. Typical Problems Addressed by Systems Modelling

The importance of the systems formalism is seen in the various problems that can be tackled, using this generic view. By posing a set of known variables, and leaving others to be estimated, a wide range of important problems are amenable to the use of modelling. These modelling problems include:

1. ***Steady state analysis and simulation***: Given the model of the system **S**, and a fixed operating state $\mathbf{x}_{ss}$ compute the outputs **y**, knowing the inputs **u,** the disturbances **d** and the system parameters **p**. Here the system is regarded as being in 'steady state', or at a particular operating point. Time varying or dynamic behaviour is not being considered here. These types of problems could be identified with standard process flowsheeting applications, or for the use of steady state models in control applications.
2. ***Dynamic analysis and simulation***: Given a model structure for **S**, predict the outputs **y,** knowing the time varying behaviour of the inputs **u**, disturbances **d** and the parameters **p**. This is similar to the previous problem,

except that time varying behaviour is assumed. This type of problem is often focused on assessing the effects of input or disturbance changes, on the outputs of the system. In many cases the combination of steady-state and dynamic models provides profound insights, and even unexpected behaviours, in complex systems.

3. ***The design problem***: In its simplest form: estimate the set of parameters **p**, for a given fixed structure of **S**, desired outputs **y** and specified inputs **u**. Here the situation can be either dynamic or steady state. This problem seeks to find, for example, the size of equipment to give a desired behaviour. There are more complex design problems that require **S** to be found within synthesis problems.
4. ***The optimisation problem***: Estimate the optimum values of the states **x,** for a given objective function F_{obj} involving the states, parameters and inputs. This is a very common application where the 'best' operating point to maximize or minimize some objective, is sought. Unit optimisers in petroleum refineries are a well-known example of such modelling practices.
5. ***Regulatory control or state driving applications***: Estimate the input **u** for a given **S**, **y**, **d** and **p**. This is a standard control issue to obtain the values of the inputs, needed to maintain the system at some specified operating point, or to drive the system from one operating point to another, such as done in batch polymerisation reactors.
6. ***System identification***: Find a structure for the system **S,** with its parameters **p,** using inputs **u** and outputs **y**. This is often done to generate a model to be used for control applications, where the resultant model is embedded into a control algorithm, such as model predictive control (MPC).
7. ***State estimation problem***: Find the internal states **x** of the system **S,** knowing inputs **u** and outputs **y**. This problem is often addressed when there is no direct way to measure the internal state of a system. Through the use of a model and known input and output data, estimates can be obtained using such approaches as Kalman filters.

A number of such applications will be seen, in the case studies presented in subsequent chapters.

1.2. MODEL USE AND DECISION MAKING

Models are widely used in capturing, understanding, investigating and exploiting the aspects of a real-world situation. In the area of product design and development, as well as the invention of processes to make those products, the use of models as decision informing and making tools abound.

1.2.1. Life Cycle Perspectives on Modelling

The extensive nature of model use can be viewed in many ways. A helpful representation that emphasises the breadth and depth of modelling, is seen in

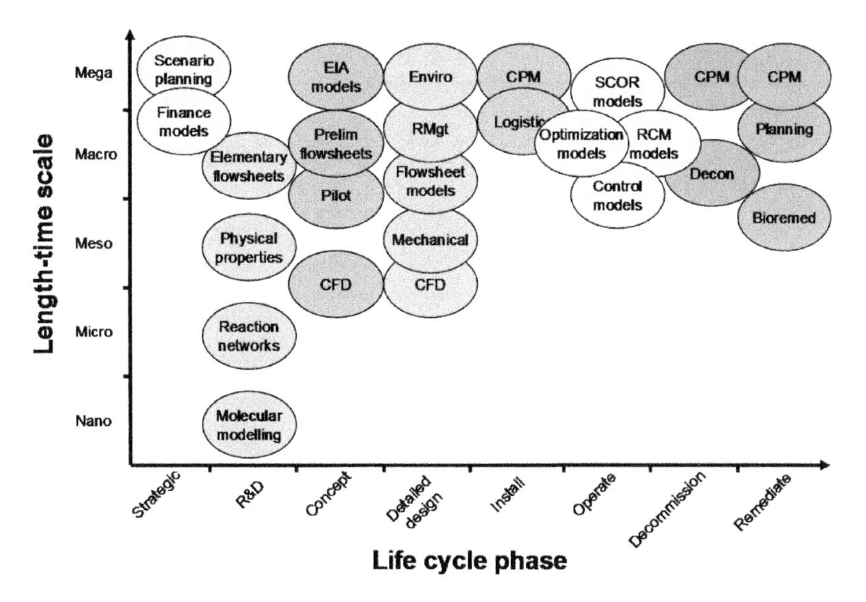

FIGURE 3 **Modelling across the life cycle and scale domains.**

Figure 3, which gives a selection of model applications across the product-process life cycle, emphasising not only the individual life cycle phases, but also the relevant scales of the system being addressed. This emphasises the multi-scale, multi-form nature of current modelling practice.

What becomes evident from Figure 3 is the very wide spread nature of product and process modelling, across all life cycle phases. Scenario planning and economic assessment models are used for long term strategic planning of companies and organizations. Research and development employs a wide range of models, potentially applied to understanding the fundamental aspects of the products, their properties and the processes that will produce them. In conceptual and detailed design, many decisions are informed through model use, whilst the rest of the life cycle phases use numerous models, for decision making purposes.

A major development over the last 20 years has been the ever increasing scale expansion, whereby modelling is addressing quantum and nanoscales, to mega or global scale issues related to supply chains and global climate variability. It is now a 'model-centric' world of product and process engineering. Table 2 gives a summary of the life cycle phases, some model applications, forms and approaches.

Figure 3 also shows the scale focus of concern, often associated with specific life cycle phases. Some phases are focussed on large-scale issues, whereas other phases reach deep into the lower scales, to resolve uncertainties that might be

TABLE 2 A Summary of Life Cycle Phase Modelling Characteristics

Process life cycle phase	Modelling applications foci	Modelling approaches
Strategic planning	Market potential. Resource assessment.	Purpose, goal, mission models. Issue based planning. Scenario models. Self-organising models.
Research & Development	Resource characterization. Basic chemistry. Reaction kinetics. Catalyst activity behaviour. Physico-chemical behaviour. Pilot scale design and operation.	Reaction systems models. Catalyst deactivation models. PFR, CSTR reactor models. Elementary flowsheet models. Fluid-phase equilibria models. Physical property models. Quantum chemistry and molecular simulation
Initial process feasibility	General mass & energy balances Alternate reaction routes Alternate process routes Input output economic analysis Preliminary risk assessment	Flowsheeting models. Semiquantitative risk models Financial analysis models
Conceptual design	Mass & energy balances. Plant or site water balances. Initial environmental impact. Detailed risk assessment. Economic modelling.	Flowsheeting models. Environmental impact models Social impact assessment models. Technical risk models. Computational fluid dynamics Discrete event modelling (DEM)
Detailed design	Detailed mass & energy balances. Vessel design and specifications. Sociotechnical risk assessment. Risk management strategies. Project management.	Flowsheeting models. Unit dynamic simulation models . CFD modelling and simulation. Mechanical simulation (FEM) 3D plant layout models. Fire and explosion models. Fault tree and event tree models Airshed models for dispersion of gases and particulates. Noise models.
Commissioning	Start-up procedures. Shutdown procedures. Emergency response Critical path analysis	Ladder logic and Grafcet models. Safety instrumented assessment models. Risk assessment models. Petri-Net/Critical path models.

(continued)

TABLE 2 *(continued)*

Process life cycle phase	Modelling applications foci	Modelling approaches
Operations	Process optimization. Process batch scheduling. Supply chain design and optimization.	Scheduling models. Unit and plant wide optimization models (LP, NLP, MILP, MINLP). Queuing models. Real-time expert system models. Artificial neural nets and variants. Empirical models (ARMAX, BJ). Maintenance models (CPN, RBM).
Retrofit	Constraint and debottle-necking studies. Plant redesign.	Flowsheeting models. Detailed dynamic simulation. CFD.
Decommissioning	Disposal processes/ strategies. Decontamination of equipment, approaches and optimal policies.	Specialised models for the processes involved.
Remediation and restoration	Geotechnical. Contaminant extraction options.	3D physical extraction pilot plants. Soil processing models for decontamination.

important in product design and process development. There is, of course, in life cycle phases such as 'operate' often a necessity to do further lower scale investigations when production or product problems persist, due to the impact of unexpected and unknown phenomena that affect outputs.

1.2.2. Global Perspectives on Model Development and Use

Some global perspectives on industrial modelling were captured in a survey by Cameron & Ingram (2008), of industrial model use and practice. An earlier survey by Foss and co-workers (Foss *et al.* 1998), also addressed aspects of industrial modelling. The key messages from these surveys were that:

- The principal life cycle phases where modelling dominated were in R&D, conceptual design, detailed design and operations
- The majority of those carrying out modelling activities, had between 5 and 10 years modelling experience

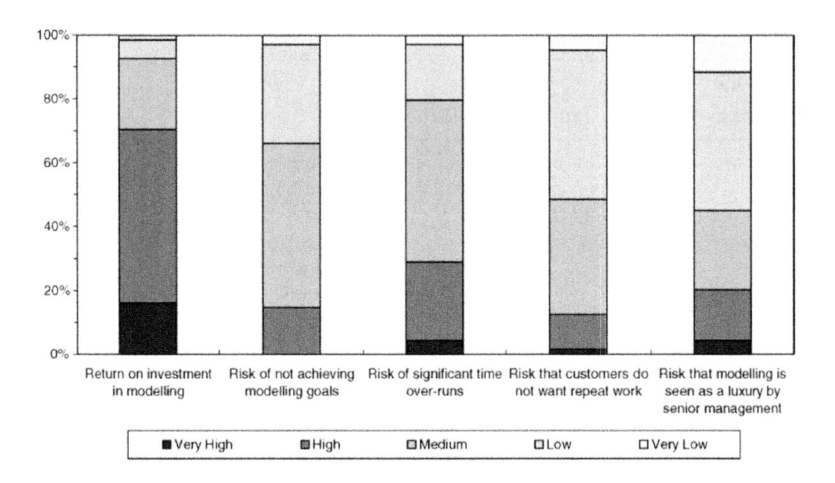

FIGURE 4 Value and risks in industrial modelling.

- The value of modelling and the risk associated with that activity showed that the return on investment was very worthwhile, as seen in Figure 4
- Despite the returns, the risk in achieving the goals of modelling were moderately high
- The biggest factors against the increased use of modelling in organizations were: cost in time and tools, tool complexity plus lack of awareness of benefits and expertise to do the work
- The post-mortem reviews of modelling activities were frequent when the outcomes were poor, but little interest was shown in learning, what went well when outcomes were good
- Modelling goals frequently changed during model development and
- Documentation of modelling assumptions and changes, were not generally easy to make for current tools and methodologies

The information in these surveys clearly shows that industry values the important role that modelling plays in scientific and economic terms. Yet, this value is certainly not without its problems and challenges, as set out more fully, in the key references.

1.3. MULTISCALE, MULTIFORM NATURE OF PRODUCT-PROCESS MODELLING

1.3.1. Multiscale Phenomena and its Impact on Modelling

A growing phenomenon in modelling product and process issues is that of multiscale modelling. In many applications, it becomes crucial to incorporate into the model a range of length scales, and the phenomena operating over those

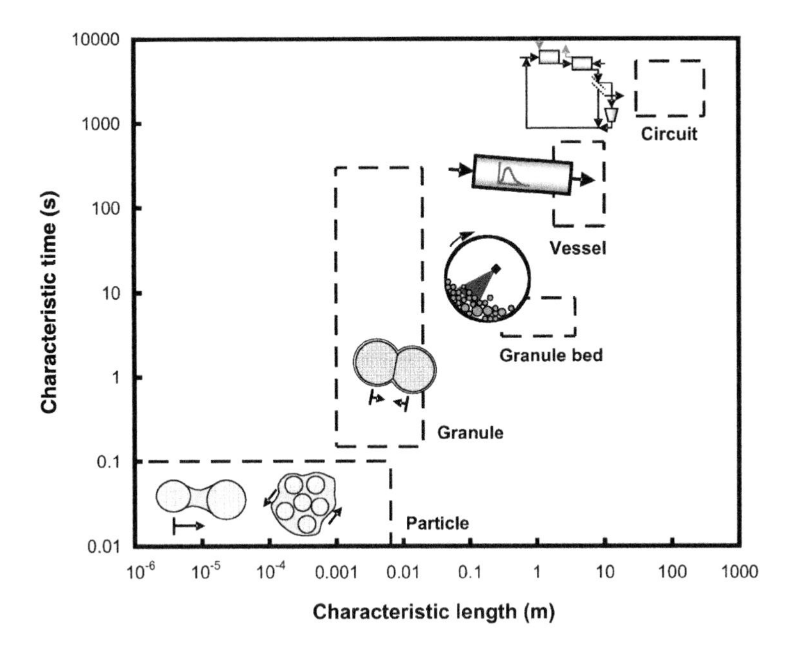

FIGURE 5 **Scale map of granulation processes (Ingram 2005).**

length scales. The focus on complex multiscale phenomena and the capture of this in models is one of the most significant developments in modelling methodology in the last 25 years. Much continues to be done in this area around the conceptual frameworks, tools and applications (Villermaux 1996, Pantelides 2001, Li and Kwauk 2003, Ingram *et al.* 2004).

Along with the idea of multiscale modelling is also the fact that at any particular scale, there is often a multitude of model forms or types that can be used to capture the important phenomena. The choice regarding the scales to incorporate and the model form at each scale is challenging, without the consideration of how the sub-models at each scale are to be connected.

Figure 5 illustrates part of the challenges in multiscale modelling, as it relates to industrial granulation processes.

This is typical of many multiscale applications. Here, a scale map from work by Ingram (2005) shows phenomena at various time-length scale regions, with the dotted boxes showing the typical extent of the time-length phenomena, operating at each level.

At the smallest level of concern are the phenomena of nucleation, particle formation, coalescence and consolidation. Beyond the micro-scale, there are important aspects of granule growth to consider, and different challenges at the meso-scale granulation bed length, where a range of phenomena such as granule motion, mixing and segregation, play a role in granule growth. Finally, there are other important dynamics at the complete circuit scale, where control stabilization and production performance, play a major role.

Characteristic lengths span 10 orders of magnitude, whilst characteristic time extends over 7 orders of magnitude. These levels of length-time coverage are not unusual in most multiscale applications. Handling the multiscale model development is the real challenge.

These multiscale challenges can be summarised, by the following important issues:

- Deciding which length scales are appropriate, for a specific application or model
- Developing or selecting appropriate models, at the scales of interest and
- Choosing suitable frameworks, to link or integrate the partial models

In many applications the choice of length scales is often dictated by–

- The key physico-chemical phenomena, related to atoms, molecules, particles, thin films, etc.
- Phases, that exist in the system
- Process equipment and unit operations, that are considered
- Complete plant analysis
- Company sites for integrated studies and
- Business enterprise considerations, across national and multinational operations

The integration of sub-models across the scales is also non-trivial, with a range of integration frameworks possible. The choice can be straight forward, but when multiple options are present, final choices are often not clear. This can be a vexing issue. Figure 6 shows a recent classification of multiscale integration

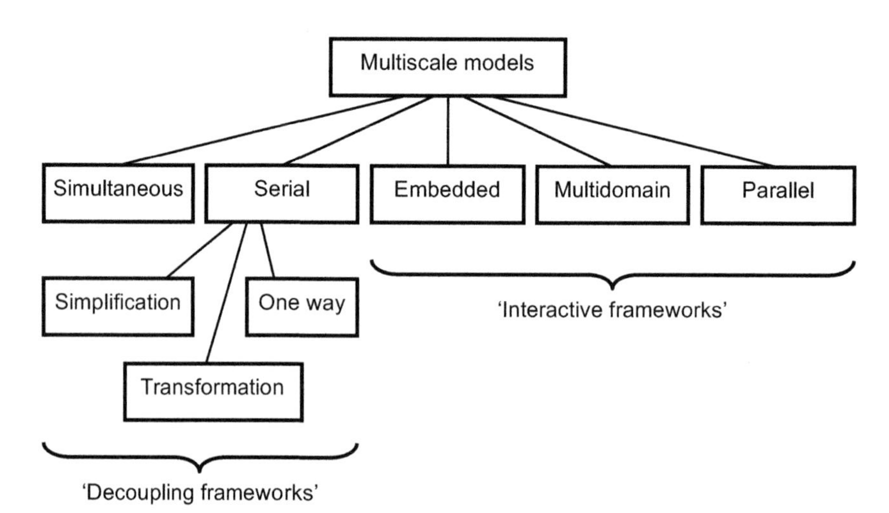

FIGURE 6 Multi-scale integration frameworks for modelling (Ingram *et al.* 2006).

frameworks that describe most application areas. Further details, of the nature of those frameworks, and the particular information flows between sub-models, give insights into the potential complexity of the framework, and the computational challenges (Ingram et al. 2006).

Over the last 15 years there has been a growing academic interest in multiscale modelling but this interest is not necessarily reflected in industry practice where uptake has been much slower (Cameron and Ingram 2008). There still remain a significant number of hurdles to effective and efficient multiscale modelling, not least, being excellent tools to facilitate model development, at individual scales and most importantly integration tools, for easy computation.

1.3.2. Multiform Nature of Modelling

The previous section, as well as Section 1.1.2 indicated that along with issues of scale in product and process modelling, there is also the idea of modelling type and form, within the scales. Figure 7 gives a simple taxonomy of some model

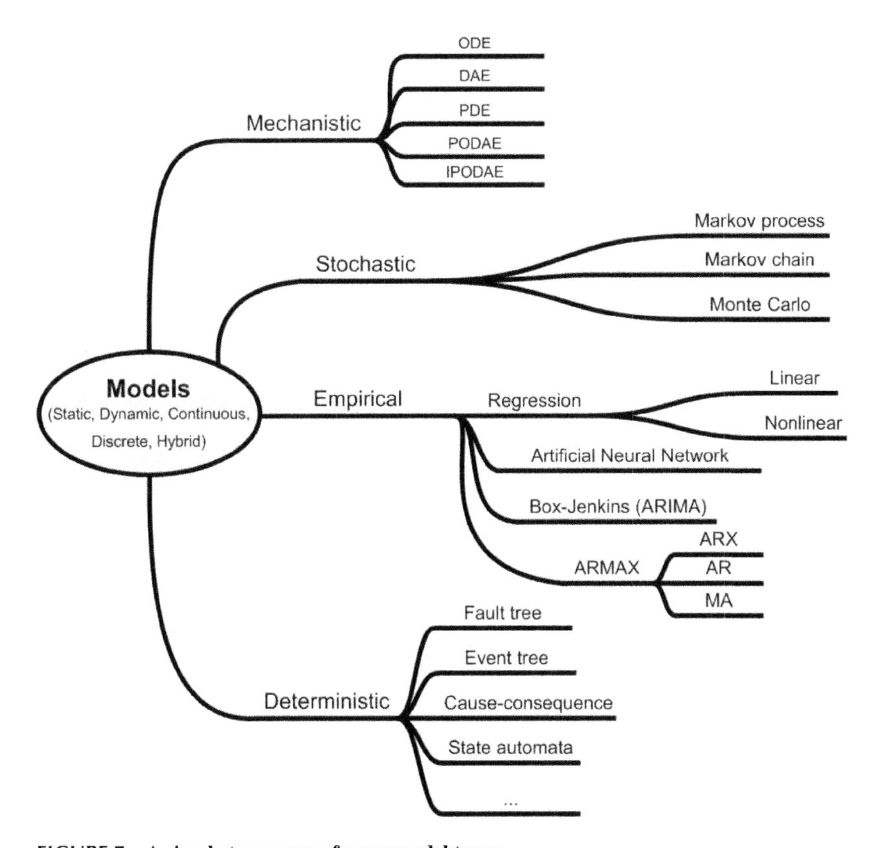

FIGURE 7 A simple taxonomy of some model types.

types, centred on the main classes of: mechanistic, stochastic, empirical and deterministic models. These categories cover most of the models developed in the product-process space. The taxonomy is not a comprehensive coverage of models types, nor unique, as some models can co-exist under different categories. It does illustrate the diversity of the models that pervade modern practice.

Many more are available, and new models continue to appear to address specific areas of science and engineering.

Some comments can be made on Figure 7 with regard to the categories:

- *Mechanistic models* capture the underlying mechanisms linked to the fundamental sciences, such as physics, chemistry and biology that are active in the real-world system. Models are often expressed in mathematical terms, as they often arise from the application of fundamental thermodynamics, related to the conservation of mass, energy and momentum.

Conservation balances can also be applied to particle systems, leading to population balance equations (Ramkrishna 2000).

These mathematical models give rise to such sub-classes of dynamic models as ordinary differential equations (ODE), differential-algebraic equations (DAE) and partial differential equations (PDE) and combinations, such as partial-ordinary differential-algebraic equations (PODAE). If integral equations appear in the model, then this can lead to IPODAE systems.

- *Stochastic models* incorporate random or probabilistic elements within the model. Numerous stochastic models exist, the most well-known being the Monte Carlo models. Others, such as the Markov processes are widely used and assume that the future state depends only on the current state
- *Empirical models* form a large and important class of models, widely used in product and process applications. They are based on observations from experiments. These models are typically based on developing relationships between the input and output data of a system that does not consider the underlying mechanisms. They are often called 'black box' models. Where some insights into mechanisms are contained in the relationships, then 'grey box' models can be developed.

The most prominent simple models are regression models that generate linear or nonlinear relationships to describe the input-output data sets. As well, artificial neural networks (ANN) have also been widely used.

Other sub-classes of dynamic models exist that include an auto-regressive (AR) part as well as a moving average (MA) part, in order to provide estimates in time series. The AR part relates to the past values in the data, whilst the MA part relates to the changing meaning of the past values in the data. An 'integrated' (I) aspect is often present, when the data is differenced to generate stationary data sets.

- *Deterministic models* are those where there is a clear use of causality, captured in the model. It assumes that the current state of the system is determined, by the previous states and the parameters. The model should generate identical outcomes for the same initial condition. In Figure 7 some representative deterministic models have been chosen, such as fault and event trees, popular in risk management applications.

Finite state automata (FSA) or finite state machines (FSM) are representations of the states in which a system can reside, and the transitions which move from one to state to another. If the state transition function is deterministic, then the FSM is deterministic. The FSM can also clearly be non-deterministic.

These multiform types that have been briefly discussed show the huge diversity of forms that contribute to the building of models, related to products and processes.

The scale and forms to be used in any model building activity will be determined by the application area, an understanding of the behaviour of the system, and the necessity to incorporate key factors of the system into a model. All this should facilitate the ability of users of the model to answer the major questions for which the model was built.

1.4. SUMMARY

In this opening chapter, we have set out some of the rationale, use and characteristics of modelling, across the product and process life cycle. It is clear that modelling has played, and will continue to play, an ever increasingly important role in decision making for a wide range of application areas.

It is also important that modelling is seen through a 'systems perspective', since this formalism provides a powerful representational, synthesis and analysis framework, that underpins modelling and model use.

This book takes a case study approach in illustrating the application of basic model-building principles to a wide range of application areas, in product and process engineering. As well, these applications have a wide range of characteristics in the underlying problem, as well as illustrating the use of multi-scale and multi-form modelling, in seeking to answer important questions for which the model was created.

Chapter 2 takes some of the concepts introduced in this chapter, and extends them to discussing some of the details related to modelling practice, as a prelude to the case studies. In those case studies, many of the ideas presented here and in Chapter 2 are 'put to work'.

1.5. GLOSSARY

A	Application
ANN	Artificial neural network
AR	Auto-regressive
ARIMA	Auto-regressive integrated moving average

ARMA	Auto-regressive, moving average
ARMAX	Auto-regressive, moving average exogenous
ARX	Auto-regressive exogenous
BJ	Box-Jenkins
CFD	Computational fluid dynamics
CPM	Critical path method
CSTR	Continuous stirred tank reactor
DAE	Differential-algebraic equations
DEM	Discrete element method
EAI	Environmental impact assessment
FEM	Finite element method
FSA	Finite state automata
FSM	Finite state machine
IPODAE	Integral partial ordinary differential-algebraic equations
LP	Linear programming
M	Model
MA	Moving average
MILP	Mixed integer linear programming
MINLP	Mixed integer nonlinear programming
NLP	Nonlinear programming
ODE	Ordinary differential equations
P	Purpose
PDE	Partial differential equations
PFR	Plug flow reactor
PODAE	Partial-ordinary differential-algebraic equations
PODE	Partial-ordinary differential equations
RBM	Risk based maintenance
RCM	Reliability centred maintenance
RM	Risk management
S	System
SCOR	Supply chain operations reference

REFERENCES

Aris, R., 1999. Mathematical Modelling: A Chemical Engineer's Perspective. London: Academic Press.

Cameron, I.T., Ingram, G.D., 2008. A survey of industrial process modelling across the product and process lifecycle. Computers & Chemical Engineering 32 (3), 420–438.

Foss, B.A., Lohmann, B., Marquardt, W., 1998. A field study of the industrial modelling process. Journal of Process Control 8, 325.

Hangos, K.M., Cameron, I.T., 2001. Process modelling and model analysis. London: Academic Press.

Hollnagel, F.E., 2009. The ETTO Principle: the efficiency-thoroughness trade-off. UK: Ashgate Publishing.

Ingram, G.D. 2005. Multiscale modelling and analysis of process systems. PhD Thesis. The University of Queensland, Australia.

Ingram, G.D., Cameron, I.T., Hangos, K.M., 2004. Classification and analysis of integrating frameworks in multiscale modelling. Chemical Engineering Science 59, 2171–2187.

Li, J., Kwauk, M., 2003. Exploring complex systems in chemical engineering – the multiscale methodology. Chemical Engineering Science 58 (3–6), 521–535.

Pantelides, C. C. 2001. Conference Proceedings *ESCAPE-11* (R. Gani and S. B. Jørgensen, Eds.), 15–26. Kolding, Denmark.

Ramkrishna, D., 2000. Population Balances: Theory and Applications to Particulate Systems in Engineering. Academic Press.

Villermaux, J., 1996. New horizons in chemical engineering, 5thWorld Congress on Chemical Engineering. San Diego, July 14.

Chapter 2

Modelling Practice

Modelling practice involves a range of people with specific knowledge areas, practical skills and competencies in application, in order to reach a final, usable model for an end purpose. These modelling practice issues span problem and model identification to model deployment and maintenance. It is also clear from industrial surveys that modelling is usually the domain of modelling experts within the organisation or done in academic research or consulting institutions (Foss *et al.* 1998, Cameron & Ingram 2008).

In this chapter we briefly review those practical issues that involve modelling for product and process applications, noting that there is significant available literature, methodologies and tools to help the modeller achieve their desired goals.

2.1. MODELLING METHODOLOGIES

Modelling practice involves a wide range of tasks, knowledge and skills in order to generate an appropriate model 'fit for purpose'. Those tasks are associated with insights within modelling methodologies or modelling steps, which are given in Figure 1. This particular modelling methodology covers the major tasks that need to be done in completing most of the modelling projects. It is noteworthy that many of these tasks can be, and usually are, repetitive due to various testing or confirmation activities that occur through the model development cycle.

In what follows, we investigate and elucidate those major tasks and some of the details within those tasks, so that readers might be made aware of practical issues that are not immediately obvious from looking at Figure 1.

2.1.1. Problem and model definition

Modelling clearly arises because of a need. Those needs can be very diverse, but are essentially associated in the product and process area with aspects of the life cycle as seen in Figure 3 of chapter 1. A detailed coverage of modelling across the product and process life cycle is given in Cameron & Newell (2008). What becomes clear from closer examination of life cycle modelling is the huge variety of model end-uses, model types and application areas. However, the initial considerations around problem definition and the necessity of

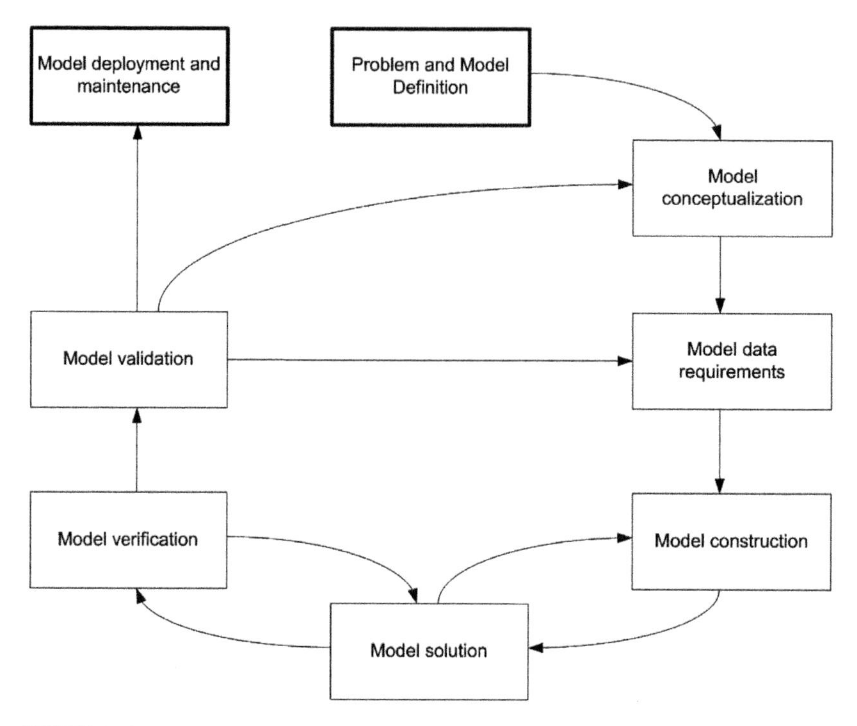

FIGURE 1 Model development framework.

model-based approaches to decision making have common elements despite the life cycle phase.

Of prime importance is the discussion around what goal or goals the modelling should address. This issue often arises from challenges posed by the life cycle phase, around how to resolve particular issues. Those issues have numerous origins such as:

- financial and feasibility assessments;
- generating insights around product and process design;
- optimisation of processes;
- risk management issues;
- operator support systems;
- operator training.

These areas of modelling dominated industrial survey responses in a 2008 study, and indications are that they still remain the top priorities in modelling practice. Of lesser importance were:

- the need to make decisions around commercial strategic planning;
- human factor assessments;
- fulfilling regulatory requirements by government agencies.

Together with addressing the goal issue is also the factor of who will be the intended customer. This can range from internal company scientists, engineers, managers and operations staff who dominate the customer base, to external personnel and agencies. Current modelling practice is mainly focussed on internal rather than external demands.

In goal setting, some of the general discussion questions that need to be asked include:

- What decisions will be addressed through the use of this model?
- Who or what will make those decisions? Will it be a person or will it be some technology in which the model is embedded such as a control system, plant or unit optimiser?
- What principal states, properties or attributes of the application area will be the focus of decision making?
- What is the anticipated uncertainty that can be tolerated in the application of the model or within the decision-making framework?
- Who will take eventual ownership of the model and be responsible for its maintenance?
- What time frame exists for model development?
- What general cost limits will apply?
- What acceptance criteria will apply to the modelling? Under what circumstances will the development cycle stop?

These questions help scope out some general aspects of the goal, which, as mentioned in Section 1.1.2, can be described in broad terms as the combination of a <model>, with an <application> applied to a <system>. At this point in the modelling activity, the general aspects surrounding the modelling activity have normally been discussed and the broader properties have been established.

A further step can generate a specification that includes statements about the required accuracy of prediction, the range of disturbances the model must handle and the nature of those disturbances and, in some cases of real-time application, the maximum execution time required.

2.1.2. Model conceptualisation

Conceiving the model is an extremely important activity in the modelling methodology. In this book we have paid particular attention to the task of initially conceptualising the model for model development. Chapter 8 in particular has some industrial examples of how the models were conceived that are typical of most practice. As well, we do not address black-box modelling in this book. It is clearly a very important area in industrial applications. In the case of empirical modelling different approaches are needed, as the internal mechanisms and controlling factors are not addressed explicitly.

Key issues and questions to be addressed for model conceptualisation include:

- What are the overall boundaries of the system under study?
- Where are the major mass, energy, momentum or population hold-ups in the system?
- Are there clearly identifiable sub-systems, and what are their boundaries and interactions with other sub-systems?
- Can we identify clear sub-system demarcations where physical and chemical phenomena might be distinct?
- What identification of specific phases can be seen in the actual or conceived system? Are there specific solid, liquid and vapour phases that need individual treatment? Are there dispersed phases in continuous phases that should be treated as such or lumped together?
- What dominant time, length and detail scales should be initially considered?
- Is spatial variation of states an important issue or can a lumped model be assumed initially?
- What states are absolutely crucial in the model?
- What constraints can be identified in the system? These might include operational constraints on various states of the system or equipment constraints.

This conceptualisation activity is best carried out through brainstorming processes with a wide range of stakeholders to generate an initial conceptualisation. It is essential that the stakeholders include people very familiar with the process or product, the modelling team, an experienced modeller with skills to decide the authenticity or applicability of positions taken by the stakeholders.

The principle of parsimony should be exercised (Occam's Razor), to ensure that the model is the simplest for the intended application.

It is important that the initial conceptualisation lead to a tractable problem and model description that can be tackled within the time, cost and implementation constraints. It is no use generating a conceptualisation that is essentially a 3 year PhD, when design or operations staff have to make decisions on tighter time frames.

The types of outcomes from such a process might include:

- Sketches of the balance volumes and their interconnections
- Clear inputs and outputs of the overall system and identification of key disturbances that might apply
- An idea of the principal states that are captured in the model
- Identification of dominant mechanisms, such as heat, mass and momentum transfer, phase behaviour and the like, applicable to the identified systems or sub-systems.
- A clearly stated set of organised and documented assumptions that describe how the conceptualisation arises, and also the arguments and positions that

stakeholders took in the adoption of the initial conceptualisation. This is vital in revisiting model concepts during validation activities.

- Clearly stated procedures to address the initial development of the model, through mass balances, energy, momentum and population balance techniques. This might lead to the development of a mass-only balance model initially, followed by the coupling of energy balances to generate the model and further enhancements as required.

The industrial survey of 2008 indicated that despite the development of process or product modelling conceptualisation tools, there was little interest (<20% surveyed) in the use of such tools to aid this stage of modelling.

2.1.3. Model data requirements

In this stage of model development there are numerous data issues to be identified that are of immediate use in the model development, as well as use in the longer-term issue of model validation. In fact, model data can be one of the greatest challenges in producing 'fit-for-purpose' models.

The types of data considerations that need to be resolved for the initial model development can include:

- Thermo-physical properties of the substances within the system, including such aspects as fixed and state-dependent properties. Are key properties in formulating the balance equations and constitutive relations a strong function of temperature, pressure and concentrations? What are the relevant applications or predictive ranges?
- What predictive models for the properties are relevant to the application? Are they readily available or are laboratory or plant investigations required to elucidate the properties?
- Phase equilibria data, from either predictive models or direct experimental look-up tables
- Rate data that might include reaction kinetics, mass and heat transfer coefficients and the sub-models that could be used. This might include appropriate thermodynamic and reaction prediction packages
- Issues around equipment specifications that might include geometric properties such as volumes, surface areas, internal details that affect phase formation and hold-ups in the system, such as flights within rotating devices, or internal stages in separation systems.
- Equipment and control constraints that might operate in the system. Are they available from the process or are they purely predictive or associated with the current state of the design?

There is an extensive thermodynamic literature that covers thermo-physical data in various fields from petroleum and petrochemical applications through to polymer systems, minerals processing applications and food processing. Some areas are clearly better served than others in provision of good-quality data and predictive models. See the following chapters on these thermo-physical data issues.

Another set of data is also necessary and that is for validation purposes. This can be a problematic area in modelling. It is especially the case when a new, potentially unique product or process is being produced. It can be equally challenging when modelling existing processes when process data is needed over a significant range for validation activities.

Associated with this is the frequent need to estimate key parameters within the model. These might be kinetic parameters, heat or mass transfer coefficients and the like. Here, experimental data is required to carry out such parameter estimates, along with parameter confidence intervals. Standard approaches for steady state and dynamic situations apply.

For products and processes that do indeed exist, the issues include:

- Development of data acquisition regimes to extract operational data from processes. This can be a difficult task, especially to generate dynamic data compared with 'steady-state' data.
- In the case of products that are primarily particulates, the challenge of sampling points and sampling methods cannot be underestimated.
- Some sampling for model validation will require invasive techniques, and preparations within the plant might be needed for insertion of specialised measurement devices.
- Access to operating processes to extract steady-state data always requires data handling from distributed control systems (DCS) or programmed logic controllers (PLC) systems, or extraction from data historians such as PI.
- Negotiation with operating divisions to access data or to perform dynamic tests on a system whilst in production can be very challenging, requiring perseverance, mutual benefit and excellent relations with engineering staff, process operators and supervisors. Nothing is ever easy!
- There is a need to apply data reconciliation techniques to ensure one has a consistent set of data that can aid validation purposes; including dealing with missing data, problems of instrument calibration, biases, outliers and the like. There exist a range of excellent written and software resources for such activities (Romagnoli & Sanchez, 2000).
- Multiple sets of data are essential but can be difficult to generate and obtain, particularly for dynamic testing on operating plant. Replication is often very difficult on operating plant.

As can be seen from the discussion here, the access and generation of important model-related data is no trivial matter. There are many sources of data needed and, in some important cases, the need for experimental planning to extract data from operating systems under production conditions. Care is required in data selection and particularly in the appropriate choice of thermo-dynamic predictive models that are fit for purpose. Inappropriate choices can spell disaster for the quality of simulation predictions.

2.1.4. Model construction

Model construction deals with the description in appropriate form of an abstracted description of the system to generate an analogue or model of the real behaviour. There are a significant number of model types as seen in Figure 6, chapter 1. The challenge is to take the understanding from the goal deliberations, combined with the model conceptualisation, and generate a model that can encompass both the conceptualisation and the end-goal uses.

Although Figure 6, chapter 1 includes empirical model types, the emphasis here is on mechanistic models. There is significant literature on modelling empirical model forms (see Ljung, 1999).

In translating the verbal description contained in the conceptualisation stage, conservation principles and accompanying constitutive relations are mainly represented in some mathematical form. In this book, the case studies have been done using integrated conceptualisation, modelling and simulation systems. Examples of these systems are ICAS (CAPEC-DTU, 2011), gPROMS (PSE, 2011), Daesim Dynamics (Daesim 2011) and ASPEN Custom Modeler (ASPEN 2011).

These systems allow the conceptualisation of the system through definition of balance volumes and then the generation or development of the governing equations to describe the dynamic or steady-state behaviour of the system.

However, it is clear that from the surveys already mentioned that about 40 per cent of the time, approaches to building models are linked to the use of MS Excel, commercial flowsheeting packages, MATLAB, computational fluid dynamics (CFD) tools and direct coding. However, there is clear diversity in approaches and tools used for other applications.

Much of the approaches depend on the exact application, and for small, simple models, the ubiquity and cost of MS Excel often dictates the choice. Once the characteristics have more substance such as model size, complexity in model and thermo-physical properties, possessing a distributed parameter nature or hybrid form, more sophisticated construction tools become inevitable. As well, if the model is to be tested and deployed in a real-time environment, then construction, testing and possibly model reduction tools will be essential.

Multiscale modelling poses particular challenges, with few, if any, environments that can handle the construction of such systems to enable integration frameworks to be properly enabled (see Pantelides 2001, Ingram *et al.* 2004).

Most model development in industry occurs in specialised teams who have expertise in model development and access to appropriate tools already mentioned. The majority of the development is done by individuals. Whatever the approach, there is a necessity for well-structured and well-written models accompanied by excellent documentation.

2.1.5. Model solution

The majority of model solutions are via the application of numerical methods. Those methods will vary in relation to the equation sets being solved. These range across:

- Algebraic equation (AEs) systems: representing steady-state behaviour, can be linear or more likely nonlinear in nature.
- Ordinary differential equations (ODEs): representing dynamic behaviour.
- Differential-algebraic equations (DAEs): representing the combination of dynamic behaviour and various non-linear algebraic relations in the model.
- Partial and ordinary differential algebraic equations (PODAEs): where both lumped and distributed parameter systems are modelled.
- Integral PODAEs: where there can be integral terms, such as those from population balances, combined with dynamic, non-linear behaviour models.
- Hybrid systems, where continuous models possess discrete event behaviour. This can be the case in developing start-up and shutdown simulations where discrete events such as valve closures take place, at specific times or under specific state conditions. These require special numerical treatment.
- Stochastic systems, where probabilistic elements exist, require special treatment.

The solution of many of these model types is routinely undertaken by modelling and simulation tools. The solution of AEs, ODEs and DAEs is now routinely performed within standard packages such as flowsheeting packages, the ICAS system, gPROMS (PSE, 2011), ICAS (CAPEC-DTU, 2011ICAS-DTU, 2011) and other such tools. MATLAB (Math works, 2011) provides a suite of numerical methods, which can handle many of the model types. Other more complex models involving PDEs can now be handled through automatic discretisation using finite difference, finite element or polynomial approximation techniques (see Hangos & Cameron, 2001).

One of the biggest challenges is the use of appropriate numerical methods to solve models that contain phenomena of widely separate time constants. These have been termed 'stiff' problems. They require numerical techniques that are implicit in nature. If the underlying phenomenon that is captured in the model has a wide range of natural time constants, then the model will have this property. This can arise from widely varying reaction kinetics, or the presence of slow heat transfer and very fast kinetics. Efficient solution methods can handle these situations.

What is essential prior to solution is the analysis of the equation set. There are several issues that need to be resolved in order to generate solutions for specific circumstances. These include:

- Degrees of freedom analysis, to ensure the set is well-posed, otherwise solution difficulties can be encountered. In this book, we show in many cases the structure of the equation set, given the chosen degrees of freedom.

These facilities allow one to see the effect of choosing different algebraic variables to satisfy the degrees of freedom.

- High-index issues, where certain choices of the variables that satisfy the degrees of freedom might lead to structural issues that make solution extremely difficult. Here we encounter 'high-index' systems, which might require reformulation of the model or application of advanced numerical methods.
- Consistent initial conditions, where the initial choice of differential variables and the algebraic variables should be such that the equation system is satisfied. Many simulation tools provide means for generating these 'consistent' conditions.
- Defining variable bounds, where it is often necessary to limit the range of a specific variable or define bounds on variables that represent physical or chemical limits.

It is important for the user to be aware of these really important issues, since many modelling problems can be understood via these structural and operational concepts. It often leads to some rethinking of the model description and reformulation of the model for effective and efficient solution.

2.1.6. Model verification

The model verification step should be clearly distinguished from model validation. In this book model 'verification' refers to the checking of the model implementation (or equation code) against the conceptual description (or original model equations). Is the model an accurate representation of the ***conceptualisation***? It is essentially a debug activity. There is significant literature on this issue (e.g. Thacker *et al.* 2004).

This is particularly important in the case where new models are being written as opposed to the use of pre-existing models in such tools as flow sheeting packages.

There are concerns on at least two levels:

- Conceptual implementation errors where underlying modelling concepts have not been correctly represented in the coded model.
- Errors of coding, where there are errors in the description, despite the concept being correctly incorporated. These can be incorrect signs, powers, wrong variable use and the like.

To help initially address the conceptual implementation issues, one aspect of checking is to look at asymptotic behaviour of the implemented model and assess whether the general behaviour is in accordance with accepted understanding. This will require running the model under a range of test scenarios to possibly determine long-term (steady-state) behaviour or short-term behaviour for a range of forcing functions on the dynamic models. In the case of batch models, this might mean looking at general patterns of dynamic transitions under varying inputs to check conceptual consistency. These tests can also be done on coded sub-models as they are developed. This approach relates to degeneracy testing in software engineering.

For coding errors and testing, there should be an adoption of modular code and avoidance of monolithic code, so as to enhance debug operations. There are approaches available that use structured one-step analysis of the code through checking line-by-line veracity. Most systems now provide detailed debug facilities that set stop-points and show variable values after each code step. They can help isolate areas of coding error.

Documentation of the code is essential when writing new models, especially when the code is handed to the customer. Lack of comprehensive documentation describing the code intent leads to significant time delays, confusion about code purpose and difficulties in maintainability and maintenance. It is clear from both industrial modelling surveys that documentation is generally regarded as insufficient and that significant improvement needs to be made on this issue.

Debugging code can be difficult and this requires code modules to be executed and tested independently as well as in their final integrated form. If there is significant use of logic structures in the code, then it is essential to check and exercise the logical pathways to ensure correct behaviour.

The message from verification is that coding should be highly organised, modular in approach, well documented with modules tested individually and then in an integrated fashion. Address the conceptual translation issues first, then sort out the coding errors. Fully test the logical structures through a range of forcing functions on the model, as well as a wide range of model parameters. Check coding interfaces with other programs.

2.1.7. Model validation

As previously mentioned, model validation is a distinct activity from model verification, yet clearly linked to it. A model that does not pass the verification test can have serious validation problems.

Validation refers to the question: Is the model a reasonable representation of the ***actual*** system? In answering this question there are a number of factors that have an impact:

- The underlying assumptions that have been made in conceptualising the model. How good are they? Do they serve to address the goals of the modelling?
- The model inputs and disturbance ranges or distributions. What are the expected ranges of these model inputs, under which the model must perform adequately in relation to the actual system?
- The outputs of the model and their relationship to the actual system.
- The internal parameters of the model. Are they derived from real data through parameter estimation, are they accurate, accepted values or 'best'

estimates? What is the output sensitivity in respect of the parameter variations?
- What is regarded as 'reasonable representation' to pass the validation testing? How much error between model outputs and system outputs can be tolerated?

Before model validation should be contemplated, it is important that a set of sensitivity analyses are performed to investigate:

- The sensitivity of model outputs to changes in inputs
- The sensitivity of model outputs to disturbances
- The sensitivity of model outputs to model parameters

These sensitivity tests give vital information on the effect of changes on model predictions from a variety of sources. Importantly, this activity will help identify the most sensitive parameters in the system in determining output changes. The information can inform issues around parameter estimation which might be used to aid validation of the model.

In industrial surveys, many industries had well-established validation procedures. Around 70 per cent of respondents stated that steady-state models were readily validated, but this dropped to 35 per cent for validation of dynamic models. As well, only 40 per cent of customers were aware of the validity range of models produced. This indicates some issues require addressing in the case of dynamic model validation and communication.

In the case where the actual system does not exist, expert opinion is often required for model validation. This is usually done by domain experts familiar with the phenomena and the process systems. The judgements required are often around gross dynamic behaviour as well as asymptotic behaviour for a given range of input functions or disturbance functions. It is important to test the model and its outputs for the full range of anticipated inputs to make judgements on validity.

In the case of data gathered from the actual process, it is clearly important to gather data over as wide an intended operational range as possible that relates to the model use. In some cases this might incorporate a significant range for the intended model application. In other cases it might be dynamic behaviour around a particular operating point. Steady-state models can be statistically validated initially through parameter estimation methods using reconciled data (see Romagnoli & Sanchez 2000 or Narasimhan & Jordache 2000).

In the case of dynamic models with data from actual systems, approaches initially via statistical parameter estimation provide a starting point for validation. Once parameter estimation has been performed, using the parameter sensitivity studies, further testing using other data sets can be done to provide model validation. Again statistical methods can be used to assess the model performance using some form of least squares estimator (Hangos & Cameron, 2001).

The focus on the model performance will be on the key state variables of the system and their dynamic behaviour. It is important to check issues around states representing mass/mole variations, pressures, temperatures, flows, heat and mass fluxes. Anomalies in the behaviour of certain states and associated properties can be clues in resolving further investigations.

Where validation is poor and substantial parameter estimation has been performed on the existing model it is time to return to conceptual and model structural issues by looking carefully at the set of assumptions made in conceptualising the model. Here it is necessary to review the assumptions around:

- Conservation balance volumes: are they complete? Do they truly represent the actual system? Are they over-simplified or too complex?
- Are the interconnections between balance volumes complete? Are certain flows or fluxes missing or incorrectly identified?
- Are there errors within the constitutive relations?
- Where control systems are modelled, are the loop elements correctly modelled and are control parameters correctly set?

These and other questions will often lead to the conceptualisation being enhanced or simplified. The rule is to start with the simplest model that does the job (Occam's Razor or the principle of parsimony). Beyond this is another cycle of the modelling methodology until convergence to a fit-for-purpose model is achieved.

2.1.8. Model deployment and maintenance

Getting the model for the application is just the beginning of the model's life. There can often be a long life for the model through the operational phase of the life cycle. It is here that the original practices in giving birth to the model play an ongoing role in its long-term use. In other cases the model will be used for decision making in early phases of the life cycle and then archived. This can have its problems.

Industry views (75 per cent agreement) suggest that there is significant loss of information and expertise when people leave the organisation due to inadequate archiving of the model, its data, assumptions and associated verification and validation information. The risk of having to repeat previous work is very high (~60 per cent response). All this leads to the need to be organised and to document and comment effectively around the development and deployment of models. Many time and cost pressures mitigate against this, and often modelling tools do not easily facilitate the capture of vital information and knowledge that can be subsequently used by future developers and operations personnel.

In order to avoid significant rework the message:

- Fully document the individual steps of the modelling methodology
- Make sure all assumptions and their justifications are captured

- Ensure any coding is well documented and written in modular form
- Indicate where changes were made to the initial conceptualisation and why
- Provide all relevant data, values and data treatment processes
- Archive validation runs, validation data and model performance.

This should go a long way to addressing model archiving and maintenance issues.

2.2. SUMMARY

This chapter has considered some practical aspects of product and process modelling via commentary on the modelling methodology set out in Figure 1. Model building for an end-goal is a purposeful activity that requires clearly defined steps and specific considerations at each of those steps. There are many challenges at each step and inevitably the modeller or team return to previous steps to refine their approaches based on a range of checks around verification and validation.

It is vital to have a clear view of the end-use of the model and the decisions that will be made via the model. This informs the conceptualisation and subsequent steps to be taken and just what will be regarded as a 'reasonable representation' of the actual system.

It is clear that there are many challenges within the modelling methodology that cut across highly abstract, conceptual thinking through to application of conservation principles in deriving underlying models, to data acquisition and the solution, simulation and validation of the developed model. Often little thought is given to the long-term history of the model if it is embedded in operations. In developing a model it is vital that the underlying assumptions and their justification be clearly stated, and that the model development be accompanied by documentation suitable for others to use, without resorting to complete rework of the model.

The following chapters will set out a series of case studies that illustrate a wide range of applications and the approaches taken to develop relevant models for those purposes.

REFERENCES

ASPEN 2011. Aspen Custom Modeler, http://www.aspentech.com/products/aspen-custom-modeler.aspx accessed February 27.

Cameron, I.T., Ingram, G.D., 2008. A survey of industrial process modelling across the product and process lifecycle. Computers and Chemical Engineering. 32, 420–438.

Cameron, I. T., and Newell, R. B. 2008. 'Modelling in the Process life-cycle, Computer Aided Process & Product Engineering' (L. Puigjaner and G. Heyen, Eds.). Wiley-Verlag, Germany.

CAPEC-DTU 2011. Computer Aided Process Engineering Center, ICAS and Tools, http://www.capec.kt.dtu.dk/Software/ICAS-and-its-Tools/ accessed February 27.

Daesim, 2011. Daesim Dynamics, http://www.daesim.com/software/daesim/dynamics/ accessed February 27.

Foss, B.A., Lohmann, B., Marquardt, W., 1998. A field study of the industrial modelling process. Journal of Process Control. 8, 325.

Hangos, K.M., Cameron, I.T., 2001. 'Process modelling and model analysis.'. London: Academic Press.

Ljung, L., 1999. 'System Identification – Theory for the User.', 2nd edit. NJ, USA: Prentice Hall.

Mathworks 2011. MATLAB, http://www.mathworks.com/products/matlab/ accessed 27 February.

Narasimhan, S., Jordache, C., 2000. 'Data reconciliation and gross error detection'. Houston, Texas: Gulf Publishing.

Pantelides, C. C., 2001. 'Conference Proceedings ESCAPE-11' (R. Gani and S. B Jørgensen, Eds.), pp. 15–26. Kolding, Denmark.

PSE, 2011. gPROMS, Process Systems Enterprise, http://www.psenterprise.com/gproms/index.html accessed 27 February.

Romagnoli, J.A., Sanchez, M.C., 2000. 'Data Processing and Reconciliation for Chemical Process Operations' 2, Process Systems Engineering. San Diego, USA: Academic Press.

Thacker, B.H., et al. 2004. 'Concepts of Model Verification and Validation. Report LA-14167-MS. 'USA: Los Alamos National Laboratory.

Chapter 3

Computer-Aided Modelling Methods and Tools

Modelling is needed in some form in almost all disciplines. In this book, we have highlighted the mathematical form. Models of this form are needed to generate information (knowledge) in the form of data through which various types of problems (modelling context) can be solved for the system under consideration. The system could be a process/product, an operation, a production system, a business and so on, while the problem could be related to design, production, planning, analysis and many more. The model provides all or part of the information needed to solve the specific problem when the equations representing the model are solved. For the same problem and system, models of different complexity (levels) and with different perspectives of interest (views) may be needed. For example, in a chemical or biochemical process, simulations at various levels of complexity may be needed for the same set of chemicals being processed. In the conceptual (process) design stage, process simulation models including sub-models for constitutive variables (such as physical properties), are needed. These process simulation models may be used from within process synthesis/design methods to generate the optimal design and/or used for verification of the synthesis/design results. In either case, the process simulation models may be of different perspectives of interest or view and having different levels of complexity.

Although mathematical models are routinely used in process-product design, analysis, planning, forecasting and so on, it is not sure whether the necessary models are being developed and used in the most efficient manner. Certainly the process of developing and using models can be improved in many ways. As pointed out by Stephanopoulos *et al.* (1990), multifaceted modelling is necessary for modelling activities with different objectives for the same process. On the other hand, the life of a model could be viewed as cyclic (see Figure 1), where a number of modelling steps are repeated until a desired model is obtained. By changing the model objectives in any cycle, a different model for the same process is obtained, leading to the concept of life cycle process modelling (Marquardt *et al.*, 2000) where different facets of process models are considered together with their relationships and the evolution of modelling artefacts in the sense of a work process. According to this concept, the evolution of a model from its initial creation, through a number of application and

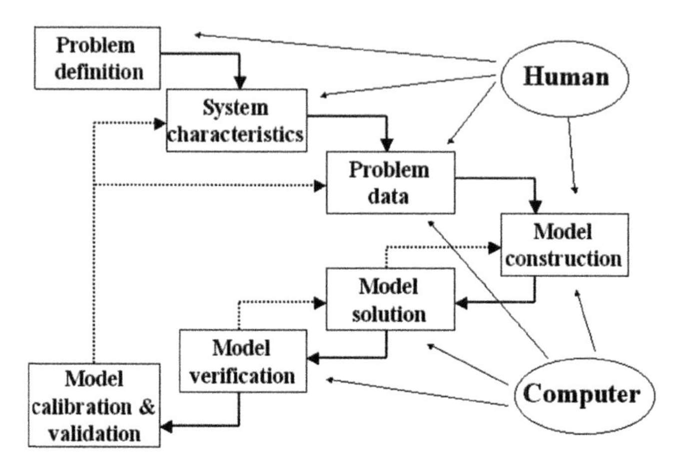

FIGURE 1 Computer-aided model development framework.

modification cycles, is monitored and different versions and views of a model are related to each other. This is certainly true when the models describe the same physical equipment or operation or chemical system. Once the common features at various levels or views have been identified, starting from a reference model frame, various versions of a model may be generated. Figure 2 illustrates this concept for the generation of various process models where the chemical

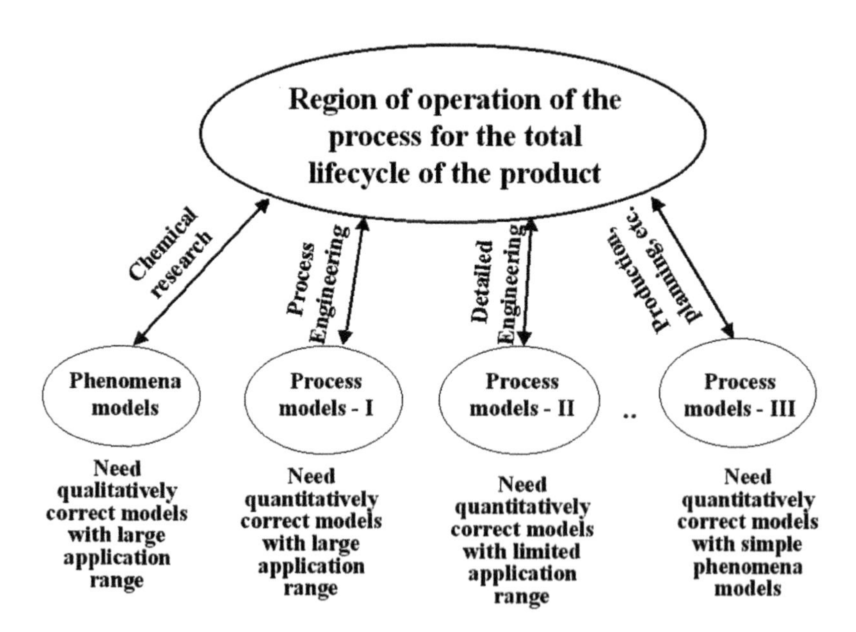

FIGURE 2 Multifaceted modelling needs.

system is fixed. This figure also highlights the fact that a considerable amount of model development work is needed for various modelling activities related to a single process. Therefore, use of systematic methods and their implementation as computer-aided modelling tools that can assist in the model development work is highly recommended.

Computer-aided modelling tools are designed to guide/help the model developer and/or user to perform the modelling tasks in a systematic and efficient manner. In this respect, Sales-Cruz and Gani (2003) and more recently, Gernaey and Gani (2010) proposed a computer-aided modelling framework (see Figure 1) where different modelling tasks are assigned to the model developer or to the computer, based on who can perform them better. For example, tasks such as modelling problem definition and model data requirements are assigned to be performed by the model developer, while model solution, model verification and model validation are assigned to be performed by the computer. Model conceptualisation, model construction and model deployment are assigned to be performed by both. Here, the model developer will decide-select options while the computer will help-guide the model developer/user in making decisions and then implement the selected options. In this way, the time-consuming and calculation/analysis-oriented tasks are performed by the computer through a set of tools while the selection-decision making tasks are performed by the model developer and/or user.

In this chapter, three types of modelling methods and tools are discussed. These tools perform a combination of the tasks shown in Figure 1. For example,

- *Methods and tools for model representation*: These tools help to define the modelling problem, in model conceptualisation, in model construction and in model deployment.
- *Methods and tools for model generation*: These tools help to generate the model equations representing the system being modelled. The typical tasks performed by these tools are model conceptualisation, model construction and model analysis.
- *Modelling toolbox*: These tools translate the mathematical model for the computer, perform model analysis, interface the model with an appropriate solver and report the results.

We start, however, with a brief overview of the structure of mathematical models.

3.1. MATHEMATICAL MODELS

Mathematical models for a process may be derived by applying the principle of conservation of mass, energy and/or momentum on a defined system volume (representing the process) and its connections to the surroundings. A process may be divided into a number of sections (sub-systems) where each section is defined by a boundary (sub-system volume) and connections with other sections and the surroundings. In this way, models for different sections of a process may be aggregated together into a total model for the process. In general, the model equations may be divided into three main classes of equations:

- balance equations (mass, energy and/or momentum equations);
- constitutive equations (equations relating intensive variables such as temperature, pressure and/or composition to constitutive variables such as enthalpies, reaction rates, heat transfers and so on);
- connection and conditional equations (equations relating surroundings–system connections, defined relations such as summation of mole fractions and so on).

The appropriate equations for each type of model may be derived based on the specific modelling objectives. These objectives are transformed into a set of model assumptions and needs, which help to describe the boundary and its connections. Therefore, based on this description, different versions of a model for the same process may be derived. For example, a simple process model may include only the mass balance equations and the connection/conditional equations because the energy and momentum balance effects are assumed to be negligible and the constitutive variables are assumed to be invariant with respect to the intensive variables. A more rigorous model may include the mass and energy balance equations, the connection/conditional equations as well as the constitutive equations. There could be two modes of these models (steady state or dynamic). In the steady-state mode, the rate of change of accumulation is assumed to be zero (or negligible), while in the dynamic mode, they vary with respect to time, the independent variable. A more rigorous model may add the distribution of the intensive variables as a function of space, in one or more dimensions, leading to a distributed model. An even more rigorous model may consider the presence of a population of (catalyst) particles on the surface of which reactions may take place. This leads to a multi-scale model of the process. Figure 3 illustrates the relationship between the three main types of model equations.

In general, the balance equations in a process model are based on the laws of conservation and take one or more of the following forms:

$$0 = f(y, z, t) \tag{1}$$

$$dy/dt = f(y, z, t) \tag{2}$$

$$\delta y/\delta t + \delta y/\delta u = f(y, u, z, t) \tag{3}$$

In the above equations, Eq. 1 represents a set of AEs (algebraic equations) and models the steady-state behaviour; Eq. 2 represents a set of ODEs (ordinary differential equations) and models the dynamic behaviour when the independent variable is time; while Eq. 3 represents a set of PDEs (partial differential equations) and may be used to model both steady-state and dynamic

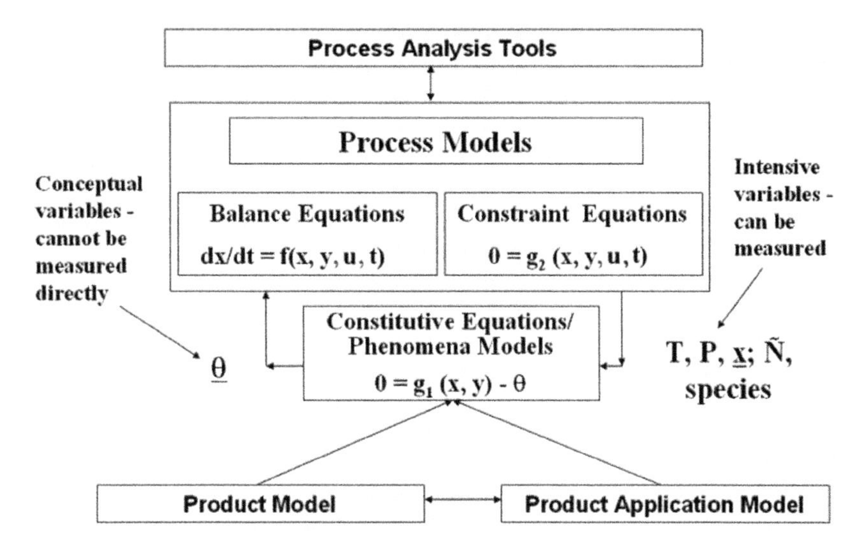

FIGURE 3 Structure of mathematical models.

behaviours, depending on the dimension of the problem and the type of the corresponding independent variables. $\boldsymbol{y}$ represents a vector of state variables (it also includes the sub-set of state variables $\mathbf{x}$), z represents a vector of specified variables while $\boldsymbol{u}$ and t are independent variables (note: u represents another independent variable in a two-dimensional PDE system).

The constitutive equations are usually algebraic (but could also be of the ODE and/or PDE type). In the algebraic form, they may be written as:

$$0 = \theta - g(T, P, n, \beta) \tag{4}$$

where θ is a constitutive variable, which may be a function of temperature (T), pressure (P), composition ($\boldsymbol{n}$) and property model parameters (β). Usually, T, P and/or $\boldsymbol{n}$ are represented by $\mathbf{y}$ in the balance equations. These models are usually explicit in nature, that is, knowing T, P, $\boldsymbol{n}$ and/or β, it is possible to compute θ.

The connection and/or conditional equations (to be called henceforth connection equations) are also usually algebraic and may be represented as:

$$0 = \mathbf{x} - h(\mathbf{y}, \mathbf{z}, \mathbf{x}, \boldsymbol{t}) \tag{5}$$

As Eq. 5 implies, some connection may be explicit (if they are not functions of $\boldsymbol{x}$) while others may be implicit (if they are functions of $\boldsymbol{x}$).

The nature of Eqs. 4 and 5 may define the complexity of the set of equations representing a process model. For example, is the set of equations a linear AE-set; or a DAE-set of index 1; or a higher index DAE or PDAE system? If the different sections (sub-systems) have different scales of size or time, they will lead to a multi-scale model.

3.2. METHODS-TOOLS FOR MODEL REPRESENTATION

As stated earlier, we are dealing with mathematical models where the system is represented by a set of equations for a purpose. The equations consist of some combination of algebraic equations, ordinary differential equations, partial differential equations, logical functions and so on, and need to be solved through an appropriate solution method, numerical or analytical. A model representation system, among others, helps to describe the system being modelled; to conceptualise the modelling process and to identify the model contents. The starting point in all cases is a method for model classification.

3.2.1. Model classification

One way to classify models is through the system it represents, which is the commonly employed method in most process or property simulators. For example, in a process simulator, all unit operation (process) models are classified according to the operation and the simulation mode (for example, steady state or dynamic) they represent, while in a property simulator (for example, a tool for molecular simulation) models are classified according to the property and scale they present. Consider the following examples: constitutive models (for prediction of properties of a polymer); unit operation models (for determining the composition profiles in a distillation column); and, model for a cylindrical catalyst pellet (for evaluating the performance of the catalyst during reaction).

Models for prediction of properties of a polymer

Let us consider the following problem: prediction of permeability of gases through a polymeric membrane. Here, the system is the molecular structure of a polymer consisting of a specified number of repeat units. Each repeat unit is described by atoms and bonds (see Figure 4). The model classification could be

Property type (Which property? Permeability)

→ *Chemical system* (What is the molecular structure? Structure of a polymer—repeat unit and its arrangement in a polymer chain)
→ *Model type* (What type of property model? A micro-scale model that can represent a polymer chain of a fixed number of repeat units, or a group contribution-based model giving only the average property of the repeat unit)
→ *Model equation and parameters* (model parameters available? The typical parameters needed for the micro-scale model are bond angle, bond length, number of atoms, types of atoms and so on; while for a group contribution model they are contributions of each group present in the polymer repeat unit for the desired property)

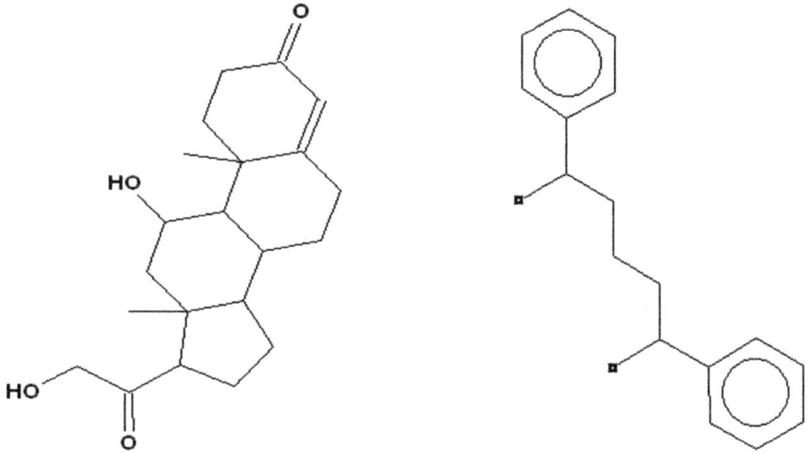

4a: 2D molecular structure of Corticosterone

4a: 2D structure of polystyrene with 2 repeat units

FIGURE 4 Molecular structure of a chemical and a polymer.

Property prediction software, such as ICAS-ProPred, use this structure, where the user needs to decide on the property type, the chemical system and the model type, while the software tool checks for parameter availability and, if available, estimates the property. This example highlights the model usage through model representation according to a specific model classification. A model development (identification) feature could be added to the above scheme if the model parameters are not available. This feature is described in Chapter 11.

Unit operation modelling—steady state or dynamic

The process model classification is highlighted from the point of view of simulation (model application) and from the point of view of model development (construction and/or generation). The first-level classification is the same for both cases. For example,

Process type (Which unit operation? Distillation column)

→ Process *system* (What is the system volume? Are there sub-volumes? N equilibrium stages; each stage has perfectly mixed liquid and vapour in equilibrium)

→ *Chemical system* (What is the chemical system and what properties are needed? Models for equilibrium constants, vapour pressures, liquid heat capacities and heats of vaporisation)

→ *Model type* (Simulation mode? Steady state or dynamic)

→ *Model form* (What type of equations represent the system? The model includes algebraic equations representing the mass and energy balance equations for every stage in the column; to calculate the property terms in these equations, constitutive models for the sub-systems need to be added; result is a large set of linear or nonlinear algebraic equations depending on whether the constitutive models are temperature/composition dependent or not for a steady-state model and a set of ordinary differential and algebraic equations for a dynamic model)

→ *Model parameters* (Does the model have parameters? If yes, how to supply them? Two types of parameters need to be specified—process unit parameters, such as number of stages, feed stage location and so on; constitutive model parameters, such as group contributions for the selected property models)

Using a process simulator, the model developer will simply define the process type, the process system, the chemical system and the model type. The process simulator will then retrieve the relevant equations and the constitutive model parameters from its model libraries (process and property) as well as the constitutive model parameters. The model developer will need to define the process system parameters (number of stages, number of compounds, feed location and so on) before the model equations can be solved.

A model generation system is also able to use the same model classification system to generate the necessary model equations. This is explained in Section 3.2.

Model for a cylindrical catalyst

Consider a cylindrical catalyst pellet as shown in Figure 5, where conduction of mass and energy during reaction takes place for a single-phase system with no accumulation.

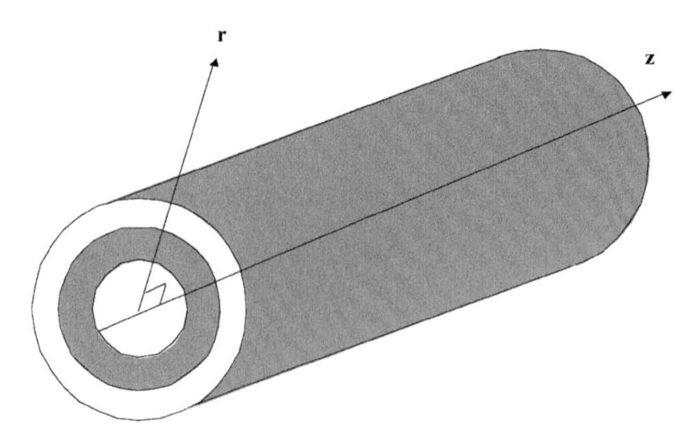

FIGURE 5 Cylindrical catalyst pellet.

Process type (Which unit operation? Which scale? Catalyst pellet within a reactor; micro-scale for catalyst pellets)

→ Process *system* (What is the system volume? Are there sub-volumes? The catalyst pellet (see Figure 5) is the process system with a well-defined volume)

→ *Chemical system* (What is the chemical system and what properties are needed? It is assumed that all the necessary constitutive models exist for this system)

→ *Model type* (Simulation model? Distributed steady-state model—radial and angular directions)

→ *Model equations and parameters* (Type and number of model equations representing the process system and sub-systems, and chemical systems? The model is represented by a two-dimensional partial differential equation (with respect to time and radial direction). Assuming all the constitutive variables are known, the model needs to be discretised in the radial direction and results in a set of ordinary differential equations. The model description in terms of system volume (control shell description) and model equations are given below:

System volume—number of control shells: 1; balance type: microscopic; balance equations: mass and energy; accumulation: no; source: reaction; conduction models: Fick's and Fourier's laws; boundary conditions: $\lambda(r = R) = \lambda_0$, $\delta\lambda/\delta r(r = 0) = 0$; correlations: momentum (correlation for pressure); coordinate system: cylindrical; finite coordinates: r, θ; start and end values of coordinates: (0, 0, 0) and (R, 2π, -); accumulation: no; source: reaction; number of phases: 1; equilibrium reaction: no.

Model equations—No bulk flow means that the velocity terms are not used. Due to the finite coordinates specification, conduction is important in the radical and angular directions, but from the geometric form and the boundary conditions it is given that there is no gradient in the angular direction. Therefore, conduction in this direction is omitted. Moreover, it is specified that accumulation is insignificant and that the only generation term is reaction. After eliminating the negligible terms, the resulting equations are transferred to cylindrical coordinates and the mass-based variables are converted to molar variables. Fick's and Fourier's laws are inserted and the resulting balance equations are given by Eqs. 6 and 7. These balance equations will, together with a call to a pressure correlation routine and the equations for the boundary conditions, define the model for a cylindrical catalyst pellet.

$$0 = \mathrm{D_{im}} \cdot \left(\frac{1}{r}\right) \cdot \left(\frac{\delta}{\delta r(r \cdot \delta c_i/\delta r)}\right) + R_i \quad i = 1, NC \tag{6}$$

$$0 = k\left(\frac{1}{r}\right)\left(\frac{\delta}{\delta r(r\delta T/\delta r)}\right) - f(P) \tag{7}$$

Other examples of distributed models—their system description and the resulting models can be found in Hangos and Cameron (2001). In this book, case studies 8.3 and 12.2 also highlight modelling issues related to distributed and multi-scale modelling.

Examples highlighting the representation of unit operations with library models from a process simulator are given in Chapter 6. An example of property modelling through library models is given in Chapter 14. While this method is very convenient, difficulties arise if different forms of the same unit operation (process) model need to be classified, or none of the available classification terms match the system being modelled. In this case, a knowledge representation system, based on ontology, for example, could be used.

Yang and Marquardt (2009) has developed such ontologies for model representation that are closely connected to model development and solution. An example of such an ontology is illustrated in Figure 6, where the different system description terms are organised such that the resulting model can be retrieved. If properly developed, this system can also help the model developer to derive the model equations (discussed in Section 3.3).

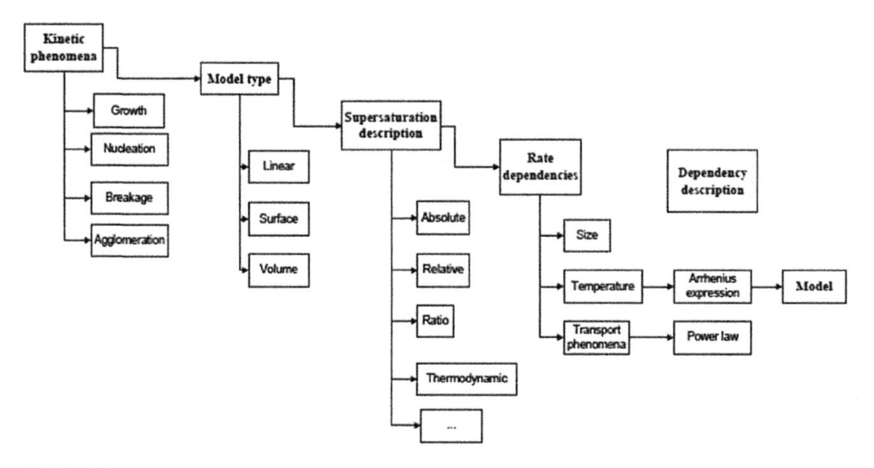

FIGURE 6 **Ontology for representing a modelling system (kinetic models for crystallization).**

3.3. MODEL GENERATION

Until the late 1990s, most of the model development work was done without too much involvement of computers. Only at the solution stage, the computers were used and are still routinely used. For example, process simulators are routinely used to perform mass and energy balance calculations of processes as well as detailed analysis of process operations. With the development of a number of computer-aided modelling tools, such as ModDev (Jensen and Gani, 1999), Modeller (Preisig 1995), Model.La (Bieszczad, 2000), ICAS-MoT (Russel and

Gani, 2000; Sales-Cruz and Gani 2003; Heitzig *et al.*, 2011), ModKit (Bogusch *et al.*, 2001), Modelica (2011), DAESIM (2011), gPROMS (2011) and ASPEN Custom Modeller (2011), this has now changed. Increasing use of these computer-aided modelling tools means a bigger role for computers in the cycle of modelling activities (see Figure 2). In this chapter, the ModDev model generation tool is presented in detail.

ModDev is a knowledge-based modelling system that is able to generate process models through a set of model building blocks and a reference model frame, using user-provided descriptions of the process, boundary, phenomena and so on. The generated model is then translated into a solvable form and linked to a solver or a simulation engine through ICAS-MoT.

3.3.1. Reference model

A reference model is a model representing a system and stored in a model library. The three main types of model equations (balance, constraint and constitutive), representing a system, can be further decomposed into sub-systems until no further decomposition is possible (or desired). The model for the final sub-system then becomes a reference model. These models can be modified to create new reference models or building blocks. In this way, describing a system through the decomposition scheme of Figure 7 helps to

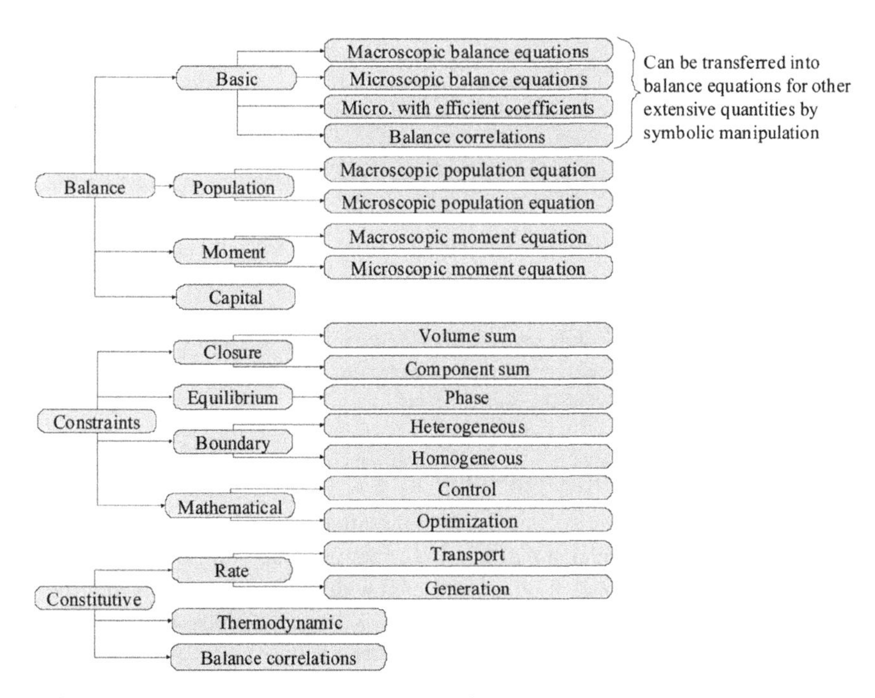

FIGURE 7 Hierarchical decomposition of the reference model.

establish the basic set of model equations. Equations 8–11 are examples of reference models at different levels or scale.

$$\text{Accumulations} = \text{Transport} - \text{Generation} \tag{8}$$

$$\delta\psi/\delta t = -\Delta\varphi + \sum \chi_j \tag{9}$$

$$\delta(V\psi)/\delta t = -\sum(\varphi_j \cdot A_j) + \sum V \cdot \chi_j \tag{10}$$

$$\text{Flux} = \text{Convection} + \text{Conduction} \tag{11a}$$

$$\varphi = \psi\nu + \kappa \tag{11b}$$

3.3.2. Control shell

In order to apply the principles of conservation of mass, energy and momentum, system boundaries need to be established. The system boundaries define a chosen region of space, often called the system volume (control shell). To model the various types of behaviour of a process, the control shells have to be defined such that the partial gradients within the boundaries can be established. Different ways to set up control shells over a tray can be incorporated through a user interface or overall flux model. Through a collection of control shells representing specific regions, it is possible to decompose a system (to be modelled) into elements where first principles modelling is applicable.

As an example, consider the modelling of a multi-tray distillation column. If the phases on the tray are assumed to be in equilibrium (assuming no spatial distribution with respect to temperature, pressure and composition) the control shell is defined for the entire tray and only one set of balance equations (mass, energy and momentum) need to be generated for the tray. If the phases are not in equilibrium, but the flux between the phases can be described by an interface or overall flux model, the control shell is defined for each phase, and, two sets of balance equations (one for each phase) need to be generated. If we now assume that the mixtures in each phase are not perfectly mixed, we will need multiple (infinitesimally small) control shells for each phase. Figure 8 illustrates the 'visual' description of different control shells.

3.3.3. Model generation algorithm

In ModDev, model generation implies the generation of different versions of an available (reference) model for different system and sub-system description. Each control shell and connection is associated with a set of equations (building blocks) generated from the reference model. A building block may also be created through supplied model equations, when a suitable reference model is not available. Schemes such as the hierarchical model decomposition scheme (see Figure 7) may be used to describe a control shell. Starting from a branch

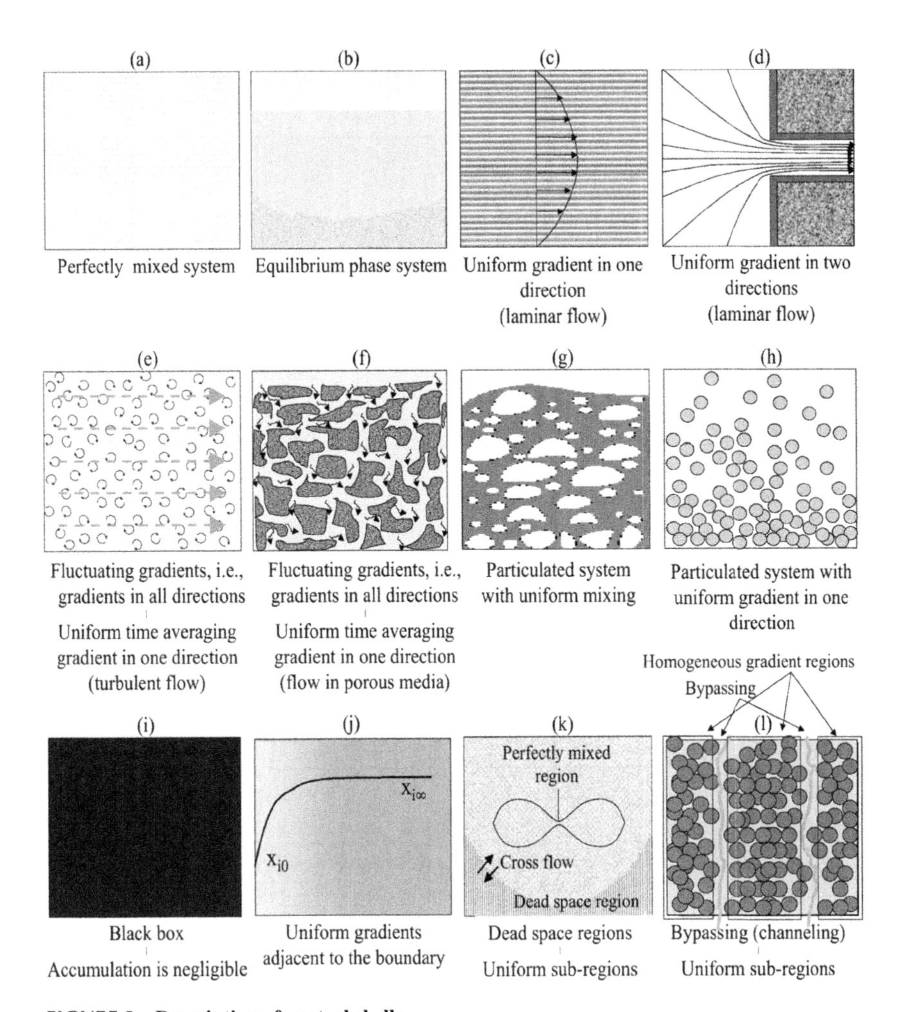

FIGURE 8 Description of control shells.

(e.g. balance) one follows a path including a sub-branch (e.g. basic) and sub-sub-branch (e.g. macroscopic balance equations), until a leaf (reference model) is reached. At this point, the reference model equations corresponding to this leaf is retrieved from the reference model library and manipulated to define a building block. Repeating this procedure for all control shells and associated phenomena generates the final model. The idea is to decompose the control shells into the smallest desired scale; retrieve the corresponding equations from a model library or provide them to create the building blocks; and then aggregate the building blocks to obtain the final set of model equations. Figure 9 highlights this procedure, while Figure 10 illustrates the decomposition scheme for a two-phase flash model. Figure 11 highlights the application of the decomposition scheme to generate a flash model.

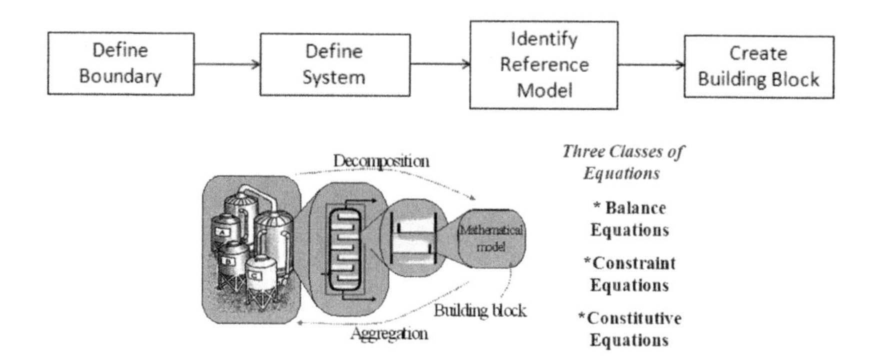

Define Boundary → Describe System → Identify Building Block

FIGURE 9 Model generation procedure.

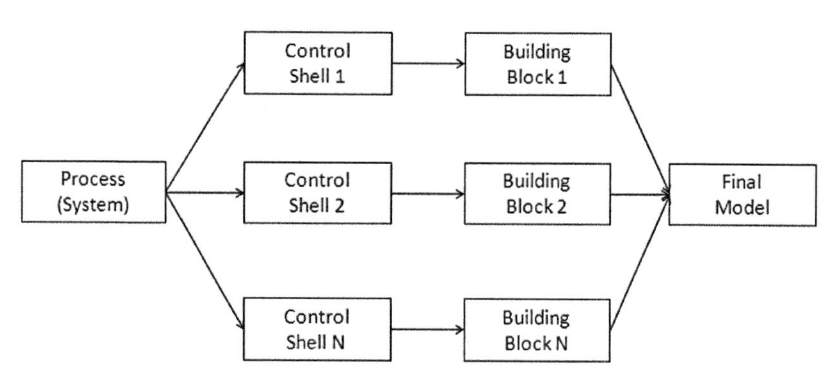

FIGURE 10. Decomposition scheme for model generation.

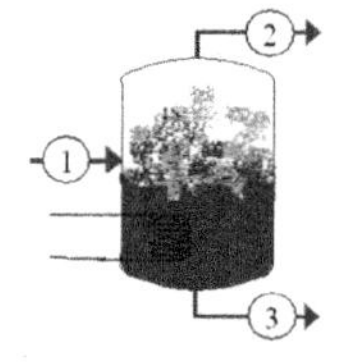

STREAM CONNECTION OBJECT
Name: *3*
Models for quantities:
Energy (enthalpy): H_3=@FUNC_E(2,$f_{3[]}$,T_3,P_3)
Models for the "from"-connection: *(equilibrium)*
Energy connection: $T_3=T_{flash}$
Momentum connection: $P_3=P_{flash}$

SHELL OBJECT
Name: *flash*
Assumed phase condition: *Calculate (VL)*
Equilibrium model: $0=f_{2i}/F_2-K_{flash}*f_{3i}/F_3$, @KEQ($T_{flash}$, P_{flash}, f_2, f_3, @K_{flash}), no accumulation, include mass & energy balance

SHELL CONNECTION OBJECT
Name: *heater*
Connection models:
Energy connection: $Q_{heater}=Q_{flash}$

FIGURE 11 Application of the model generation procedure.

A process separating a stream into two phases through a separator (see Figure 11) will now be used as an example to highlight the different forms and versions of models that can be generated. The starting point is a simple mass balance model. Description of the system boundary (shell object) and its connections is shown in Figure 12.

Model	Equation	Index	Type
M1: Simple mass balance model (steady state); S_i is split factor for component i; NC is number of components; MB is mass balance; C1 is constitutive equation; C2 is connection equation	$0 = Fz_i - Vy_i - Lx_i$	$i = 1, NC$	MB
	$0 = S_i - (Vy_i)/(Fz_i)$	$i = 1, NC$	C1
	$0 = 1 - \sum S_i$	$i = 1, NC$	C2
M2: Simple mass balance model (steady state); replace S_i with equilibrium constant K_i for component i; selected C1 model for K_i assumes ideal system; p_i is vapour pressure of component i	$0 = Fz_i - Vy_i - Lx_i$	$i = 1, NC$	MB
	$0 = K_i - p_i(T)/P$	$i = 1, NC$	C1
	$0 = P_i - f(T)$	$i = 1, NC$	C1
	$0 = y_i - x_i K_i$	$i = 1, NC$	C2
	$0 = 1 - \sum y_i$	$i = 1, NC$	C2
M3: Mass balance model (steady state); select another model for K_i for component i that does not assume ideal system; γ_i is activity coefficient for component i in liquid phase; φ_i is fugacity coefficient of component i in vapour phase	$0 = Fz_i - Vy_i - Lx_i$	$i = 1, NC$	MB
	$0 = K_i - (p_i\gamma_i)/(P\varphi_i)$	$i = 1, NC$	C1
	$0 = P_i - f(T)$	$i = 1, NC$	C1
	$0 = \gamma_i - f(T, x)$	$i = 1, NC$	C1
	$0 = \varphi_i - f(T, P, y)$	$i = 1, NC$	C1
	$0 = y_i - x_i K_i$	$i = 1, NC$	C2
	$0 = 1 - \sum y_i$	$i = 1, NC$	C2
M4: Rigorous steady-state model with mass and energy balance (EB); H_k is enthalpy for stream k; different constitutive (enthalpy) models for liquid and vapour streams	M3 plus		
	$0 = Q - (FH_F - VH_V - LH_L)$		EB
	$0 = H_k - f(T, P, \text{composition})$		C1
M5: Rigorous dynamic model with mass and energy balance; same model as M4 except new MB and EB (only the extra equations are shown); assume negligible vapour holdup; n_i is molar liquid holdup for compound i; E is energy holdup	$dn_i/dt = Fz_i - Vy_i - Lx_i$	$i = 1, NC$	MB
	$dE/dt = Q - (FH_F - VH_V - LH_L)$		EB
	$0 = x_i - n_i/(\sum n_i)$		C2
	$0 = L - f(n, A, T, P)$		C2
	$0 = V - f(n, T, P)$		C2
M6: Rigorous two-phase reactor model; add reaction terms to MB and EB and add kinetic (constitutive) model; d are kinetic parameters; ΔH_R is heat of reaction; k is reference reactant	Reaction term (mass) = $(Lx_k)\text{Rate}_i$		
	Reaction term (energy) = $\Delta H_R L$		
	$0 = \text{Rate}_i - f(T, P, n, d)$		

From the above models, it can be noted that when the chemical system in a process is fixed, the C1 (constitutive) models also become fixed and will not change but the MB, EB and C2 models may change as one moves from one form of the model to another. On the other hand, for each form of the process model, the EB and MB models become fixed and the C1 and C2 models may be changed

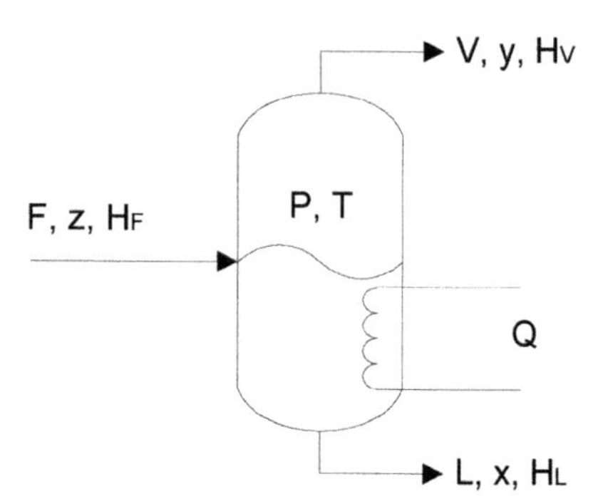

FIGURE 12 **A two-phase separation process. F, V and L are the flow rates of feed, vapour and liquid streams, respectively. *z*, *y* and *x* are the composition (mole fractions) of feed, vapour and liquid streams, respectively. H_F, H_V and H_L are the enthalpies of feed, vapour and liquid streams, respectively. *P* is the pressure, *T* is the temperature and *Q* is the heat duty.**

for different chemical systems. This has been illustrated in Chapter 10. Model M5 may be further simplified by assuming rate of change of energy holdup (this will convert the EB to an algebraic equation) or the vapour holdup may not be assumed to be negligible (making the model more complex as now the holdup n will be sum of liquid and vapour holdups). Alternatively, models M3, M4 and M5 may be simplified by selecting simpler C1 model equations for the liquid and vapour phase fugacities. Note that the nonlinearity of EB and MB equations becomes clear by inserting the corresponding C1 equations into them. Note also that models M1 and M2 are linear because the corresponding constitutive variables are independent of composition. Models consisting of C1 and C2 are commonly used for equilibrium saturation point calculations (does not need MB and EB) that enable the generation of phase diagrams. In a similar fashion, kinetic models represented by C1 and their corresponding C2 equations generate reaction yield diagrams (attainable regions).

In Appendix, incremental model generation through ModDev is highlighted.

3.4. MODELLING TOOLBOX

The idea of a modelling toolbox is to take a derived model, translate it to a form that can be analysed through a computer, analyse the model in terms of degrees of freedom, variables specification, singularity and so on, and then solve the model equations through an appropriate solver. If the model satisfies the modelling needs, the model can be saved as a model object, which can be used (deployed) for various applications. If the model performance is not acceptable, model parameters may be estimated or fine-tuned through a model parameter estimation (model identification) option. All accepted available models may also be employed for process optimisation and other related studies where the process model is used in an inner loop. In this way, the modelling toolbox performs a number of modelling tasks (as outlined in Figure 1). A number of tools exist to perform the tasks. The tools vary in the number of tasks they are able to perform. Well-known software such as MatLab (2011) fall under this category. Also, process simulators such as gPROMS and ASPEN provide tools for some of the modelling tasks. In this chapter, the ICAS-MoT is described because it is the tool used in most of the case studies presented in Part II of this book.

3.4.1. ICAS-MoT

ICAS-MoT takes a model represented by a set of equations as input and generates either a solution to the model equations or exports the model as a COM-object for use in a simulation engine or external software. The model equations can be imported from a model generation tool, a text file containing the equations, a model written in XML or a model directly added by the user through an editor option in MoT. The work process in MoT is divided into two main activities—model definition and model solution. In addition to the above, MoT also allows model transfer in terms of COM-objects for link to the ICAS simulation engine and/or use in external software. The connection between ICAS-MoT and ICAS-ModDev is shown in Figure 13.

The current version of ICAS-MoT has four main options, and within each option, many sub-options are available to help the model developer.

Model definition: This refers to the transfer of a derived model to ICAS-MoT. Different options are available, for example generation of model with ModDev and model analysis and solution with ICAS-MoT. Models may be imported to ICAS-Mot as a text file or as an XML file, or the equations can be directly typed or 'pasted'.

Model (numerical) analysis: This refers to the analysis of an uploaded model in terms of degrees of freedom, singularity, choice of variables to specify and so on. In this step, the model equations are also ordered to generate a calculation order

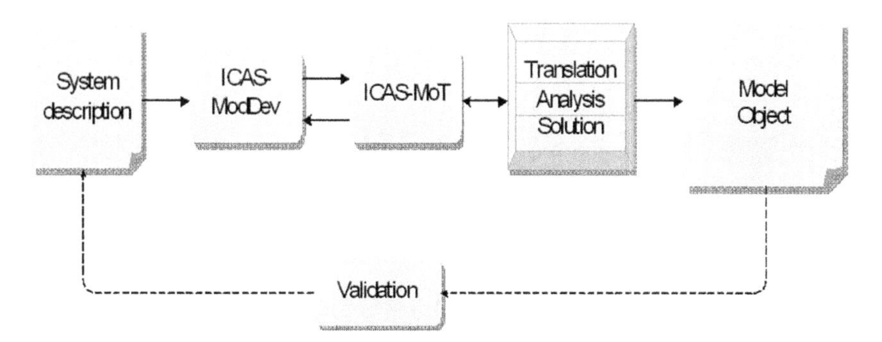

FIGURE 13 Connection between ICAS-MoT and ICAS-ModDev.

for the equations. The classification of equations and variables take place here. Appropriate solvers are connected to solve the set of model equations for a specified set of known variable values.

Model solution: This refers to the actual solution of the model equations.

Analysis tools: These tools refer to options to perform sensitivity analysis, parametric sensitivity analysis, eigen-value analysis, generating correlation statistics and many more.

Model definition

As shown in Figure 14, model definition includes model creation (that is, writing a derived set of model equations according to the syntax rules of MoT), model translation (translates and expands the created text-based model), import (reads models introduced through external text files and/or XML files) and modify (allows the modification of the text-based model). The important step here is translation that dissects the text-based equations using a Reverse Polish Notation (RPN) algorithm and classifies equations and variables using a multi-layered classification system for equations and variables. First, MoT identifies the equations and variables according to a set of simple syntax rules.

Model analysis

The analysis step invokes the second-analysis layer for the classification of variables. In this layer, all variables, identified previously as those appearing

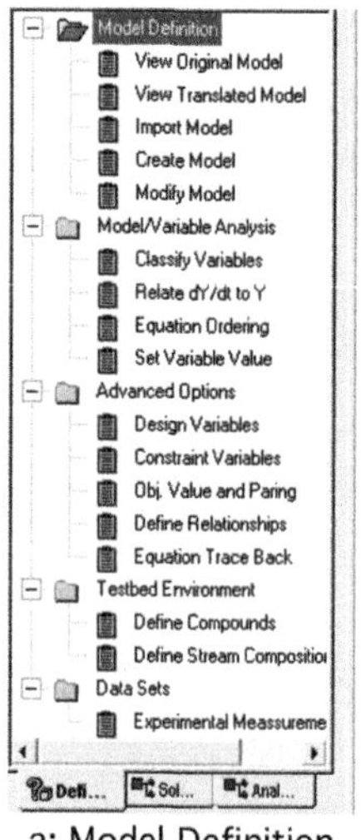

a: Model Definition

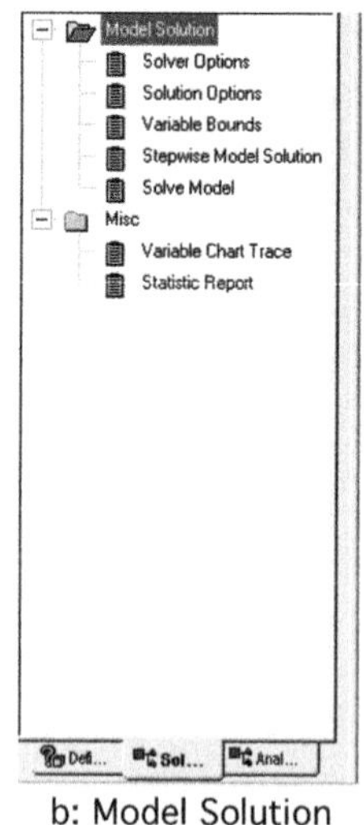

b: Model Solution

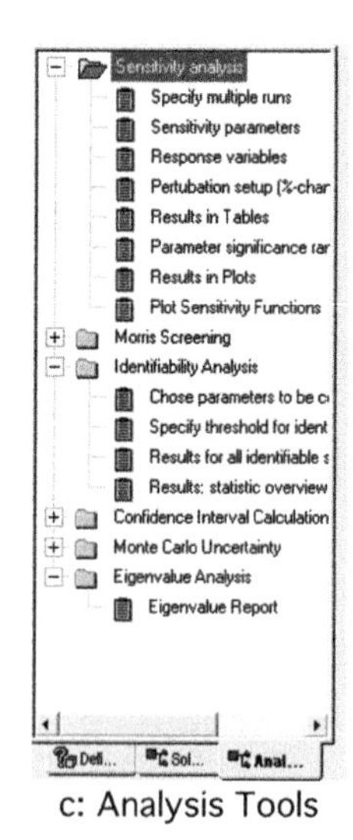

c: Analysis Tools

FIGURE 14 Task options available in ICAS-MoT. (a) Model definition, (b) model solution, (c) analysis tools.

on the RHSs of equations and classified as parameters, are now re-classified as follows:

- parameter (variables with known values);
- explicit (variables that are functions only of parameters and/or dependent-prime variables);
- implicit-unknown (variables related to algebraic equations where there are more than one unknown variables per equation);
- dependent (variables appearing with the differential operators on the LHSs of ODEs and/or PDEs);
- dependent-prime (the derivative operator related to the dependent variable).

Note that according to this classification, the LHS variables can only be unknown, dependent and/or dependent-prime. Once the classification of the variables according to the rules of the analysis layer has been done, ICAS-MoT offers a number of analysis options:

- generation/analysis of incidence matrix (with equations as row index and variables as column index);
- check for singularity of matrix (identifies equations with no unknown variables);
- equation trace-back (examines the model equations).

A typical screenshot from the model analysis task is shown in Figure 15.

- Equation-by-equation solution mode (solving residuals of each translated equation—this is an equation debug option to check whether the values

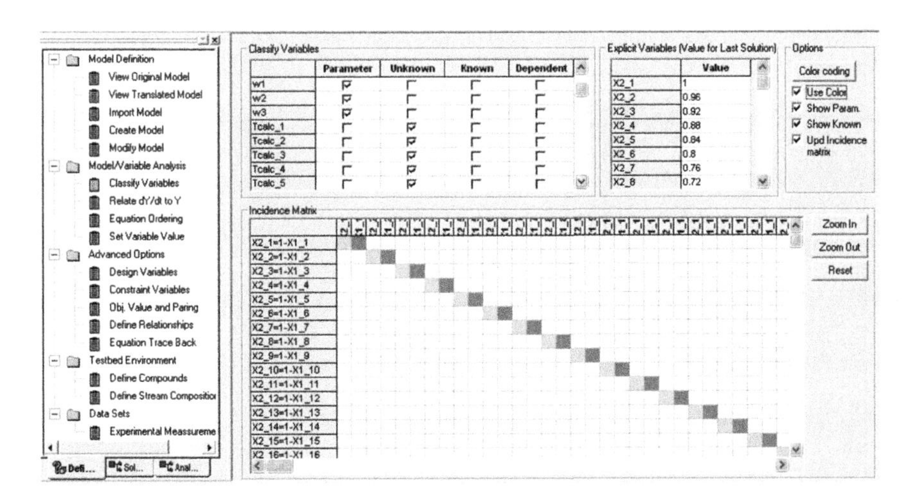

FIGURE 15 Screen-shot from ICAS-MoT for model analysis.

passed to the equation are correct or the derived function is correct). Figure 16 shows this feature of ICAS-MoT.

- Decomposition, partitioning and ordering of the model equations (identifies the sub-sets of equations that need to be solved simultaneously).
- Degree of freedom analysis is automatically performed to ensure that the problem is not ill posed (number of equations do not match the number of unknown variables) before going to the solution step.

Model solution

The solution step in MoT involves the following:

- Admin: administration of the solution procedure (drives the numerical solution procedure);
- Solver-link: connection to the solver specified by Admin;
- Resid: computes the function values for the translated equations.

The Admin is divided into a global administrator and one or more local administrator(s), as needed by the solution strategy. Each local administrator instantiates one or more solvers from the solver library. The possible local administrators are

	Break	Equation	Result
1		ntot=((n_0)+(n_1))	32
2		hVr_0=(((((EDippr104_0*0.2*Tref+DDippr104_0*0.25)*Tref+CDippr104_0/3.0)*Tref+BDippr104_0*0.5)*Tref+ADippr104	6167.39
3		hVr_1=(((((EDippr104_1*0.2*Tref+DDippr104_1*0.25)*Tref+CDippr104_1/3.0)*Tref+BDippr104_1*0.5)*Tref+ADippr104	18575.4
4		hLr_0=(((((EDippr104_0*0.2*Tref+DDippr104_0*0.25)*Tref+CDippr104_0/3.0)*Tref+BDippr104_0*0.5)*Tref+ADippr104_	6167.39
5		hLr_1=(((((EDippr104_1*0.2*Tref+DDippr104_1*0.25)*Tref+CDippr104_1/3.0)*Tref+BDippr104_1*0.5)*Tref+ADippr104_	18575.4
6		zTank_0=n_0/ntot	0.40625
7		zTank_1=n_1/ntot	0.59375
8		V=ValveV*(Pout-Pmin)	0.07
9		DenV=Pout/(0.08314*Tout)	0.0601395
10		hL_1=(((((EDippr104_1*0.2*Tout+DDippr104_1*0.25)*Tout+CDippr104_1/3.0)*Tout+BDippr104_1*0.5)*Tout+ADippr104_	7940
11		hL_0=(((((EDippr104_0*0.2*Tout+DDippr104_0*0.25)*Tout+CDippr104_0/3.0)*Tout+BDippr104_0*0.5)*Tout+ADippr104_	2757.9
12		hV_1=(((((EDippr104_1*0.2*Tout+DDippr104_1*0.25)*Tout+CDippr104_1/3.0)*Tout+BDippr104_1*0.5)*Tout+ADippr104	7940
13		hV_0=(((((EDippr104_0*0.2*Tout+DDippr104_0*0.25)*Tout+CDippr104_0/3.0)*Tout+BDippr104_0*0.5)*Tout+ADippr104	2757.9
14		hVap_1=(ADippr103_1*(1-(Tout/DB_Tc_1))^(BDippr103_1+CDippr103_1*(Tout/DB_Tc_1)+DDippr103_1*(Tout/DB_Tc_1	43668.8
15		hVap_0=(ADippr103_0*(1-(Tout/DB_Tc_0))^(BDippr103_0+CDippr103_0*(Tout/DB_Tc_0)+DDippr103_0*(Tout/DB_Tc_0	46951.3
16		Psat_1=(10^(DB_AntoineA_1-DB_AntoineB_1/(Tout-273.15+DB_AntoineC_1)))/760	4.67668e-005
17		K_1=Psat_1/Pout	4.67668e-005
18		x_1=zTank_1/(1+phi*(K_1-1))	1.18744
19		y_1=x_1*K_1	5.5533e-005
20		Psat_0=(10^(DB_AntoineA_0-DB_AntoineB_0/(Tout-273.15+DB_AntoineC_0)))/760	2.53064e-006
21		K_0=Psat_0/Pout	2.53064e-006
22		Residual_2=((zTank_0*(1-K_0)/(1+phi*(K_0-1)))+(zTank_1*(1-K_1)/(1+phi*(K_1-1))))	
23		x_0=zTank_0/(1+phi*(K_0-1))	
24		Hl=(((hL_0-hLr_0)*x_0)+((hL_1-hLr_1)*x_1))	
25		y_0=x_0*K_0	
26		Hv=(((hV_0-hVr_0+hVap_0)*y_0)+((hV_1-hVr_1+hVap_1)*y_1))	
27		Residual_3=(Hl*ntot*(1-phi)+Hv*ntot*phi-Htank)/1000	
28		Ttank=Tout	
29		dL_1=ADippr101_1/BDippr101_1^(1+(1-Ttank/CDippr101_1)^DDippr101_1)	
30		dL_0=ADippr101_0/BDippr101_0^(1+(1-Ttank/CDippr101_0)^DDippr101_0)	
31		DenL=1/((x_0/dL_0)+(x_1/dL_1))	
32		Level=ntot*(1-phi)/(Area*DenL)	
33		Residual_1=ntot*(1-phi)/DenL+ntot*phi/DenV-Vol	
34		Pl=Pout	
35		Pv=Pl	
36		Tl=Tout	

FIGURE 16 **Debug solution mode—equation by equation.**

- algebraic administrator;
- integration administrator;
- optimisation administrator.

The global administrator combines all the local administrators needed to solve a given problem and handles the overall solution sequence. The local administrators are only connected to the model via the global administrator. From the model classification information, ICAS-MoT detects and lists the needed local administrators (Figure 17). If more than one local administrator is needed, the variable classification rules for the third-solution layer are invoked. Here, variables are classified as follows:

- design/manipulated (parameters whose values may be changed in the outer loop);

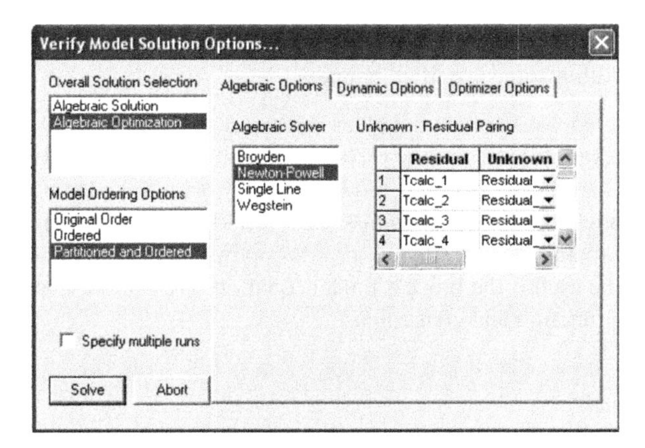

FIGURE 17 (A) **Different solver options in ICAS-MoT (AE-solver).**

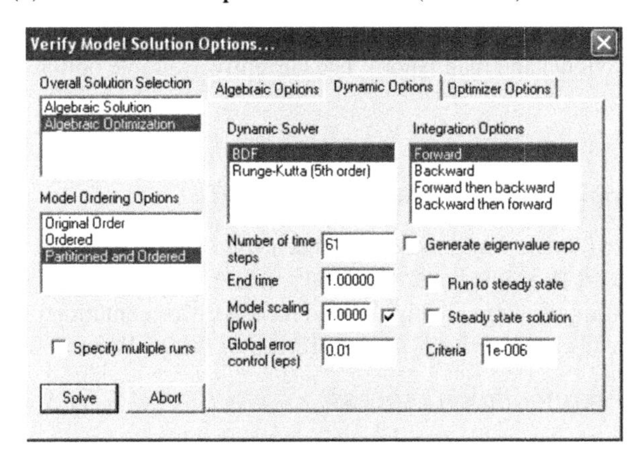

FIGURE 17B **Different solver options in ICAS-MoT (DAE-solver).**

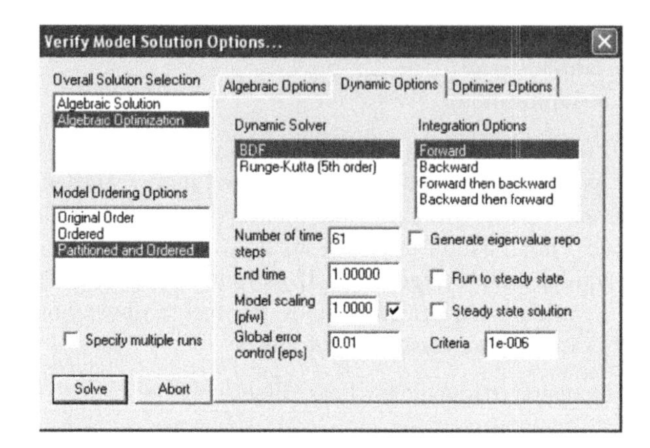

FIGURE 17C Different solver options in ICAS-MoT (optimisation-solver)

- constraints (unknown, dependent and/or dependent-prime whose values must be within some specified bounds).

If the problem solution also involves parameter estimation based on supplied data, the data variables need to be classified as real, integer and/or binary. Once the variables have been identified, the appropriate solver linked to ICAS-MoT is invoked through the global administrator. Note that in problems related to optimisation (parameter estimation, process optimisation and so on), multiple solvers may be used if the process model equations are solved separately from the objective function and constraints.

Analysis tools

These tools are developed within ICAS-MoT to help with the numerical solution of the problem, to analyse the optimisation (also parameter estimation) results, to analyse the simulation model (in terms of eigenvalues for model reduction or response behaviour) and many more. The currently available options are shown in Figure 14c.

3.4.2. Example of application of ICAS-MoT

The ICAS-MoT modelling toolbox has been used in most of the case studies presented in Part II of this book. Here only the steps/options used in solving a typical modelling problem are highlighted for a beer fermentation process. The model has already been derived (Andres-Toro *et al.*, 1996, Woinaroschy, 2002).

3.4.2.1 *Beer fermentation process*

These models are generally highly nonlinear and subject to many parameters that need to be identified. In this example, all the parameters have been taken from the published example. The objective here is not to analyse this model if it is correct, but to highlight the ICAS-MoT options for the tasks needed in its testing and analysis.

The model is described by the following set of ODEs:

$$\frac{dX_{\mathrm{lag}}(t)}{dt} = -\mu_{\mathrm{lag}} \cdot X_{\mathrm{lag}}(t) \tag{12}$$

$$\frac{dX_{\mathrm{act}}(t)}{dt} = m_{\mathrm{x}} \cdot X_{\mathrm{act}}(t) - k_{\mathrm{m}} \cdot X_{\mathrm{act}}(t) + m_{\mathrm{lag}} \cdot X_{\mathrm{lag}}(t) \tag{13}$$

$$\frac{dX_{\mathrm{dead}}(t)}{dt} = k_{\mathrm{m}} \cdot X_{\mathrm{act}}(t) - \mu_{\mathrm{D}} \cdot X_{\mathrm{dead}}(t) \tag{14}$$

$$\frac{dS(t)}{dt} = -\mu_{\mathrm{S}} \cdot X_{\mathrm{act}}(t) \tag{15}$$

$$\frac{dE(t)}{dt} = -\mu_{\mathrm{a}} \cdot f \cdot X_{\mathrm{act}}(t) \tag{16}$$

$$\frac{dA(t)}{dt} = \mu_{\mathrm{eas}} \cdot \frac{dS(t)}{dt} \tag{17}$$

$$\frac{dD(t)}{dt} = k_{\mathrm{dc}} \cdot S(t) \cdot X_{\mathrm{act}}(t) - k_{\mathrm{dm}} \cdot D(t) \cdot E(t) \tag{18}$$

where, X_{act}, X_{lag}, X_{dead} are active, suspended and dead biomass concentration, and S, E, A and D are substrate, ethanol, ethyl acetate and diacetyl concentrations, respectively. The translated ICAS-MoT model is shown in Figure 18. The model has 17 equations (7 ODEs and 10 explicit AEs) and 21 variables (2 parameters, 2 known, 7 dependent and 10 explicit). It is solved with the BDF-integration method in ICAS-MoT with the initial values of the dependent variables listed in Figure 19.

A sample of the dynamic simulation results is shown in Figure 20, which shows the transient behaviour of ethanol concentration in the fermenter.

3.4.3. Using ICAS-MoT from Excel

Microsoft Excel is an extremely powerful tool, which is used by millions of people everyday. Functions tailored to a specific task can be programmed into Excel to extend its capabilities with customised analysis tools. A simple method of customising Excel is to create a macro. An Excel macro file for using models developed in ICAS-MoT through the COM[1] technology is

[1]COM is an acronym for Component Object Model, and it is the widely accepted standard for integration of external functionality into Microsoft Office applications, such as Excel.

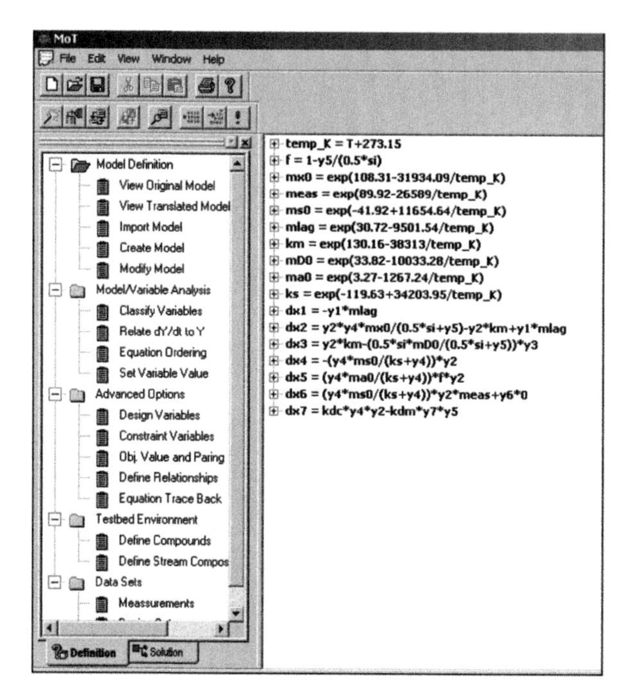

FIGURE 18 **Translated model of beer fermentation model.**

available. The Excel macro is customised such that it is able to execute a sequence of modelling-related activities: reading of input data, execution of the MoT model and writing of output (results) data, as shown in Figure 21. The building process that follows involves code generation, compiling, linking and evaluation of the MoT model.

In this way, the process of evaluation and use of a MoT export model is completely automatic. The main advantage is that users can prepare data for a MoT export model, get output results from the MoT export model and store results from the same model using different parameters, directly through an

	Dependent
y5	1
y1	1
y2	1
y4	120
y3	1
y6	0
y7	0

FIGURE 19 **Initial conditions for the beer fermentation model.**

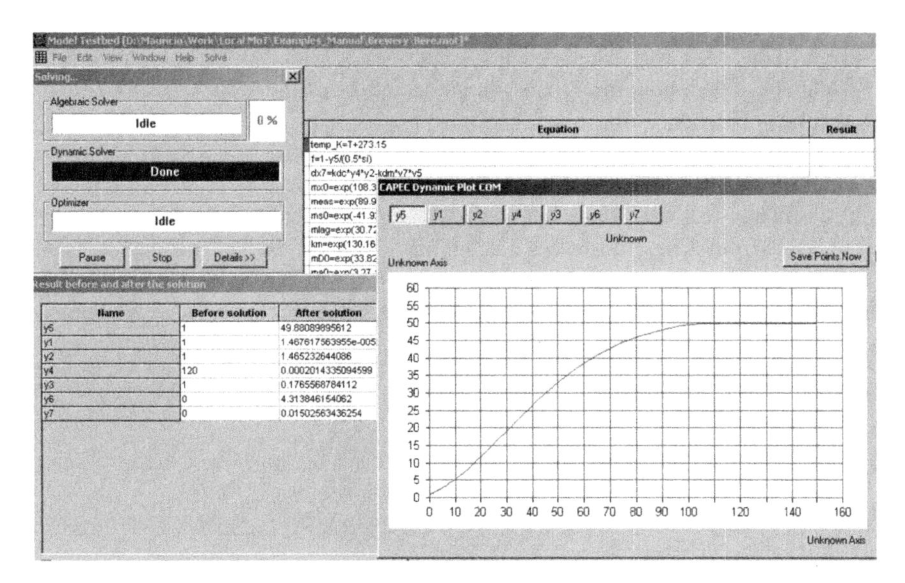

FIGURE 20 Beer fermentation model solution completed.

Excel worksheet environment. This makes the model use easier for those not familiar with the modelling toolbox environment. A scheme for using ICAS-MoT for model development is highlighted in Figure 21.

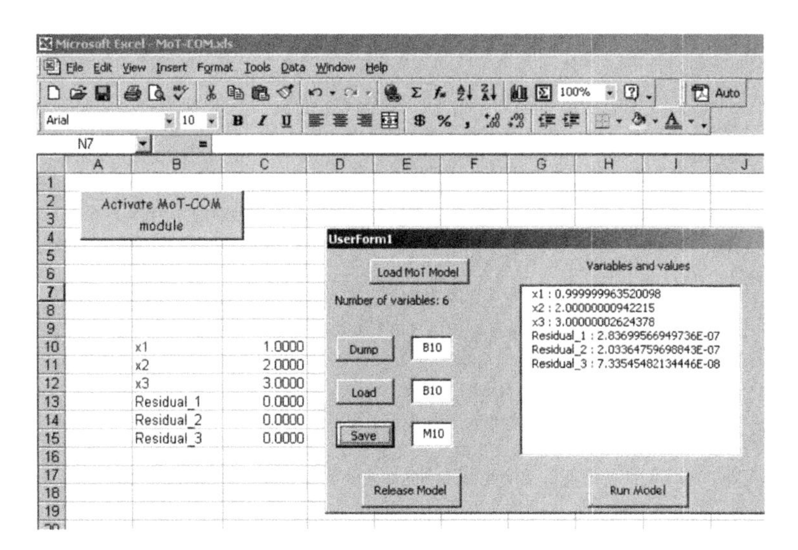

FIGURE 21 Excel macro interface.

3.5. SUMMARY

Computer-aided modelling tools can guide and assist the model developer and significantly reduce the time and resources used in model development. Through a computer-aided framework that allows the use of various modelling methods and tools, it is possible to perform the modelling tasks in a systematic and efficient manner. A number of very useful modelling methods are already available from different sources — commercial as well as from academia. Two of these have been highlighted to illustrate some of the modelling concepts. Interested readers should try as many of the tools as possible to decide which ones match their requirements best.

REFERENCES

Andres-Toro, B., Giron-Sierra, J. M., Lopez-Orozco, J. A., and Fernandez-Conde, C., 1996. Optimisation of a batch fermentation process by genetic algorithms, 183–188.

ASPEN 2011. Aspen Custom Modeler, http://www.aspentech.com/products/aspen-custom-modeler.aspx accessed February 27.

Bieszczad, F. J. 2000. A framework for the language and logic of computer-aided phenomena-based process modelling. PhD Thesis, Massachusetts Institute of Technology.

Bogusch, R., Lohmann, B., Marquardt, W., 1997. Computer-aided process modelling with ModKit. Computers and Chemical Engineering. 21, 1105–1115.

Bogusch, R., Lohmann, B., Marquardt, W., 2001. Computer-aided process modelling with MODKIT. Computers & Chemical Engineering. 25, 963–995.

Daesim 2011. Daesim Dynamics, http://www.daesim.com/software/daesim/dynamics/accessed February 27.

Eggersmann, M., Hackenberg, J., Marquardt, W., Cameron, I.T., 2002. Applications of modelling: a case study from process design. B. Braunschweig., R. Gani (Eds.), *In Software Architectures and Tools for Computer Aided Process Engineering*. Vol. 11, Elsevier CACE Series, pp. 335–372.

Gani, R., Hytoft, G., Jaksland, C., Jensen, A.K., 1997. An integrated computer-aided system for integrated design of chemical processes. Computers and Chemical Engineering. 21, 1135–1146.

Gernaey, Krist V., Gani, Rafiqul., 2010. A model-based systems approach to pharmaceutical product-process design and analysis. Chemical Engineering Science. 65 (21), 5757–5769.

Hangos, K., Cameron, I.T., 2001. Process Modelling and Process Analysis. London: Academic Press.

Jensen, Anne Krogh., Gani, Rafiqul., 1999. A computer aided modeling system. Computers & Chemical Engineering. 23 (Supplement 1), S673–S678.

Heitzig, M., Sin, G., Sales Cruz, M., Glarborg, P., Gani, R., 2011. A computer-aided modeling framework for efficient model development, analysis and identification: Combustion and reactor modeling. Industrial & Engineering Chemistry Research Puigjaner Special Issue. 50, 5253–5265.

Mathworks 2011, MATLAB, http://www.mathworks.com/products/matlab/ accessed February 27.

Marquardt, W., von Wedel, L., and Bayer, B. 2000. Perspectives on lifecycle process modelling. In Foundation of Computer Aided Process Design. In: M.F., Malone, J.A., Trainham, B., Carnahan, (Eds.), Vol. 96, 192–214. AIChE Symposium Series 323.

Modelica, 2011. Modelica Association, https://www.modelica.org/, accessed February.

Preisig, H. A., 1995. MODELLER—An object-oriented computer-aided modelling tool. In: L.L. Biegler and M.F. Doherty, Editors, Foundations of computer-aided process design, AIChE symposium series, Vol. 91, No. 304, pp. 328–331.

PSE, 2011. gPROMS, Process Systems Enterprise, http://www.psenterprise.com/gproms/index.html accessed February 27.

Russel, B. M. R., and Gani, R. 2002. MoT—a modelling test-bed. In ICAS Manual (CAPEC Report), Technical University of Denmark.

Sales-Cruz, M., Gani, R., 2003. A modelling tool for different stages of the process life. S. P. Asprey., S. Macchietto (Eds.), *Computer-Aided Chemical Engineering, vol. 16: Dynamic Model Development*. Amsterdam: Elsevier, pp. 209–249.

Skiadas, I.V., Gavala, H.N., Lyberatos, G., 2000. Modeling of the Periodic anaerobic baffled reactor based on the retaining factor concept. Water Research. 34, 3725–3736.

Stephanopoulos, G., Henning, G., Leone, H., 1990. MODEL.LA—A Modelling Framework for Process Engineering—II. Multifaceted Modelling of Processing Systems. Computers and Chemical Engineering. 14, 813–846.

von Wedel, L., Marquardt, W., Gani, R., 2002. Modelling frameworks. B. Braunschweig., R. Gani (Eds.), *In Software Architectures and Tools for Computer Aided Process Engineering*. Vol. 11, Elsevier CACE Series, pp. 89–126.

Woinaroschy, A., 2002. Personal communication. Romania: Department of Chemical Engineering, University 'Politehnica' of Bucharest, 1–5 Polizu Str., 78126-Bucharest.

Yang, A., Marquardt, W., 2004. An ontology-based approach to conceptual process modeling. A. Barbarosa-Povoa., H. Matos (Eds.), *European symposium on computer aided process engineering*. Vol. 14, Amsterdam: Elsevier, pp. 1159–1164.

APPENDIX: APPLICATION OF MODDEV

This tutorial concentrates on the basic features of ModDev in order to guide the user through the different steps of the model generation procedure followed in ModDev. Simple examples are only considered here.

The model generation procedure followed in ModDev consists of providing descriptions of the process (to be modelled) in terms of shells, streams and connections. Based on the descriptions, ModDev generates the corresponding equations. In case the corresponding equations of a shell/stream/connection are not available, ModDev allows the user to define new equations.

Process modelling with ModDev

Before starting with the first example, let us first review the various features available in ModDev and how to start with ModDev in ICAS. Figure A1 shows the screen when ICAS is started.

Close the 'task manager' and click on ModGen ('model generation'). Figure A2 highlights the various tools available in the starting screen of ModGen.

We will start every example from this screen (minus the model and equations shown in Figure A2) and generate our own process model. There are five detailed examples starting from a simple steady-state tank mixer and ending with a dynamic reactor where the reaction is defined through a kinetic model.

Tutorial example 1—steady-state tank (mixer)

In this example, we have two streams coming in and one stream going out. We will perform both mass and energy balance.

Step1: Describe the process through a flow diagram—Draw a shell connected by the inlet streams and one outlet stream. This is the necessary description of the shell (tank), streams (inlet/outlet) and connection (shell–stream). As the shells and streams are drawn, the default component mass balance and energy balance equations come on the right half of the screen.

Step2: Describe the shell—Double click on the shell to enter the 'shell description' tool. For this simple model, the default equations for mass and energy balance are acceptable. Therefore, no further action is needed. Exit by clicking on 'close'.

Step3: Describe streams—Double click on stream 1 to enter the 'stream description' tool. Introduce new variables by clicking on 'variables'. Search for temperature and select it as a variable for addition. In the same manner, add also pressure as a variable. Click on 'energy-temperature' to introduce a model for enthalpy as a function of temperature (for the simple example, we will ignore the effect of pressure). Figure A3 shows the screenshot at this stage.

Close the 'model search' tool and double click on the 'enthalpy' in the 'model definition' tool. A list of models will be shown; select the DIPPR100

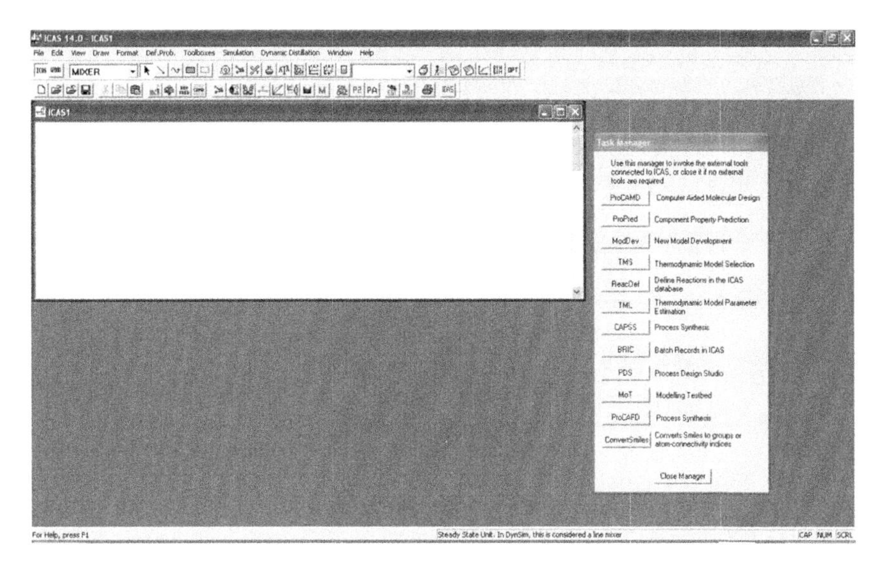

FIGURE A1 Starting screen in ICAS.

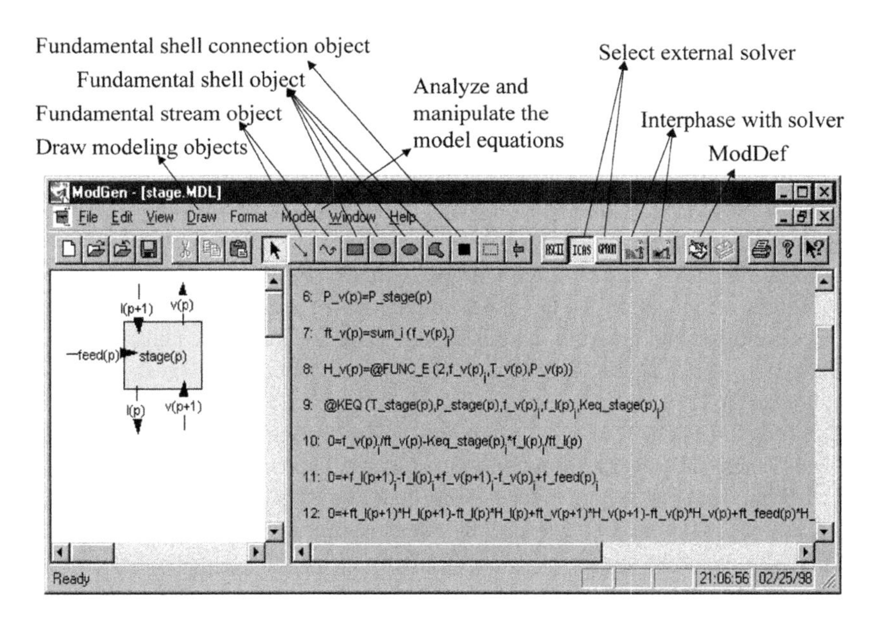

FIGURE A2 Starting screen of ModGen with available tools highlighted.

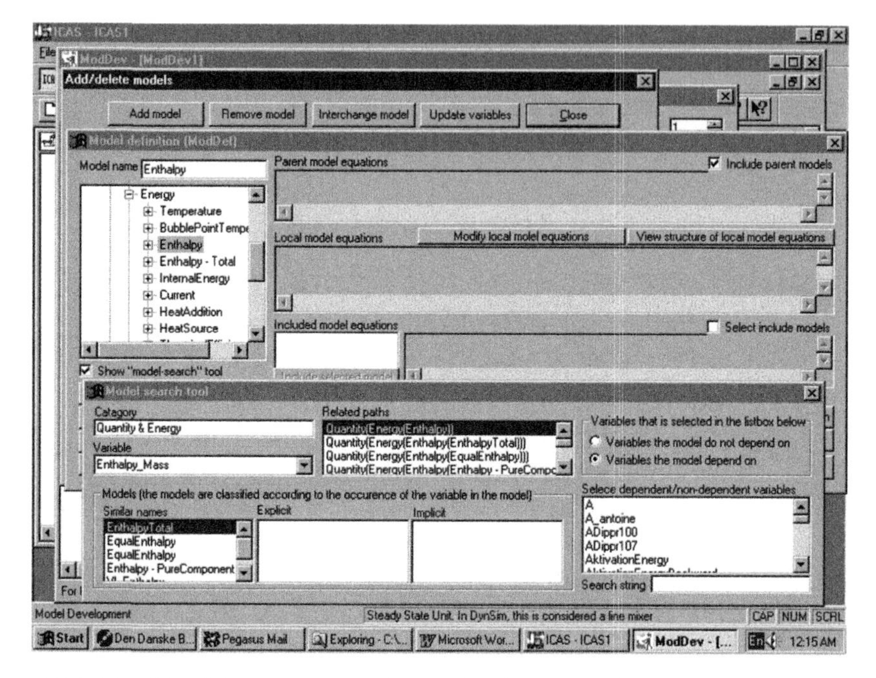

FIGURE A3 **'Add/delete models' tool plus associated tools—'model definition' and 'model search'.**

correlation for pure component enthalpy by clicking on DIPPR100. The correlation will be shown under 'local model equations'. Now drag down the 'model definition' tool and click on 'add model' in the 'add/delete models' tool. Exit by clicking on 'close' in the 'add/delete models' tool. Note that it is not possible to exit from 'model definition' when 'add/delete models' is still active. Now the main screen will show the enthalpy correlation for components in stream 1.

Repeat the above procedure for streams 2 and 3. For stream 3, also click on 'equilibrium connection' on the main 'stream description' tool. This will make the temperature and pressure of stream 3 the same as that of the tank.

At this point, on the main 'ModGen' screen, seven equations will be shown (see Figure A4).

Step4: Definition of 'closure equations'—The model is not yet complete because there are several terms on the right-hand side of Eq. 7 that have not been defined but that are not independent equations. These equations are termed as 'closure equations' or constraints. In order to define them, double click on the stream (for example, stream 1), select 'related variables' on the 'stream description' tool and then add the 'mass' and 'energy' related constraints. An example is shown in Figure A5.

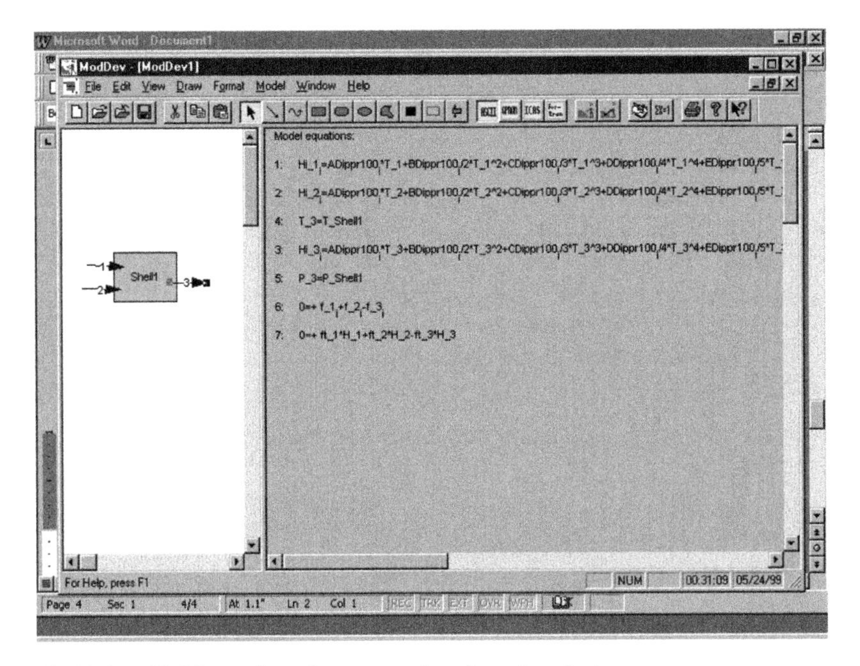

FIGURE A4 Model equations after streams have been described.

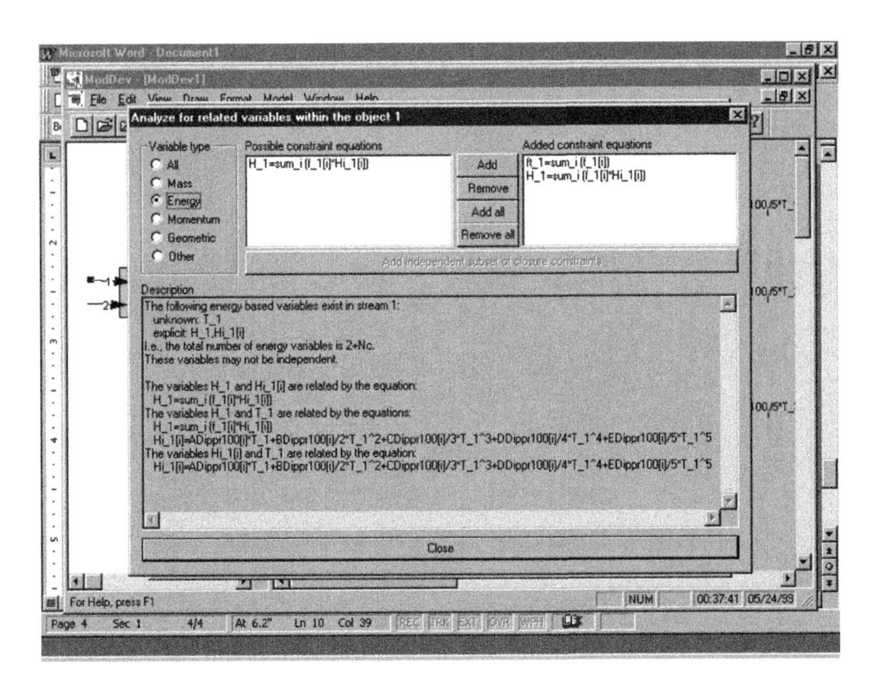

FIGURE A5 Tool for adding 'closure equations'.

Repeat this procedure for streams 2 and 3.

When 'closure equations' for all three streams have been defined, a process model consisting of a total of 13 equations is generated. This is the steady-state mass and energy balance model for a tank mixer where the effect of pressure is neglected. Figure A6 shows the final model equations.

Note that this tutorial stops at the model generation step. Before a code is generated for connection to a solver, it is necessary to analyse the model equations (see Part II of the tutorial examples).

Step5: Saving of a generated model—Select 'file' on the toolbar and then 'save as'. As name, give any preferred name. As part of this tutorial, the same model file is enclosed and it is called model1a.mdl (note that the extension will always be mdl).

Tutorial example 2—dynamic tank mixer model with negligible energy holdup

We start with the previous model (model1a.mdl) as the base model and add/change the process description. Since we would like a dynamic model, Eq. 12 is not acceptable. We need to add an 'accumulation' term.

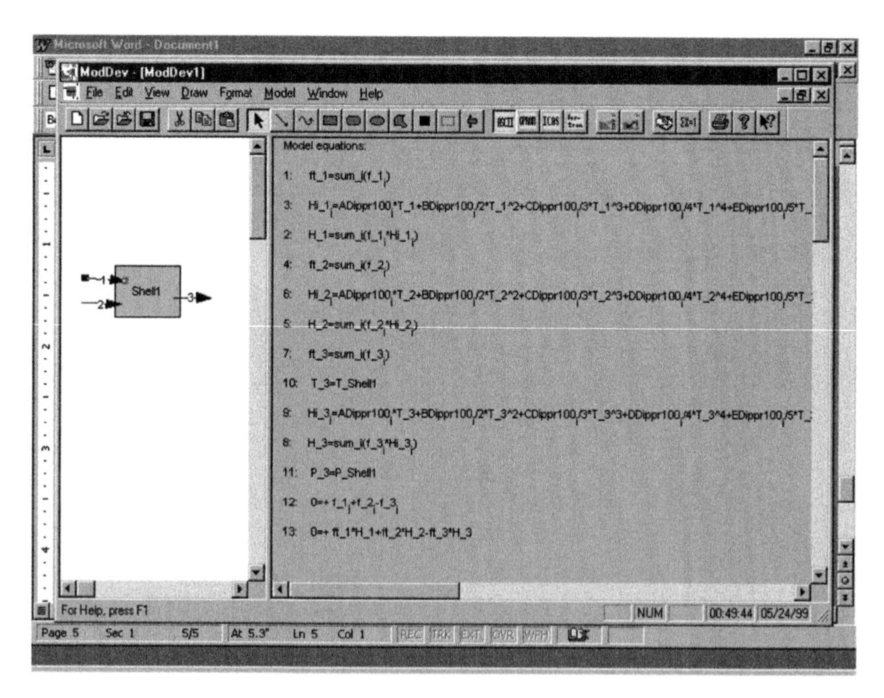

FIGURE A6 **Process model for a steady-state tank mixer.**

Step1: Change component balance equation—Double click on the shell to enter the 'shell description' tool. Click on 'component balance', select 'accumulation' in the 'component balance' tool and exit by clicking on 'close' in the 'component balance' tool. This will add a molar holdup term to the left-hand side of Eq. 12 (replacing the '0' from the steady-state version). The molar holdup term will now have to be related to the flow out of the tank (stream 3). This is done in the next step, before exiting from the 'shell description' tool.

Step2: Relate molar holdup to stream 3 flow—Click on 'other models' and double click on 'rootmodel', 'geometric', 'height' and single click on 'fluidheight' to generate the corresponding liquid level model equation (as shown in Figure A7). The model is added by clicking on 'add model' on the 'add/delete models' tool.

The next step is to relate the liquid height to the flow out (stream 3). To do this, double click on stream 3 and then click on 'other models' in the 'stream description' tool. This will again start the 'add/delete models' tool with 'model definition' and 'model search' tools. The path to finding the level-flow model is shown in Figure A8 (from 'rootmodel', 'transport' is chosen, followed by 'interphase' and, ultimately, valve3). Note that before clicking on 'add model'

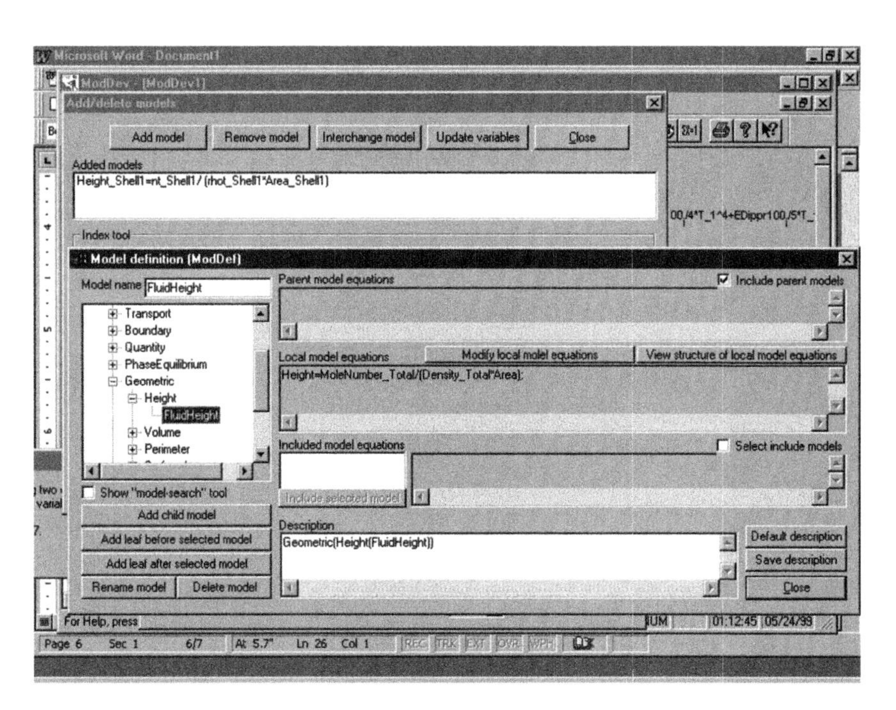

FIGURE A7 **Add the equation for liquid height in the tank.**

in 'add/delete models' tool, define the index for 'drum' as 'shell1' (shown in Figure A8).

Before leaving the 'add/delete models' tool, the equation relating the total flow out from stream 3 to the component flows for stream 3 needs to be generated. The path for finding this equation is shown in Figure A9. As in the total flow equation, before clicking on the 'add model', the 'from' index should be specified as 'shell1' on the 'add/delete models' tool.

At the end of this step, three equations are added. The first relates the total molar holdup of liquid to the height of the liquid (introducing tank geometric parameters), then the height is related to the total flow out (stream 3) and then the total flow is related to the component flow. Two additional terms—total molar liquid holdup and liquid mole fraction in the tank—are introduced by the three new equations. These additional terms are dependent on the other variables and therefore need 'closure equations'.

Finally, after adding the above three equations, ModDev gives a warning that two equations have same explicit variables. This means that one of the equations has to be removed.

Examining the model equations on the main ModDev screen shows that the two equations both have the total flow out from stream 3 as the unknown

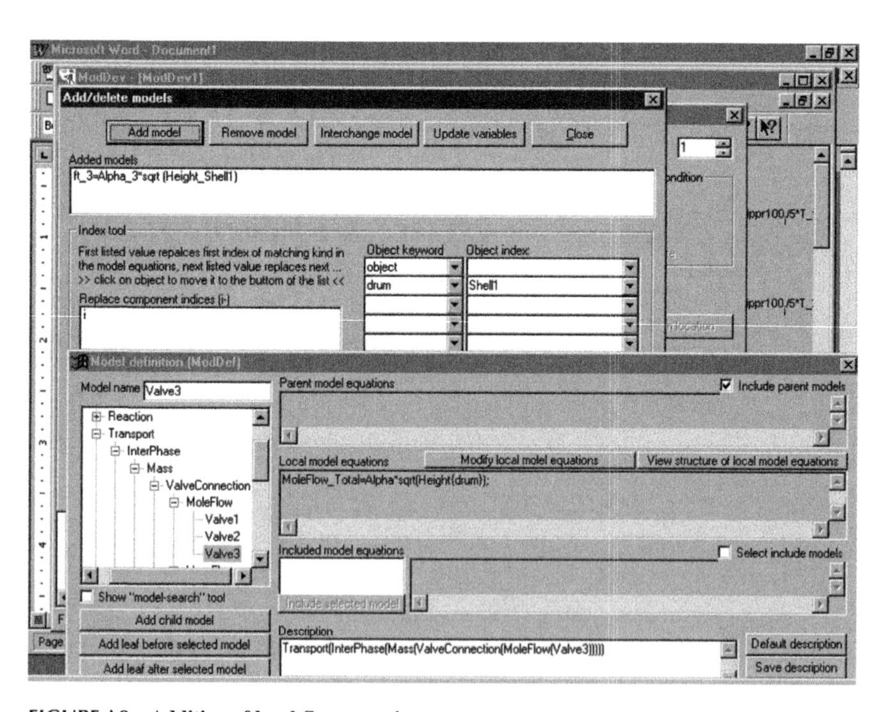

FIGURE A8 Addition of level-flow equation.

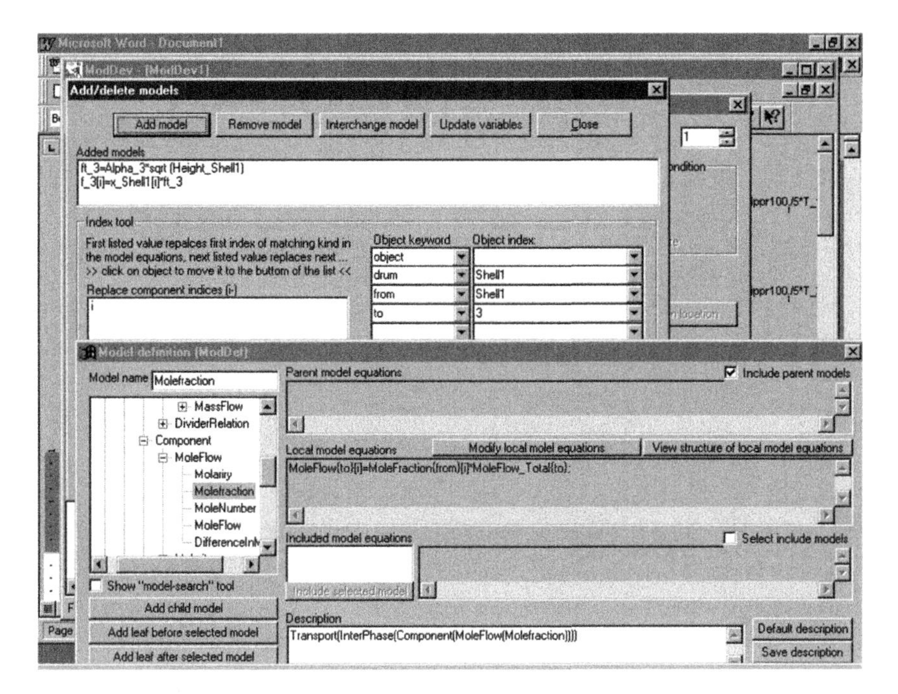

FIGURE A9 **Addition of component flow equation.**

variable (on the left-hand side of the equation). One of these equations needs to be removed, that is,

$$ft_3 = sum_3(f_{3_i}) \quad \text{A1}$$

need to be deleted. Addition of the necessary 'closure equations' and deletion of the above equation is done in the next step (step 3).

Step3: Addition and deletion of 'closure equations'—Click on the shell and then select 'model' in the toolbar. From the 'model' window items, select 'closure equations' as shown in Figure A10. From the 'closure equations' tool, remove the Eq. A1 from the list of closure equations and add the following equations,

$$xShell1_i = nShell1_i/ntShell1 \quad \text{A2}$$

$$ntShell1 = sumi(n\ Shell1_i) \quad \text{A3}$$

to the list of added closure equations (as shown in Figure A11).

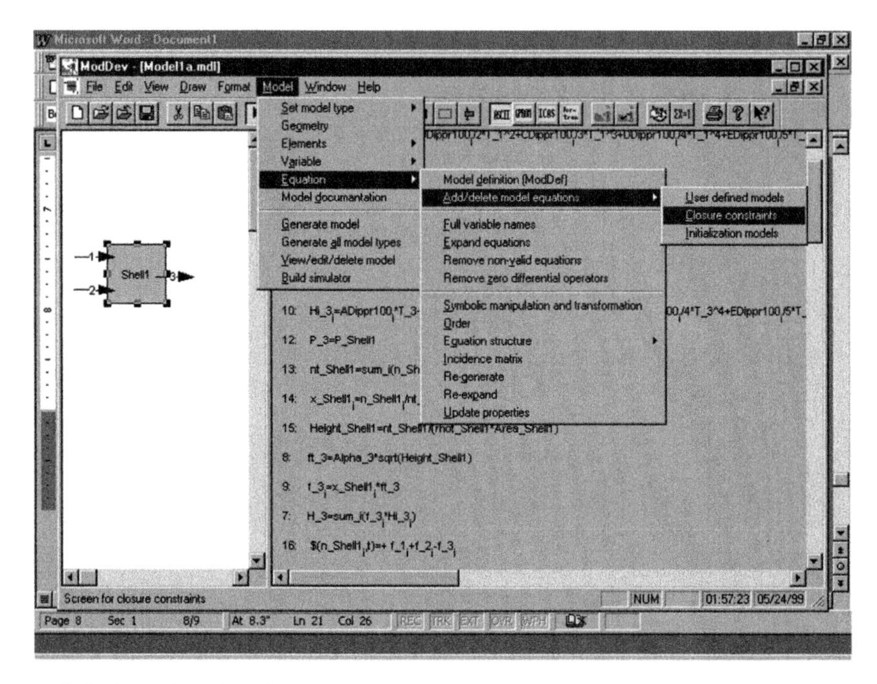

FIGURE A10 Selection of 'closure equations' from toolbar.

Exit from the 'closure equations' tool and now the warning will disappear and we will have a dynamic tank model with 17 equations as shown in Figure A12 (note that Eqs. A1 and A3, which is not shown in the figure, is the same as in Figure A6).

Step4: Save the model—Before exiting from ModDev or starting another problem, save the model under the 'save as' option. In the files related to this tutorial, the file model1b.mdl corresponds to this model.

Tutorial example 3—dynamic tank with mass and energy holdup

We start with the model from example 2 as the basis. We only need to add the energy holdup and related equations. Open ModDev with the model1b.mdl file.

Step1: Add energy holdup—Double click on the shell and then select 'energy balance'. In the 'energy balance' tool, select 'accumulation' and exit.

Step2: Relate the stream 3 enthalpy to the tank liquid enthalpy—Double click on stream 3 and select 'other models'. The 'add/delete models' tool opens and adds the model equation as shown in Figure A13.

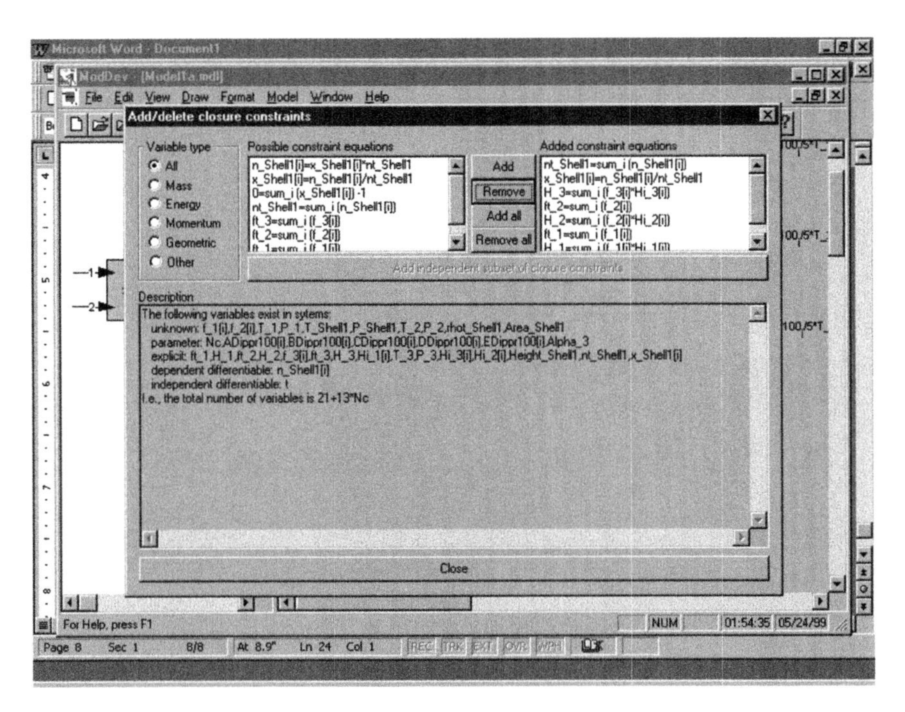

FIGURE A11 Addition and removal of closure equations.

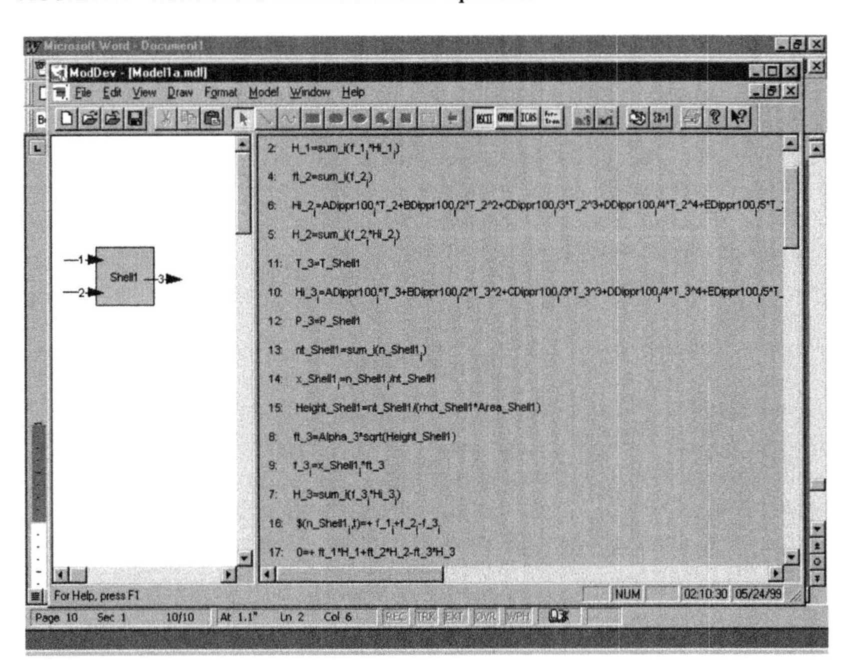

FIGURE A12 Dynamic model for a tank mixer with negligible energy holdup.

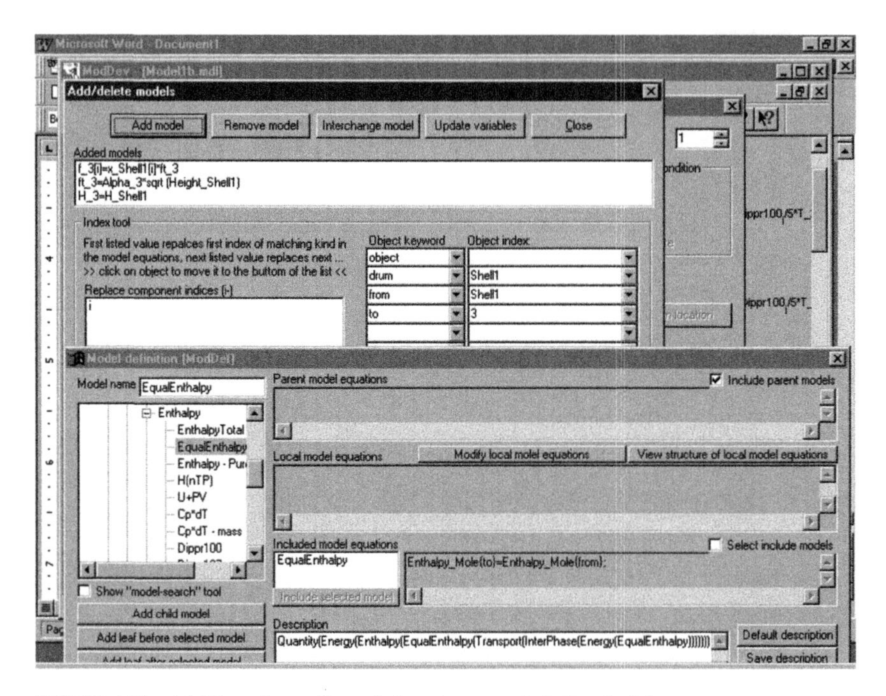

FIGURE A13 Addition of equation relating stream and shell enthalpies.

ModDev will now give a warning that the same explicit variable appears in two equations. This means one closure equation must be removed. Also, the shell enthalpy needs to be related to the total enthalpy through a closure equation.

Step3: Addition and removal of closure equations—Highlight the shell and enter the 'closure equations' tool as shown in Figure A11. Add Eq. A4 and remove Eq. A5 as shown in Figure A14.

$$\mathrm{HShell1} = \mathrm{HtShell1}/\mathrm{ntShell1} \qquad \text{A4}$$

$$H_3 = \mathrm{sumi}(f_3[i]^*\mathrm{Hi}_3[i]) \qquad \text{A5}$$

Now the warning will disappear and we will have the full model. Note that this dynamic model considers the existence of only one phase, that is, liquid. The final model equations are shown in Figure A15.

Step4: Save the model—Save this model before starting a new problem or exiting from ModDev. Use the 'save as' option. In the files related to this tutorial, the file model1c.mdl corresponds to this model.

Tutorial example 4—tank with reaction (CSTR)

For this example, any of the above three models can be used. For simplicity, model1a.mdl will be used. Models model1b.mld and model1c.mdl will need

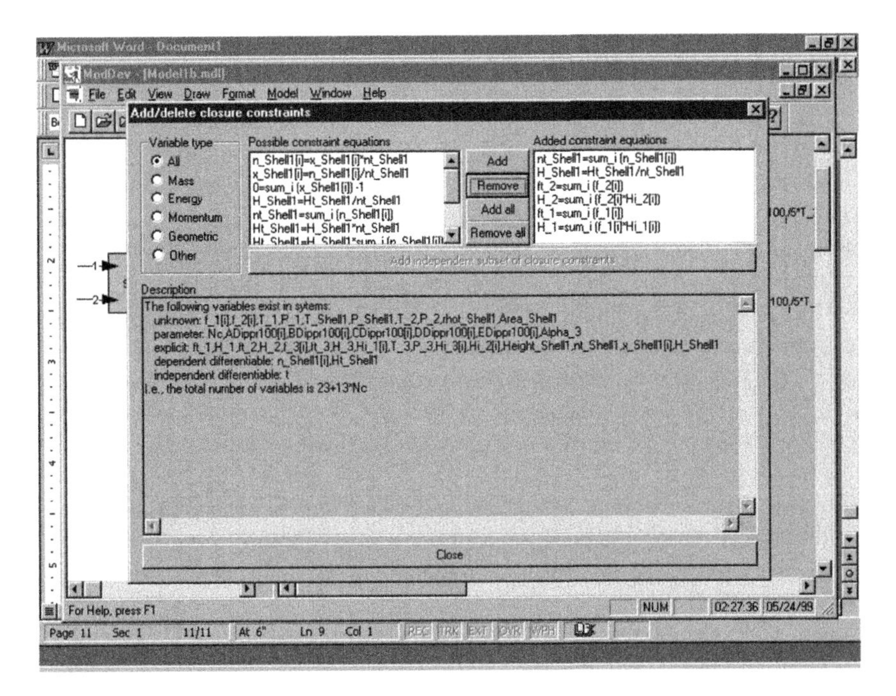

FIGURE A14 Addition and removal of closure equations.

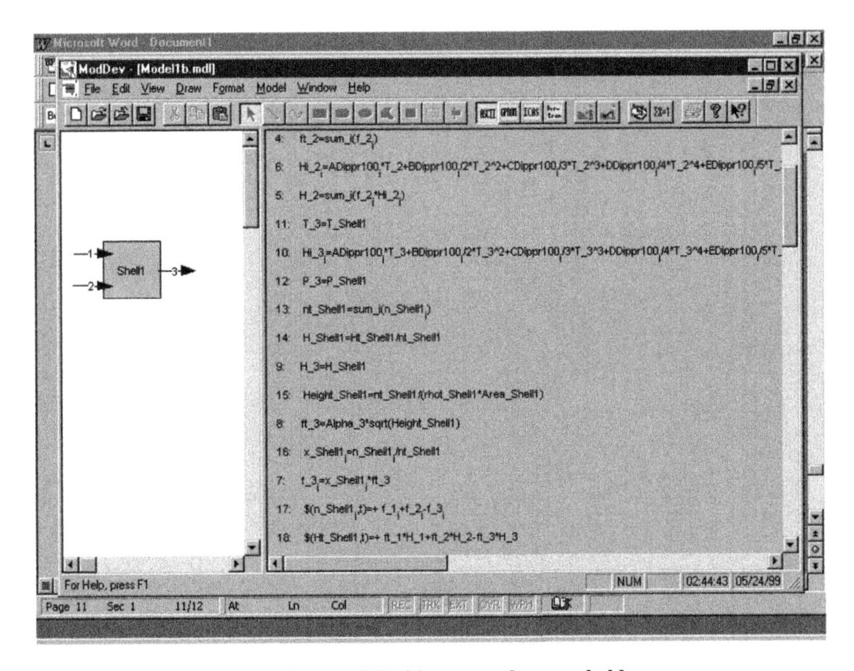

FIGURE A15 Dynamic tank mixer model with mass and energy holdup.

exactly the same steps. In the first example, a stoichiometric reaction with specified degree of reaction will be used. In the second example, a kinetic model will be used. The starting point is model1a.mdl—start ModDev by openning the file model1a.mdl.

Step1: Add the reaction term—Double click on the shell and click on 'reactions' in the 'shell description' tool. In the 'reaction' tool, double click on 'root' until 'stoichiometric' is shown. Type a name for the reaction and click on 'add child reaction' and then click on 'define components in the reaction'. Fill out the reaction details as shown in Figure A16. Back on the 'reaction' tool, select for 'add the current reaction' as shown in Figure A17.

Step2: Define equation for degree of reaction—Exit from the 'reaction' definition tool and click on 'other models'. This will open the 'add/delete models' tool. Follow the path (as shown in Figure A18) to choose the degree of reaction equation.

Note that before clicking on 'add model', the 'from' index should be set to 3 indicating the component flow from stream 3. Exit from the 'shell description' menu. The final version of the model is shown in the main screen of ModDev (see Figure A19).

Note that adding the same set of equations to the dynamic tank models generates the corresponding dynamic CSTR models (one-phase systems).

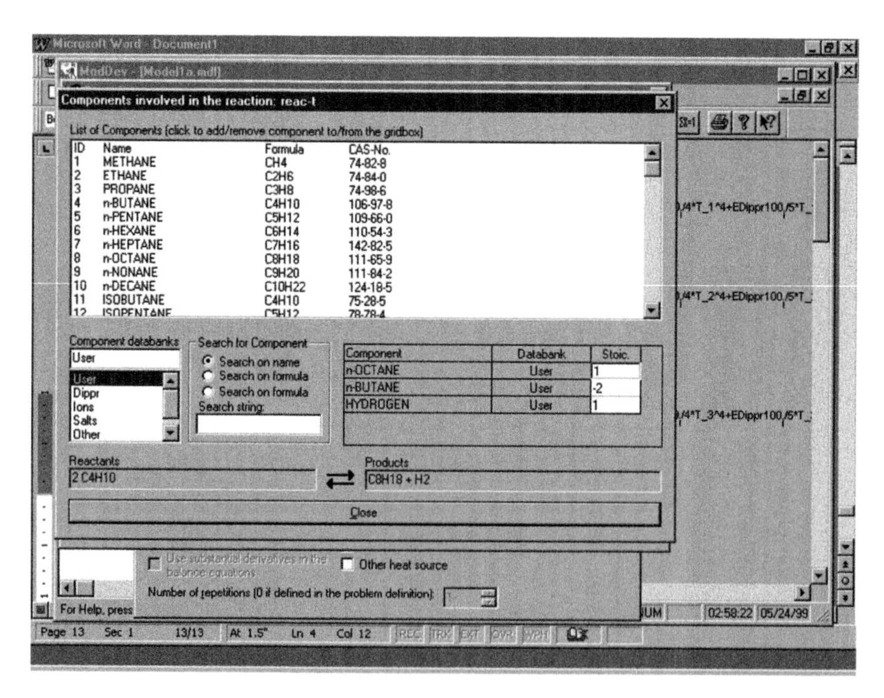

FIGURE A16 **Reaction definition.**

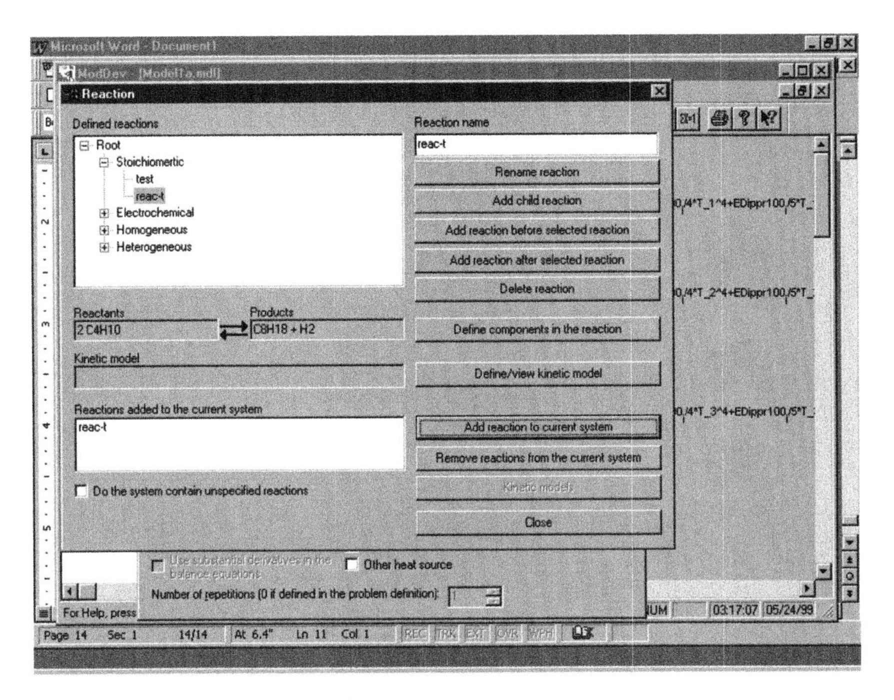

FIGURE A17 Adding reaction to the current system.

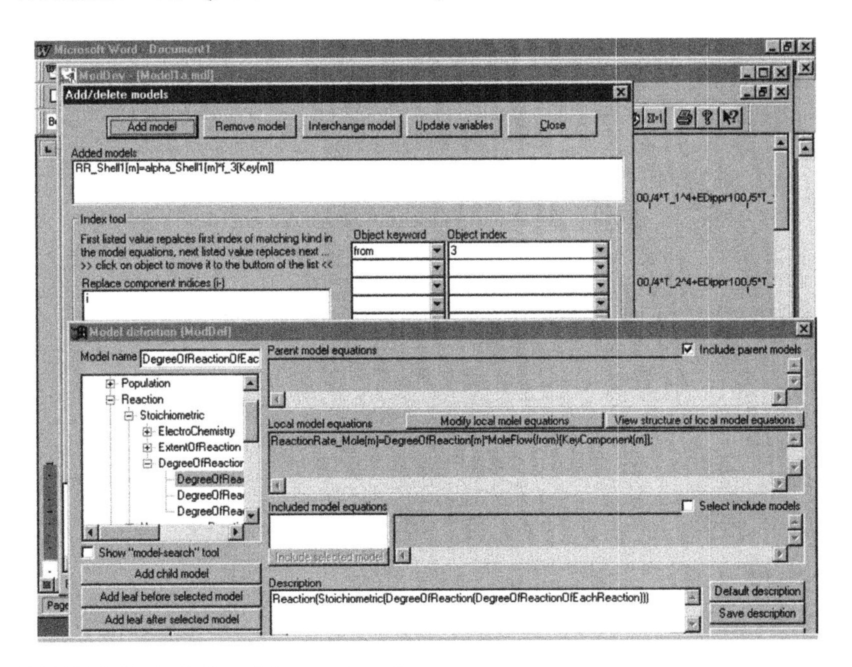

FIGURE A18 Defintion of degree of reaction.

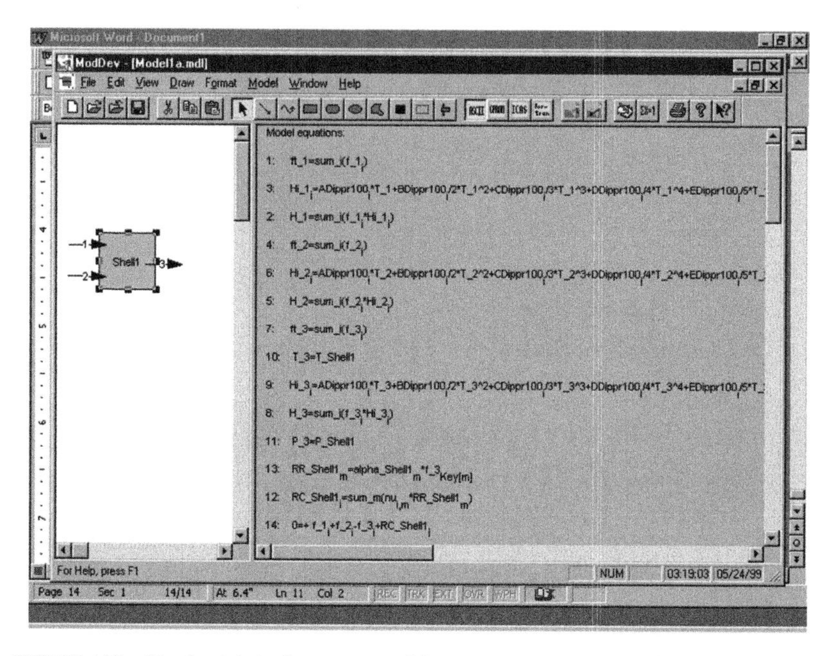

FIGURE A19 Steady-state tank-reactor model.

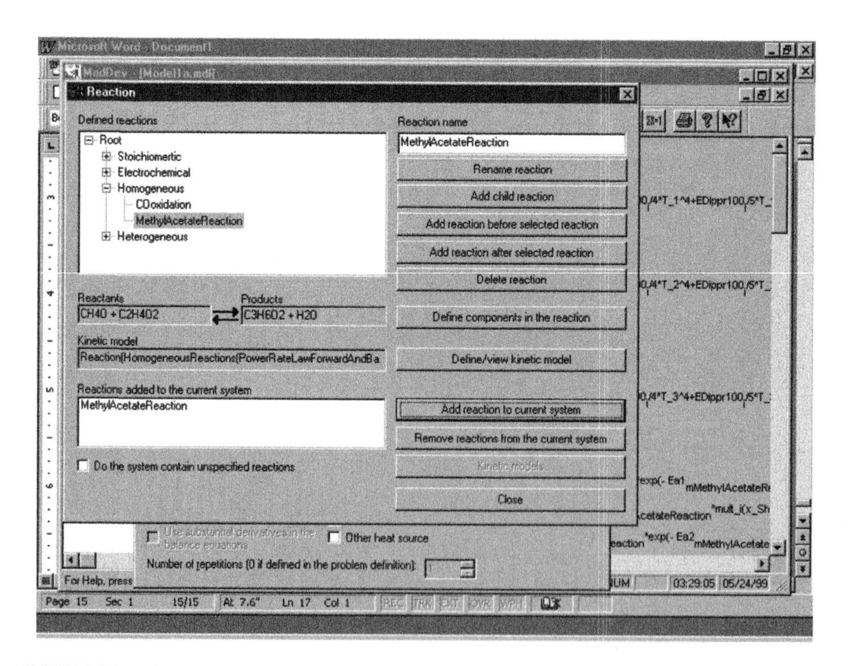

FIGURE A20 Selection of methylacetate reaction.

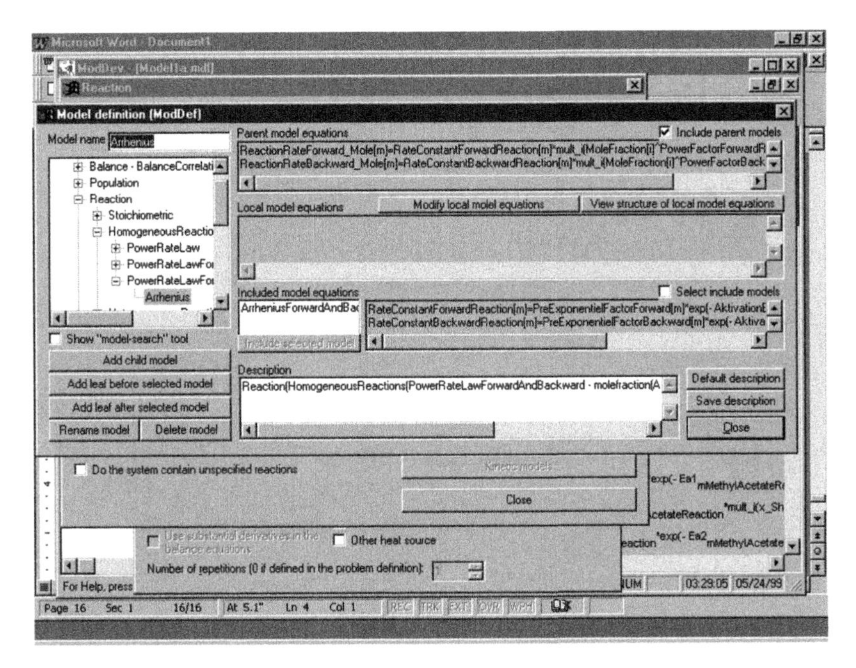

FIGURE A21 Kinetic model definition through the 'model definition' tool.

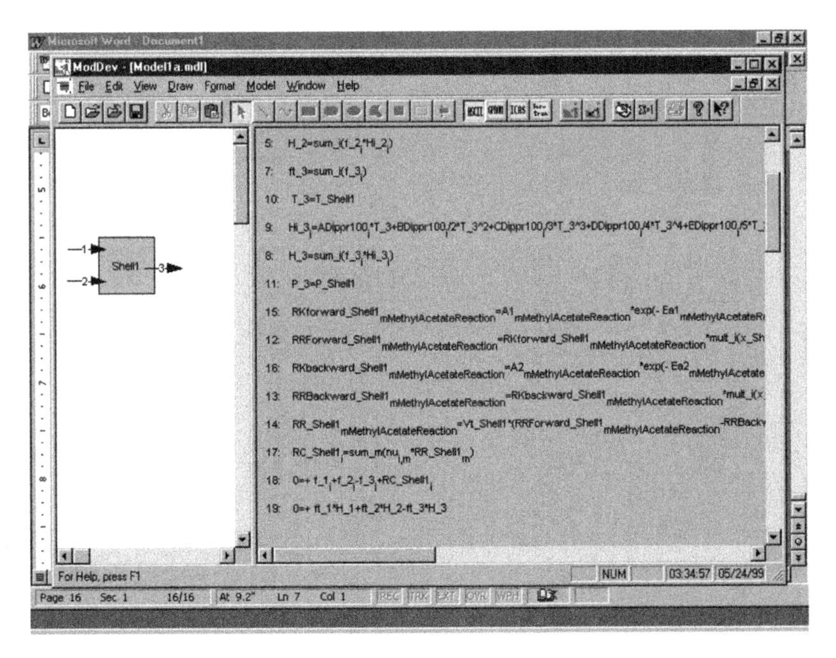

FIGURE A22 CSTR model with reaction kinetics.

Save this model as model2a-r.mdl

Tutorial example 5—tank-reactor model where kinetic model is included

The starting point again is the steady-state tank model (for simplicity). Enter the 'reaction' tool; double click on the root, the 'homogeneous' and the 'MethylAcetateReaction'. Figure A20 shows the 'reaction' tool with the various selected options. Figure A21 shows the details of the predefined kinetic model. Figure A22 shows the final process model equations. Note that when defining/viewing the kinetic model, the 'model definition' tool is opened without the 'add/delete models' tool. The final model is saved as model3a-r.mdl.

Other examples

Other examples dealing with reactors with jacketed heating, flash, evaporator and distillation columns and many more can be obtained from the authors.

Chapter 4

Overview of the Case Studies

In Part II of this book, we are trying to highlight the modelling practice through a number of modelling case studies. In each case, we have provided detailed descriptions of the models used in terms of model objectives, assumptions, model description (including model equations), model analysis and numerical solution. The data used for the numerical solution has also been given so that the readers may try to compare and/or validate their solutions with ours. We have not tried to provide a detailed derivation of the model, as in most of the case studies the models have been taken from published literature. Unless otherwise indicated, all the models are analysed and solved through the ICAS-MoT modelling tool (Sales-Cruz, 2006; Heitzig et al. 2009). The model equations in each case are listed in Appendix-1 and the corresponding ICAS-MoT file can be downloaded from a website for the book, established by Elsevier (http://www.elsevierdirect.com/companion.jsp?ISBN=9780444531612). Also, a list of other interesting models that may be downloaded from the same Website is given in Appendix-2.

An overview of the models used in the case studies of this part is given through Table 1. Examples of the workflows and data flows associated with the modelling procedure are shown in Figures 1-3. The text below gives a brief overview of the contents of the chapters of part-II.

Chapter 5

The chapters with the case studies have been organised into themes and, within them, into specific topics. Chapter 5 deals with constitutive (mainly property) models, their identification (parameter estimation) and their application. Through Sections, 5.1, 5.2 and 5.3, we have tried to highlight the use of an incremental modelling approach. That is, first we develop the constitutive property models needed to study the evaporation of a chemical from a water droplet. Then we identify the models in terms of estimating their parameter values by matching a set of collected property-related data. Then we use these identified models within the framework of an evaporation model. In addition, other examples of parameter estimation involving solid–liquid data and electrolyte systems are also given. Finally, in Section 5.3.4, the use of constitutive (kinetic) models to generate attainable region diagrams is highlighted. This is similar, in principle, to the use of constitutive (property) models to generate phase equilibrium diagrams for chemical systems.

Chapter 6

Chapter 6 deals with process simulation and optimisation. Here, we highlight the use of the equation-oriented approach as well as the sequential modular

TABLE 1 List of Case Studies Covered in Part II

Theme	Case study	Short Description	Chapter	Section
Constitutive models (development)	Pure component property	Modelling of temperature dependent correlations. The example shows the modelling of the Antoine correlation for pure component vapour pressures as a function of temperature.	5	5.1.1
		Modelling of PVT relations through an equation of state. How to model the molar volume of a chemical at a given temperature and pressure through a cubic equation of state	5	5.1.2
	Mixture property	Modelling of liquid phase activity coefficients through the Wilson model	5	5.1.3
Constitutive models (parameter estimation)	Pure component property	Regression of Antoine correlation parameters through supplied data	5	5.2.1
	Mixture properties	The regression of the Wilson model parameters by matching supplied data of activity coefficients	5	5.2.2
	Mixture properties	Regression of elec-NRTL model parameters by matching supplied solubility data of an ionic liquid	5	5.2.3
Constitutive models (application)	VLE data fit	Parameter estimation for the Wilson model parameters with vapour–liquid equilibrium data (uses also the Antoine correlation)	5	5.3.1
	SLE calculation	Use of the NRTL model within a model for SLE calculation	5	5.3.2
	Evaporation from a droplet	Use of Antoine correlation, Wilson model and a VLE calculation model within a model for calculating evaporation from a droplet	5	5.3.3

	Attainable region diagram	Use of kinetic models within model for attainable region diagram calculation	5	5.3.4
Steady-state flowsheeting	Equation oriented	Derivation of simple mass balance models and their use in process flowsheet simulation	6	6.1
	Process simulator	Use of model library for process flowsheet modelling and simulation	6	6.2
Modelling of dynamic systems	Blending tank	Derivation and solution of a complex model for a blending tank	7	7.1
	Fuel cell (DMFC)	Analysis and solution of a mutli-scale dynamic model and for a direct methanol fuel cell	7	7.2
	Fluidised bed	Analysis and solution of a multi-scale model for a fluidised bed	7	7.3
	Chemical reactor	Analysis and solution (through ICAS-MoT) of a simple dynamic model for a CSTR	7	7.4
	Polymerisation reactor	Analysis and solution of a dynamic model for a polymerisation reactor (multiple solutions)	7	7.5
Distributed parameter modelling	Processing of oil-shale	Modelling of a retorting device for the processing of oil-shale (discretisation along spatial direction)	8	8.1
	Granulation process	Modelling of a commercial granulator for producing industrial-grade phosphate fertilisers (discretisation along spatial direction)	8	8.2
	Short-path evaporation	Steady-state modelling of a short-path evaporation process (discretisation in radial direction, integration in axial direction)	8	8.3

(continued)

TABLE 1 *(continued)*

Theme	Case study	Short Description	Chapter	Section
Process modelling and simulation	Tennessee Eastman	Analysis and solution of the simplified version of the Tennessee Eastman Challenge process model	9	9.1
	Tennessee Eastman	Analysis and solution of the complete Tennessee Eastman Challenge process model	9	9.2
Batch process operations	Batch crystallisation	Development, analysis and solution of models of different types of batch cooling crystallisation	10	10.1
	Batch distillation	Uses of a dynamic model for batch distillation combined with event modelling	10	10.2
Parameter estimation	Reaction kinetics	Regression of model parameters for different types of kinetic models	11	11.1–11.3
	Reaction kinetics	Model discrimination (least squares fitting)	11	11.4
	Reaction kinetics	Model discrimination (maximum likelihood principle)	11	11.5
	Reaction kinetics	Uses of orthogonal collocation and finite difference methods in parameter estimation	11	11.6
Bio-, agro-, pharma-applications	Controlled release	Analysis and solution of a model for controlled release from a microcapsule	12	12.1
	Fermentor	Analysis and solution of a dynamic model for fermentor (for conversion of glucose to ethanol)	12	12.2
	Milk pasteurisation	Analysis and solution of a model for a milk pasteurisation process	12	12.3
	Milling	Analysis and solution of a model for a milling process (used as part of tablet manufacturing process)	12	12.4
	Granulation	Analysis and solution of a model for a granulation process (used as part of tablet manufacturing process)	12	12.4
	Tablet pressing	Analysis and solution of a model for a tablet pressing process (used as part of tablet manufacturing process)	12	12.4

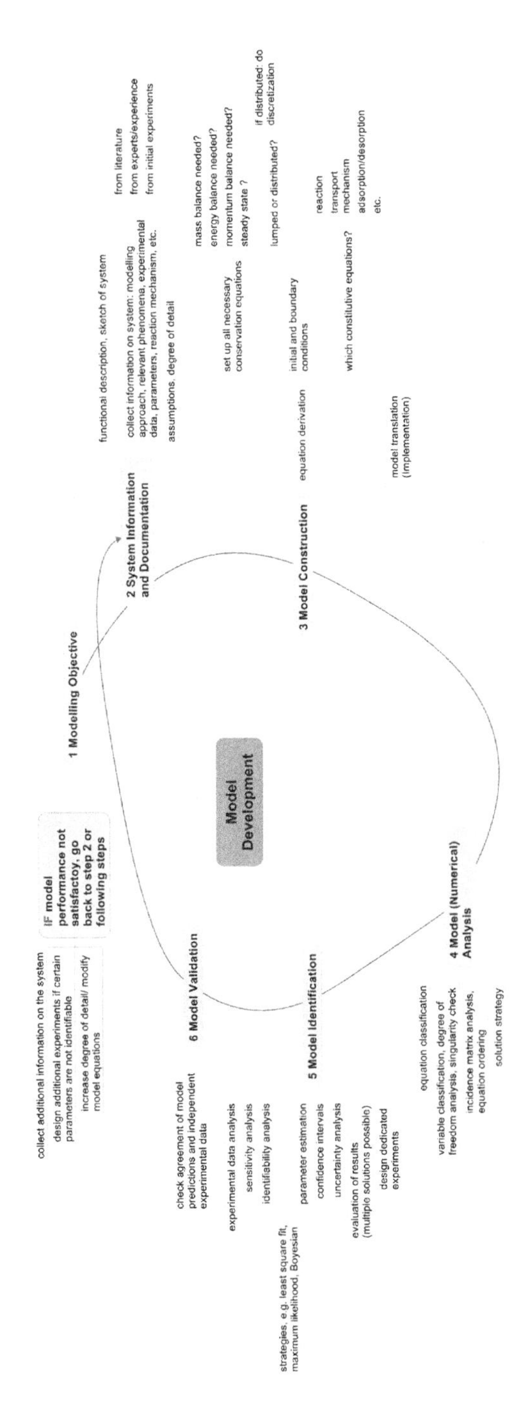

FIGURE 1 Workflow for single-scale model development.

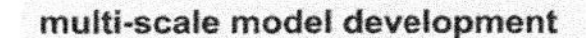

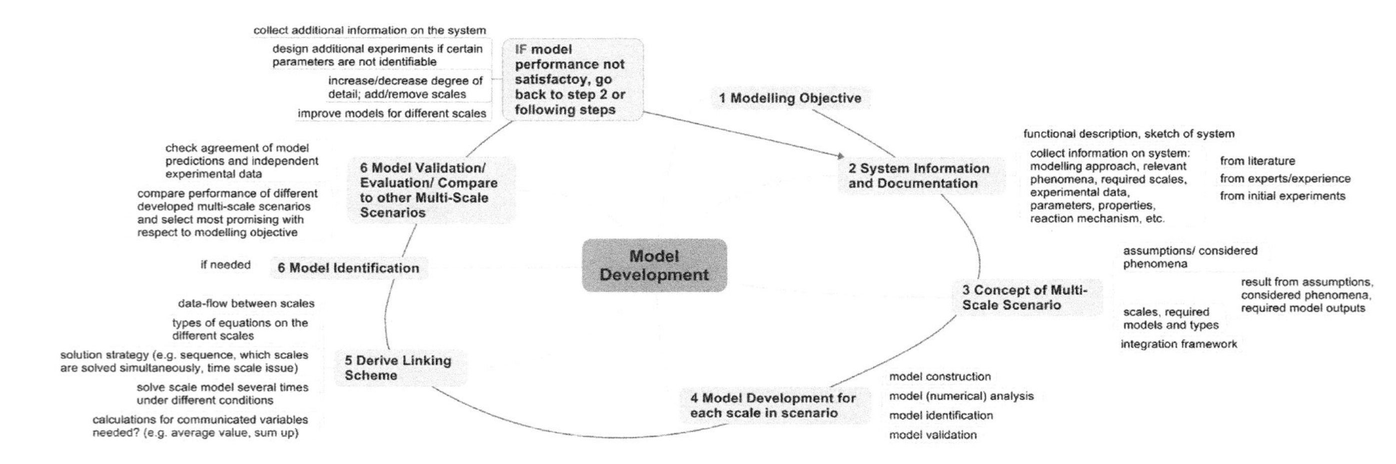

FIGURE 2 **Workflow for multi-scale model development.**

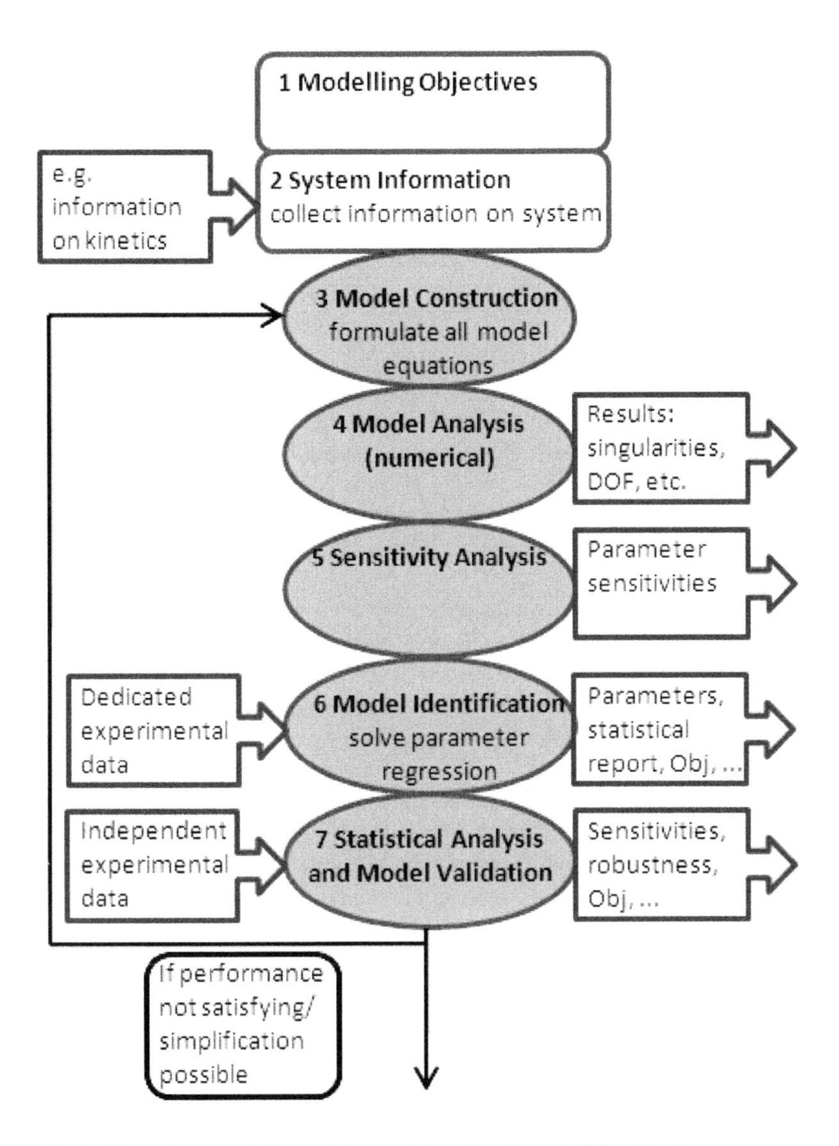

FIGURE 3 Flow diagram of the workflow and data flow in model identification.

approach. The use of two modelling approaches, for example deriving model equations and solving them or using already developed models from a model library, is highlighted. In the former, we use the ICAS-MOT modelling tool to develop, analyse and solve the model equations, whilst in the latter, we use a commercial simulator (PRO/II) to perform the simulations.

Chapter 7

Chapter 7 highlights models for different (dynamic) applications. Five detailed case studies are presented. The first deals with the modelling of a complex unit operation (blending tank). The second involves the multi-scale dynamic modelling of the performance of a direct methanol fuel cell. The third also involves multi-scale modelling, but this time studies a fluidised bed reactor. The fourth presents a typical dynamic model for a chemical reactor. The final case study in this chapter involves the dynamic modelling of a polymerisation reactor.

Chapter 8

Through chapter 8, we highlight some aspects of distributed modelling. In the first case study, an oil-shale pre-heating cooling unit is modelled. In the second case study, the dynamic performance of a granulator has been modelled, whilst in the third case study, the operation of a short-path evaporator has been modelled. In each case the model equations have been discretised in one (spatial) direction and integrated in the other spatial direction (Cases 8.1 and 8.3) or with respect to time (case 8.2).

Chapter 9

The modelling issues related to the well-known Tennessee-Eastman challenge problem are highlighted in Chapter 9 through two case studies, involving different forms of the models proposed in the published literature. Both these models are quite complex, and the use of a systematic step-by-step solution strategy helps to get the solutions, in these cases multiple solutions.

Chapter 10

Chapter 10 covers aspects of modelling, and simulation of batch operations. Two case studies are presented. In the first case study, a detailed step-by-step procedure to develop a model for the study of batch crystallisation operations is presented; then a model is developed and analysed using this procedure, and finally solved for a specific chemical system (sucrose crystallisation). In the second case study, an example of events modelling is provided through the multiple running of a batch distillation operation. Here, using an available dynamic model for a process, an events modelling feature is added to evaluate the multiple operations associated with batch distillation.

Chapter 11

Chapter 11 returns to the topic of constitutive models, in this case, different types of parameter estimation problems related to identification of kinetic models. Examples cover simple steady-state least squares regression to non-linear least squares fitting, to model selection, to the use of maximum likelihood principle and to finally, use of orthogonal collocation for dynamic optimisation.

Chapter 12

The final set of examples covered in chapter 12 highlight the theme modelling, for bio, agro and pharma applications. We start with an example from the agro-chemical sector and highlight the modelling of the performance of a microcapsule for controlled release. The second example deals with a bio-process, that is, a bio-fermentation process for the conversion of glucose to ethanol. This model is able to monitor the performance of the fermentor with respect to a number of key operational variables. The last three examples come from the pharma sector. They deal with the operational steps in the production of drug tablets. First the model of a milling process is given, then a model for a granulation process and finally, a model for a tablet pressing operation. In all three cases, the models have been developed with the objective to support process-product monitoring through application of PAT (process analytical technology). The models supplement to the collected data for monitoring the product quality under a specific condition of operation.

Chapter 5

Constitutive Models*

In this chapter, we present different types of constitutive models and examples of their application. Mainly, constitutive models related to modelling of physical properties are covered here. Another type of constitutive models, related to reaction kinetics, is covered in Chapter 11, where identification aspects of modelling are highlighted. Each modelling case is organised in terms of –

- Modelling objectives
- Model description
- Model analysis
- Numerical solution and analysis of results

The chapter is divided into three parts. The first part deals with creation and solution of property models, the second part deals with parameter estimation, and the third part highlights the application of constitutive models. These examples also illustrate model reuse and incremental modelling issues that are important in model-based studies. In most of these examples, the property (constitutive) models are analysed with respect to their numerical solution using the systematic procedure of Gani *et al.* (2006), which consists of the following steps:

Step 1. List the independent set of equations representing the model for the desired property and all the variables found in the equations; classify the variables as scalars, vectors and matrices.

Step 2. Determine the degrees of freedom (DOF) by subtracting the number of equations from the number of variables. Based on the DOF, select the variables that need to be specified and classify them as those fixed by the problem, fixed by the system, fixed by the property model and adjustable (regressed) model parameters. The remaining variables would typically be the remaining state conditions, properties and intermediate values.

Step 3. Establish an incidence matrix of equations and all variables (except those fixed by the system and the property model). If the case has relatively few variables, this may be done visually by putting equations in columns and variables in rows and putting an * on the column-row index where a variable appears in an equation.

Step 4. Eliminate the columns for variables that have been selected as specified variables (that is, variables set by the problem). The incidence matrix must then become square.

*Written in conjunction with Prof. Mauricio Sales-Cruz, Dr. Chiara Piccolo and Miss Martina Heitzig.

Step 5. Order the equations so that a lower tri-diagonal form will appear, if possible. (Standard equation ordering techniques exist in computer-aided modelling tools, such as ICAS-MoT (Sales-Cruz and Gani, 2003; Sales-Cruz, 2006), to obtain the ordered equation set.)

- **a.** If the incidence matrix shows a lower tri-diagonal form, then all the equations can be solved sequentially (one unknown for each equation) corresponding to the order giving the lower tri-diagonal form. Expect some of the equations to be nonlinear, requiring iterative solution.
- **b.** If there are elements in the upper tri-diagonal portion of the matrix, equations will need to be solved simultaneously and/or iteratively.

Note that steps 3–5 may also be combined by directly generating the ordered incidence for the square system.

Additional information on property models and how to use them for process and product design calculations is given by Kontogeorgis and Gani (2004).

5.1. CONSTITUTIVE MODELS – PHYSICAL PROPERTIES

5.1.1. Modelling of temperature-dependent pure compound physical properties

The objective here is to create a model object for use on a stand-alone basis as well as for use in other modelling applications, such as, as part of other models and/or use of the models to generate chemical system information. Here, the objective is to develop a temperature-dependent pure component property model. Typical examples for these temperature-dependent properties are vapour pressure, heat capacity, density, viscosity and many more. In this example, we will use the case of vapour pressure modelling with the Antoine correlation.

5.1.1.1 Model Description

The Antoine correlation for vapour pressure, in mmHg, at a given temperature (in°C) is given by the following equation.

$$Log_{10}Y = A - B/(T + C) \qquad (1)$$

where, Y is the vapour pressure; A, B and C are compound-specific regressed parameters, and T is the temperature.

5.1.1.2 Model Analysis

We have only one equation with one unknown variable (Y). This means that A, B, C and T must be specified in order to calculate Y.

Solution strategy: Equation 1 is regarded as an explicit algebraic equation. That is, given all the variables on the right-hand side of the equation, the variable on the left-hand side can be calculated.

5.1.1.3 Numerical Solution

To calculate the vapour pressure for any compound at a specific temperature, the coefficients A, B and C needs to be specified. For methanol, according to the CAPEC database (Nielsen, *et al.*, 2001), $A = 8.09058$; $B = 1583.72600$ and $C = 239.16200$. This correlation is valid for T_{min} (K) = 175.47 and T_{max} (K) = 512.64.

At $T = 30\,°C$, Y can be calculated with a calculator or any equation solver to be 160.941 mmHg.

Discussion: A good way to check if the vapour pressure correlation is correct or not is to check the calculated vapour pressure at the normal boiling point. The value should be the atmospheric pressure (1 atm or 760 mmHg). Another way to calculate the vapour pressure of a pure compound is to use an equation of state, such as the SRK equation of state (Soave, 1972). Saving this model as a 'model object', will make it possible to be used as parts of other models.

5.1.2. Modelling of a pure component property with an equation of state

In this example, first a property model (the SRK equation of state) is analysed and then the calculation of a pure component property is highlighted (from Gani *et al.*, 2006).

5.1.2.1 Model Description

The SRK EOS relates the pressure (P), molar volume (V) and temperature (T) of a single pure compound and is written as

$$P = RT/(V - b) - a/V(V + b) \tag{2}$$

In the above equation, R is the universal gas constant while a, and b are model parameters. Therefore, given values of R, a, b, T and P, Eq. 2 can be used to calculate V. However, the value of the parameter a depend on T as well as the 'mixing' model (only for mixtures) and pure component properties of the involved chemical, while parameter b depends only on the 'mixing' model (only for mixtures), and the pure component properties of the involved chemical.

5.1.2.2 Model Analysis

The five-step property model analysis procedure of Gani *et al.* (2006) is applied here.

Step 1: The independent equations and variables are listed in Table 1a.

Step 2: From Table 1, the total number of equations is four and the total number of variables is 12, meaning that the DOF is 8. This means that eight variables need to be specified. These are the following: three set by the acentric factor, critical temperature and critical pressure of the system compound (ω_i, T_{ci}, P_{ci}), three set by the property model (R, ψ_A, ψ_B) and two set by the problem (any two from P, V, T).

Step 3: An incidence matrix for all the equations and variables minus those set by the system and property model is given in Table 1b.

Step 4: To make the matrix square, two columns must be removed by specifying two variables from those set by the problem (any 2 from P, V, T). There are three possible problem cases: (a) specify T, P and find V; (b) specify T, V and find P; and (c) specify P, V and find T. In all the cases, two of the last three columns of the matrix are eliminated.

Step 5: The incidence matrix for case (a) is shown in Table 1c after ordering is done to achieve a lower tri-diagonal form.

Problem (b) gives the same result when P is substituted for V in the matrix. However, problem (c) does not give a lower tri-diagonal form, as shown in Table 1d. This means that, for problem (c), equations 2 and 5 need to be solved simultaneously for T.

TABLE 1a List of Property Model Equations for Pure Compound Property Calculation with the SRK EOS

Eq. No.	Equation	Number of equations
2	$P = RT/(V - b_i) - a_i/V(V + b_i)$	1
3	$m_i = 0.48 + 1.574\ \omega_i - 0.176\ \omega_i^2$	1
4	$b_i = \psi_B R\ T_{ci}/P_{ci}$	1
5	$a_i = \psi_A (R^2 (T_{ci})^2/P_{ci})[1 + m_i (1 - T/T_{ci})^{1/2}]^2$	1

Total number of equations = 4

Total number of variables = 12 [R, T, P, V, ψ_A, ψ_B, a_i, b_i, m_i, ω_i, T_{ci}, P_{ci}]

TABLE 1b Incidence Matrix for SRK EOS for Pure Components

Equation	Variables					
	a_i	b_i	m_i	P	V	T
2	*	*		*	*	*
3			*			
4		*				
5	*		*			*

TABLE 1c Ordered Incidence Matrix for Calculating the Molar Volume of a Pure Compound at Specified T, P with the SRK EOS

Equation	Variables			
	b_i	m_i	a_i	V
4	*			
3		*		
5		*	*	
2	*		*	*

TABLE 1d Ordered Incidence Matrix for Calculating Pure Compound Volume at Specified P, V with SRK EOS

Equation	Unknown Variables			
	b_i	m_i	a_i	T
4	*			
3		*		
5		*	*	*
2	*		*	*

Numerical Solution

According to the model analysis given above, to compute the pure compound molar volume (or density) at a specified temperature and pressure, it is necessary to specify the following variables:

- Property model parameters: R, ψ_A, ψ_B.
- Properties of the compound: acentric factor (ω_i), critical temperature (T_{ci}) and critical pressure (P_{ci}).

Let us consider the compound n-decane for which we will calculate the molar volume at temperature, $T = 373$ K and pressure, $P = 1$ atm. The SRK property model parameters are: $[\psi_A, \psi_B, R] = [0.42747, 0.08664, 82.0575]$. The needed properties are: $\omega_i = 0.4923$; $T_{ci} = 617.700$; $P_{ci} = 20.8240$

Solving equations 2–5 in the order given in Table 1c gives the (molar) volume, $V = 1.45$ cm^3/g.

Discussion: In principle, any other equation of state could be used in the same way as shown above. If the model equations can be ordered in the lower tri-diagonal form, a non-iterative solution strategy can be devised. Gani *et al.* (2006) has analysed more complex equations of state such as the CPA EOS

(Kontogeorgis, *et al.* 1996) and the PC-SAFT EOS (Gross and Sadowski, 2002). These models (listed in Appendix-2) are stored in a library of ICAS-MoT model objects, which can be downloaded by the interested reader.

5.1.3. Modelling of liquid-phase activity coefficients

The objective here is to create a model object for use on a stand-alone basis as well as for use in other modelling applications, such as, as part of process models and/or use of the models to generate chemical system information (phase equilibrium calculations). Here, the objective is to develop a model to calculate the activity coefficient of a compound i in a binary liquid mixture where the mixture composition and temperature are specified. In this example, we will use the simple Wilson activity coefficient model (Wilson, 1964).

5.1.3.1 Model Description

The equations for the Wilson activity coefficient model (to be called the Wilson model) can be found in many textbooks on thermodynamic properties. The standard form of the Wilson model for a binary mixture is given by the following equations:

Binary interaction terms

$$D_{1,1} = (V_1/V_1)^*\exp(-A_{1,1}/T) \tag{6}$$

$$D_{1,2} = (V_2/V_1)^*\exp(-A_{1,2}/T) \tag{7}$$

$$D_{2,1} = (V_1/V_2)^*\exp(-A_{2,1}/T) \tag{8}$$

$$D_{2,2} = (V_2/V_2)^*\exp(-A_{2,2}/T) \tag{9}$$

Internal variable $E\ (i)$

$$E_1 = x_1{}^*D_{1,1} + (1 - x_1)^*D_{1,2} \tag{10}$$

$$E_2 = x_1{}^*D_{2,1} + (1 - x_1)^*D_{2,2} \tag{11}$$

Activity coefficients (γ_i)

$$Ln(\gamma_1) = 1 - \ln(E_1) - (x_1{}^*D_{1,1}/E_1 + (1 - x_1)^*D_{2,1}/E_2) \tag{12}$$

$$Ln(\gamma_2) = 1 - \ln(E_2) - (x_1{}^*D_{1,2}/E_1 + (1 - x_1)^*D_{2,2}/E_2) \tag{13}$$

In the above equations, the vector $\underline{V}$ and the matrix $\underline{\boldsymbol{A}}$ are the model parameters, which are mixture specific, T is the temperature and the vector x_1 is the mole fraction of compound 1 in the binary mixture. Matrix $\underline{\boldsymbol{D}}$ and the vector $\underline{E}$ are internal variables.

5.1.3.2 Model Analysis

For a binary system, we have eight equations with eight unknown variables ($\underline{\boldsymbol{D}}$, $\underline{E}$, $\underline{\gamma}$). The degree of freedom is 8, which means that the model parameters ($\underline{\boldsymbol{A}}$, $\underline{V}$) and the mixture condition (x_1 and T) need to be specified.

Solution Strategy: Equations 6-13 may be organised in the lower tri-diagonal form, indicating the order of the equations to be solved. That is, solve the equations sequentially, starting from the top row. The incidence matrix showing the ordering of the equations in the lower tri-diagonal form is given in Table 2. Note that the specified variables are not included. Rows indicate equations, columns indicate variables and * indicate the variables found in each equation. The variables appearing in the diagonal are assigned as the unknown variable for the respective equation.

5.1.3.3 Numerical Solution

To calculate the activity coefficients, In ($\underline{\gamma}$), the mixture specific parameters, $\underline{V}$ and $\underline{\boldsymbol{A}}$ need to be known. The vector $\underline{V}$ refers to the volume parameters for the two compounds, while the matrix $\underline{\boldsymbol{A}}$ refers to the interaction between the two compounds (the self-interactions, $A_{i,i}$ are ignored, that is set to zero). So, only values for $A_{1,2}$ and $A_{2,1}$ are needed.

For the mixture methanol–water, the values of V and A can be found in the CAPEC database (Nielsen et al. 2001). They are ($V_1 = 0.0805$, $V_2 = 0.018$; $A_{1,2} = -22.6$, $A_{2,1} = 298.3$).

TABLE 2 Ordering of the Wilson Activity Coefficient Model Equations

Equations	Variables							
	$D_{1,1}$	$D_{1,2}$	$D_{2,1}$	$D_{2,2}$	E_1	E_2	$ln\ \gamma_1$	$ln\ \gamma_2$
6	*							
7		*						
8			*					
9				*				
10	*	*			*			
11			*	*		*		
12	*		*		*	*	*	
13		*		*	*	*		*

For $T = 298$ K and $x_1 = 0.4$, the calculated values for activity coefficients, ln (γ), are,

$$\ln(\gamma_1) = 0.0904$$

$$\ln(\gamma_2) = 0.1174$$

See Appendix-1 for a copy of the model equations as implemented and solved in ICAS-MoT (chapter-5-1-2-wilson.mot).

Discussion: A good way to check calculations is to generate γ values as a function of composition. At the two ends (mole fraction = 1 for each compound), the respective γ-values should be unity, and as the mole fraction decreases, the activity coefficient increases. Also, as an exercise, the models for NRTL (Renon and Prausnitz, 1968, UNIFAC (Fredenslund *et al.* 1972, Kang *et al.* 2003) and elec-UNIQUAC (Thomsen *et al.*, 1996) may also be tried (these models can be downloaded from the model library - see Appendix-2).

5.2. PROPERTY MODEL PARAMETER ESTIMATION

5.2.1. Pure component property

The objective here is to use experimentally measured data to regress the property model parameters. In this example, we will use the Antoine correlation model for calculating pure component vapour pressures that we have developed in Section 5.1.1. The model is the same as before, but the compound-specific model parameters are now unknown and need to be regressed using the available experimental data.

5.2.1.1 Model Description

We are going to use Eq. 1 as our model.

5.2.1.2 Model Analysis

Now we have one equation, but three unknown parameters (in addition to the vapour pressure) and 20 experimental data points. We can set up an optimisation problem where we find the best parameters corresponding to a minimum of the sum of squares of the difference between the calculated and measured values of the vapour pressures. The optimisation problem is formulated as,

Minimise,

$$Fobj = \left(\textstyle\sum_k (P_{\exp}[k] - P_{calc}[k])^2\right)/N \quad (14)$$

Subject to A, B, C

$$P_{calc}[k] = A - B/(T[k] + C) \quad (15)$$

TABLE 3 Incidence Matrix for Vapour Pressure Model Parameter Estimation

Variable/Equation	$\underline{P}_{calc}$	F_{obj}	$dF/d\theta$	A	B	C
Eq. 14	*			+	+	+
Eq. 15	*	*				
Eq. 16		*	*	+	+	+

N is the number of data points. A local solution to the above problem is obtained when $d_{Fobj}/d\theta_i$, where the vector $\underline{\theta}$ represents parameters A, B and C, is equal to zero,

$$d_{Fobj}/d\ \theta_i = 0, \quad i = 1,2,3 \tag{16}$$

Solution strategy: The optimisation problem (Eqs. 14-15) can be represented by the incidence matrix given in Table 3.

According to the incidence matrix of Table 3, it is necessary to employ an iterative solution strategy where values for A, B and C are assumed and Eqs. 14-15 are solved and, if the conditions (Eq. 16) are not satisfied, then the previous steps are repeated with another set of values for the parameters. Another alternative is to directly solve Eq. 16 (solve three equations for three unknown variables) by inserting Eq. 15 into Eq. 14 to obtain analytical expressions for Eq. 16.

5.2.1.3 Numerical Solution

Column 1 in Table 4 gives a set of measured vapour pressures for methanol. Column 2 gives the corresponding calculated values for the converged solution.

The corresponding regressed parameter values are $A = 5.1665$, $B = 1583.2$, $C = -35.282$; $F_{obj} = 6.25 \times 10^{-6}$. Appendix-1 gives the model equations as implemented in ICAS-MoT (chapter-5-2-1-antoine-fit.mot).

5.2.2. Activity coefficient model-1

The objective here is to use experimentally measured data to regress the model parameters for a model to estimate the liquid-phase activity coefficients of compounds present in a liquid solution. We will use the Wilson model (see Section 5.1.2) for calculating the activity coefficient values. The model is the same as presented in Section 5.1.2, but the compound-specific model parameters are now unknown, and needs to be regressed using the available experimental data.

5.2.2.1 Model Description

We are going to use Eqs. 6-13 as our model.

TABLE 4 Measured and Calculated Data Corresponding to Regressed Model Parameters

Data-point number	Variable values		
	P_{expt}	P_{calc}	T
1	−1.409	−1.405	273.15
2	−1.193	−1.191	281.15
3	−0.991	−0.991	289.15
4	−0.802	−0.802	297.15
5	−0.624	−0.626	305.15
6	−0.456	−0.459	313.15
7	−0.298	−0.301	321.15
8	−0.149	−0.152	329.15
9	−0.008	−0.011	337.15
10	0.125	0.122	345.15
11	0.252	0.249	353.15
12	0.372	0.369	361.15
13	0.487	0.484	369.15
14	0.596	0.594	377.15
15	0.700	0.698	385.15
16	0.800	0.798	393.15
17	0.895	0.894	401.15
18	0.986	0.985	409.15
19	1.073	1.073	417.15
20	1.157	1.157	425.15
21	1.238	1.238	433.15
22	1.315	1.315	441.15
23	1.390	1.389	449.15
24	1.462	1.461	457.15
25	1.532	1.530	465.15
26	1.600	1.597	473.15

Note: The measured and calculated pressure values are given in terms of P = Log 10 (Y), where Y is the measured or calculated value

5.2.2.2 Model Analysis

Now we have eight equations, but four unknown parameters (in addition to the eight unknown variables) and N experimental data points. Again, we can set up an optimisation problem where we find the best parameters corresponding to a minimum of the sum of squares of the differences between the calculated and measured values of the activity coefficients. The optimisation problem is formulated as,

Minimise,

$$Fobj = (\textstyle\sum_k(\gamma_{\exp}[k] - \gamma_{calc}[k])^2)/N \tag{17}$$

Subject to V_1, V_2, $A_{1,2}$, $A_{2,1}$, and Eqs. 6-13

TABLE 5 Incidence Matrix for Wilson Model Parameter Estimation

Variable/Equation	γ_{calc}	F_{obj}	$dF/d\theta$	$A_{1,2}$	$A_{2,1}$
Eqs. 6-13	*			+	+
Eq. 17	*	*			
Eq. 18		*	*	+	+

N is the number of data points. A local solution to the above problem is when $d_{Fobj}/d\theta_i$, where the vector $\underline{\theta}$ represents parameters V_1, V_2, $A_{1,\,2}$, $A_{2,\,1}$, is equal to zero. The parameters V_1 and V_2 can be calculated from the pure compound information. Therefore, it is only $A_{1,\,2}$, $A_{2,\,1}$, which need to be regressed.

$$d_{Fobj}/d\theta_i = 0, \quad i = 1, 2 \tag{18}$$

Solution Strategy: The optimisation problem (Eqs. 17-18) can be represented by the incidence matrix given in Table 5.

According to the incidence matrix of Table 5, it is necessary to employ an iterative solution strategy where values for $A_{1,\,2}$, $A_{2,\,1}$ are assumed and Eqs. 6-13, 17-18 are solved and, if the conditions (Eq. 18) are not satisfied, repeat with another set of values. Another alternative is to directly solve Eq. 18 (solve two equations for two unknown variables) by inserting Eqs. 6-13. into Eq. 18.

5.2.2.3 Numerical Solution

Table 6 gives a set of measured activity coefficient values for methanol. In principle, the measured values for water should also have been given.

The result of the regression is shown in Figure 1, where the regressed parameter values for A (1, 2) and A(2, 1) are given together with plots of γ_i as a function of x_i. Note that compound 1 is methanol and values for V_1 and V_2 are fixed to 0.08050 and 0.01809, respectively.

Discussion: From Figure 1, it can be seen that a good fit of the data has been obtained. This regressed set of parameters can now be used as initial estimates for 'fine-tuning' the same model parameters with respect to vapour–liquid phase equilibrium data (see Section 5.3.1). In the same way, model parameters for any activity coefficient model can be regressed by providing a set of measured (or generated) data of activity coefficients. As an exercise, try to estimate the NRTL- and UNIFAC-model parameters using the dataset given in Table 6.

5.2.3. Activity coefficient model-2

The objective here is to use experimentally measured data to regress the model parameters for the elec-NRTL model to estimate the molecular interaction parameters involving ionic liquids in water.

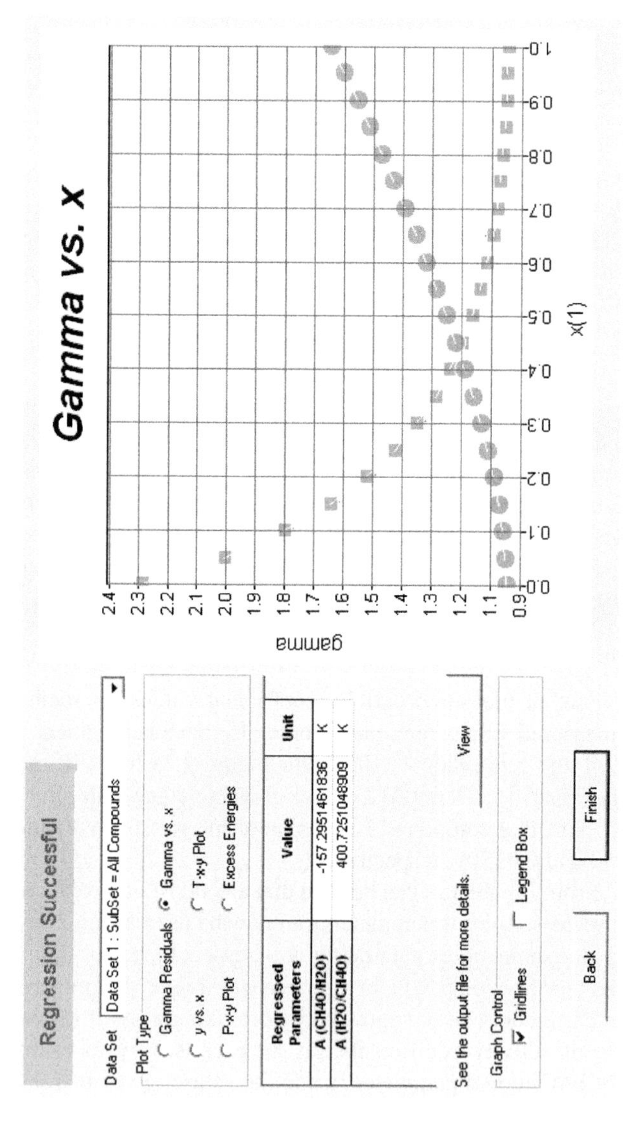

FIGURE 1 **Results of regression for the Wilson model parameters.**

TABLE 6 Generated Activity Coefficient Values at $T = 298.15$ K as a Function of Composition

Data-point number	Variable values		
	T	x_i	$\gamma_{expt}(i)$
1	298.15	0.0000	2.2447
2	298.15	0.0500	1.9576
3	298.15	0.1000	1.7484
4	298.15	0.1500	1.5912
5	298.15	0.2000	1.4702
6	298.15	0.2500	1.3753
7	298.15	0.3000	1.2997
8	298.15	0.3500	1.2388
9	298.15	0.4000	1.1893
10	298.15	0.4500	1.1488
11	298.15	0.5000	1.1157
12	298.15	0.5500	1.0885
13	298.15	0.6000	1.0664
14	298.15	0.6500	1.0484
15	298.15	0.7000	1.0339
16	298.15	0.7500	1.0226
17	298.15	0.8000	1.0139
18	298.15	0.8500	1.0075
19	298.15	0.9000	1.0032
20	298.15	0.9500	1.0008
21	298.15	1.0000	1.0000

5.2.3.1 Model Description

The elec-NRTL model (Chen and Evans, 1986) equations for a binary mixture are given below (subscripts *a,* and *c* indicate anion and cation, respectively).

Definition of mole fractions

$$v = v_c + v_a \tag{19}$$

$$X_a = v_a{*}m/[(1000/Ms) + (v{*}m)] \tag{20}$$

$$X_c = v_c{*}m/[(1000/Ms) + (v{*}m)] \tag{21}$$

$$X_m = 1 - X_a - X_c \tag{22}$$

Internal parameters - $\underline{\tau}$ and **G**

$$\tau_{ca-m} = A_{ca-m}/(R^*T) \quad (23)$$

$$\tau_{m-ca} = A_{m-ca}/(R^*T) \quad (24)$$

$$G_{m-ca} = \exp(-C_{m-ca}{}^*\tau_{m-ca}) \quad (25)$$

$$G_{ca-m} = \exp(-C_{ca-m}{}^*\tau_{ca-m}) \quad (26)$$

Total ion activity coefficients are written as,

$$\ln\gamma_{a-total} = \ln\gamma_a + \ln\gamma_{DH-a} \quad (27)$$

$$\ln\gamma_{c-total} = \ln\gamma_c + \ln\gamma_{DH-c} \quad (28)$$

where NRTL term for ions are given by,

$$\begin{aligned}\ln\gamma_c = &\{[(X_m)^2]^*\tau_{ca-m}{}^*G_{ca-m}\}/(X_a{}^*G_{ca-m} + X_c{}^*G_{ca-m} + X_m)^2 \\ &+ (\tau_{m-ca}{}^*Z_c{}^*X_m{}^*G_{m-ca})/(X_a + X_m{}^*G_{m-ca}) \\ &- (\tau_{m-ca}{}^*Z_a{}^*X_a{}^*X_m{}^*G_{m-ca})/[(X_m{}^*G_{m-ca} + X_c)^2] \\ &- \tau_{ca-m}{}^*G_{ca-m} - \tau_{m-ca}{}^*Z_c\end{aligned} \quad (29)$$

$$\begin{aligned}\ln\gamma_a = &\{[(X_m)^2]^*\tau_{ca-m}{}^*G_{ca-m}\}/(X_a{}^*G_{ca-m} + X_c{}^*G_{ca-m} + X_m)^2 \\ &+ (\tau_{m-ca}{}^*Z_a{}^*X_m{}^*G_{m-ca})/(X_c + X_m{}^*G_{m-ca}) \\ &- (\tau_{m-ca}{}^*Z_c{}^*X_c{}^*X_m{}^*G_{m-ca})/[(X_m{}^*G_{m-ca} + X_a)^2] \\ &- \tau_{ca-m}{}^*G_{ca-m} - \tau_{m-ca}{}^*Z_a\end{aligned} \quad (30)$$

where Debye–Hückel term for ions are given by,

$$\begin{aligned}\ln\gamma_{DH-a} = &- [(1000/Ms)^{1/2}]^*A_\varphi{}^*[(2/\rho)^*(Z_a)^2]^*[\ln(1+\rho^*I^{1/2})] \\ &+ (Z_a)^{2*}(I^{1/2} - 2^*I^{1/2})/(1+\rho^*I^{1/2})\end{aligned} \quad (31)$$

$$\begin{aligned}\ln\gamma_{DH-c} = &- [(1000/Ms)^{1/2}]^*A_\varphi{}^*[(2/\rho)^*(Z_c)^2]^*[\ln(1+\rho^*I^{1/2})] \\ &+ (Zc^2)^*(I^{1/2} - 2^*I^{1/2})/(1+\rho^*I^{1/2})\end{aligned} \quad (32)$$

where Debye–Hückel parameter (A_φ) and ionic strength (I) are given by,

$$A_\varphi = 1/3^*(2^*3.1415^*N_a{}^*d/1000)^{0.5*}[(e^2)/(\varepsilon_0{}^*\varepsilon^*\kappa^*T)]^{3/2} \quad (33)$$

$$I = 0.5^*[X_a{}^*(Z_a)^2 + X_c{}^*(Z_c)^2] \quad (34)$$

Mean molal activity coefficient is written as,

$$\ln \gamma_m = (1/v)*(v_c*\ln \gamma_{c-total} + v_a*\ln \gamma_{a-total}) - \ln(1 + 0.001*Ms*m*v) \quad (35)$$

Note: in the above Eqs. 19-35, m is molality (moles ions/kg solvent); Z_a is the absolute value of charge of anion; Z_c is the absolute value of charge of cation; Ms is the solvent molecular weight; and v is the salt stoichiometric coefficient.

5.2.3.2 Model Analysis

Now we have 17 equations (Eqs. 19-35) that are needed to compute the mean molal activity coefficients; the elec-NRTL model parameters are: A_{ca-m} and A_{m-ca}. We have N experimental data points for the mean molal activity coefficients for the system IL (ionic-liquid) in water. Again, we can set up an optimisation problem where we find the best parameters corresponding to a minimum of the sum of squares of the difference between the calculated and measured values of the vapour pressures. The optimisation problem is formulated as,

Minimise,

$$Fobj = (\textstyle\sum_k (\gamma_{m-expt}[k] - \gamma_{m-calc}[k])^2)/N \quad (36)$$

Subject to A_{ca-m}, A_{ca-m} and Eqs. 19-35

N is the number of data points. A local solution to the above problem is when $d_{Fobj}/d\theta_i$, where the vector $\underline{\theta}$ represents the parameters A_{ca-m}, A_{ca-m}, is equal to zero.

$$d_{Fobj}/d\theta_i = 0, \quad i = 1, 2 \quad (37)$$

Solution Strategy: The incidence matrix for the optimisation problem (Eqs. 36-37) is similar to the one given in Table 5. In Table 7, the incidence matrix for solving Eqs. 19-35 is given.

The degrees of freedom for the elec-NRTL model (Eqs. 19-37) related to the calculation of the mean molal activity coefficient for a binary mixture is 19. This means that the following variables need to be specified: constants (Na; d; e; ε_0; ε; κ; Ms, R); system variables (ρ, m, Z_a, Z_c, v_a; v_c; T); NRTL model parameters (C_{m-ca}; C_{ca-m}; A_{m-ca}; A_{ca-m}).

TABLE 7 Incidence Matrix for Calculating the Mean Molal Activity Coefficients with the elec-NRTL Model

Variable/Equation	γ_{m-calc}	F_{obj}	$dF/d\theta$	A_{ca-m}	A_{m-ca}
Eqs. 19-35	*			+	+
Eq. 36	*	*			
Eq. 37		*	*	+	+

Similar to the optimisation problem in Section 5.2.2, it is necessary to employ an iterative solution strategy where the starting values for $A_{ca\text{-}m}$, $A_{ca\text{-}m}$ are guessed, Eqs. 19-35 are solved and, if the conditions (Eq. 37) are not satisfied, the procedure is repeated with another set of parameter values. Another alternative is to directly solve Eq. 37 (solve two equations for two unknown variables) by inserting Eqs. 19-35 into Eq. 37 and obtaining the analytical expressions for Eq. 37.

5.2.3.3 Numerical Solution

Measured mean molal activity coefficients of an IL (tributylammonium chloride) in water (Lindenbaum and Boyd, 1964) are given in Table 8. The parameter estimation progress in terms of objective function value and the parameters as a function of iteration number is shown in Figures 2a–2c.

Values of the specified variables are: $Na = 6.0221E+23$; $d = 1000$; $e = 1.6022E\text{-}19$; $\varepsilon_0 = 8.8542E\text{-}12$; $\varepsilon = 80$; $\kappa = 1.3807E\text{-}23$; $C_{m\text{-}ca} = 0.2$; $C_{ca\text{-}m} = 0.2$; $T = 298.15$; $R = 8.3145$; $Ms = 18$; $Za = 1$; $Zb = 1$; $\rho = 14.9$; $v_a = 1$; $v_b = 1$.

The result of the regression is shown in Figures 2b-2c. The optimal parameter values in terms of $\tau_{ca\text{-}m}$ and $\tau_{m\text{-}ca}$ are: $\tau_{ca\text{-}m} = -4.16$ $\tau_{m\text{-}ca} = 8.12$, which are quite close to those reported by Belveze *et al.* (2004). The implementation of this model in ICAS-MoT is given in Appendix-1 (chapter-5-2-3-elec-nrtl-parfit.mot).

Discussion: The same procedure can be repeated for other 'electrolyte' binary systems involving water as the solvent. Replacing Eqs. 19-35 for another activity coefficient model would allow one to estimate the parameters of this model. The fitted model can be applied in process models where liquid-phase activity coefficients are needed, as well as for generating phase equilibrium diagrams.

5.3. CONSTITUTIVE MODEL APPLICATIONS

5.3.1. Vapour-Liquid Equilibrium Calculation

The objective here is to use the property models (pure component as well as activity coefficient) developed above in a new model for the calculation of vapour–liquid equilibrium (VLE) calculations. That is, for a given pressure and liquid-phase composition, calculate the bubble point temperature and the corresponding vapour-phase composition.

5.3.1.1 Model Description

We are going to use Eqs. 1, 6-13 as the constitutive models to be embedded for the calculation of a single point of a VLE data series. The VLE condition at a given pressure and liquid-phase composition is given by (assuming ideal vapour phase and using the gamma-phi approach),

TABLE 8 Mean Molal Activity Coefficient Values at T = 298.15 K as a Function of Molality of IL

Data-point number	Variable values		
	T	M	$\gamma_{m\text{-}expt}$
1	298.15	0.1	0.752
2	298.15	0.2	0.701
3	298.15	0.3	0.670
4	298.15	0.4	0.650
5	298.15	0.5	0.637
6	298.15	0.6	0.629
7	298.15	0.7	0.626
8	298.15	0.8	0.625
9	298.15	0.9	0.625
10	298.15	1.0	0.627
11	298.15	1.2	0.629
12	298.15	1.4	0.638
13	298.15	1.6	0.639
14	298.15	1.8	0.640
15	298.15	2.0	0.640
16	298.15	2.5	0.636
17	298.15	3.0	0.627
18	298.15	3.5	0.614
19	298.15	4.0	0.597
20	298.15	4.5	0.583
21	298.15	5.0	0.574
22	298.15	5.5	0.568
23	298.15	6.0	0.564
24	298.15	7.0	0.562
25	298.15	8.0	0.564
26	298.15	9.0	0.574
27	298.15	10.0	0.587
28	298.15	11.0	0.603
29	298.15	12.0	0.621
30	298.15	13.0	0.644
31	298.15	14.0	0.667
32	298.15	15.0	0.688
33	298.15	18.0	0.752

$$K_i = \gamma_i p_i^S / P \tag{38}$$

$$y_i = K_i x_i \tag{39}$$

$$\mathrm{R}_{esid} = \left[\sum_i (1 - y_i)\right] \tag{40}$$

$$T = f(d\mathrm{R}_{esid}/dT,\ T) \tag{41}$$

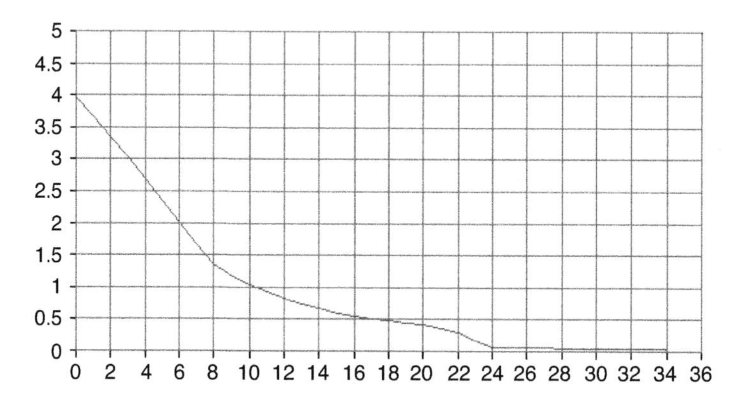

FIGURE 2a **Results of regression for the elec-NRTL model parameters (y-axis is the objective function, and x-axis is the number of iterations).**

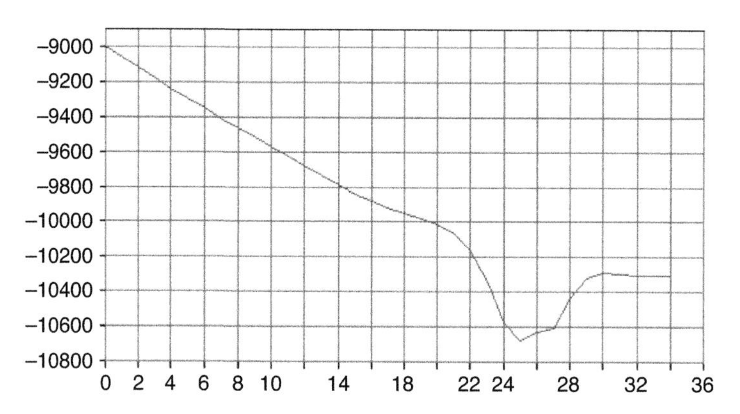

FIGURE 2b **Results of regression for the elec-NRTL model parameters (y-axis is A (ca, m), and x-axis is the number of iterations).**

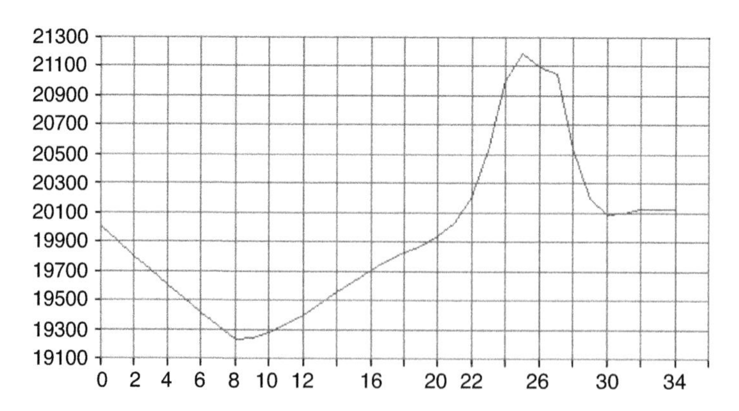

FIGURE 2c **Results of regression for the elec-NRTL model parameters (y-axis is A (m, ca), and x-axis is the number of iterations).**

In the above equations, K_i is the equilibrium constant, p_i^S is the vapour pressure at a specific temperature, P is the system pressure, x_i and y_i are the equilibrium liquid and vapour mole fractions of compound i, respectively, and R_{esid} is the residual value, which should be equal to zero (within a convergence criteria) at equilibrium.

5.3.1.2 Model Analysis

We now have two equations for vapour pressure (Eq. 1), eight equations for activity coefficients (Eqs. 6-13), two equations for the equilibrium constants (Eq. 38), two equations for the vapour-phase mole fractions (Eq. 39) and one equation for R_{esid} (Eq. 40). For the five new equations (counting only Eqs. 38-40) we have $\underline{K}$, $\underline{y}$ and R_{esid} as the unknown variables with P and x_i as the specified variables. Note that Eq. 41 is used to generate new values of T in an iterative solution scheme.

Solution Strategy: The bubble point calculation (Eqs. 1, 6-13 and 38-41) can be represented by the incidence matrix given in Table 9, where equations 1, 6-13, 38-39 and 40-41 are represented as a set. Again, only the unknown variables are used in the incidence matrix.

According to the incidence matrix of Table 9, it is necessary to employ an iterative solution strategy where for an assumed value of T, Eqs. 1, 6-13 are solved and based on this, if the condition for equilibrium (Eq. 40-41) is also satisfied, then the assumed temperature is the bubble point temperature and the corresponding y_i are the equilibrium vapour mole fractions. Otherwise, Eq. 41 is used to generate new value for T and the procedure repeated. The incidence matrix also shows that since a lower tri-diagonal form cannot be found, all the equations need to be solved either simultaneously or sequentially in an iterative manner (as indicated above). Note that in the former, Eq. 41 is not included. Also, the incidence matrix shows that for each iteration two sets of constitutive model equations (models for vapour pressure and activity coefficients) need to be solved.

TABLE 9 Incidence Matrix for Bubble Point Calculation

Variable/Equation	$\underline{p}^{sat}$	$\underline{\gamma}$	K_i	y_i	dR_{esid}/dT	T
Eq. 1	*					+
Eqs. 6-13		*				
Eq. 38	*	*	*			
Eq. 39			*	*		
Eq. 40-41				*	*	*

5.3.1.3 Numerical Solution

Consider two cases:

i) Consider the following bubble point temperature calculation – what is the bubble point temperature for a mixture of methanol–water having 0.4 methanol mole fraction in the liquid phase and for a system pressure of 1 atm? Using the following Antoine correlation parameters for methanol and water,
Methanol: $A = 8.09058$; $B = 1583.72600$ & $C = 239.16200$
Water: $A = 8.08131$; $B = 1730.63$; & $C = 233.426$
and the Wilson model parameters given in Figure 1, the following values are obtained – bubble point temperature = 350.42 K, $y_1 = 0.7225$

ii) In a second problem, multiple measured data of pressure, temperature, liquid and vapour mole fractions are given. This data is used to fine-tune the activity coefficient model. The model developed above for single point calculation is now repeated for N points and then an objective function is added to minimise the sum of squares of the residuals to obtain a fine-tuned set of Wilson model parameters. Table 10 gives a set of measured VLE data.

TABLE 10 Measured VLE Data for Methanol–water Mixture (CAPEC Database (Nielsen *et al.*, 2001)

Data-point number	Variable values			
	P	T	x_1	y_1
1	101.3000	369.5600	0.0190	0.1370
2	101.300	362.7200	0.0770	0.3680
3	101.300	358.3800	0.1330	0.4920
4	101.300	357.7600	0.1400	0.5080
5	101.300	354.5100	0.2050	0.5940
6	101.300	352.4500	0.2600	0.6470
7	101.300	350.8600	0.3100	0.6830
8	101.300	350.5800	0.3200	0.6900
9	101.300	349.5400	0.3580	0.7130
10	101.300	348.0700	0.4200	0.7460
11	101.300	347.0900	0.4620	0.7680
12	101.300	346.3800	0.5010	0.7870
13	101.300	344.7000	0.5860	0.8260
14	101.300	343.8300	0.6380	0.8480
15	101.300	343.5900	0.6510	0.8520
16	101.300	343.1100	0.6790	0.8640
17	101.300	342.4100	0.7170	0.8820
18	101.300	340.8900	0.8030	0.9220
19	101.300	339.6600	0.8790	0.9520
20	101.300	338.4600	0.9510	0.9810

The result of the regression is shown in Figure 3 (see also the chapter-5-3-1-wilson-vle-fit.mot file in Appendix-1).

Discussion: From Figure 3, it can be seen that a good fit of the data has been obtained, but they are different from the ones fitted with the generated activity coefficient data. Having established the model parameters for VLE calculations, this model can now be used in process calculations such as a single-stage PT-flash, equilibrium-based distillation column and many more where a gamma-phi approach for VLE calculations need to be used.

5.3.1.4 Development of a Process Model

A PT-flash model (mass balance only) is obtained by combining the following equations:

Eq. 1, Eqs. 6-13, Eqs. 38-39 and the mass balance equation,

$$F z_i = L x_i + V y_i \quad i = 1, NC \tag{42}$$

where F is the flow rate of the stream entering the flash unit; L is the liquid stream leaving the flash unit; V is the vapour stream leaving the flash unit, x_i and y_i are the equilibrium compositions of compound i in the liquid and vapour streams, respectively, leaving the flash unit.

Eq. 42 can be rearranged to the following form, where $\beta = V/F$.

$$z_i = \beta y_i + (1 - \beta) x_i \quad i = 1, NC \tag{43}$$

In addition to Eq. 40, the liquid-phase compositions $\underline{x}$ and the feed compositions $\underline{z}$ are constrained to sum to unity.

$$Resid_x = \sum_i x_i - 1; \quad Resid_y = \sum_i y_i - 1; \quad Resid_y = \sum_i z_i - 1 \quad i = 1, NC \tag{44}$$

Since the $Resid_x$ and $Resid_y$ both must be zero (or smaller than a convergence criteria) at solution of the model equations, this means that inserting Eq. 44 into Eq. 8 results in the total mass balance equation,

$$F = V + L \tag{45}$$

To have a PT-flash model (mass and energy balance), the following additional equations (energy balance plus associated property model equations) are needed,

$$F H_F = L H_L + V H_V + Q \tag{46}$$

where H_j are the enthalpies of the corresponding streams (subscript j refers to subscripts F, L and V in Eq. 46) and Q is the heat added or removed. Note that the addition of enthalpies to the set of equations means that we now need constitutive models for calculating them. Different versions of the PT-flash model, as implemented in ICAS-MoT, are given in Appendix-2. The ordered incidence matrix for the PT-flash model is given in Table 11.

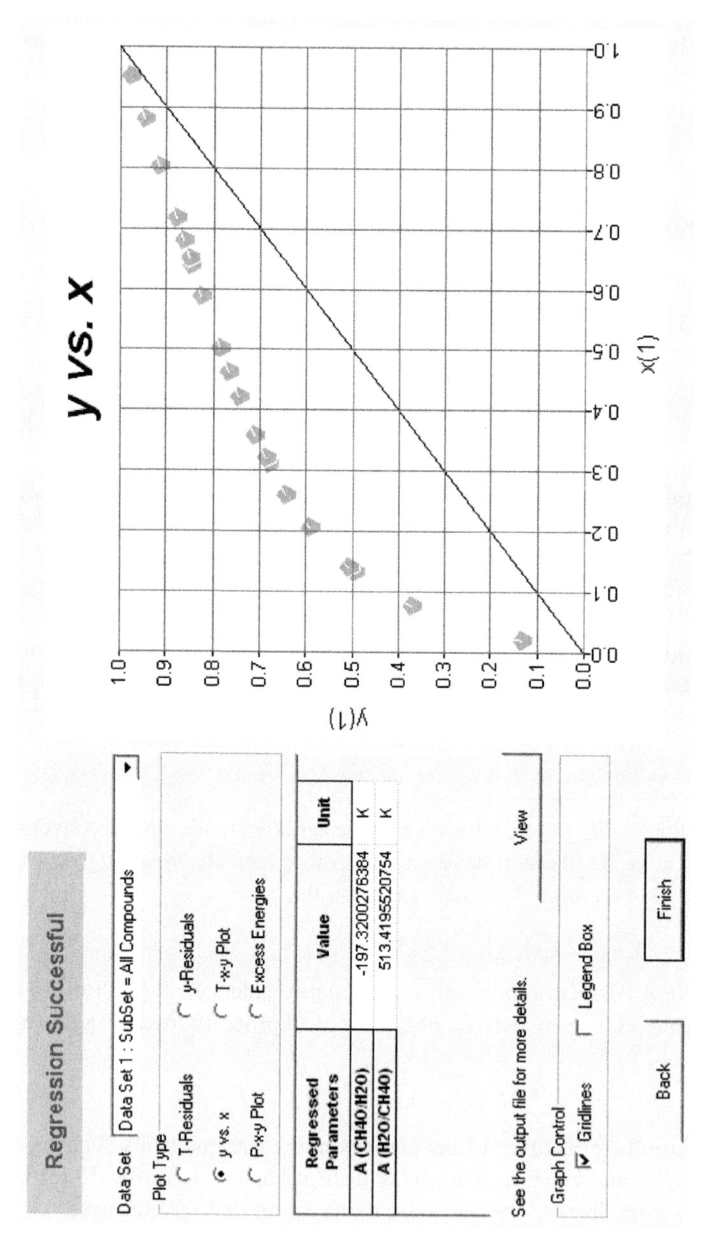

FIGURE 3 **Results of regression for the Wilson model parameters.**

TABLE 11 Incidence Matrix for a *PT*-flash Model with Gamma-phi Approach

Equations	Known variables				Unknown variables							
	F	$\underline{Z}$	P	T	$\underline{P}^{sat}$	$\underline{\gamma}$	$\underline{x}$	$\underline{y}$	β	V	L	Q
1				*	*							
6-13				*		*	*					
43		*	*		*	*	*		*			
38-39			*		*	*	*	*				
44								*	*			
$\beta=V/F$	*								*	*		
45	*									*	*	
46[1]	*									*	*	*

[1] *It is assumed that the enthalpy values of each stream are available after Eq. 45 has been solved.*

From the above analysis (incidence matrix), it is clear that a sub-set of equations (Eqs. 6-13, 43, 38-39, 44) need to be solved simultaneously for $\underline{\gamma}$, $\underline{x}$, $\underline{y}$ and β. Also, by assuming values for the two variables lying in the upper tri-diagonal ($\underline{x}$ and β), an iterative solution strategy can be generated. Note that in this case not all the property model equations belong to the partition within the repetitive cycle of equations (or the set of simultaneous equations). Note also that the energy balance equation can be solved after the mass balance–related equations have been solved (for the case when P and T are specified).

To generate an equilibrium-based distillation column model, the flash model developed above is repeated *NS* times, where *NS* is the number of equilibrium stages. Therefore, the incremental modelling steps are as follows: (a) develop and validate the constitutive models to be used; (b) develop the phase equilibrium models to be used; (c) develop the single-stage (process) model; (d) combine a-c to develop the multistage model. Note that this model structure is generic, for example, changing models a-b allows different property models for the same process model.

5.3.2. Solid–Liquid Equilibrium

In this example, the use of the NRTL model (Renon and Prausnitz, 1968) for the calculation of liquid-phase activity coefficients is discussed for the computation of solid–liquid equilibrium. The modelling objective is to develop a model that can be used for generating a solid saturation curve by calculating the saturation compositions of a solute as a function of temperature (in a binary mixture) and as a function of the composition of a second solvent at a constant temperature (for a ternary mixture).

5.3.2.1 Model Description

The solid–liquid equilibrium is defined by,

$$z\gamma^S = x\gamma^L \exp[(\Delta H_f/(R\ T_m))(T_m - T)/T] \tag{47}$$

In the above equation, z and x are mole fractions of the solute in the solid and liquid phases, respectively; γ^S and γ^L are the activity coefficients of the solute in the solid and liquid phases respectively; ΔH_f is the heat of fusion of the solute; T_m is the normal melting point of the solute, R is the universal constant, and T is the temperature. Note that Eq. 47 assumes that the entropic terms related to solid formation are negligible. Also, it is assumed that pressure has no influence.

Equation 47 is simplified (subscript s refers to the solute) to the following form when only one solute precipitates to the solid phase.

$$1 - x_s\gamma^L \exp[(\Delta H_f/(RT_m))(T_m - T)/T] \tag{48}$$

$$x_{NC} = 1 - \sum_i x_i \quad i = 1, NC - 1 \tag{49}$$

In Eqs. 48-49, $\underline{x}$ is the vector of liquid-phase mole fractions while x_s is the mole fraction of the solute that will precipitate from the solution. For γ^L, a constitutive model for the calculation of activity coefficients in the liquid phase is needed. In this example, we will use the NRTL model and the corresponding model equations are given below,

$$t_{ij} = b_{ij}/(R{*}T) \quad i = 1,\ NC;\ j = 1, NC \tag{50}$$

$$G_{ij} = \exp(-a_f{*}t_{ij}) \quad i = 1, NC; j = 1, NC \tag{51}$$

$$\begin{aligned}\ln\gamma_i = {} & [\textstyle\sum_k(\tau_{ki}G_{ki}x_k)/\sum_k(G_{ki}x_k)] \\ & + \textstyle\sum_j\{[\sum_k(\tau_{ki}G_{ki}x_k)/\sum_k(G_{ki}x_k)]{*}[G_{ij}x_j/\sum_k(G_{ki}x_k)] \\ & \times [\tau_{ij} - [\textstyle\sum_k(\tau_{ki}G_{ki}x_k)/\sum_k(G_{ki}x_k)]\} \\ & \times i = 1, NC\end{aligned} \tag{52}$$

In the above equations, $\tau_{ii} = 0$ and $G_{ii} = 1$. The model parameters are a_f (usually = 0.3) and b_{ij} (the interaction between molecules i and j).

5.3.2.2. Model Analysis

For a binary mixture, the model consists of two SLE-related equations (Eqs. 48-49) and 3*NC NRTL-model equations (Eqs. 50-52), not counting the $i = j$ equations and variables – for a total of 3*NC +2 equations. The unknown variables are NC for each of $\underline{\tau}$, $\boldsymbol{G}$, $\underline{\gamma}$; and 2 for x_s and x_{NC}. The variables to be specified are 1 defined by the problem (T), 8 defined by the model (R, a_f, b_{ij}) and 2 defined by the properties of the solute (T_m, H_f). Note that $b_{ii} = 0$; consequently, only b_{ij} for $i \neq j$ needs to be specified (note also that τ_{ii} and G_{ii} are not counted as they are set to 0 and 1, respectively).

Solution Strategy: The incidence matrix for model discussed above is given in Table 12 (specified or known variables are not included).

It can be noted that a lower tri-diagonal system is not found. This means that Eqs 49, 52 and 48 need to be solved simultaneously or iteratively (with respect to x_s). The incidence matrix structure is the same if x_s is replaced by T (that is, specify the saturation composition and calculate the saturation temperature).

Numerical Solution

Results from the solution of the above model equations are given for the binary systems (benzene–naphthalene) and (morphine–ethanol) and the ternary system (benzene–toluene–naphthalene). The model equations as implemented in ICAS-MoT tool are given in Appendix (see file chapter-5-3-2-nrtl-sle.mot).

Benzene–naphthalene: The precipitation of naphthalene is considered. Therefore, we need the corresponding pure component property data for naphthalene: $\Delta H_f = 18980$ KJ/Kmole; $T_m = 353$ K. The model parameters are: $R = 8.314$ J K^{-1} mole^{-1}; $a_f = 0.3$; $b_{ii} = 0$; $b_{12} = 0.01$; $b_{21} = 0.01$.

For a single point calculation, T is specified (= 333 K) and the x_1 is calculated (=0.68). Repeating the calculations for a series of temperatures, the T versus x diagram is obtained, as shown in Figure 4. Note that before the eutectic point, it is benzene that will precipitate out.

Benzene–toluene–naphthalene: Again, only the precipitation of naphthalene is considered. However, now we have an extra unknown variable (the composition of toluene = x_2). Since the model equations are the same, in addition to T we have to specify either x_2 or x_3 (mole fraction of benzene). The calculated ternary diagram is shown in Figure 5.

Morphine–ethanol: The precipitation of morphine is considered. Therefore, we need the corresponding pure component property data for morphine: $\Delta H_f =$ 18980 KJ/Kmole; $T_m = 353$ K. The model parameters are: $R = 8.314$ J K^{-1}

TABLE 12 Incidence Matrix for SLE Calculation with the NRTL Model (Binary System)

Equations	Variables				
	$\underline{\tau}$	$\underline{G}$	x_{NC}	γ^L	x_s
50	*				
51		*			
49			*		*
52	*	*	*	*	*
48				*	*

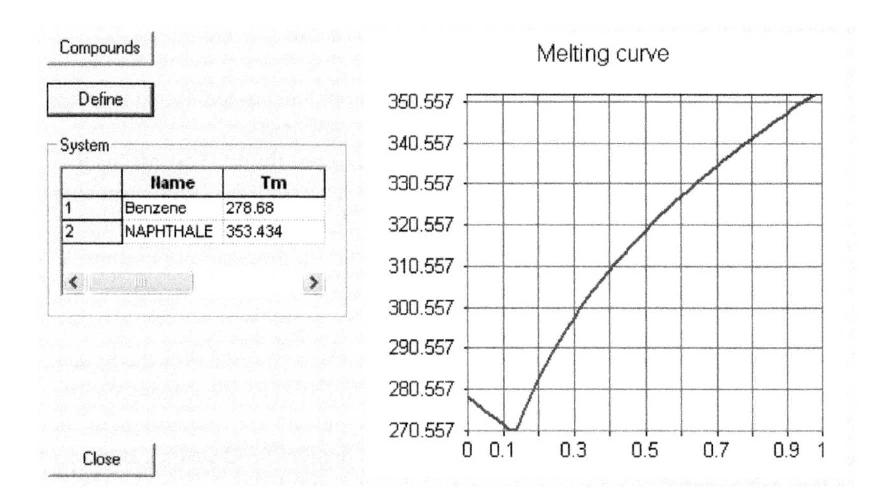

FIGURE 4 Calculated SLE phase diagram for benzene–naphthalene (y-axis is temperature, and x-axis is mole fraction of naphthalene).

mole^{-1}; $a_f = 0.3$; $b_{ii} = 0$; $b_{12} = 0$; $b_{21} = 0$. Note that by setting all the $\underline{b}$-parameters to zero, the calculation is performed for an ideal system ($\underline{\gamma} = 1$).

For a single point calculation, T is specified (=333) and the x is calculated (=0.68).

Discussion: The above model and calculations can be repeated for other chemical systems by simply providing the property model parameters and the two pure component properties. The NRTL model can be simplified to an ideal solution model by simply setting the $\underline{b}$-parameters to zero. The use of other models may be investigated by replacing Eqs. 50-52 with the equations for another property model for liquid-phase activity coefficients. For example, replace the NRTL model equations with the UNIFAC model equations (see Gani *et al.* 2006 and also models in the library; see list in Appendix-2).

5.3.3. Evaporation from a Droplet

5.3.3.1 Problem Definition (Modelling Scenario)

The models reuse and model aggregation features are highlighted through the modelling of evaporation from a water droplet with a solvent. A small droplet of water and methanol (solvent) is heated (for example, through a heat source such as sunlight or artificial light) through a constant heat supply, which causes evaporation of components from this droplet into the environment until the whole droplet is evaporated.

5.3.3.2 Controlling Factors/Mechanisms

- Mass transfer in droplet
- Heat transfer in droplet

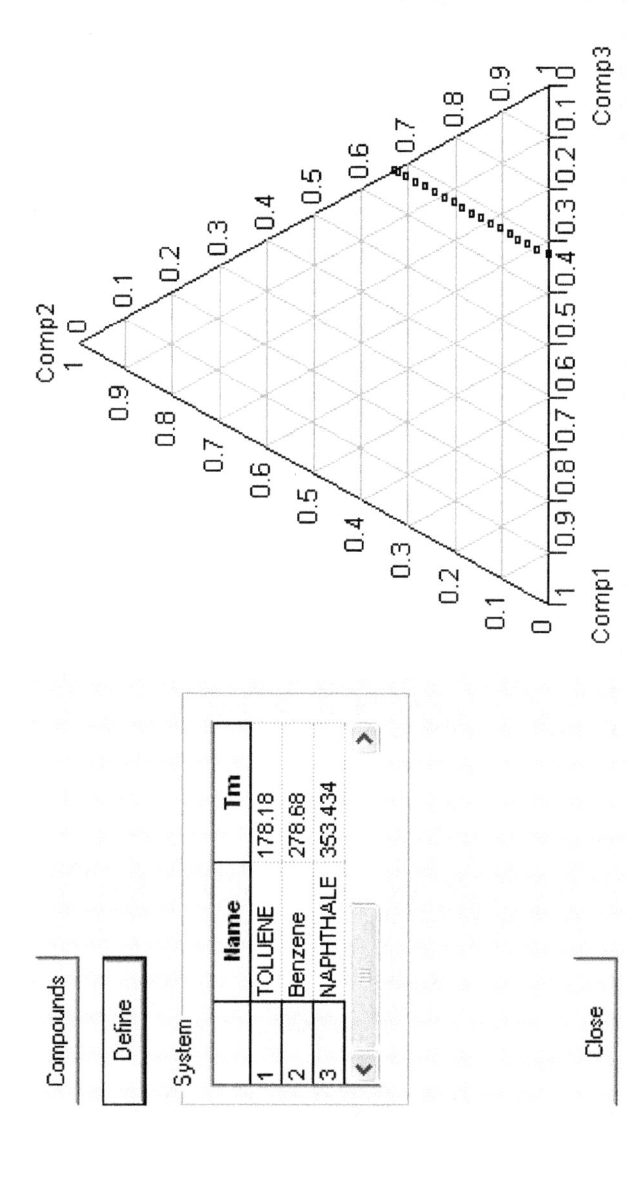

FIGURE 5 Calculated SLE for ternary mixture of benzene–toluene–naphthalene at 333 K.

5.3.3.3 Assumptions

- Two phases (liquid and vapour).
- Instantaneous equilibrium at the interface (of liquid and vapour), no transfer resistances in the phases.
- Well-mixed ideal liquid.
- Two-component system: methanol (1) and water (2).
- Equimolar amount of water and methanol in the initial droplet.
- Constant heat Q_R from the environment due to a radiation source.
- Uniform decrease of the droplet in all directions.
- Evaporated components are instantaneously removed from the droplet and go into the environment where no amount of methanol and water can be found.
- Constant physicochemical properties (heat capacities, heat of vapourisation,...).
- A lumped dynamic model is used.

5.3.3.4 Model Description

Balance volume: The balance volume to be used in developing the model equations is shown in Figure 6.

Balance equations – Conservation of moles: The total moles of the droplet decreases over time because of evaporation into the environment.

$$\frac{dn_i}{dt} = -Vy_i \tag{53}$$

Balance equations – Conservation of energy: The change of energy in the droplet is because of the assumption of thermal and mass insulation only, related to the transferred energy to the droplet and the leaving energy due to the evaporation.

$$\frac{dE}{dt} = -VH^V + Q_R \tag{54}$$

The left-hand side of Eq. 54 can be transformed in order to derive an equation for the temperature in the droplet over time, assuming that energy stored in the droplet hold-up depends on the enthalpy. Furthermore, it can be assumed that

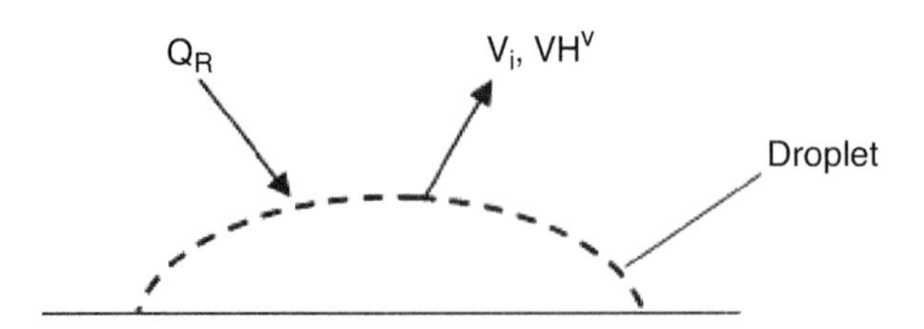

FIGURE 6 Balance volume for droplet.

the specific heat capacity of the liquid is constant, and therefore leading to the following form using the chain-rule for differential equations.

$$\frac{dE}{dt} = \frac{d(ncp^L_{mix}T)}{dt} = ncp^L_{mix}\frac{dT}{dt} + cp^L_{mix}T\frac{dn}{dt} \tag{55}$$

The enthalpy of the vapour in Eq. 54 is expressed through:

$$H^V = cp^V_{mix}T + \Delta H^V \tag{56}$$

where ΔH^V is the heat of vaporisation.

Inserting Eqs. 52, 54 and 55 into Eq. 53 and reordering of the terms leads to,

$$\frac{dT}{dt} = \frac{V(cp^L_{mix}T - cp^V_{mix}T - \Delta H^V) + Q_R}{(n_{Meth} + n_W)cp^L_{mix}} \tag{57}$$

<u>*Constitutive models – The equilibrium equation at the interphase*</u>:

$$y_i = x_i k_i \tag{58}$$

The closing condition for the equilibrium,

$$0 = \sum_{i=1}^{NC} y_i - 1 \tag{59}$$

The equilibrium constant model where the vapour phase is assumed to be ideal is given by,

$$k_i = \frac{P^S_i \gamma_i}{P} \tag{60}$$

The vapour pressure can be calculated using the Antoine equation (same as Eq. 1),

$$P^S_i = 10\left(A^{ps}_i - \frac{B^{ps}_i}{C^{ps}_i + T}\right) \tag{61}$$

<u>*Conditional equations – Define mole fractions of each compound*</u>:

$$x_i = \frac{n_i}{n_{meth} + n_W} \tag{62}$$

<u>*Consititutive model – The Wilson model for activity coefficients*</u>: Written in short form (compare with Eqs. 6-13).

$$\ln \gamma_i = -\ln \sum_{l=1}^{NC} x_i A^W_{i,l} - \sum_{n=1}^{NC} x_i A^W_{n,i} / \sum_{l=1}^{NC} x_l A^W_{n,l} \tag{63}$$

where,

$$A_{l,n} = \frac{V^W_i}{V^W_i} \exp\left(-\frac{\Delta\lambda_{l,n}}{T}\right) \tag{64}$$

Constitutive model – Heat capacity of the liquid mixture:

$$cp^L_{mix} = \sum_{i=1}^{2} x_i cp^L_i \tag{65}$$

Constitutive model – Heat capacity of the vapour mixture:

$$cp^V_{mix} = \sum_{i=1}^{2} y_i cp^V_i \tag{66}$$

Constitutive model – Heat of vaporisation:

$$\Delta H^V = \sum_{i=1}^{2} x_i \Delta H^V_i \tag{67}$$

The vapour is dependent on the pressure difference in the system to the environment. Only if the pressure in the system is greater than that of the environment and the droplet is still existing, a vapour flow can be observed and is modelled through event modelling for $P \geq P_0$.

$$V = \kappa(n_1 + n_2)(P - P_0) \tag{68}$$

The transfer coefficient κ can be fitted through experimental data. However, since this data is not existent, $\kappa = 1$.

5.3.3.5 Model Analysis

Analysis for ODE-System (constitutive and conditional equations): An analysis of the ordinary differential equations (Eq. 52) and Eq. 57 gives three equations, three dependent variables (n_{meth}, n_W, T), one independent variable time and seven variables (V, y_i, cp^L_{mix},cp^V_{mix}, ΔH^V, Q_R). Hence, the degree of freedom for the ODEs is 7. One variable, Q_R, is fixed by the problem. Therefore, either the constitutive equations should be used to calculate the remaining 6 variables, or fixed values need to be given (for variables: V, y_i, cp^L_{mix}, cp^V_{mix}, ΔH^V).

Analysis for AE-System (constitutive and conditional equations): Analysing the derived algebraic (constitutive and conditional) equations gives 19 equations (Eqs. 58-68) and 39 extra variables y_i, x_i, k_i, P^S_i, P, γ_i, A^{ps}_i, B^{ps}_i, C^{ps}_i, V^W_i, $A_{l,n}$, $\Delta\lambda_{l,n}$, cp^L_{mix}, cp^L_i, cp^V_{mix}, cp^V_i, ΔH^V, ΔH^V_i, V, κ, P_0. Hence, the degree of freedom for the AE system is 20, meaning that 20 variables need to be specified. Note that solving Eqs. 58-68 provides the values of (V, y_i, cp^L_{mix}, cp^V_{mix}, ΔH^V). The 20 variables to specify are arranged in terms of: one variable fixed by problem – P_0; six variables fixed by system – cp^L_i, $cp^V_i \Delta H^V_i$; and 13 adjustable model parameters – A^{ps}_i, B^{ps}_i, C^{ps}_i, V^W_i, $\Delta\lambda_{l,n}$, κ.

Incidence Matrix: In order to check how the choice of parameters affects the solution procedure of the system and to determine the order of the equations to solve, the incidence matrix for the ODEs as well as for the AEs is given in Table 13. Because the incidence matrix of the AE system cannot be

TABLE 13 Incidence Matrix for AE System Including RHS of ODE's of Evaporation Problem

Equations	x_1	x_2	$A_{i,j}$	γ_1	γ_2	P_1^S	P_2^S	cp_{mix}^L	V	P	k_1	k_2	y_1	y_2	ΔH^V	cp_{mix}^V	RHS1	RHS2	RHS3
AE 62	X																		
AE 62		X																	
AE 64			X																
AE 63	X	X	X	X															
AE 63	X	X	X		X														
AE 61						X													
AE 61							X												
AE 65	X	X						X											
AE 68									X	X									
AE 60						X				X	X								
AE 60							X			X		X							
AE 58	X										X		X						
AE 58		X										X		X					
AE 59													X	X					
AE 67	X	X													X				
AE 66													X	X		X			
ODE 53									X				X				X		
ODE 53									X					X				X	
ODE 57								X	X						X	X			X

transformed into a lower triangular form, Equations 68 to 60 need to be solved simultaneously (that is, as shown in Table 13, this set defines an implicit set of AEs).

5.3.3.6 Numerical Solution

Problem data: The data is collected depending on the model and assumptions (see Tables 14–16). Also property model parameters are retrieved from the ICAS database. Note also that before performing any simulations of the described system, the property model parameters could be fitted/verified against available experimentally measured data (as shown in Sections 5.2 and 5.3.1). The Antoine correlation parameters are taken in this example from ICAS database and listed in Table 14.

TABLE 14 Fitted Antoine Parameter

	A_i^{PS}	B_i^{PS}	C_i^{PS}
Methanol	5.203082	1585.500436	−33.101218
Water	5.148	1731.681	−40.517

TABLE 15 Data Collection

κ	Q_R	P_0	cp_1^L	cp_2^L	cp_1^V	cp_2^V	ΔH_1^V	ΔH_2^V
s atm-1	J min-1	atm	J mol-1 K-1	J mol-1 K-1	J mol-1 K-1	J mol-1 K-1	J mol-1	J mol-1
1	500	1	88.4	75.3	46.2	33.7	35800	42600

TABLE 16 Initial Values for ODE System

n_1	n_2	T
mol	Mol	K
1	1	330

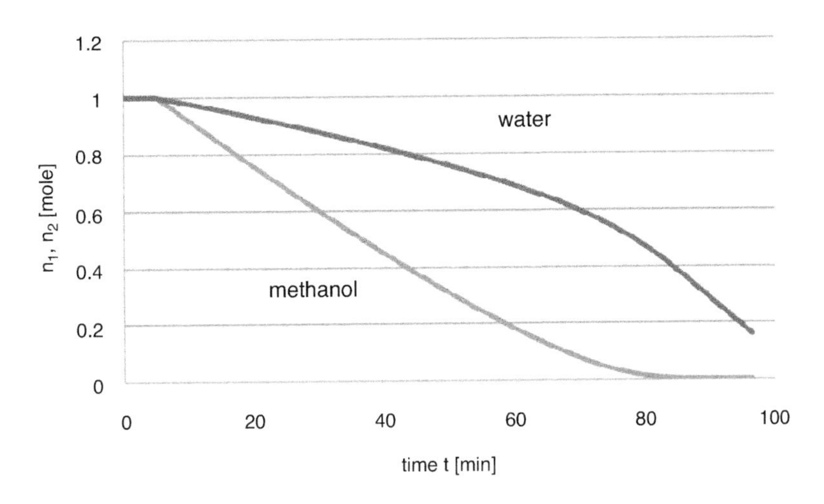

FIGURE 7a Evaporation of methanol and water from droplet.

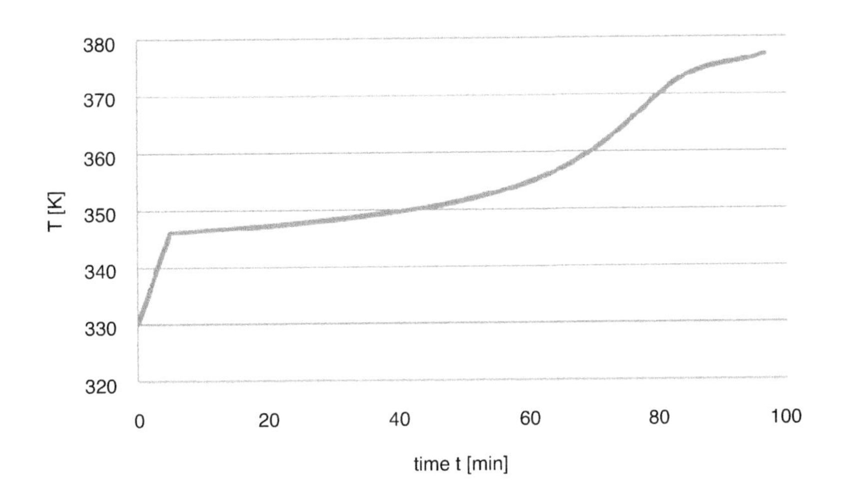

FIGURE 7b Temperature change in the liquid droplet during the evaporation and heating.

Wilson model parameters used are: $\Delta\lambda_{1,2} = -165.9$, and $\Delta\lambda_{2,1} = 492.4$

The values for the fixed property variables are retrieved from the ICAS database (calculated at $T = 330\ K$) and given in Table 15. Note that the values of κ, Q_R, P_0 were not calculated but assumed.

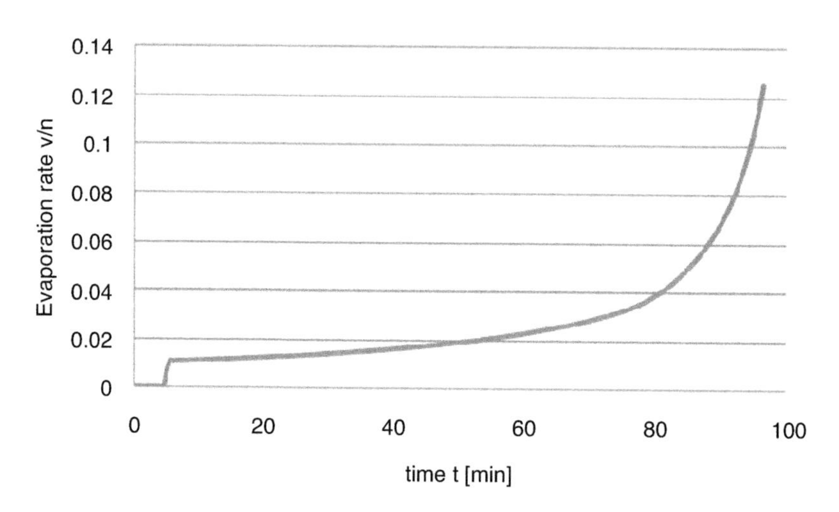

FIGURE 8 **Evaporation rate over time.**

Finally, initial values of the dependent variables are needed to start the initial value integration. Table 16 lists these values.

Simulation results: Results of the evaporation of the droplet problem are given in Figures 7a, 7b, 8 and 9. Starting from a temperature of the droplet of 330 K, the boiling point of the mixture is not reached and all transferred heat into the droplet is used to heat the droplet (Figure 7b). Therefore, the amount of moles in the mixture remains constant (Figure 7a) until $t = 8$min When the boiling point of the mixture is reached, methanol is evaporated faster as it is the component with the lower boiling point (Figure 7a). After 85 minutes, almost all the methanol is evaporated.

The temperature profile (Figure 7b) increases linearly at the beginning where nothing evaporates. During the evaporation of the mixture (between 8 min – 85 min), the slope of the temperature changes from almost constant at $t = 8$ min to a high slope at the end of the evaporation.

The evaporation rate is zero until evaporation starts (see Figure 8) and increases over time due to decreasing amount of droplet left while the simulated value of the evaporation vapour flow V shows a constant value.

The comparison between a simulation with and without activity coefficient model (that is, assuming ideal liquid and vapour phases), here presented for the temperature change in the droplet, clearly shows a big difference between the calculated values from these two models (Figure 9). Therefore, it is important to

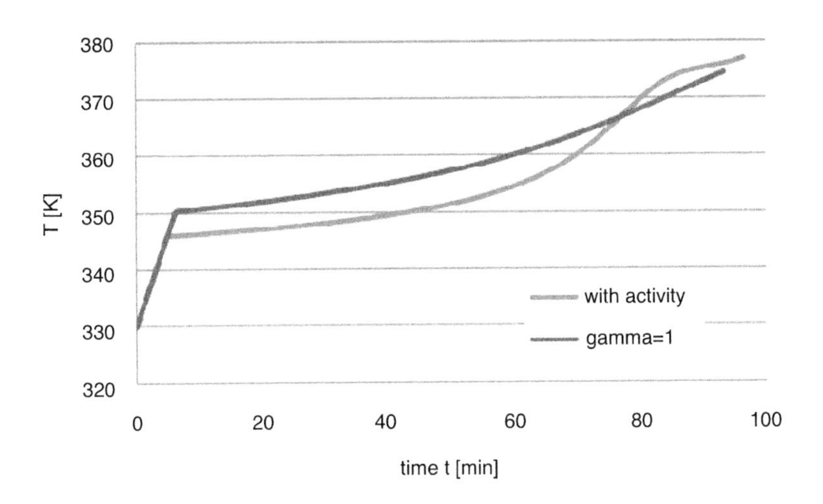

FIGURE 9 Comparison of the simulated temperatures in the droplet during evaporation using an activity coefficient model (Wilson) and assuming ideal liquid ($\gamma_i = 1$).

consider the use of an activity coefficient model when the liquid phase non-ideality cannot be ignored.

The model equations, as implemented in ICAS-MoT, are given in Appendix-1 (see file chapter-5-3-3-drop-evap.mot).

5.3.4. Kinetic model use – Computation of attainable region diagram

The objective here is to use validated kinetic models to calculate the attainable region (Biegler *et al.* 1997). We will consider the isothermal liquid-phase system of elementary series, parallel and reversible van de Vusse reactions, which involves four species,

$$A \underset{k_{-1}}{\overset{k_1}{\rightleftharpoons}} B \overset{k_2}{\longrightarrow} C$$

$$2A \overset{k_3}{\longrightarrow} D$$

The feed to the system is pure A, with $C_{A0} = 1$ $kmol/m^3$. Consider that we have a feed stream of *A* and we are considering a reactor system with only one exit (*i.e.*, product) stream. What we would like to compute and plot is the composition of the desired product B (on Y-axis) against the composition of the limiting reactant A (X-axis).

TABLE 17 Reaction Rate Constants

Reaction	Constant	value	Units
$A \xrightarrow{k_1} B$	k_1 =	0.01	s^{-1}
$B \xrightarrow{k_{-1}} A$	k_{-1} =	5	s^{-1}
$B \xrightarrow{k_2} C$	k_2 =	10	s^{-1}
$2A \xrightarrow{k_3} D$	k_3 =	100	m^3/kmol s

5.3.4.1 Model Description

Let us consider a PFR (Plug-flow reactor). We define $C = [C_A, C_B]^T$ where:

$$r(C) = \left[-k_1 C_A + k_{-1} C_B - k_2 C_A^2, \quad k_1 C_A - k_2 C_B - k_{-1} C_B\right]^T \tag{69}$$

$$\frac{dC_A}{d\tau} = -(k_1 C_A + k_3 C_A^2 - k_{-1} C_B) \tag{70}$$

$$\frac{dC_B}{d\tau} = -(k_2 C_B + k_{-1} C_B - k_1 C_A) \tag{71}$$

Dividing Eq. 71 by Eq. 70, we obtain

$$\frac{dC_B}{dC_A} = \frac{k_1 C_A - (k_{-1} + k_2) C_B}{k_{-1} C_B - k_1 C_A (1 + {}^{k_3}/_{k_1} C_A)} \tag{72}$$

5.3.4.2 Model Analysis

In order to generate the data for plotting an attainable region, we need to solve the ordinary differential equation (Eq. 72) with C_A as the independent variable and C_B as the dependent differential variable, and integrate backwards from $C_A = 1$ to $C_A = 0$. The data needed to solve Eq. 72 is given in Table 17.

5.3.4.3 Numerical Solution

The calculated values of C_B as a function of C_A are plotted in Figure 10. The corresponding ICAS-MoT file for this calculation can be found in Appendix-1 (chapter-5-3-4-ar-pfr.mot).

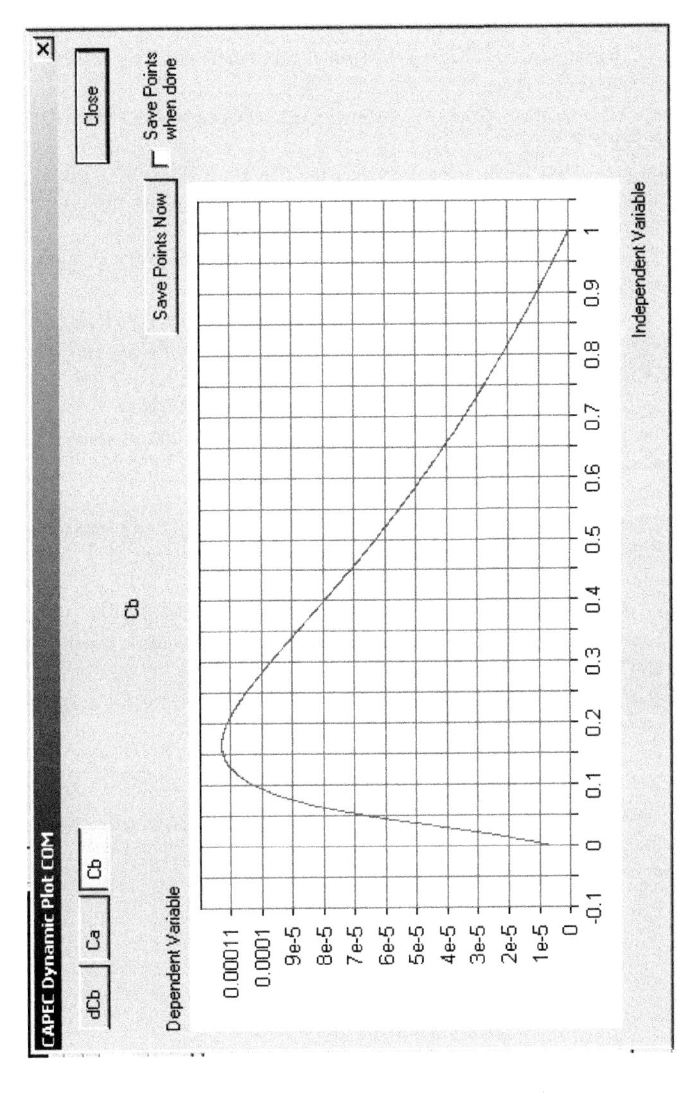

FIGURE 10 Calculated attainable region with PFR model (Eq. 72) – C_B (y-axis) versus C_A (x-axis).

REFERENCES

Belveze, L.S., Brennecke, J.F., Stadtherr, M., 2004. Industrial and Engineering Chemistry Research. 43, 815–825.

Biegler, L.T., Grossmann, I.E., Westerberg, A.W., 1997. Systematic Methods of Chemical Process Design. Prentice Hall.

Chen, C.C., Evans, L.B., 1986. AIChE Journal. 32, 444.

Fredenslund, Aa., Rasmussen, P., Gmehling, J. Vapor-Liquid Equilibria Using UNIFAC. Elsevier Scientific Publishing Company.

Gani, R., Muro-Suñé, N., Sales-Cruz, M., Leibovici, C., O'Connell, J.P., 2006. Fluid Phase Equilibria. 250, 1–32.

Gross, J., Sadowski, G., 2002. Industrial and Engineering Chemistry Research. 41, 1084–1093.

Kang, J.W., Abildskov, J., Gani, R., Cobas, J., 2002. Industrial and Engineering Chemistry Research. 41, 3260–3273.

Kontogeorgis, G.M., Yakoumis, I.V., Meijer, H., Hendriks, E.M., 1990. Fluid Phase Equilibria. 158–160, 201–211.

Kontogeorgis, G.M., Gani, R. Computer Aided Property Estimation for Product and Process Design. In: G. M. Kontogeorgis. R. Gani (Eds.), *Computer Aided Chemical Engineering, CACE-19*. Elsevier B. V..

Lindenbaum, S., Boyd, G.F., 1964. Journal of Physical Chemistry. 68, 911–917.

Nielsen, T.L., Abildskov, J., Harper, P.M., Papaeconomou, I., Gani, R., 2001. Journal of Chemical and Engineering Data. 46, 1041–1044.

Renon, H., Prausnitz, J.M., 1968. AIChE Journal. 14, 135–144.

Sales-Cruz, M., Gani, R., 2003. A Modelling Tool for Different Stages of the Process Life. In: S. P. Asprey., S. Macchietto (Eds.), *Computer-Aided Chemical Engineering, Vol. 16: Dynamic Model Development*. Amsterdam: Elsevier; pp. 209–249.

Sales-Cruz, M. 2006. 'Development of a computer aided modelling system for bio and chemical process and product design, PhD-thesis.' Technical University of Denmark, Lyngby, Denmark.

Soave, G., 1972. Chemical Engineering Science. 27, 1197–1203.

Thomsen, K., Rasmussen, P., Gani, R., 1996. Chemical Engineering Science. 51, 3675–3683.

Wilson, G.M., 1964. Journal of the American Chemical Society. 86, 127–130.

Chapter 6

Steady-State Process Modelling

In this chapter, two types of process modelling approaches are highlighted

- Approach-1: Derive models for individual unit operations of a process, then combine them to represent a specific process flowsheet and then solve the model equations, according to a defined solution strategy
- Approach-2: Use process simulation software where pre-defined models of unit operations already exist in a model library, represent a process flowsheet with models selected from the library and then perform simulations of the process flowsheet

To highlight these two types of modelling and simulation approaches, we will use four well-known process simulation problems – the Williams-Otto plant (Biegler et al., 1997); the process for hydrodealkylation of toluene (Douglas, 1985); process for conversion of ethylene to ethanol (Biegler et al. 1997); and a bioethanol production process (Alvarado-Morales, 2009).

6.1. PROCESS DESCRIPTIONS

6.1.1. Williams-Otto Plant

The process flow diagram for the Williams-Otto plant is shown in Figure 1. In the reactor, the following three reactions take place.

$$A + B \rightarrow C$$

$$C + B \rightarrow P + E$$

$$P + C \rightarrow G$$

P is the main product, C is an intermediate and E is a by-product. After the reaction, the reactor effluent is cooled (not shown in the flow-diagram) and then by-product G is removed as solid waste through a decanter. The remaining liquid is sent to a distillation column where the product P is removed as the top product and the un-reacted A and B together with C is recycled back to the reactor. A purge is also used.

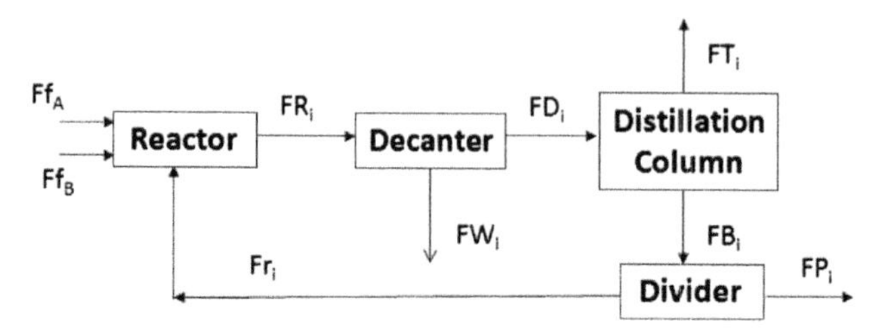

FIGURE 1 **Flow-diagram for the Williams-Otto plant (adapted from Biegler et al 1997).**

6.1.2. Hydrodealkylation (HDA) Process

The flow diagram for the HDA process is shown in Figure 2. Two reactions take place in the reactor.

$$\text{Toluene} + \text{Hydrogen} = \text{Benzene} + \text{Methane}$$

$$\text{Benzene} = \text{Diphenyl} + \text{Hydrogen}$$

The reaction takes place in the gas phase at high temperature and pressure. The reactor effluent is cooled (not shown in the flow-diagram), and the liquid and vapour streams are separated in a flash operation. The vapour containing mainly hydrogen and methane is divided into a purge stream and a recycle stream. The liquid stream from the flash passes through a train of distillation columns to separate the remaining gases, the benzene product and the unreacted toluene for recycle. More information on this process can be found in Douglas (1988).

6.1.3. Conversion of Ethylene to Ethanol

In this process a feed mixture containing 96% ethylene, 3% propylene and 1% methane is reacted at high temperature (between 535 K–575 K) and pressure (68 atm), and is converted to ethanol based on the following reactions,

$$\text{Ethylene} + \text{Water} \rightarrow \text{Ethanol}$$

$$\text{Propylene} + \text{Water} \rightarrow \text{isopropanol}$$

$$2\text{Ethanol} \rightarrow \text{Diethyl Ether} + \text{Water}$$

The first two reactions are kinetically controlled, whilst the third reaction is equilibrium controlled. According to the available reaction data, the conversion of ethylene is only between 5 and 7%, whilst the conversion of propylene is around 10% of the ethylene conversion. Since the reaction takes place in the vapour phase, a PT-flash is used after the reactor to create two phases and to

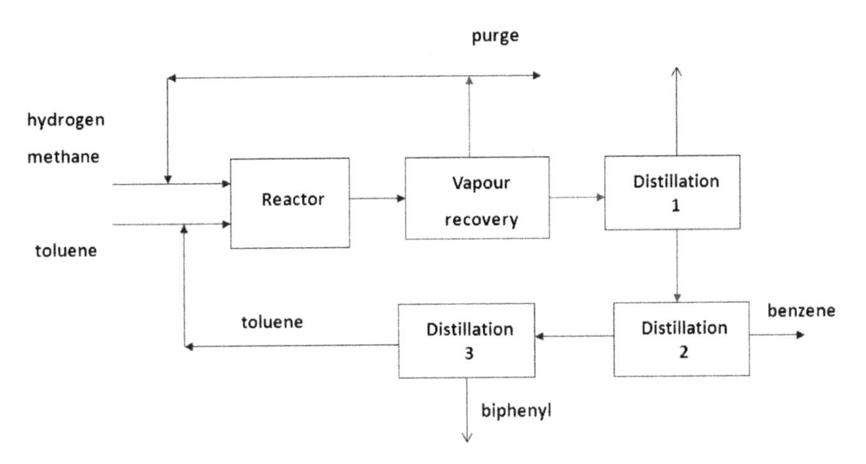

FIGURE 2 Flow-diagram for hydrodealkylation of toluene

remove the gases (ethylene, propylene and methane) from the rest. However, as diethyl ether and some ethanol also go to the vapour phase, an absorber with water as the solvent is used to recover the ethanol and diethyl ether. This stream is then combined with the liquid product stream from the flash and it passes through a network of distillation columns to separate the ethanol product, the waste water (with by-products) and a purer water stream, which is recycled. Figure 3 shows the flow diagram of the flowsheet. More details of this process can be found in Biegler et al. (1997).

6.1.4. Bio-Ethanol Process

A base case design of the bioethanol production process is based on the one given by NREL (Wooley et al., 1999). The main operations of the process are highlighted in Figure 4.

Feedstock handling: The feedstock, in this case hardwood chips, is delivered to the feed handling area for storage and size reduction.

Pre-treatment: The main operation of the pre-treatment processing step is the 'pre-treatment reactor', which converts most of the hemi-cellulose portion of the feedstock to soluble sugars (primarily, xylose, mannose, arabinose and galactose) by hydrolysis using dilute sulphuric acid and elevated temperature. The hydrolysis under these conditions also solubilises some of the lignin in the feedstock. In addition, acetic acid is released from the hemi-cellulose hydrolysis. Degradation products of pentose sugars (primarily furfural), and hexose sugars (primarily hydroxymethylfurfural), are also formed. Following the pre-treatment reactor, the hydrolysate (consisting of a mixture of liquid and solid particles) is flash cooled. This operation

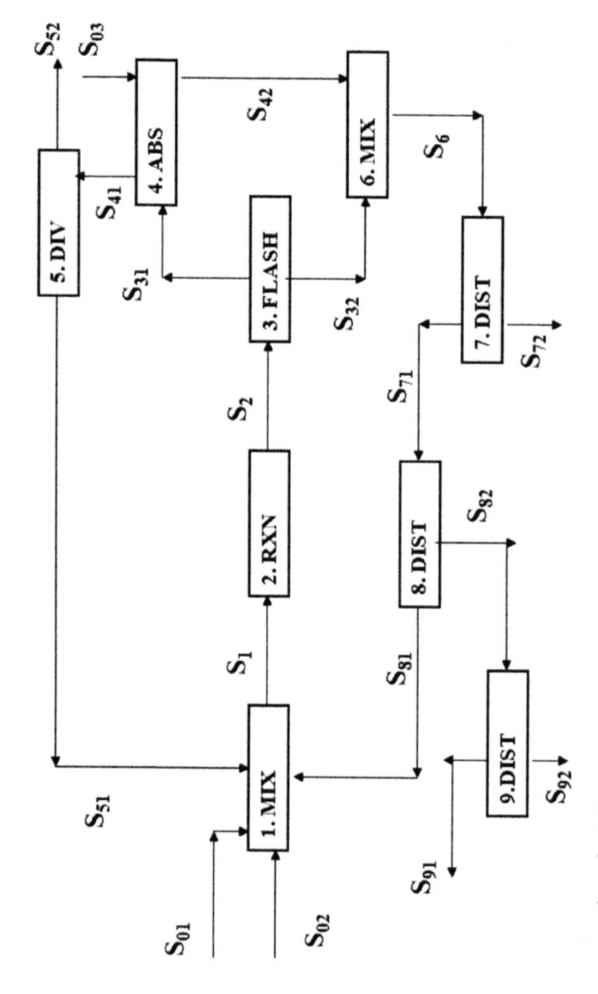

FIGURE 3 **Process for conversion of ethylene to ethanol**

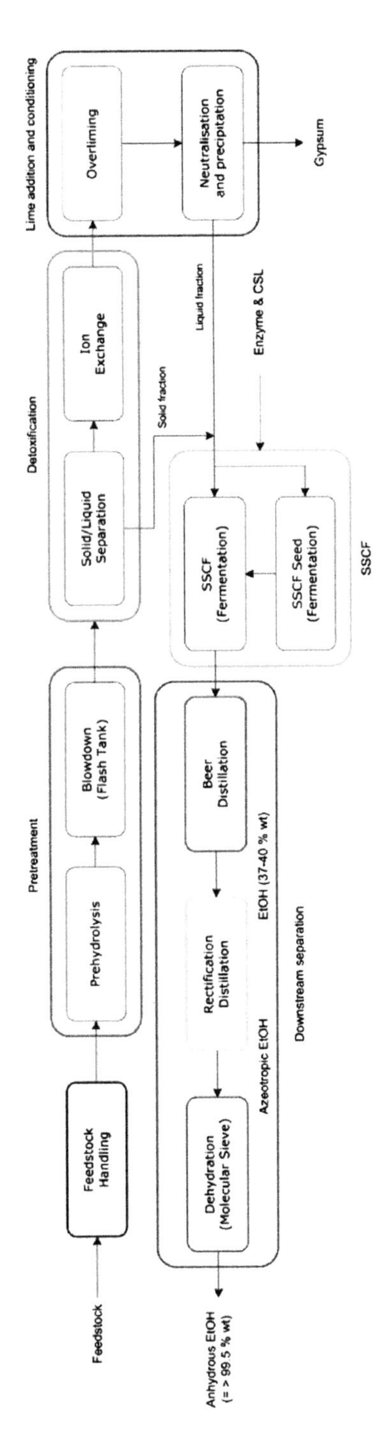

FIGURE 4 Production of bio-ethanol from cellulose (Alvarado-Morales et al 2009).

vapourises a large amount of water, a portion of acetic acid and much of the furfural and hydroxymethylfurfural.

Detoxification, lime addition and conditioning: The unreacted solid phase is separated from the liquid hydrolysate. The latter contains sulphuric acid and other inhibitors in addition to the hemicellulose sugars. Before fermentation, detoxification of the liquid hydrolysate is performed to remove the inhibitors, formed during the pre-treatment of biomass. Ion exchange is used to remove acetic acid and sulphuric acid that will be toxic to the microorganisms in the fermentation. After ion exchange the pH is raised to 10 (by adding lime) and held at this value for a period of time. Neutralisation and precipitation of gypsum then takes place. The gypsum is removed via filtration and the hydrolysate is mixed again with the solid fraction (from the solid-liquid detoxification separation unit) before being sent to SSCF.

SSCF: Following the lime addition, a small portion of the detoxified slurry is diverted to the simultaneous saccharification and co-fermentation (SSCF) seed process area for microorganism production (*Zymomonas mobilis*), whilst the bulk of the material is sent to the (SSCF) process area. Two different operations are performed here: saccharification (hydrolysis) of the remaining cellulose to glucose using cellulase enzymes, and fermentation of the resulting glucose and other sugars, to ethanol. For the fermentation, the recombinant *Zymomonas mobilis* bacterium is used, which will ferment both glucose and xylose to ethanol. The resulting ethanol broth is collected and sent to the downstream separation area.

Downstream Separation: After the SSCF, distillation and molecular sieve adsorption are used to recover the ethanol from the fermented beer and nearly 100% pure ethanol is obtained. The first distillation column (beer column) removes the dissolved CO_2 and most of the water, and the second distillation column concentrates the ethanol to near azeotropic composition. Subsequently, the residual water from the nearly azeotropic mixture is removed by vapour phase molecular sieve adsorption.

6.2. MODEL DESCRIPTION

The modelling/simulation of the four process flowsheets are highlighted in this chapter through two solution approaches: equation oriented (with the solver of ICAS-MoT) and modular (with process simulators). To highlight the main concepts and principles, simple mass balance models are used in the equation-oriented approach. In the process simulation part, rigorous simulation covering mass and energy balances are used, after an initial mass balance modelling/simulation.

6.2.1. Equation-Oriented Approach (Model Derivation)

For mass balance calculations, almost all chemical process flowsheets can be represented by a combination of mass balance models of mixers (representing all kinds of mixing operations), stoichiometric (conversion) reactors (representing all types of reactors), component splitters (representing all types of separation operations) and dividers (representing all types of stream splitting operations). They can also be represented by in-house library models available in many process simulators. In order to highlight the concepts and principles, these models will be derived and their use highlighted through the four well-known process flowsheet examples.

6.2.1.1. Assumptions

For all four mass balance models, it is assumed that the mixture is perfectly mixed and homogeneous. The temperatures and pressures are perfectly controlled and, therefore, do not change. Under these conditions, the separations can be modelled through the concept of separation factors, whilst for the reactions, either conversions of the limiting reactant or the reaction rate model will need to be supplied. The assumptions also decompose the mass balance models from the energy balance model for each of the four unit operations.

6.2.1.2. Model Derivation

The balance volumes and the stream connections are shown in Figures 5a–5c.

Mixer model: NC is the number of compounds; NM is the number of input streams to the mixer – see Figure 5a.

$$f_{i,NM+1} = \sum_j f_{ij} \qquad \text{for } i = 1, NC;\ j = 1, NM \tag{1}$$

Reactor model: k is the limiting reactant; j is the stream entering the reactor – see Figure 5b.

$$f_{ij+1} = f_{ij} + \gamma_i\, \eta_k f_{kj} \qquad \text{for } i = 1, NC \tag{2}$$

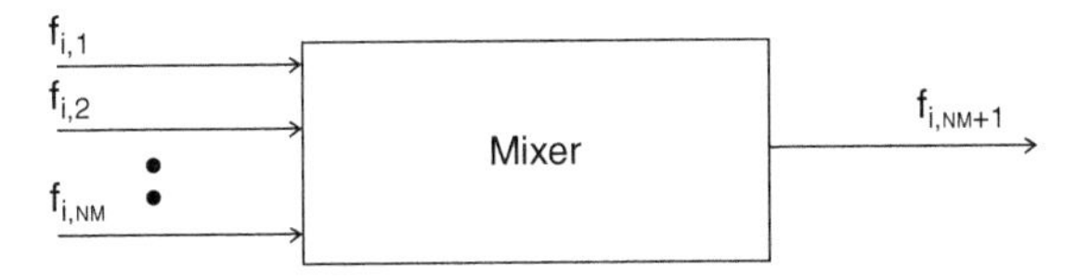

FIGURE 5a The flow-diagram for a mixer.

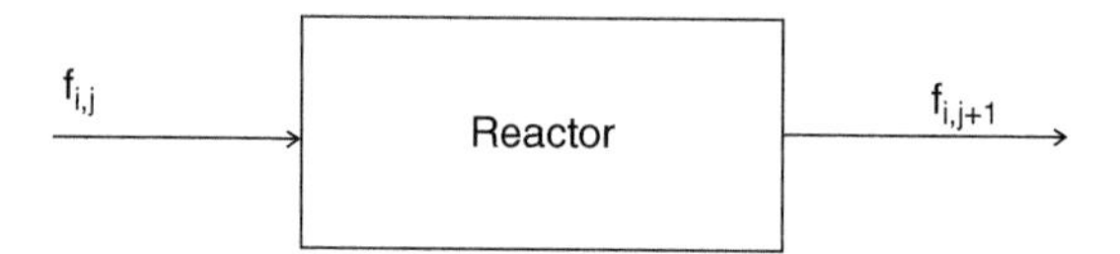

FIGURE 5b Simple reactor.

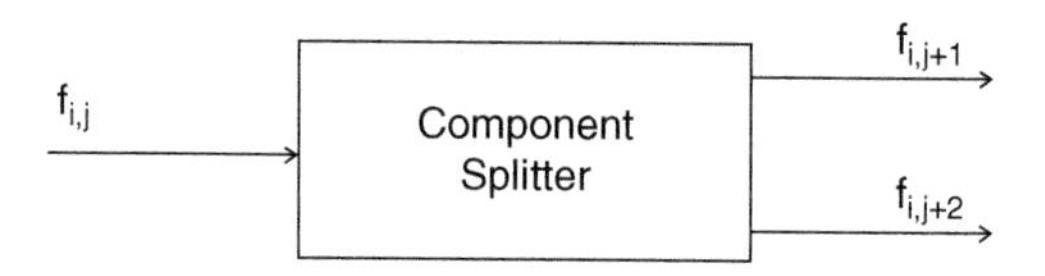

FIGURE 5c Component splitter or divider (note that the balance volumes are the same for the component splitter and divider).

<u>*Component splitter model*</u>: for one feed stream j and two product streams, $j+1$ and $j+2$) – see Figure 5c.

$$f_{ij}+1=\xi_{iS}f_{ij} \qquad \text{for } i=1, NC \tag{3}$$

$$f_{ij+2}=(1-\xi_{iS})f_{ij} \tag{4}$$

<u>*Stream divider model*</u>: for one feed stream j and two product streams, $j+1$ and $j+2$ – see Figure 5c.

$$f_{ij+1}=\beta f_{ij} \qquad \text{for } i=1, NC \tag{5}$$

$$f_{ij+2}=(1-\beta)f_{ij} \qquad \text{for } i=1, NC \tag{6}$$

In the above equations, f_{ij} is the component flow rate of compound i in stream j; γ_i is the stoichiometric coefficient of compound i; η_k is the conversion with respect to the limiting reactant k; ξ_{iS} is the separation factor of component i in the component splitter; β is the separation factor of component i in the divider.

Note that the difference between a component splitter and a stream divider is that in the stream divider, all the streams have the same compositions, whilst in the component splitter, they are all different. In this way, the stream divider is a special case of the component splitter. Note that instead of the conversion, a reaction rate expression could also be used.

Based on the above mass balance model, energy balance equations (one per unit operation) can easily be added to compute the energy added/removed from each unit operation. For example, in the case of a component splitter, assuming that the inlet stream is a liquid, and the outlet streams are saturated vapour and liquid streams, the following energy balance equation is obtained,

$$H^F(T^F, \underline{z})F_j = H^L(T^L, \underline{x})F_{j+2} + H^V(T^V, \underline{y})F_{j+1} + Q \tag{7}$$

In Eq. (7)7, H^J is the enthalpy for stream j; x, y, and z are the mole fractions of the liquid, vapour and feed streams, respectively; T^J is the temperature of stream j; and Q is the heat added or removed.

Following this procedure, more complex models can be developed for each unit operation. The model complexity needs to match the model objectives and the model assumptions. In the next higher level of the model, for example, the objective could be to replace a component splitter, a single-stage two-phase flash (assuming vapour–liquid equilibrium) or a distillation column, depending on the separation task. To obtain dynamic models, the left-hand sides of Eqs. 1-7 are replaced with a time derivative term. Examples of these models can be found in chapters 5 and 10, and in appendix-2. Note that because the separation factors and reaction conversions, etc., are specified, the constitutive models are not needed in these simple process models.

6.2.1.3. Williams-Otto Plant: Modelling and Simulation

The model equations for this process are derived from Eqs. 1-6, since only a mass balance model is being used. However, instead of conversion, the reaction rate is calculated through a temperature-dependent kinetic model (for a specified temperature). The model for the reactor includes the mixer, whilst two sets of equations are needed for the component splitter to represent the decanter and the distillation column (see Figure 1). The model equations as implemented in ICAS-MoT (Sales-Cruz, 2006) are given in Appendix-1 (see ch-6-1-Williams-Otto-Biegler.mot file). Here, they are represented in a compact short form to highlight the model analysis and solution strategy issues.

Reactor (mass balance for components A, B, C, E, P and G) – NC equations:

$$\begin{aligned}
FR_A &= Ff_A + Fr_A - k_1{*}X_A{*}X_B{*}V{*}\rho \\
FR_B &= Ff_B + Fr_B - (k_1{*}X_A + k_2{*}X_C){*}X_B{*}V{*}\rho \\
FR_C &= Fr_C + (2{*}k_1{*}X_A{*}X_B - 2{*}k_2{*}X_B{*}X_C - k_3{*}X_P{*}X_C){*}V{*}\rho \\
FR_E &= Fr_E + 2{*}k_2{*}X_B{*}X_C){*}V{*}\rho \\
FR_P &= Fr_P + (k_2{*}X_B{*}X_C - 0.5{*}k_3{*}X_P{*}X_C){*}V{*}\rho \\
FR_G &= Fr_G + (1.5{*}k_3{*}X_P{*}X_C){*}V{*}\rho
\end{aligned} \tag{8}$$

Definition of mole fractions – NC equations:

$$X_i = FR_i/(\Sigma FR_i) \qquad i = A,\ B,\ C,\ E,\ P,\ G \tag{9}$$

Constitutive (kinetic) models – 3 equations:

$$k_i = k_{0i}{*}\exp(-E_i/T) \qquad i = 1,\ 2,\ 3 \tag{10}$$

Decanter (component splitter) – 2*NC equations:

$$FD_i = \xi_{iS1} {}^* FRi \qquad i = A,\ B,\ C,\ E,\ P,\ G \tag{11a}$$

$$FW_i = (1 - \xi_{iS1}) {}^* FR_i \tag{11b}$$

Distillation Column (component splitter) – 2*NC equations:

$$F_{Bi} = \xi_{iS2} {}^* FD_i \qquad i = A,\ B,\ C,\ E,\ G \tag{12a}$$

$$FB_P = \xi_{PS2} {}^* FD_E \tag{12b}$$

$$FT_i = (1 - \xi_{iS2}) {}^* FD_i \qquad i = A,\ B,\ C,\ E,\ G \tag{13a}$$

$$FT_P = FD_P - \xi_{PS2} {}^* FD_E \tag{13b}$$

Flow splitter (stream divider) – 2*NC equations:

$$FP_i = \beta {}^* FB_i \qquad i = A,\ B,\ C,\ E,\ P,\ G \tag{14}$$

$$Fr_i = (1 - \beta) {}^* FB_i \qquad i = A,\ B,\ C,\ E,\ P,\ G \tag{15}$$

Model Analysis: Based on the above model, we have a total of 8NC + 3 equations, and 11NC + 3*3 + 5 variables. This means that 3NC + 8 variables need to be specified before the remaining 8NC + 3 unknown variables (F_R, $\underline{F}_W$, $\underline{F}_D$, $\underline{F}_T$, $\underline{F}_B$, $\underline{F}_P$, $\underline{F}_r$, $\underline{X}$, k_1, k_2, k_3) can be calculated. Table 1 gives the model analysis for this simple process flowsheet model, and Table 2 gives the incidence matrix. Note the ' + ' on the row for Eq. (8)8. This indicates that either the full set of equations need to be solved simultaneously, or, an iterative solution scheme as shown in Figure 6 can be considered where a guess of the stream variable (F_r) is made so that Eq. (2)2 can be solved for F_R. Repeating this for every unit operation as shown in Figure 6, one can see that after the mixer is solved, a new calculated value of F_r is obtained. If the guessed value and the calculated value are not the same (that is, the difference is not within a specified tolerance), a new iteration is started with a new guess. To accelerate the convergence, a suitable acceleration technique (Broydon, quasi Newton-Rahpson, etc.) may be used.

Numerical solution: Variables that need to be specified are: $k_{01} = 5.9755{}^*10^9$; $k_{02} = 2.5962{}^*10^{12}$; $k_{03} = 9.6283{}^*10^{15}$; $E_1 = 12000$; $E_2 = 15000$; $E_3 = 20000$; $\xi_{iS1} = 1$ for $i = A, B, C, E, P$; $\xi_{GS1} = 0$; $\xi_{iS2} = 1$ for $i = A, B, C, E, G$; $\xi_{PS2} = 0.1$; $T = 674.43$; $V = 30$; $\rho = 50$; $F_{fA} = 13179.6$; $F_{fB} = 30038.3$; $\xi_D = 0.1$.

The model equations as implemented in ICAS-MoT can now be solved for the specified variable values given above.

TABLE 1 Model Analysis

Equations:	Number
Eq. 10	3
Eq. 8	NC
Eq. 9	NC
Eq. 11 (times 2)	NC*2
Eq. 12 (a + b)	NC
Eq. 13 (a + b)	NC
Eq. 14	NC
Eq. 15	NC
Total: NE	8*NC + 3
Number of Variables:	
Component flow-rates: $\underline{F}_R$, $\underline{F}_r$, $\underline{F}_W$, $\underline{F}_D$, $\underline{F}_B$, $\underline{F}_T$, $\underline{F}_P$, F_{fA}, F_{fB}	7*NC + 2
Component mole fractions: $\underline{X}$	NC
Reactor parameters: $\underline{\gamma}$ (stoichiometric coefficients), $\underline{k}_0$, $\underline{E}$ (rate constants), $\underline{k}$	NC + 2*3 + 3
Stream calculator parameters: $\underline{\xi}_{S1}$, $\underline{\xi}_{S2}$ (*i.e.*, recovery of each compound compared to feed)	NC*2
Divider parameters: ξ_D (divide fraction compared to feed)	1
Other variables (ρ, V)	2
Total: NV	11*NC + 14
Degrees of freedom: NV – NE	3*NC + 11

Variables to specify (decisions): F_{fA}, F_{fB} (2 process variables); V, $\underline{\gamma}$, $\underline{k}_0$, $\underline{E}$, $\underline{\xi}_{S1}$, $\underline{\xi}_{S2}$, ξ_D (3NC + 7 equipment parameters); ρ (mixture properties)
*Unknown variables: $\underline{F}_R$, $\underline{F}_r$, $\underline{F}_W$, $\underline{F}_D$, $\underline{F}_B$, $\underline{F}_T$, $\underline{F}_P$, $\underline{X}$ (8*NC process variables) + $\underline{k}$ (3 rate parameters)*

TABLE 2 Incidence Matrix

	Specified variables										Calculated variables							
	F_{fA}	F_{fB}	$\underline{k}_0$	$\underline{E}$	ρ	V	$\underline{\xi}_{S\text{-}s}$	$\underline{\xi}_{S\text{-}d}$	ξ_D	$\underline{k}$	$\underline{F}_R$	$\underline{X}$	$\underline{F}_D$	$\underline{F}_W$	$\underline{F}_B$	$\underline{F}_B$	$\underline{F}_P$	$\underline{F}_r$
Eq. 10			*	*						(*)								
Eq. 8	*	*			*	*					(*)	+						+
Eq. 9											*	(*)						
Eq. 11a								*			*		(*)					
Eq. 11b								*			*			(*)				
Eq. 12							*						*		(*)			
Eq. 13							*						*			(*)		
Eq. 14									*							*	(*)	
Eq. 15									*								*	(*)

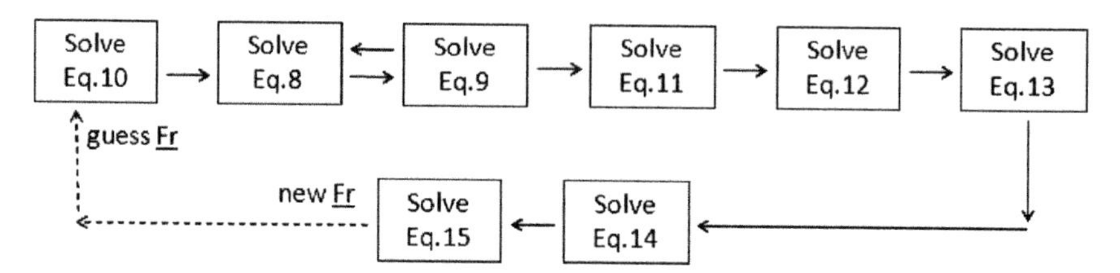

FIGURE 6 Calculation sequence, for the Williams-Otto plant model.

TABLE 3 Simulation Results

Compound	Fr	FW	FP	FT
A	41743.7	0	4638.2	
B	129423.2	0	14380.3	
C	6842.2	0	760.2	
E	128097.6	0	14233.0	
P	12809.7	0	1423.3	4648.5
G	0	3134.2	0	

Discussion: Add an objective function (Biegler et al. 1997) to the simulation model of the Williams-Otto plant and maximise the profit function subject to F_{fA}, F_{fB}, V, s_D and T as the design variables (see Table 4).

$$ROI = (2207*F_{PT}+50*F_P - 168*F_{fA} - 252*F_{fB} - 2.22*F_r - 84*F_W + 600*V*\rho - 1041.6)/(6*\rho*V)$$

The objective function and the change of the design variables as a function of iteration-index are shown in figures 7a–7d.

6.2.1.4. Conversion of Ethylene to Ethanol- Numerical Solution

The Eqs. 1-6 are used in this case study to model the mass balance for the conversion of ethylene to ethanol. The model equations, as implemented in ICAS-MoT, are given in Appendix-1 (see ch-6-ethylene-ethanol-mb.mot file). The list of specified variables and their values are given in Table 5. In addition, the conversions for reactions 1 and 2 are 0.07 for ethylene and 0.007 for propylene, respectively. For the equilibrium reaction, the amount of diethyl ether (DEE) formed can be calculated from ethanol (EA) and water (W) – as given in Biegler et al. (1997).

$$f_{DEE} = K_c (f_{EA})^2 / f_W \tag{17}$$

TABLE 4 Design Variables for the Profit Maximisation

Lower bound	*Variable*	Upper bound	Initial estimate	Optimal value
0	F_{fA}	15000	11540	13567.56
30	V	60	30	30
0	p	0	50	50
0	F_{fB}	35000	31000	30887.76
580	T	680	674.36	675.09
0.01	s_D	0.99	0.1002	0.1006

where f_i is the flow rate (moles/h) of the compound *i*, and K_c is the equilibrium constant.

The simulation results in terms of calculated stream variables for the outlet streams are given in Table 6. Using the values given in Table 5 and the model equations (Eqs. 1-6), interested readers can check their models/solver by comparing their calculated values against those given in Table 6.

6.2.2. Modular Modelling (Process Simulator)

6.2.2.1. Conversion of Ethylene to Ethanol

Here, the same process flowsheet for the ethylene conversion to ethanol is modelled through PRO/II (2006), using the models present in its model library. The simulator organises the model equations, according to its built-in solution strategy (sequential modular approach). The following models from the model library are used: mixer, stoichiometric reactor, stream calculator (for all the separators, that is, PT-flash, absorber and the three distillation columns) and splitter (for divider) – as shown in Figure 8. The data given in Table 5 is also used

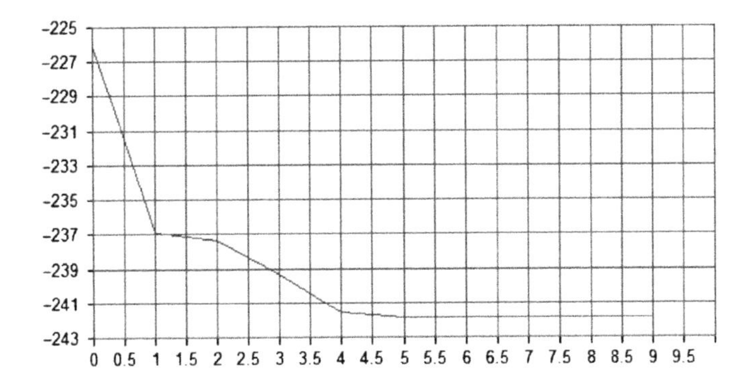

FIGURE 7a ROI (objective function) on y-axis versus iteration-index (x-axis).

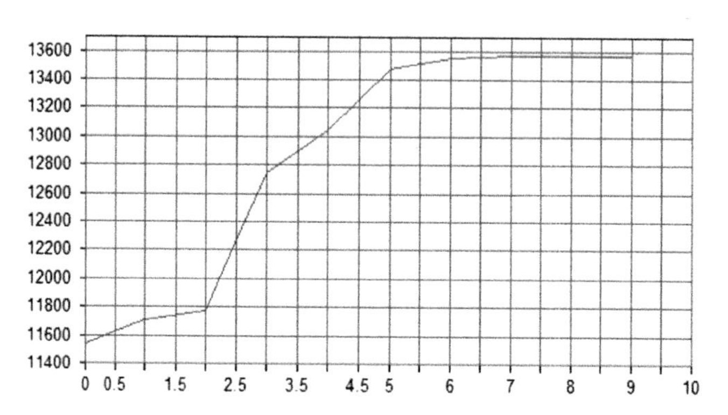

FIGURE 7b **FfA (feed flow-rate of A) on y-axis versus iteration-index (x-axis).**

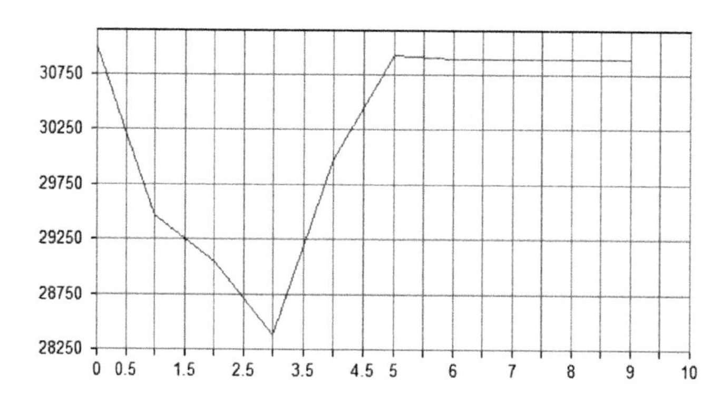

FIGURE 7c **FfB (feed flow-rate of B) on y-axis versus iteration-index (x-axis).**

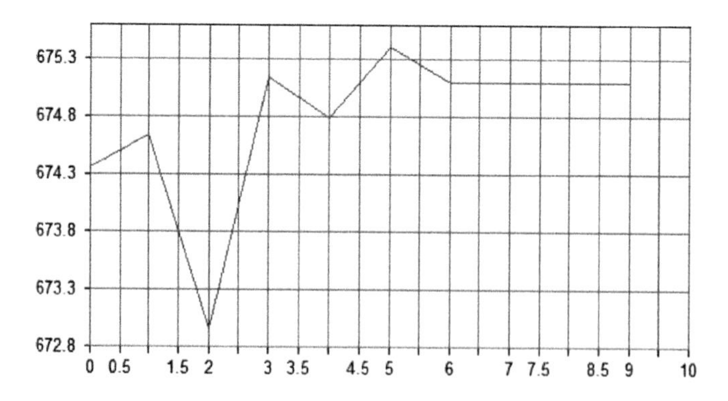

FIGURE 7d **T (reactor temperature) on y-axis versus iteration-index (x-axis).**

TABLE 5 Values of Specified Variables for Conversion of Ethylene to Ethanol

	Feed*	Solvent	ξ_{Flash}	ξ_{Abs}	ξ_{Purge}	$\xi_{Dist\text{-}1}$	$\xi_{Dist\text{-}2}$	$\xi_{Dist\text{-}3}$
Ethylene	96		0.985	0.979	0.995	1	1	1
Propylene	3		0.932	0.901	0.995	1	1	1
Methane	1		0.996	1	0.995	1	1	1
Ethanol			0.121	0.01	0.995	0.995	0.005	0.995
Isopropanol			0.083	0.0065	0.995	0.960	0.05	0.05
Water	777.4	37.7	0.054	0.0019	0.995	0.1	0	0.22
Diethyl ether			0.5	0.24	0.995	1	0.995	1

*Note: *- combined reactor feed (in gmol/s); values of $\xi_i$correspond to those given in Biegler et al. (1997).*

here, and, in principle, can be used in any other process simulator, using the same models.

However, if the separator is to represent a distillation column, simply selecting the corresponding model makes the simulator extract the corresponding model equations – as shown in Figure 9. Note, however, in this case, the user is not able to view the equations but the user needs to provide values of the known variables. The detailed simulation results are not given here. However, the corresponding PRO/II files can be downloaded from the book website (http://www.elsevierdirect.com/companion.jsp?ISBN=9780444531612). Note that now as more equations are added for more rigorous models, more variables also need to be specified. For a complete list, consult the PRO/II manual. Note also that the constitutive (property) model equations are also retrieved by simply selecting the corresponding model from the constitutive (property) model library.

6.2.2.2. *Hydroalkylation Process*

For this process, first we will simulate the process with the simple mass balance model, and then we will replace the stream calculators with distillation columns

TABLE 6 Simulation Results for the Outlet Streams in gmol/s

	Feed*	Solvent	S_{52}	S_{72}	S_{91}	S_{92}
Ethylene	96		5.78	0	0	0
Propylene	3		1.12	0	0	0
Methane	1		1	0	0	0
Ethanol			0	0.453	89.317	0.448
Isopropanol			0	0.075	0.086	1.630
Water	777.4	37.7	0	646.890	15.813	56.064
Diethyl ether			0.001	0	0.010	0

*Note: *: - combined feed (in gmol/s).*

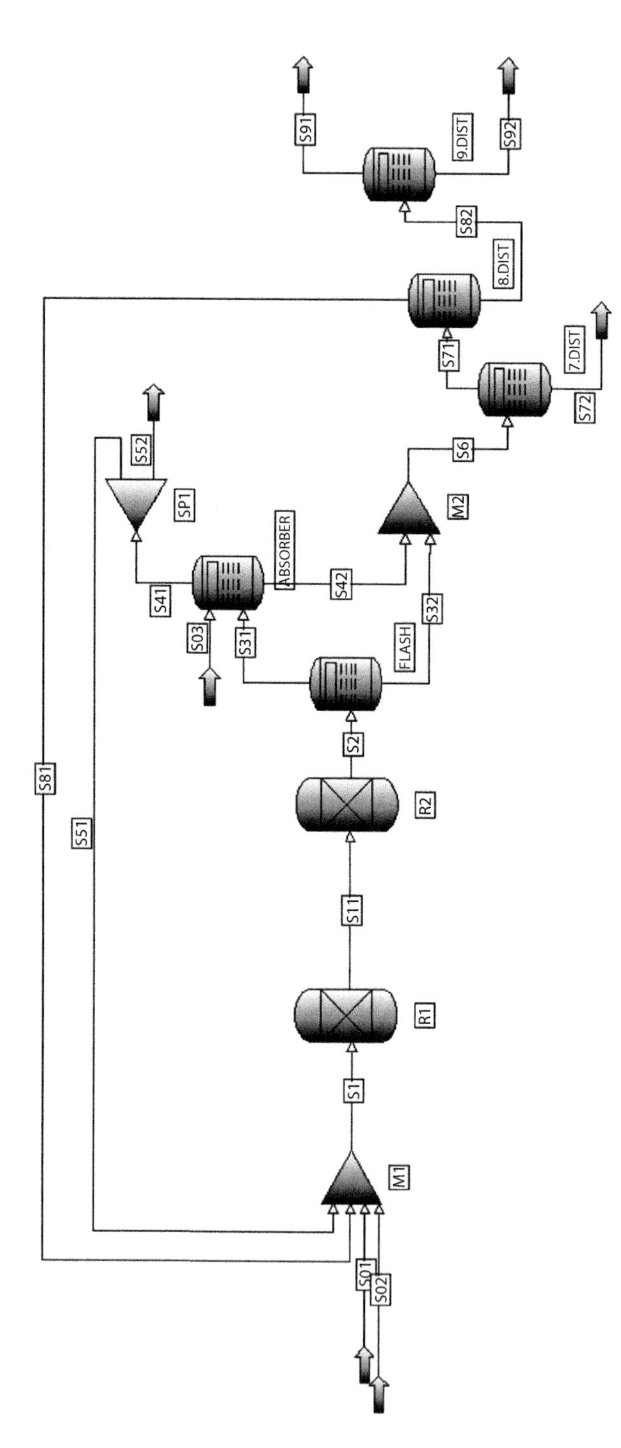

FIGURE 8 Conversion of ethylene to ethanol process flowsheet as represented with models from PRO/II.

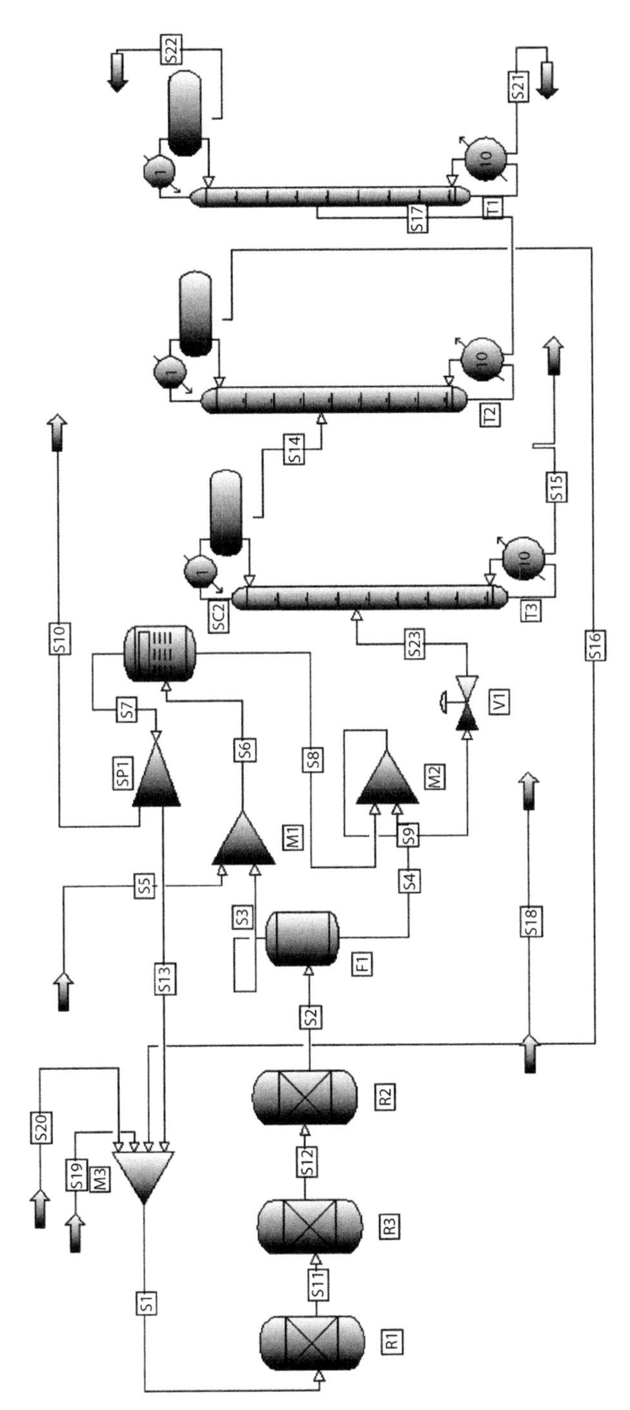

FIGURE 9 Process flowsheet, for conversion of ethylene as represented with models from PRO/II (rigorous model).

and add heat exchangers and pumps. The data for this process can be found in many published articles and books. The data used here is from Douglas (1988) – see Table 7. Using the same mass balance models given earlier in this chapter (Eqs. 1-6), this process flowsheet can be modelled for mass balance calculations. The process flowsheet, as represented by PRO/II models, is shown in Figure 10. Note that the purge is modelled as a component splitter and not as a divider (and the simulation results are given in Table 8).

In order to convert the simple flowsheet to the detailed flowsheet, the stream calculators need to be replaced by distillation columns. Also, heat exchangers, pumps, etc., need to be added. To perform the simulations, the set of variables needed to specify for each model in PRO/II is given in Table 9 and the complete flowsheet is shown in Figure 11. The simulated results are given in terms of a stream summary (see Tables 10a–10c).

6.2.2.3. *Bio-ethanol Process Simulation*

Modelling of this bio-ethanol process (Alvarado-Morales, et al. 2009), with a process simulator, requires several issues to be resolved. For example, in order to include the full list of compounds usually found in this process, new compounds, their physical properties and their constitutive (property) model parameters need to be added to the property model library (database) of the simulator. The process specifications (Alvardo-Morales, et al. 2009) are given in Table 12. The reactor models are also complicated by the lack of data – full lists of known reactions taking place in the different reactors are given in Tables 13–18. For each of these reactions, it is necessary to know either the reaction rate (constitutive) model or an overall conversion. As can be noted, the pre--treatment, hydrolysis steps are modelled as reactors followed by separation. For solid separations, a simple mass balance model is used. For a number of unit (separation) operations, stream calculators have been used because of the lack of

TABLE 7 Specifications for Hydrodealkylation Process (Simple Mass Balance Model)

Compound	γ		ξ (Separation factor)				
	Reactor		PT-Flash	Purge	Distillation		
	1	2	SC1	SC2	SC3	SC4	SC5
Methane	1	0	0.9857	0.9840	1	1	1
Hydrogen	−1	1	0.9984	1.0000	1	1	1
Benzene	1	−1	0.0409	0	0	1	1
Toluene	−1	0	0.0134	0	0	0.0009	1
Biphenyl	0	1	0	0	0	0	0

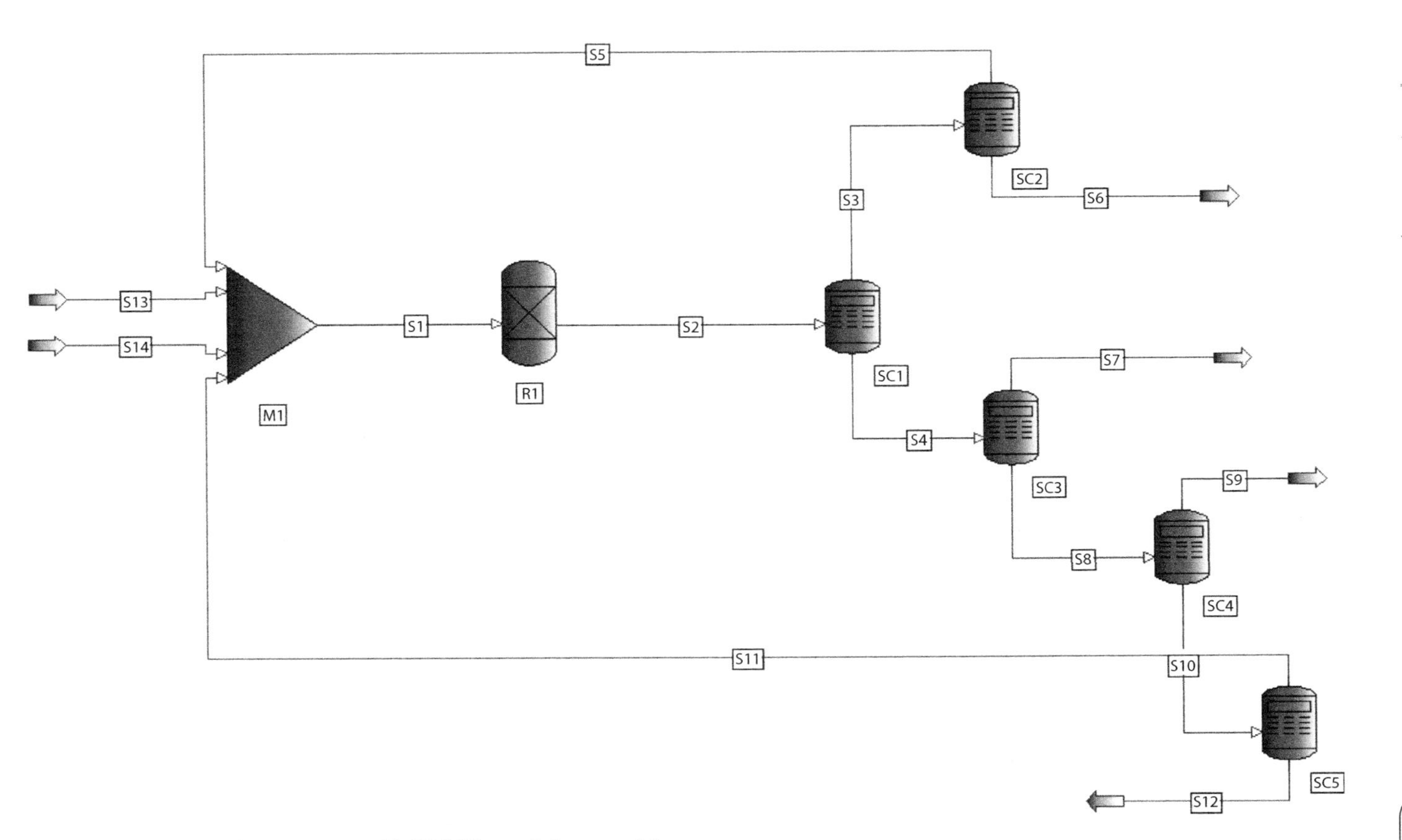

FIGURE 10 Modelling of HDA process with PRO/II mass balance model.

TABLE 8 Simulated Results from PRO/II – Mass Balance Only

Stream	S1	S2	S3	S4	S5	S6	S7	S8	S9	S10	S11	S12	S13	S14
Name														
Flow-rate (kmol/hr)	2297.35	2296.22	1800.44	495.79	1553.33	247.11	125.36	370.42	274.40	96.02	93.22	2.80	280.00	370.80
Composition														
Toluene	0.1625	0.0406	0.0000	0.1882	0.0000	0.0000	0.0000	0.2519	0.0003	0.9708	1.0000	0.0000	1.0000	0.0000
Hydrogen	0.8328	0.7121	0.8627	0.1649	1.0000	0.0000	0.6521	0.0000	0.0000	0.0000	0.0000	0.0000	0.0000	0.9709
Benzene	0.0000	0.1195	0.0000	0.5533	0.0000	0.0000	0.0000	0.7406	0.9997	0.0000	0.0000	0.0000	0.0000	0.0000
Methane	0.0047	0.1266	0.1373	0.0880	0.0000	1.0000	0.3479	0.0000	0.0000	0.0000	0.0000	0.0000	0.0000	0.0291
Biphenyl	0.0000	0.0012	0.0000	0.0056	0.0000	0.0000	0.0000	0.0076	0.0000	0.0292	0.0000	1.0000	0.0000	0.0000

TABLE 9 List of Specifications for the Rigorous Simulation with PRO/II – HDA Process

Unit Operation/ Stream	Variables to specify	Specified value
Stream 13	$T, P, F, \underline{x}$	See Table 10b
Stream 14	$T, P, F, \underline{x}$	See Table 10b
Pump –P1	Outlet pressure	
Heat exchanger – E1	Outlet temperature (process stream)	
Heat exchanger – E2	Outlet temperature (process stream)	
Heat exchanger – E3	Outlet temperature (process stream)	
Reactor – R1	Reaction, conversion, stoichiometric coefficients	See Table 7
PT Flash – F1	T, P	
Distillation column – T1	Number of stages (NP), feed stage (NF), Product rate (distillate or bottom), recovery of compound i at top (ξ_{Di}), recovery of compound i at bottom (ξ_{Bi})	
Distillation column – T2	Number of stages (NP), feed stage (NF), Product rate (distillate or bottom), recovery of compound i at top (ξ_{Di}), recovery of compound i at bottom (ξ_{Bi})	
Distillation column – T3	Number of stages (NP), feed stage (NF), Product rate (distillate or bottom), recovery of compound i at top (ξ_{Di}), recovery of compound i at bottom (ξ_{Bi})	
	Constitutive Models	
VLE	Model for fugacity coefficients	SRK equation of state
Pure component (temperature-dependent) properties	Vapour pressure, density, hat capacity, heat of vaporisation , viscosity, *etc.*	DIPPR correlaions

Note: the missing data in the last column of the table indicate the design (decision) variables

appropriate process models. However, each of these unit operations can be modelled separately, and as long as the inlet and outlet stream properties can be matched, the process flowsheet results would still be valid.

A copy of the PRO/II file corresponding to the process flow-sheet (Figure 12) and results (Table 11) can be downloaded from the book website (http://www.elsevierdirect.com/companion.jsp?ISBN=9780444531612).

6.2.2.4. Discussion

For the interested reader, an interesting exercise would be to convert the bio-ethanol process flowsheet into a simple mass balance model represented by Eqs.

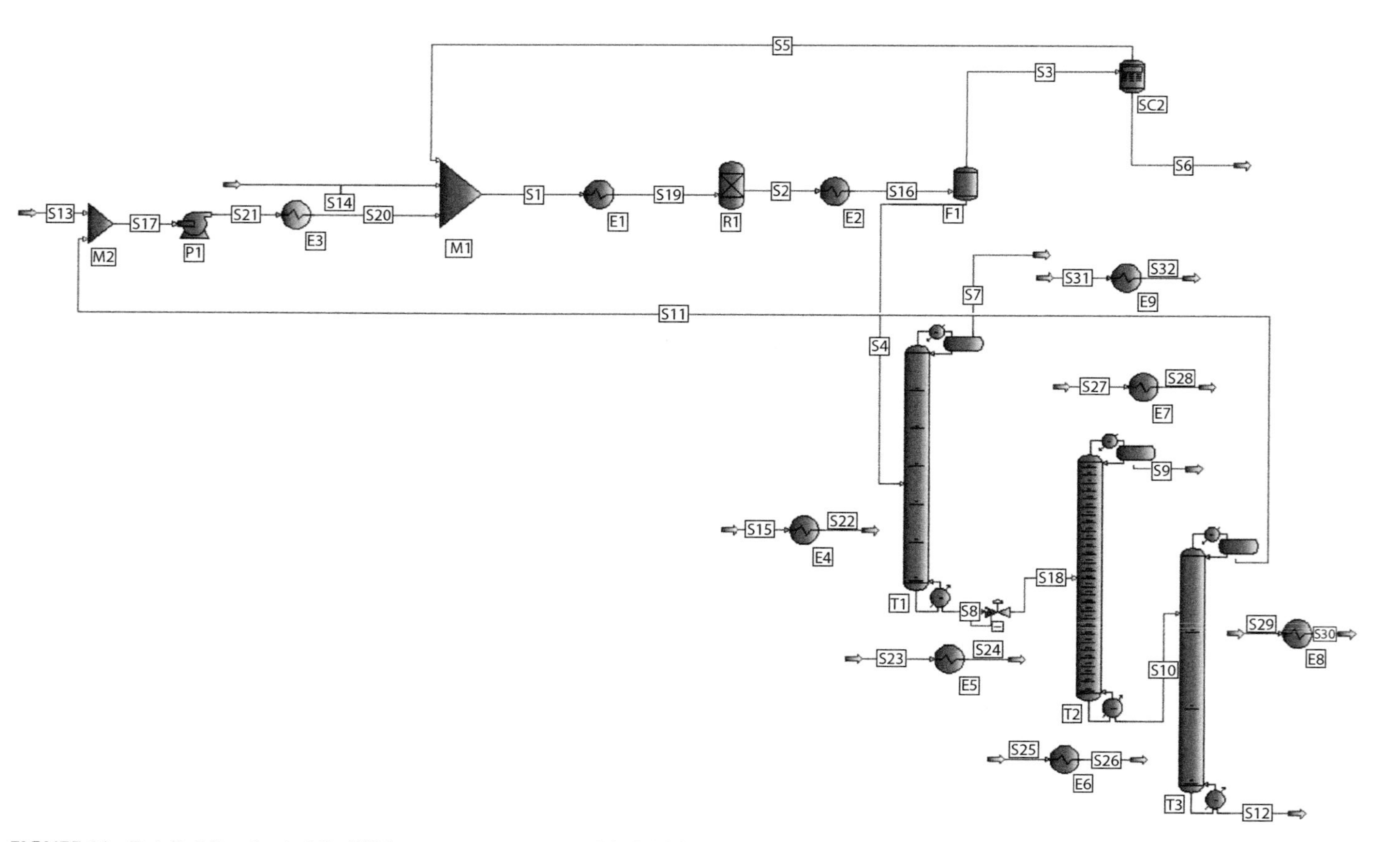

FIGURE 11 Detailed flowsheet of the HDA process, as represented in PRO/II – note that the heat integration is included as a separate flowsheet.

TABLE 10A Simulated Results from PRO/II – Mass and Energy Balance with Rigorous Models

	Streams									
Stream variable	**S1**	**S2**	**S3**	**S4**	**S5**	**S6**	**S7**	**S8**	**S9**	**S10**
Phase	Mixed	Vapour	Vapour	Liquid	Vapour	Mixed	Vapour	Liquid	Liquid	Liquid
Temperature (K)	455.550	959.817	293.150	293.150	296.952	296.952	345.690	542.360	352.920	384.855
Pressure (atm)	32.934	32.934	32.934	32.934	32.934	32.934	32.930	32.930	1.000	1.000
Flow rate (kmol/hr)	1756.851	1757.532	1383.417	374.115	1075.015	308.403	8.634	365.481	269.755	95.726
					Composition					
Toluene	0.212	0.053	0.000	0.248	0.000	0.001	0.001	0.254	0.000	0.968
Hydrogen	0.783	0.625	0.793	0.006	1.000	0.071	0.245	0.000	0.000	0.000
Benzene	0.000	0.156	0.003	0.722	0.000	0.013	0.031	0.739	1.000	0.003
Methane	0.005	0.164	0.204	0.017	0.000	0.915	0.723	0.000	0.000	0.000
Biphenyl	0.000	0.002	0.000	0.007	0.000	0.000	0.000	0.008	0.000	0.029

TABLE 10B Simulated Results from PRO/II – Mass and Energy Balance with Rigorous Models

	Stream									
Stream variable	**S11**	**S12**	**S13**	**S14**	**S15**	**S16**	**S17**	**S18**	**S19**	**S20**
Phase	Liquid	Liquid	Liquid	Vapour	Liquid	Mixed	Liquid	Vapour	Vapour	Vapour
Temperature (K)	383.798	507.332	297.039	294.261	542.360	293.150	320.415	398.162	960.000	574.089
Pressure (atm)	1.000	1.000	1.000	32.934	32.930	32.934	1.000	1.000	32.934	32.934
Flow rate (kmol/hr)	92.837	2.890	280.000	309.000	2018.900	1757.532	372.837	365.481	1757.532	372.837
Composition (mole fractions)										
Toluene	0.997	0.032	1.000	0.000	0.254	0.053	0.999	0.254	0.212	0.999
Hydrogen	0.000	0.000	0.000	0.971	0.000	0.625	0.000	0.000	0.783	0.000
Benzene	0.003	0.000	0.000	0.000	0.739	0.156	0.001	0.739	0.000	0.001
Methane	0.000	0.000	0.000	0.029	0.000	0.164	0.000	0.000	0.005	0.000
Biphenyl	0.000	0.968	0.000	0.000	0.008	0.002	0.000	0.008	0.000	0.000

TABLE 10C Simulated Results from PRO/II – Mass and Energy Balance, with Rigorous Models

Stream Variable	Stream											
	S21	S22	S23	S24	S25	S26	S27	S28	S29	S30	S31	S32
Phase	Liquid	Vapour	Liquid	Vapour	Liquid	Vapour	Vapour	Liquid	Vapour	Liquid	Vapour	Mixed
Temperature (K)	320.909	548.263	384.855	419.996	507.332	527.330	354.000	352.920	384.000	383.798	526.552	345.700
Pressure (atm)	32.934	32.930	1.000	1.000	1.000	1.000	1.000	1.000	1.000	1.000	32.930	32.930
Flow rate (kmol/hr)	372.837	2018.900	223.800	223.800	699.800	699.800	669.700	669.700	1359.900	1359.900	270.060	270.060
Composition (mole fraction)												
Toluene	0.999	0.254	0.968	0.968	0.032	0.032	0.000	0.000	0.997	0.997	0.057	0.057
Hydrogen	0.000	0.000	0.000	0.000	0.000	0.000	0.000	0.000	0.000	0.000	0.003	0.003
Benzene	0.001	0.739	0.003	0.003	0.000	0.000	1.000	1.000	0.003	0.003	0.887	0.887
Methane	0.000	0.000	0.000	0.000	0.000	0.000	0.000	0.000	0.000	0.000	0.053	0.053
Biphenyl	0.000	0.008	0.029	0.029	0.968	0.968	0.000	0.000	0.000	0.000	0.000	0.000

TABLE 11 Selected Results from Simulation with PRO/II (Alvarado-Morales, *et al.* 2009).

Stream	S1	S2	S4	S5	S7	S9	S13	S19	S22	S26
Phase	Mixed	Vapour	Liquid	Liquid	Vapour	Mixed	Mixed	Mixed	Mixed	Mixed
Temperature (°C)	25	164	25	74	268	190	101	76	76	75
Pressure (atm)	1.00	4.42	3.40	3.00	13.00	12.20	1.00	1.00	1.00	1.00
Total mass rate (kg/h)	159115	16960	921	47583	44596	269175	224041	304141	211946	213301
Total molar rate (kmol/h)	4871	941	9	2626	2475	10806	8337	15955	10961	10977
Molar fraction composition										
Cellulose	0.045	0	0	0	0	0.019	0.024	0.000	0.000	0.000
Hemicellulose	0.025	0	0	0	0	0.003	0.003	0.000	0.000	0.000
Galactan	0.000	0	0	0	0	0.000	0.000	0.000	0.000	0.000
Mannan	0.004	0	0	0	0	0.000	0.001	0.000	0.000	0.000
Arabinan	0.001	0	0	0	0	0.000	0.000	0.000	0.000	0.000
Lignin	0.039	0	0	0	0	0.017	0.023	0.000	0.000	0.000
Glucose	0	0	0	0	0	0.001	0.002	0.001	0.001	0.001
Mannan	0	0	0	0	0	0.001	0.002	0.001	0.001	0.001
Galactose	0	0	0	0	0	0.000	0.000	0.000	0.000	0.000
Xylose	0	0	0	0	0	0.008	0.011	0.004	0.006	0.006
Arabinan	0	0	0	0	0	0.000	0.000	0.000	0.000	0.000
Ethanol	0	0	0	0	0	0	0	0.000	0.000	0.000
Water	0.873	1.000	0	0.995	1.000	0.941	0.925	0.985	0.988	0.987
Sulphuric Acid	0	0	1.000	0	0	0.001	0.001	0.000	0	0.001
Furfural	0	0	0	0.000	0	0.000	0.000	0.000	0.001	0.001
Ammonia	0	0	0	0	0	0	0	0.004	0	0
Carbon Dioxide	0	0	0	0	0	0	0	0	0	0
Hydroxymethyl Furfural	0	0	0	0.000	0	0.000	0.000	0.000	0.000	0.000
Acetic Acid	0	0	0	0.001	0	0.006	0.007	0.003	0.001	0.001
Zymomonas mobilis	0	0	0	0	0	0	0	0.000	0.000	0.000
Cellulase (enzyme)	0	0	0	0	0	0	0	0	0	0
Total enthalpy (kcal/h) x10^6	−552.3	196.5	0.1	61.5	519.6	287.9	−137.0	403.1	282.9	274.9

Temperature (°C)	77	77	77	34	33	33	93	111	93	114
Pressure (atm)	1.00	1.00	1.00	1.00	1.00	1.00	1.00	1.77	1.77	1.77
Total mass rate (kg/h)	215800	216104	213998	309369	382706	382706	365913	47193	19081	17719
Total molar rate (kmol/h)	11123	11129	11092	13672	17319	17733	17342	1987	463	388
Molar fraction composition										
Cellulose	0.000	0.000	0.000	0.013	0.011	0.003	0.003	0	0	0
Hemicellulose	0.000	0.000	0	0.002	0.001	0.001	0.001	0	0	0
Galactan	0.000	0.000	0	0.000	0.000	0.000	0.000	0	0	0
Mannan	0.000	0.000	0	0.000	0.000	0.000	0.000	0	0	0
Arabinan	0.000	0.000	0.000	0.000	0.000	0.000	0.000	0	0	0
Lignin	0.000	0.000	0.000	0.012	0.011	0.011	0.011	0	0	0
Glucose	0.001	0.001	0.001	0.001	0.001	0.001	0.001	0	0	0
Mannan	0.001	0.001	0.001	0.001	0.001	0.001	0.001	0	0	0
Galactose	0.000	0.000	0.000	0.000	0.000	0.000	0.000	0	0	0
Xylose	0.006	0.006	0.006	0.006	0.005	0.001	0.001	0	0	0
Arabinan	0.000	0.000	0.000	0.000	0.000	0.000	0.000	0	0	0
Ethanol	0.000	0.000	0.000	0.000	0.001	0.022	0.022	0.193	0.828	0.988
Water	0.988	0.988	0.989	0.960	0.956	0.925	0.945	0.799	0.158	0.012
Sulphuric Acid	0	0.000	0.000	0.000	0.000	0.000	0.000	0.000	0	0
Furfural	0.001	0.001	0.001	0.000	0.000	0.000	0.000	0.004	0.000	0
Ammonia	0	0	0	0.001	0.000	0.000	0.000	0.003	0.014	0
Carbon Dioxide	0	0	0	0	0.000	0.021	0	0	0	0
Hydroxymethyl Furfural	0.000	0.000	0.000	0.000	0.000	0.000	0.000	0	0	0
Acetic Acid	0.001	0.001	0.001	0.002	0.002	0.002	0.002	0.000	0.000	0
Zymomonas mobilis	0.000	0.000	0	0.000	0.000	0.001	0.001	0	0	0
Cellulase (enzyme)	0	0	0	0.000	0.002	0.002	0.002	0	0	0
Total enthalpy (kcal/h) x10^6	287.5	290.0	291.5	−209.6	−220.1	−47.9	285.7	406.6	93.8	81.4

TABLE 12 **Process Data for Bioethanol Production (Alvarado-Morales, et al. 2009)**

Feature	Value
Feedstock	Harwood chips
Composition (w/w)	Cellulose 22.3%, hemicellulose 9.9 %, lignin 14.5 %, moisture 48.1 %
Feed flow rate	159 ton/h
Pre-treatment reactor	
Agent	Dilute sulphuric acid
Acid concentration (w/w)	0.5%
Temperature	190 °C
Pressure	12.2 atm
Solids in the reactor (w/w)	22%
Flash vessel	
Temperature	101 °C
Pressure	1 atm
Detoxification	
Type	Ion exchange followed by neutralisation with alkali
Eluent	Ammonia
Alkali	Calcium hydroxide
SSCF	
Temperature	30 °C
Initial solids level (w/w)	20%
Enzyme	Cellulase
Biocatalyst	Zimomonas mobilis
Enzyme level	15 FPU/g of cellulose
Pressure	1 atm
Downstream separation	
Beer distillation	
Pressure	1.77 atm
Stages	32
Feed stage	4
Reflux ratio	6.1
Rectification column	
Pressure	1.77 atm
Stages	60
Feed stage	44
Reflux ratio	3.2

TABLE 13 Reaction Data for Pretreatment (Wooley et al. 1999)

Reaction	Conversion	Modelled
$C_6H_{10}O_5 + H_2O \rightarrow C_6H_{12}O_6$	Cellulose	0.065
$C_6H_{10}O_5 + 1/2H_2O \rightarrow 1/2C_{12}H_{22}O_{11}$	Cellulose	0.007
$C_5H_8O_4 + H_2O \rightarrow C_5H_{10}O_5$	Hemicellulose	0.750
$C_5H_8O_4 \rightarrow C_4H_3OCHO + 2H_2O$	Hemicellulose	0.100
$C_6H_{10}O_5 + H_2O \rightarrow C_6H_{12}O_6$	Mannan	0.750
$C_6H_{10}O_5 \rightarrow C_6H_6O_3 + 2H_2O$	Mannan	0.150
$C_6H_{10}O_5 + H_2O \rightarrow C_6H_{12}O_6$	Galactan	0.750
$C_6H_{10}O_5 \rightarrow C_6H_6O_3 + 2H_2O$	Galactan	0.150
$C_5H_8O_4 + H_2O \rightarrow C_5H_{10}O_5$	Arabinan	0.750
$C_5H_8O_4 \rightarrow C_4H_3OCHO + 2H_2O$	Arabinan	0.100
$C_2H_4O_2 \rightarrow CH_3COOH$	Acetate	1.0

TABLE 14 Reaction Data for Ion Exchange (Wooley et al. 1999)

Reaction	Conversion	Modelled
$H_2SO_4 + 2NH_3 \rightarrow (NH_4)_2SO_4$	Sulphuric Acid	1.0
$CH_3COOH + NH_3 \rightarrow CH_3COONH_4$	Acetic Acid	1.0

TABLE 15 Reaction Data for Overliming Process (Wooley et al. 1999)

Reaction	Conversion	Modelled
$H_2SO_4 + Ca(OH)_2 \rightarrow CaSO_4 2H_2O$	Lime	1.0

TABLE 16 Reaction Data for SSCF-Saccharification (Wooley et al. 1999)

Reaction	Conversion	Modelled
$C_6H_{10}O_5 + 1/2H_2O \rightarrow 1/2C_{12}H_{22}O_{11}$	Cellulose	0.012
$C_6H_{10}O_5 + H_2O \rightarrow C_6H_{12}O_6$	Cellulose	0.800
$C_{12}H_{22}O_{11} + H_2O \rightarrow 2C_6H_{12}O_6$	Cellobiose	1.000

TABLE 17 Reaction Data for SSCF fermentation (Wooley et al. 1999)

Reaction	Conversion	Modelled
$C_6H_{12}O_6 \rightarrow 2CH_3CH_2OH + 2CO_2$	Glucose	0.920
$C_6H_{12}O_6 + 1.2NH_3 \rightarrow 6CH_{1.8}O_{0.5}N_{0.2} + 2.4H_2O + 0.3O_2$	Glucose	0.027
$C_6H_{12}O_6 + 2H_2O \rightarrow C_3H_8O_3 + O_2$	Glucose	0.002
$C_6H_{12}O_6 + 2CO_2 \rightarrow 2HOOCCH_2CH_2COOH + O_2$	Glucose	0.008
$C_6H_{12}O_6 \rightarrow 3CH_3COOH$	Glucose	0.022
$C_6H_{12}O_6 \rightarrow 2CH_3CHOHCOOH$	Glucose	0.013
$3C_5H_{10}O_5 \rightarrow 5CH_3CH_2OH + 5CO_2$	Xylose	0.850
$C_5H_{10}O_5 + NH_3 \rightarrow 5CH_{1.8}O_{0.5}N_{0.2} + 2H_2O + 0.25O_2$	Xylose	0.029
$3C_5H_{10}O_5 + 5H_2O \rightarrow 5C_3H_8O_3 + 2.5O_2$	Xylose	0.002
$C_5H_{10}O_5 + H_2O \rightarrow C_5H_{12}O_5 + 0.5O_2$	Xylose	0.006
$3C_5H_{10}O_5 + 5CO_2 \rightarrow 5HOOCCH_2CH_2COOH + 2.5O_2$	Xylose	0.009
$2C_5H_{10}O_5 \rightarrow 5CH_3COOH$	Xylose	0.024
$3C_5H_{10}O_5 \rightarrow 5CH_3CHOHCOOH$	Xylose	0.014

TABLE 18 Reaction Data for SSCF Contamination Loss (Wooley et al. 1999)

Reaction	Conversion	Modelled
$C_6H_{12}O_6 \rightarrow 2CH_3CHOHCOOH$	Glucose	1.0
$3C_5H_{10}O_5 \rightarrow 5CH_3CHOHCOOH$	Xylose	1.0
$3C_5H_{10}O_5 \rightarrow 5CH_3CHOHCOOH$	Arabinose	1.0
$C_6H_{12}O_6 \rightarrow 2CH_3CHOHCOOH$	Galactose	1.0
$C_6H_{12}O_6 \rightarrow 2CH_3CHOHCOOH$	Mannose	1.0

1-6, analysing the total equation set and determining the optimal calculation order (solution strategy). Another interesting problem to try would be to convert the simple flowsheet for the conversion of ethylene to ethanol to a more rigorous process flowsheet. Since all the values of the specified variables are given (or calculated from the simulation results), any other process simulator could also be tried.

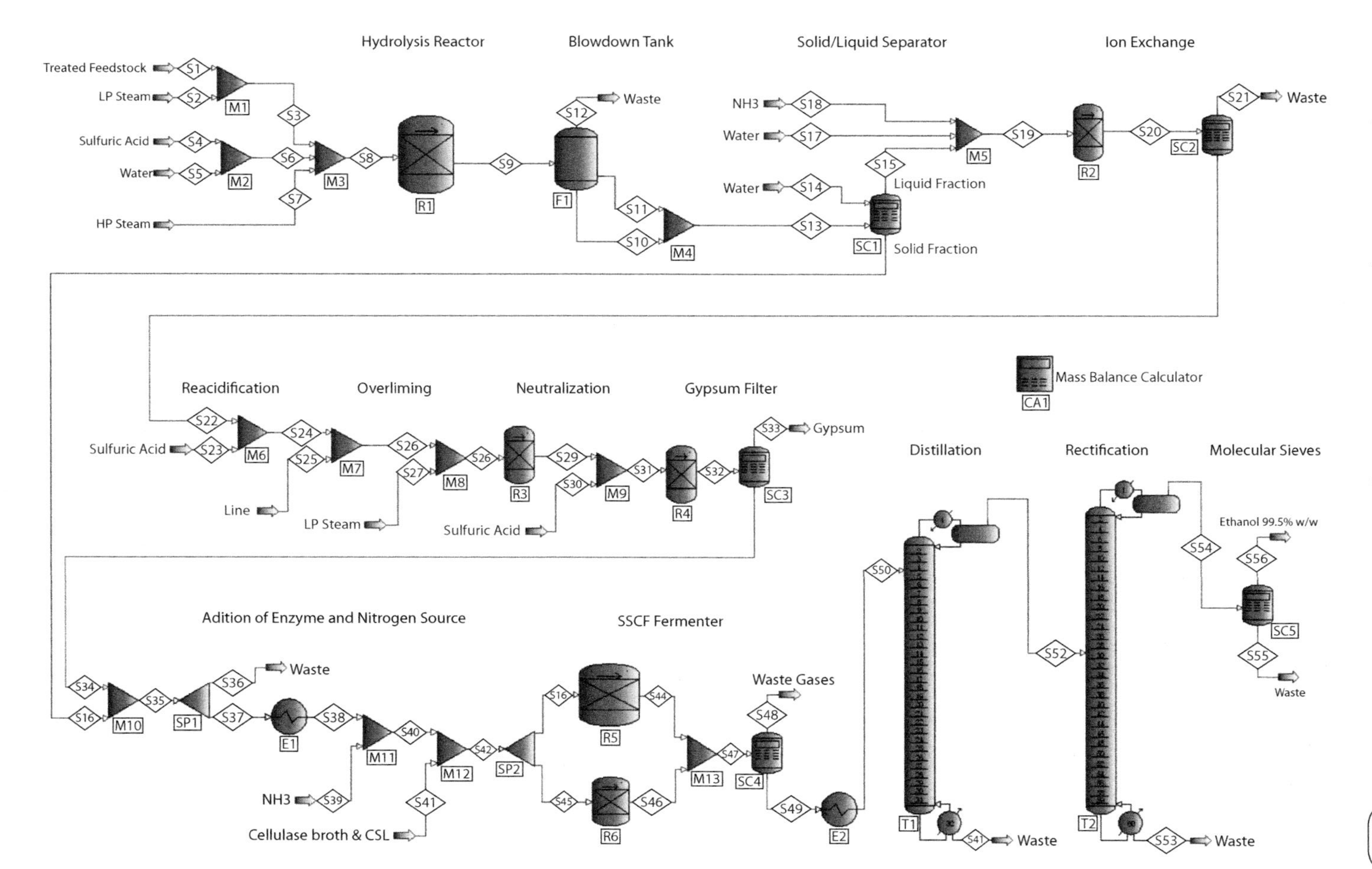

FIGURE 12 Flow-sheet of bioethanol production as represented with models from PRO/II (Alvarado-Morales, et al 2009).

REFERENCES

Biegler, L.T., Grossmann, I.E., Westerberg, A.W., 1997. 'Systematic Methods of Chemical Process Design.' Prentice Hall.

Douglas, J. M., 1988. Conceptual Design of Chemical Processes, McGraw-Hill, New York.

Alvarado-Morales, M., Terra, J., Gernaey, K.V., Gani, R., Woodley, J.M., 2009. Chemical Engineering Research and Design. 87, 1171–1183.

Sales-Cruz, M., 2006. 'Development of a computer aided modelling system for bio and chemical process and product design PhD-thesis.' Denmark: Lyngby.

PROII User's Guide (2006). Simulation Sciences Inc, Brea, USA.

Wooley, R., Ruth, M., Sheehan, J., Ibsen, K.H., Majdeski, H., Galvez, A., 1999. 'Lignocellulosic Biomass to Ethanol Process Design and Economics Utilizing Co-Current Dilute Acid Prehydrolysis and Enzymatic Hydrolysis-Current and Futuristic Scenarios.' Golden Colorado USA: National Renewable Energy Laboratory.

Chapter 7

Models for Dynamic Applications*

7.1. COMPLEX UNIT OPERATIONS: BLENDING TANK

7.1.1. Problem description: Blending tank

The blending tank is one of the most common unit operations that can be found in chemical and related industries, such as, food, pharmaceutical, refinery, agrochemicals, cosmetics, and, health-care industries. The design of this kind of process unit differs in according to their application: some may have a serpentine for heat exchange or a heat jacket around the unit, while others may be adiabatic. Most of the time, the unit is equipped with an internal mixer that allows smooth blending of chemical products inside the equipment.

7.1.1.1 System Description

Consider the process of blending two streams of liquid in a tank where all streams can vary in flow rate, composition and temperature. The uncontrolled stream F_1 consists of components A and B whilst stream F_2 is pure component B. Figure 1 shows the process and its control scheme.

7.1.2. Purpose of the model

The blending tank process model is to be used for analysis of the dynamic performance under disturbances in the feed flow rates, concentrations and temperatures. It could also be used to test an appropriate control scheme for the system.

The goal: 'Develop a model to analyse the performance of control-loops for level, composition and temperature control of a blending tank'. The model should have an accuracy of $\pm 10\%$ compared with the plant data.

* Written in conjunction with Prof Mauricio Sales-Cruz, Dr. Ricardo Morales-Rodriguez and Miss Martina Heitzig

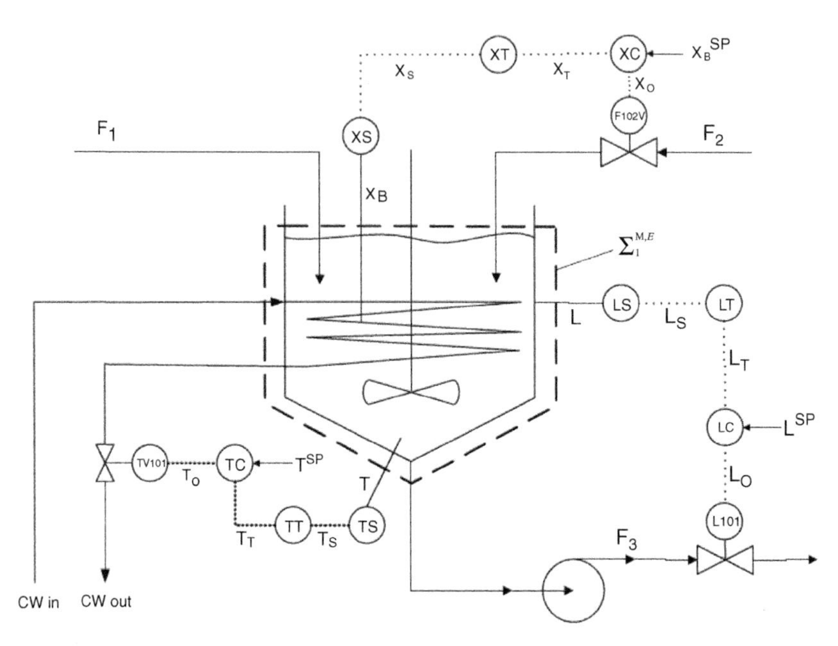

FIGURE 1 **Blending tank representation.**

7.1.3. Model development

The development of the model needs to consider different phenomena that take place during the operation of the process. This section lists some of the key mechanisms, assumptions considered in the mathematical model as well as the development of the model in terms of mass and energy balances, and constitutive equations.

7.1.3.1 Key mechanisms

The following mechanisms are considered:

- Fluid mixing in stirred tank
- Heat transfer from vessel to cooling coil
- Composition control through adjustment of blending stream through control valve
- Temperature control through adjustment of cooling water flow
- Level control via standard loop adjusting control outlet valve.

7.1.3.2 Assumptions of the mathematical model

A1: Mass as well as energy holdups are considered important.

A2: The balance volumes are considered to be the liquid holdup in the tank ($\Sigma E_1^{M,E}$) and the cooling coil holdup (ΣE_2^{E}).

A3: The liquid holdup is considered well mixed (homogeneous).
A4: There are no mass or heat transfers from the system to the surroundings.
A5: No chemical reaction occurs.
A6: Only two components (A, B) are considered.
A7: There is no heat of mixing in the system, just sensible heat changes.
A8: Thermo-physical properties are considered to be constant over the operating range.
A9: Specific energies of the streams are independent of pressure.
A10: Shaft work is considered to be insignificant.
A11: There are no physical losses from the system.
A12: There is no significant loss of heat to the environment from the vessel.
A13: There are no evaporative losses from the vessel.

7.1.3.3 Balance volume diagram

First, a sketch of the balance volume diagram (BVD) is made. BVD provides an understanding of:

- The principal balance volumes being considered in the model
- The convective flows in the system
- The diffusive or molecular flows (heat and mass fluxes) in the system.

Figure 2 gives the BVD being used for model development.

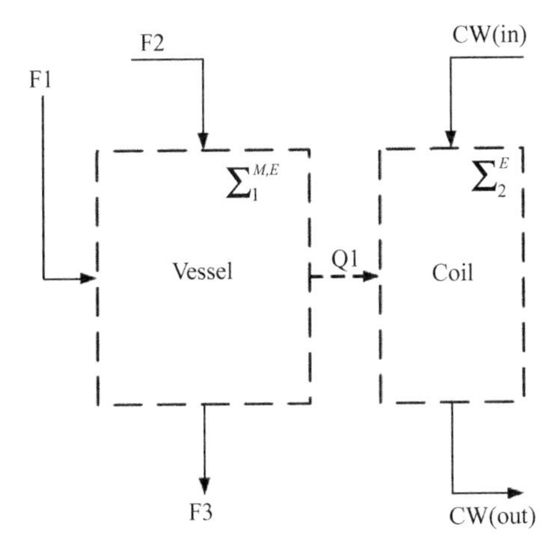

FIGURE 2 Blending system balance volume diagram (BVD).

7.1.4. Model analysis

7.1.4.1 List of Equations

Equation			No. of Eq.
Mass Balance	$\frac{dm_A}{dt} = f_{A,1} - f_{A,3}$	(1)	1
	$\frac{dm_B}{dt} = f_{B,1} + f_{B,2} - f_{B,3}$	(2)	1
Energy balance for the vessel	$\frac{dE_V}{dt} = F_1\widehat{h}_1 + F_2\widehat{h}_2 - F_3\widehat{h}_3 - Q_1$	(3)	1
Energy Balance in the cooling coil sub-system	$\frac{dE_c}{dt} = F_{CW}\left(\widehat{h}_{CW_{in}} - \widehat{h}_{CW_{out}}\right) + Q_1$	(4)	1
Heat transfer to coil	$Q_1 = UA(T - T_{CWav})$	(5)	1
	$T_{CW_{av}} = \frac{\left(T_{CW_{in}} - T_{CW_{out}}\right)}{2}$	(6)	1
Total flows for each streams	$F_j = \sum_i f_{i,j} \qquad j = 1,2,3$	(7)	3
Specific energy definitions	$\widehat{h}_1 = T_1 c_{p_A} x_{A,1} + T_1 c_{p_B} x_{B,1}$	(8)	1
	$\widehat{h}_2 = T_2 c_{p_B}$	(9)	1
	$\widehat{h}_3 = T_3 c_{p_A} x_{A,3} + T_3 c_{p_B} x_{B,3}$	(10)	1
	$\widehat{h}_{CW_{in}} = T_{CW_{in}} c_{p_{CW}}$	(11)	1
	$\widehat{h}_{CW_{out}} = T_{CW_{out}} c_{p_{CW}}$	(12)	1
Define the mass fractions	$x_{A,1} = \frac{f_{A,1}}{F_1}$	(13)	1
	$x_{B,1} = (1 - x_{A,1})$	(14)	1
	$x_{A,3} = \frac{f_{A,3}}{F_3}$	(15)	1
	$x_{B,3} = (1 - x_{A,3})$	(16)	1
Define total energy holdups	$E_v \approx M_v c_{p_v} T$	(17)	1
	$E_c \approx M_c c_{p_{CW}} T_{CW_{out}}$	(18)	1
Total holdup in the vessel	$M_v = m_a + m_b$	(19)	1
Perfect mixing relations for outlet stream F_3	$\frac{f_{i,3}}{\sum_i f_{i,3}} = \frac{m_i}{\sum_i m_i} \qquad i = A$	(20)	1
Outlet stream temperature to be tank temperature	$T_3 = T$	(21)	1
Tank level	$L = \frac{M_v}{\rho_v A_v}$	(22)	1
Composition loop in blending tank which sets flow rate (F_2) of stream 2	$\frac{dx_S}{dt} = \frac{(x_B - x_S)}{\tau_x}$	(23)	1
	$\frac{dsx}{dt} = wx$	(24)	1
	$\frac{dwx}{dt} = \frac{GxAx_O - sx}{\left(\tau_x^A\right)^2} - \frac{2xixA}{\tau_x^A} wx$	(25)	1
	$x_B = \frac{m_B}{M_v}$	(26)	1
	$x_T = x_B$	(27)	1
	$x_B^{SPC} = x_B^{SP}/100$	(28)	1

(continued)

(continued)

Equation			No. of Eq.
	$e_x(t) = \left(x_B^{SPC} - x_T\right)$	(29)	1
	$x_O = BxC + KC_x e_x(t) + \frac{KC_x}{\tau_{C,x}} \int_0^{\tau} e_x(t) d\tau$	(30)	1
	$F_2 = \rho_v CV_x sx \sqrt{\Delta P_2}$	(31)	1
Level control loop which adjusts outlet flow rate F_3	$\frac{dL_S}{dt} = \frac{(L - L_S)}{\tau_L}$	(32)	1
	$\frac{dsL}{dt} = wL$	(33)	1
	$\frac{dwL}{dt} = \frac{GLAL_O - sL}{\left(\tau_L^A\right)^2} - \frac{2xiLA}{\tau_L^A} wL$	(34)	1
	$L_T = \frac{L_s - L_{min}}{L_{max} - L_{min}}$	(35)	1
	$L^{SPC} = L^{SP}/100$	(36)	1
	$e_L(t) = \left(L^{SPC} - L_T\right)$	(37)	1
	$L_O = BLC + KC_L e_L(t)$	(38)	1
	$F_3 = \rho_v CV_L sL \sqrt{\Delta P_3}$	(39)	1
Temperature control loop which adjusts cooling water flow rate F_{CW}	$\frac{dT_S}{dt} = \frac{(T - T_S)}{\tau_T}$	(40)	1
	$\frac{dsT}{dt} = wT$	(41)	1
	$\frac{dwT}{dt} = \frac{GTAT_O - sT}{\left(\tau_T^A\right)^2} - \frac{2xiTA}{\tau_T^A} wT$	(42)	1
	$T_T = \frac{T_s - T_{min}}{T_{max} - T_{min}}$	(43)	1
	$T^{SPC} = \frac{T^{SP} - T_{min}}{T_{max} - T_{min}}$	(44)	1
	$e_T(t) = \left(T^{SPC} - T_T\right)$	(45)	1
	$T_O = BTC + KC_T e_T(t) + \frac{KC_T}{\tau_{C,T}} \int_0^{\tau} e_T(t) d\tau$	(46)	1
	$F_{CW} = \rho_w CV_T sT \sqrt{\Delta P_{CW}}$	(47)	1

7.1.4.1 List of Variables

Variables		Units
$f_{i,j}$	Flow rate of component i in stream j.	kg/s
m_i	Mass holdup	kg
E_v	Total energy holdup in the vessel	J
F_j	Total mass flow-rate in stream j;	kg/s
$\widehat{h}_j$	Specific enthalpy in stream j	kg/s
Q_1	Heat transferred to the cooling water	J/s
E_c	Total energy holdup in the coil	J
F_{CW}	Total mass flow rate through the coil	kg/s

(continued)

(continued)

Variables		Units
$\widehat{h}_{CW_{in}}$	Specific enthalpy for inlet cooling water stream	J/kg
$\widehat{h}_{CW_{out}}$	Specific enthalpy for output cooling water stream	J/kg
U	Overall heat transfer coefficient	$W/(m^2\ ^\circ C)$
S	Heat transfer surface area	m^2
A_v	Tank cross sectional area	m^2
T	Temperature in the unit operation	$^\circ C$
T_j	Temperature in stream j	$^\circ C$
T_{CWav}	Average temperature of the cooling water	$^\circ C$
T_{CWin}	Inlet cooling water temperature	$^\circ C$
T_{CWout}	Output cooling water temperature	$^\circ C$
$x_{i,j}$	Liquid concentration of component i in stream j	
M_v	Total holdup in the vessel	kg
M_c	Coil mass holdup	kg
c_{p_v}	Vessel content heat capacity	$J/kg\ K$
c_{p_A}	Heat capacity for compound A	$J/kg\ K$
c_{p_B}	Heat capacity for compound B	$J/kg\ K$
$c_{p_{CW}}$	Heat capacity for cooling water	$J/kg\ K$
x_S	Composition (x_B) sensor reading	(mass fraction)
x_B	Current composition (x_B) in the vessel	(mass fraction)
τ_x	Composition sensor time constant	s
sx	Composition control valve stem position (fraction)	
wx	Composition control valve stem velocity	s^{-1}
GxA	Composition control valve actuator gain	
x_O	Composition output from controller	
$xixA$	Composition control valve actuator damping factor	
τ_x^A	Composition control valve actuator time constant	s
x_T	Composition from transmitter	
x_B^{SP}	Composition (x_B) control set point	(%)
x_B^{SPC}	Composition (x_B) control set point	(mass fraction)
$e_x(t)$	Error calculation of composition	
BxC	Composition controller bias	
KC_x	Composition controller gain	
$\tau_{C,x}$	Composition controller integral (reset)	s
ρ_v	Vessel content density	kg/m^3
CV_x	Composition control valve coefficient	$m^3/Pa^{0.5}\ s$
ΔP_2	Pressure drop over composition control value	Pa
L_S	Level sensor reading	m
L	Current level in the vessel	m
τ_L	Level sensor time constant	s
sL	Level control valve stem position (fraction)	
wL	Level control valve stem velocity	s^{-1}
GLA	Level control valve actuator gain	
L_O	Level output from controller	m
$xiLA$	Level control valve actuator damping factor	
τ_L^A	Level control valve actuator time constant	s
L_T	Level from transmitter	
L_{min}	Level transmitter 0% height	m
L_{max}	Level transmitter 100% height	m
L^{SP}	Level control set point	(%)
L^{SPC}	Level control set point	(mass fraction)
$eL(t)$	Error calculation of Level	
BLC	Level controller bias	

(continued)

(continued)

Variables		Units
KC_L	Level controller gain	
CV_L	Level control valve coefficient	$m^3/Pa^{0.5}\ s$
ΔP_3	Pressure drop over level control value	Pa
T_S	Temperature sensor reading	$^{\circ}C$
τ_T	Temperature sensor time constant	s
sT	Cooling water control valve stem position (fraction)	
wT	Cooling water control valve stem velocity	s^{-1}
GTA	Cooling water control valve actuator gain	
T_O	Temperature output from controller	
$xiTA$	Cooling water control valve actuator damping factor	
τ_T^A	Cooling water control valve actuator time constant	s
T_T	Temperature from transmitter (fraction)	
$T_{\min}$	Temperature transmitter 0% height	$^{\circ}C$
$T_{\max}$	Temperature transmitter 100% height	$^{\circ}C$
T^{SP}	Temperature control set point	$^{\circ}C$
T^{SPC}	Temperature control set point	(fraction)
$e_T(t)$	Error calculation of Temperature	
BTC	Temperature controller bias	
KC_T	Temperature controller gain	
$\tau_{C,T}$	Temperature controller integral (reset)	S
ρ_{CW}	Cooling water density	kg/m^3
CV_T	Temperature control valve coefficient	$m^3/Pa^{0.5}\ s$
ΔP_{CW}	Pressure drop over temperature control value	Pa

7.1.4.2 Basic process unit: Degrees of freedom analysis

The model for the basic volume consists of the following equations (see list of equations given in section 7.1.4.1): mass and energy balance equations (1–4); heat transferto coil (5–6); total mass flow equations (7); specific energy definitions (8–12); mass fraction definition equations (13–16); total energy holdup definition equations (17–18); total holdup in the vessel (19); perfect mixing relations for outlet stream 3 (20); outlet stream temperature (21); and tank level (22).

- State variables (states)

  m_A, m_B, E_v, E_c

- Parameters

  U, A, A_v

- Constants

  $c_{p_A}, c_{p_B}, c_{p_{CW}}, c_{p_v}, \rho$

- Algebraic variables (algebraics)

$$f_{A,1}, f_{B,1}, f_{B,2}, f_{A,3}, f_{B,3}, F_1, F_2, F_3, F_{CW}, L, T_{CWav}$$
$$\widehat{h}_1, \widehat{h}_2, \widehat{h}_3, \widehat{h}_{CW_{in}}, \widehat{h}_{CWout},$$
$$Q_1, x_{A,1}, x_{B,1}, x_{A,3}, x_{B,3},$$
$$M_v, M_c, T_1, T_2, T_3, T, T_{CW_{in}}, T_{CWout}$$

$N_U = 41$ (states + parameters + constants + algebraics)
$N_E = 24$ (equations)
$N_{DF} = 41–24 = 17$

A natural choice of 17 variables would be all variables related to inlet streams, and the outlet flow from the tank plus the parameters. Hence we get:

Parameters U, A, A_v
Constants $c_{p_A}, c_{p_B}, c_{p_{CW}}, c_{p_v}, \rho$
Algebraics $f_{A1}, f_{B1}, f_{B2}, c_{p_A}, c_{p_B}, c_{p_{CW}}, c_{p_v}, T_1, T_2, F_3, T_{CW_{in}}, F_{CW}, M_c$

7.1.4.3 Basic process unit: Incidence matrix and calculation procedure

Incidence matrix for the set of differential algebraic equations is shown in Table 1, where it points out the equations to be solved and the variables assigned to each equation. Note that the incidence matrix has not been ordered to a lower tridiagonal form. The analysis of the incidence matrix suggests a simultaneous solution approach of the complete set of differential algebraic set of equations. The calculation procedure is as follows - at time zero, the initial condition for the states (differential variables) are known. Using these values and the 17 specified variables, the 20 algebraic equations are solved for the 20 unknown variables (see Table 1). These calculated values allow now the evaluation of the right hand sides of the ordinary differential equations (1–4). With the right hand sides of these equations known, the integration method predicts new values for the states for a new time and whole procedure is repeated until the end time is reached.

7.1.4.4 Addition of control loops: Degrees of freedom analysis

By adding the necessary equations, the basic process unit model can easily be extended to include control loops.

The model now consists of the following equations:
Ordinary differential equations – (1–4), (23–25), (32–34), (40–42)
Algebraic equations – (5–22), (26–30), (35–39), (43–47)

There are now nine new state variables (differential variables) associated with the sensor equations, that is, x_S, sx, wx, L_S, sL, wL, T_S, sT and wT.

- State variables (states)

$$m_A, m_B, E_v, E_c, x_S, sx, wx, L_S, sL, wL, T_S, sT, wT$$

TABLE 1 Incidence Matrix for the Basic Process Unit

Equations	Unknown variables																							
	m_A	m_A	E_v	E_c	F_1	F_2	$f_{A,3}$	$f_{B,3}$	M_v	T	$T_{CW_{out}}$	$T_{CW_{av}}$	Q_1	$x_{A,1}$	$x_{B,1}$	$x_{A,3}$	$x_{B,3}$	T_3	$\widehat{h}_1$	$\widehat{h}_2$	$\widehat{h}_3$	$\widehat{h}_{CW_{in}}$	$\widehat{h}_{CW_{out}}$	L
7a					*																			
7b						*																		
20	*	*					*																	
7c							*	*																
19	*	*							*															
17			*						*	*														
18				*							*													
6											*	*												
5										*		*	*											
13					*									*										
14														*	*									
15							*									*								
16																*	*							
21										*								*						
8														*	*				*					
9																				*				
10																*	*	*			*			
11																						*		
12											*												*	
22									*															*
1	*						*																	
2		*						*																
3			*		*	*							*						*	*	*			
4				*									*									*	*	*

- Parameters

The new parameters include the valve characteristics, mixer shape and the sensor time constants:

U, A, A_v
$\tau_x, GxA, xixA, \tau_x^A, x_B^{SP}, BxC, KC_x, \tau_{C,x}, \rho_v, CV_x, \Delta P_2,$
$\tau_L, GLA, xiLA, \tau_L^A, L_{\min}, L_{\max}, L^{SP}, BLC, KC_L, CV_L, \Delta P_3,$
$\tau_T, GTA, xiTA, \tau_T^A, T_{\min}, T_{\max}, T^{SP}, BTC, KC_T, \tau_{C,T}, \rho_{CW}, CV_T, \Delta P_{CW}$

- Constants

$c_{P_A}, c_{P_B}, c_{P_{CW}}, c_{P_v}$

- Algebraic variables (algebraics)

$f_{A,1}, f_{B,1}, f_{B,2}, f_{A,3}, f_{B,3}, F_1, F_2, F_3, F_{CW},$
$\widehat{h}_1, \widehat{h}_2, \widehat{h}_3, \widehat{h}_{CW_{in}}, \widehat{h}_{CW_{out}},$
$Q_1, x_{A,1}, x_{B,1}, x_{A,3}, x_{B,3},$
$M_v, M_c, T_1, T_2, T_3, T, T_{CW_{in}}, T_{CW_{out}},$
$x_B, x_T, x_O, x_B^{SPC}, e_x(t)$
$L, L_T, L_O, L^{SPC}, e_L(t)$
$T_T, T_O, T^{SPC}, e_T(t)$

$N_U = 91$ (states + parameters + constants + algebraics)
$N_E = 48$ (equations)
$N_{DF} = 91–48 = 43$

In this case we note the control system will manipulate the flows necessary to maintain a given set point for the level, concentration and temperature. The 48 variables to specify are the 38 parameters, the 4 constants and 6 of the algebraic variables that are related to the inlet streams. Hence we get:

Parameters

U, A, A_v
$\tau_x, GxA, xixA, \tau_x^A, x_B^{SP}, BxC, KC_x, \tau_{C,x}, \rho_v, CV_x, \Delta P_2,$
$\tau_L, GLA, xiLA, \tau_L^A, L_{\min}, L_{\max}, L^{SP}, BLC, KC_L, CV_L, \Delta P_3,$
$\tau_T, GTA, xiTA, \tau_T^A, T_{\min}, T_{\max}, T^{SP}, BTC, KC_T, \tau_{C,T}, \rho_{CW}, CV_T, \Delta P_{CW}$

Algebraic

$f_{A1}, f_{B1}, T_1, T_2, T_{CW_{in}}, M_c$

Constants

$c_{P_A}, c_{P_B}, c_{P_{CW}}, c_{P_v}$

7.1.4.5 Addition of control loops: Incidence matrix

Table 2 shows the incidence matrix for the blending tank mathematical model with the control loops. The incidence matrix has been ordered to indicate the

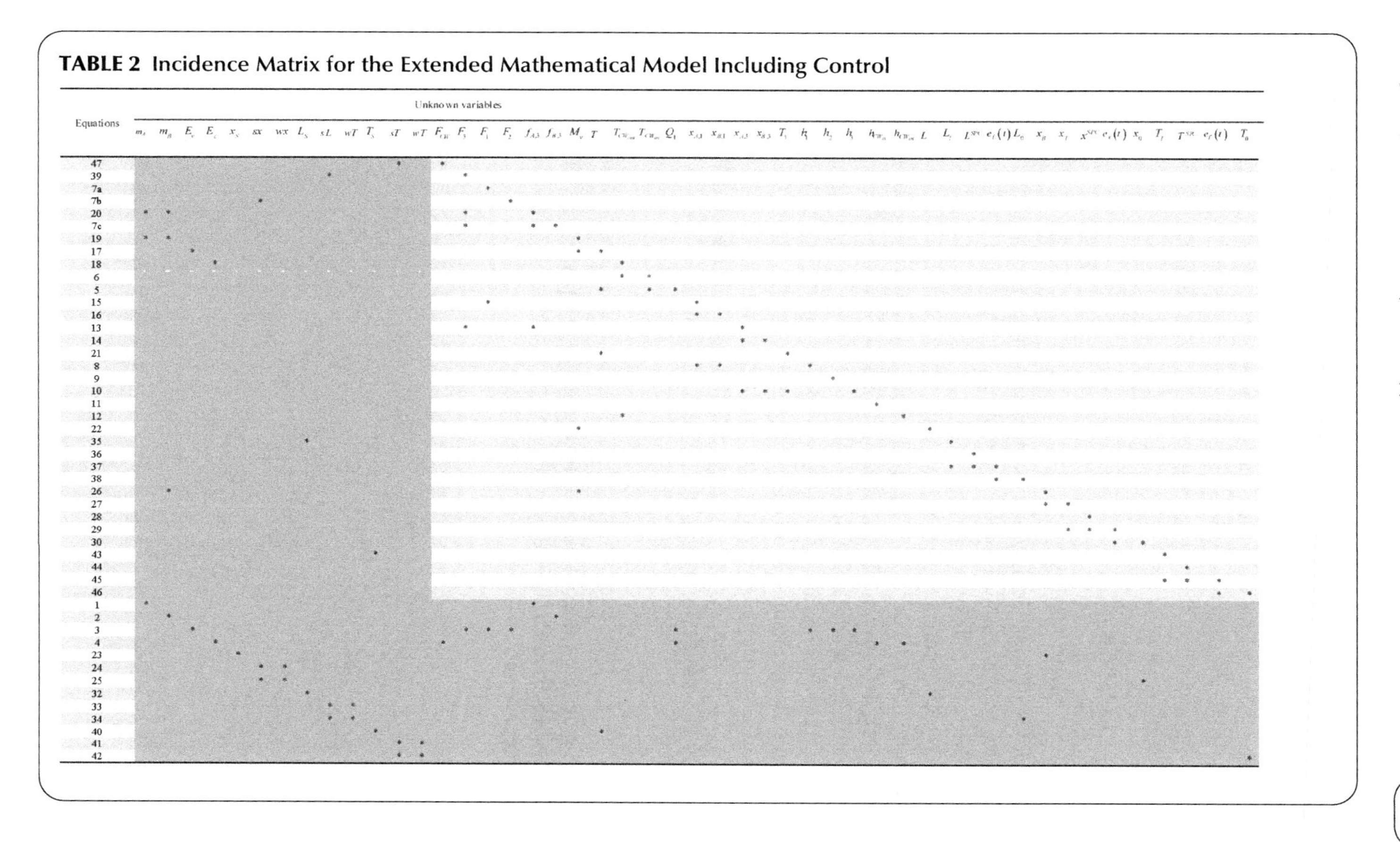

TABLE 2 Incidence Matrix for the Extended Mathematical Model Including Control

Equations	Unknown variables																																															
	m_A	m_B	E_v	E_c	x_S	sx	wx	L_S	sL	wT	T_S	sT	wT	F_{CW}	F_3	F_1	F_2	f_{A3}	f_{B3}	M_v	T	$T_{CW_{in}}$	$T_{CW_{out}}$	Q_1	x_{A3}	x_{B1}	x_{A3}	x_{B3}	T_1	h_1	h_2	h_3	$h_{CW_{in}}$	$h_{CW_{out}}$	L	L_2	L^{SP}	$e_L(t)$	L_0	x_B	x_t	x^{SP}	$e_x(t)$	x_0	T_t	T^{SP}	$e_T(t)$	T_0
47												*		*																																		
39									*						*																																	
7a																*																																
7b						*											*																															
20	*	*													*			*																														
7c															*			*	*																													
19	*	*																		*																												
17			*																	*	*																											
18				*																		*																										
6																						*	*																									
5																					*		*	*																								
15																*									*																							
16																									*	*																						
13															*			*									*																					
14																											*	*																				
21																					*								*																			
8																									*	*				*																		
9																															*																	
10																											*	*	*			*																
11																																	*															
12																						*												*														
22																				*															*													
35								*																												*												
36																																					*											
37																																				*	*	*										
38																																						*	*									
26		*																		*																				*								
27																																								*	*							
28																																										*						
29																																									*	*	*					
30																																											*	*				
43											*																																		*			
44																																														*		
45																																													*	*	*	
46																																															*	*
1	*																	*																														
2		*																	*																													
3			*												*	*	*							*						*	*	*																
4				*										*										*									*	*														
23					*																																			*								
24						*	*																																									
25						*	*																																					*				
32								*																											*													
33									*	*																																						
34									*	*																													*									
40											*										*																											
41												*	*																																			
42												*	*																																			*

ordinary differential equations in the shaded part and the algebraic equations above it. From the algebraic equations, it can be seen that a lower tridiagonal form is obtained for this set of equations. The calculation procedure is therefore, given the values of the states at a specific time, solve the algebraic equations, then calculate the right hand sides of the ordinary differential equations and then predict new values of the states at a new time and repeat the procedure until the end time is reached.

7.1.5. Model solution

7.1.5.1 Solution strategy

The solution of the blending tank mathematical model equations is carried out using the computer-aided modelling tool-box, ICAS-MoT (Sales-Cruz et al. 2003; Sales-Cruz 2006).

Figure 3 highlights the work flow that ICAS-MoT follows in order to solve any mathematical model. First the mathematical model is introduced to ICAS-MoT, followed by model translation, analysis and verification/solution. Solution of the model equations is obtained through an appropriate solution strategy that incorporates the solution order of the equations given by the incidence matrix; in this case a BDF-integrator is used to obtain the simulation results. Subsequently, validation and implementation are considered in the final step. It is important to highlight the flexibility of the modelling tool in tackling this kind of problems that is modelling of the process with/without control loops.

7.1.5.2 Basic process unit model: Values of specified variables

Table 3 gives the values of the specified variables and parameters as well as the initial values used for the ordinary differential equations for the solution of the basic process unit model without control loops.

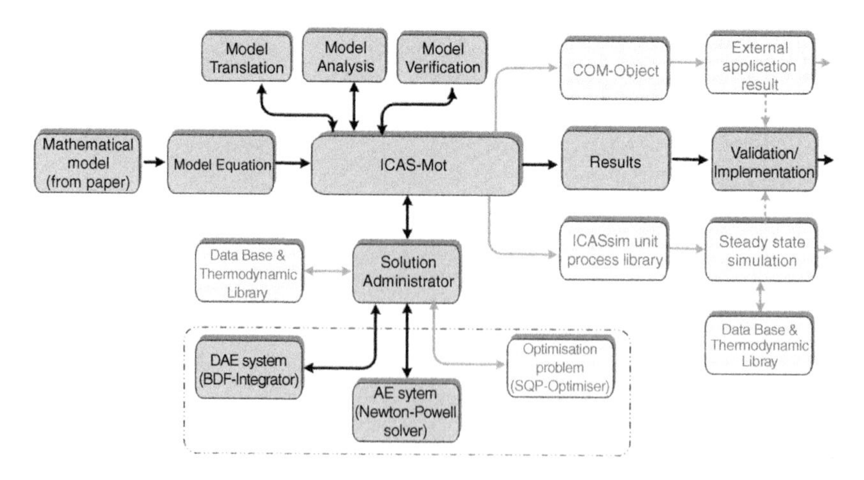

FIGURE 3 **Work flow and tools used from ICAS-MoT for the blending tank system.**

TABLE 3 Data for Simulation of the Basic Process Unit

	Property	Value
Initial values	$E_{V,initial}$	98262000 J
	$E_{c,initial}$	10909700 J
	$m_{a,initial}$	1243.1 kg
	$m_{b,initial}$	532.8 kg
Algebraic variables	$f_{A,1}$	3.85 kg/s
	$f_{B,1}$	0.95 kg/s
	$f_{A,2}$	0 kg/s
	T_1	35°C
	T_2	25°C
	F_3	5.5 kg/s
	$T_{CW_{in}}$	12°C
	F_{CW}	2.938 kg/s
	M_c	129.85 kg
Constants	c_{p_A}	2200 J/kg
	c_{p_B}	2100 J/kg
	$c_{p_{CW}}$	4200 J/kg
	c_{p_v}	2170 J/kg
Parameters	U	500 $W/m^2\ K$
	A	20.78 m^2
	A_v	1.767 m^2
	F_1	Increment 20% at the time 500 s

7.1.5.3 Basic process unit model: Solution

Employing the data given in Table 3 it is now possible to carry out open-loop simulations of the blending tank with the basic process unit model (without control loops). Values of selected variables at the simulated steady state are given in Table 4.

Figure 4 shows the dynamic responses of selected variables: the mass of the compounds inside the blending tank, mass fraction of compounds A and B in stream 3, the change in the tank level and the temperatures in different streams and in the outlet cooling water.

TABLE 4 Results for Simulation of the Basic Process Unit

Property	Value
m_a	2179.48 kg
m_b	869.72 kg
$x_{A,3}$	0.71477
$x_{B,3}$	0.28522
L	2.575 m
T_3	26.42°C
$T_{CW_{out}}$	20.55°C

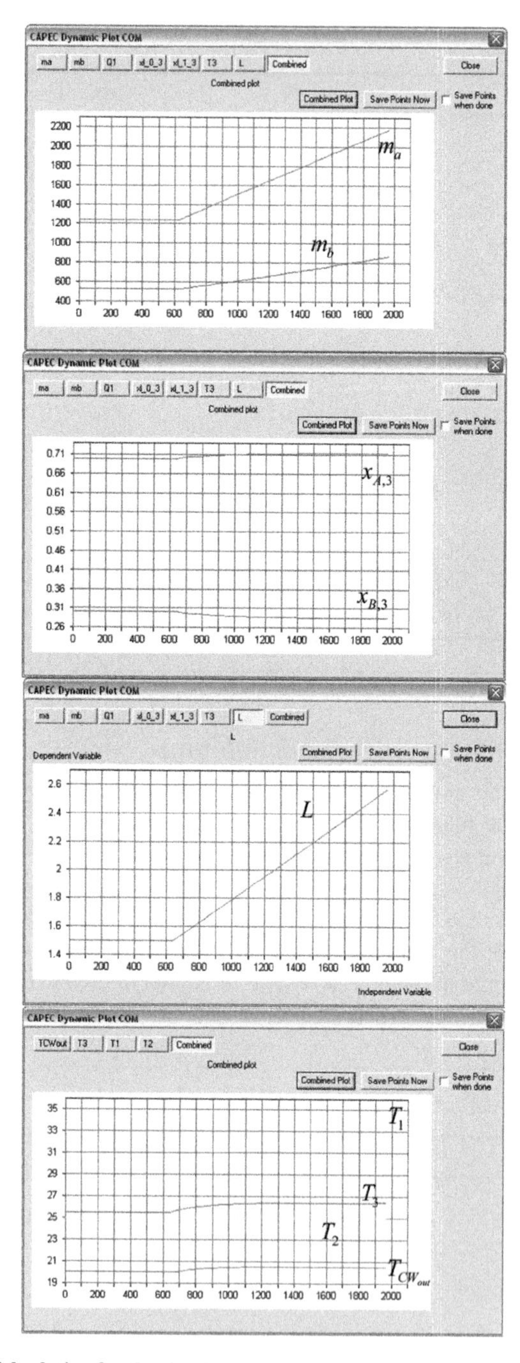

FIGURE 4 **Model solution for the simple process unit model.**

7.1.5.4 Addition of control loop: Values of specified variables

Table 5 gives the values of the specified variabes, parameters and the initial values for the states of the ordinary differential equations involving the control loops.

TABLE 5 Data for Simulation of the Basic Process Unit

	Property	Value
Initial values	$E_{V,initial}$	98262000 J
	$E_{c,initial}$	10909700 J
	$m_{a,initial}$	1243.1 kg
	$m_{b,initial}$	532.8 kg
	$L_{S,initial}$	1.5 m
	$sL_{initial}$	0.5
	$wL_{initial}$	0 s^{-1}
	$x_{S,initial}$	0.3 *mass_fraction*
	$sx_{initial}$	0.5
	$wx_{initial}$	0 s^{-1}
	$T_{S,initial}$	25.5°C
	$sT_{initial}$	0.4
	$wT_{initial}$	0 s^{-1}
Algebraic variables	$f_{A,1}$	3.85 kg/s
	$f_{B,1}$	0.95 kg/s
	T_1	35°C
	T_2	25°C
	$T_{CW_{in}}$	12°C
	M_c	129.85 kg
Constants	c_{p_A}	2200 J/kg
	c_{p_B}	2100 J/kg
	$c_{p_{CW}}$	4200 J/kg
	c_{p_v}	2170 J/kg
Parameters	A_V	1.767 m^2
	U	500 $W/m^2\ K$
	A	20.78 m^2
	τ_x	60 s
	GxA	1
	$xixA$	0.9
	τ_x^A	5
	x_B^{SP}	30%
	BxC	0.5
	KC_x	20
	$\tau_{C,x}$	150 s
	CV_x	6.608e–6 $m^3/Pa^{0.5}\ s$
	ΔP_2	100e3 Pa
	τ_L	2 s
	GLA	1
	$xiLA$	0.9
	τ_L^A	5
	L_{min}	0 m
	L_{max}	3 m
	L^{SP}	50%
	BLC	0.5
	KC_L	−1

(continued)

TABLE 5 *(continued)*

Property	Value
CV_L	5.192e–5 $m^3/Pa^{0.5}s$
ΔP_3	100e3 Pa
τ_T	5 s
GTA	1
$xiTA$	0.9
τ_T^A	5
T_{min}	0°C
T_{max}	100°C
T^{SP}	25.5°C
BTC	0.4
KC_T	20
$\tau_{C,T}$	60 s
CV_T	2.322e-5 $m^3/Pa^{0.5}s$
ΔP_{CW}	100e3 Pa
ρ_{CW}	1000 kg/m^3
ρ_v	670 kg/m^3
F_1	Increment 20% at the time 500 s

7.1.5.5 Addition of control loops: Solution

For the data given in Table 5, closed loop simulations have been made with the model equations given in Table 2. Selected results are given in Table 6.

The dynamic responses corresponding the data of Table 5 and results of Table 6 are shown in Figures 6–10. These figures show the transient of the process when different control scenarios are implemented.

TABLE 6 Results for Simulation Involving the Control Loops

Property	Value
m_a	1489.67 kg
m_b	638.42 kg
F_1	5.76 kg/s
F_2	0.84 kg/s
F_3	6.59 kg/s
$F_{CW_{out}}$	0 kg/s
T_1	35°C
T_2	25°C
T_3	33.42°C
$T_{CW_{out}}$	54.55°C
L	1.79 m
$x_{A,1}$	0.8
$x_{A,3}$	0.7
x_O	0.59
T_O	0.59
L_O	0

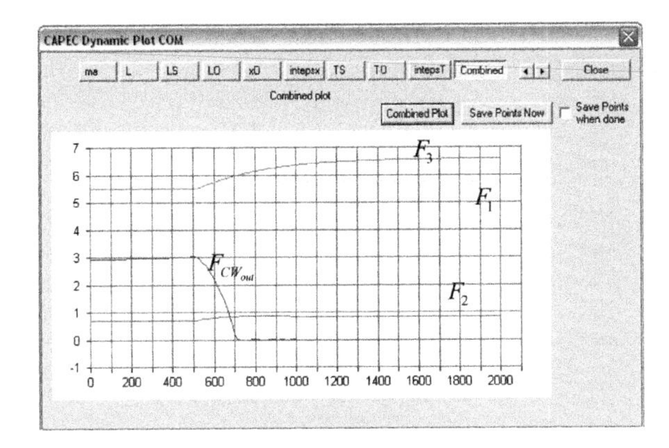

FIGURE 5 **Model solution for F_1, F_2, F_3 and F_{CW}.**

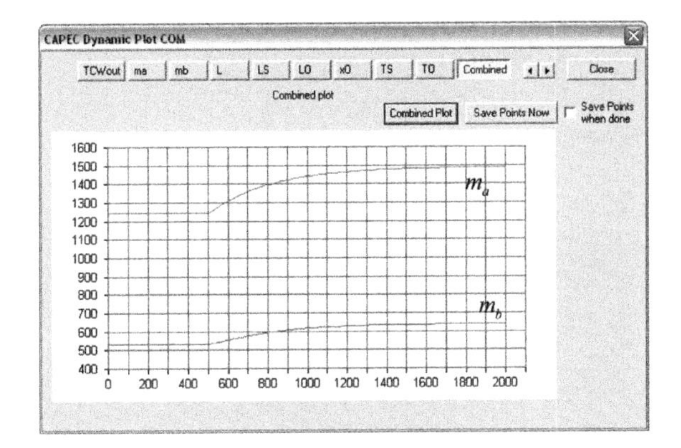

FIGURE 6 **Model solution for the m_a and m_b.**

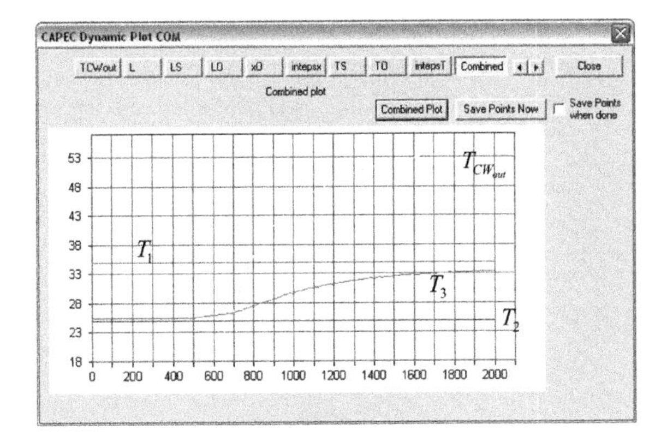

FIGURE 7 **Model solution for T_1, T_2, T_3 and $T_{CW_{out}}$.**

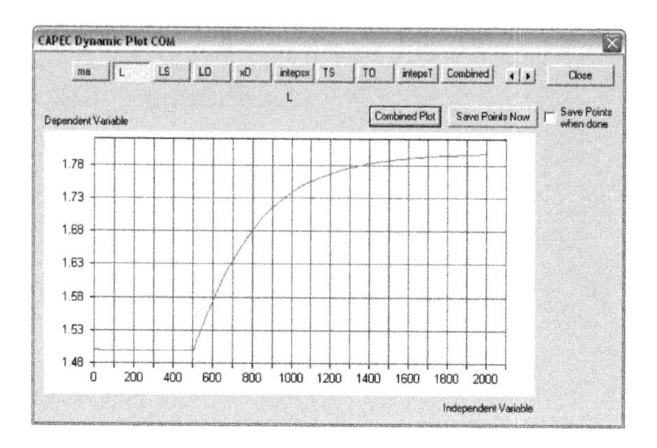

FIGURE 8 **Model solution for level tank *L*.**

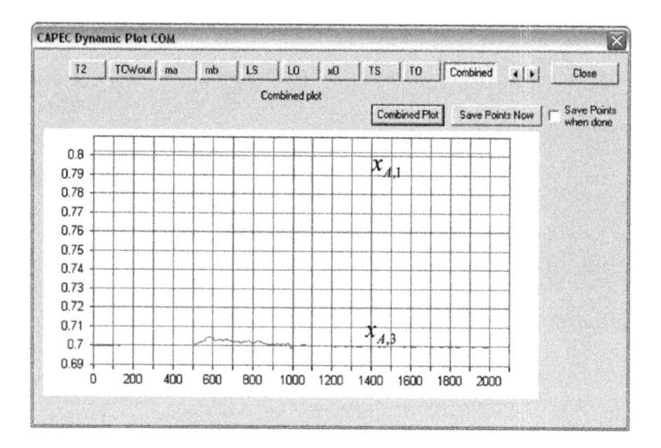

FIGURE 9 **Model solution for mass composition in stream 3.**

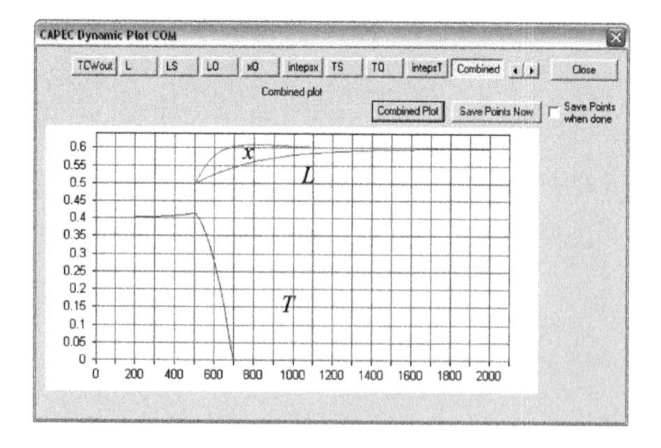

FIGURE 10 **Controller output for level (*L*), temperature (*T*) and composition(*x*).**

7.2. COMPLEX INTEGRATED OPERATIONS: DIRECT METHANOL FUEL CELL

The search for alternative energy sources to reduce the global demand and use of oil, coal and gas, as well as the concern for climate changes on earth, has motivated research for new energy supplies, production, design and efficient use of resources.

Among these new alternatives, renewable energy sources such as, wind power, tidal power, geothermal, solar power, wave power, biomass, bio-fuels and hydropower are being investigated. The increasing use of nuclear energy and the need for improvement of energy sources makes fuel cells an interesting alternative.

In this case study a model-based analysis of the fuel cell has been performed (Morales-Rodriguez, 2009). The objective here is to highlight the use of multiscale modelling rather than improvements in the design of the fuel cell. A fuel cell is a device that is based on an electrochemical reaction to produce energy where the energy released from the chemical reaction is converted directly into electric power. Most recently, research in fuel cells has promoted it as an alternative energy source, and storage for different applications, such as automobiles, buses, cell phones and computers, has been investigated (Cooper, 2007). Some advantages related to fuel cells as a future energy source are:

- high efficiencies;
- reduced air emission;
- extremely reliability;
- quiet operation;
- remote status monitoring.

7.2.1. System description

Fuel cells consist of an electrolyte sandwiched between two electrodes (see Figure 11). The fuel used in the operation is located on the anode side, and chemical reaction is carried out in the catalyst layer of this side; the oxidant (usually oxygen) is fed to the cathode side, which reacts with the ions allowed to cross through the electrolyte membrane. The electrons generated in the anode part of the cell are not allowed to pass through the membrane to the cathode side. So, these electrons must move through an electrical circuit to reach the other side of the cell.

Different types of fuel cells can be designed in terms of the electrolyte employed, the kind of reaction taking place in the cell, the operation temperature range, the kind of catalyst required, the fuel required and the mechanism of the charge conducted in it. Table 7 lists the different types of fuel cells.

There are five main sections in fuel cells: anode-channel compartment, anode catalyst layer, cathode-channel compartment, cathode catalyst layer and ion-conducting membrane.

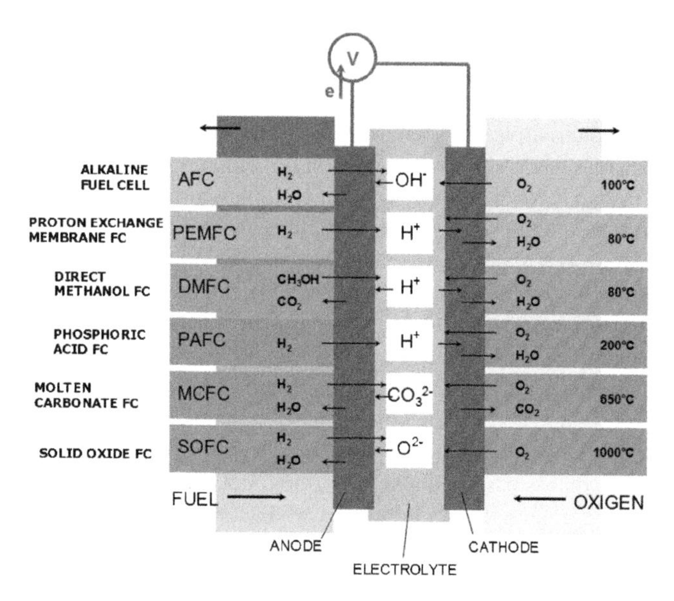

FIGURE 11 **Fuel cell diagrams.**

7.2.1.1 Anode-Channel Compartment

The fuel for the performance of the device is supplied through this compartment. When the fuel is crossing the channel, part of the fuel is delivered to the catalyst layer of the fuel cell.

7.2.1.2 Anode Catalyst Layer

Due to the chemical reaction that is taking place and because of the presence of the catalyst, the generation of electrons and ions is performed in this component of the fuel cell. Mass transfer phenomena also play an important role in this section.

7.2.1.3 Cathode-Channel Compartment

The oxidant is passed through the cathode-channel compartment; it is transported to the cathode catalyst layer where the oxidant performs a chemical reaction.

7.2.1.4 Cathode Catalyst Layer

The oxidant undergoes a chemical reaction in this section of the fuel cell. Most of the time, the oxidant has a chemical reaction with the protons that are generated in the anode compartment.

TABLE 7 Classification of Different Kind of Fuel Cells (Cooper, 2007)

		Efficiency	Operating temperature	Ion movement	Electrolyte	Electrodes/ Catalyst	Thermal output	Advantages	Disadvantages
Low-Temperature	Proton Exchange Membrane	40-47%	50-100 °C	H^+ from anode to cathode	Solid organic polymer	Porous carbon coated with Pt catalyst	Warm water	Solid electrolyte reduces corrosion and electrolyte management problems. Low temperature Quick start	Requires expensive catalyst High sensitivity to fuel impurities Low temperature waste heat Waste heat temperature not suitable for combined-cycle.
	Alkaline	50-60%	25-90 °C	OH^- from cathode to anode	Alkaline solution (aqueous solution of potassium, etc.)	Porous carbon coated with non-precious-metal catalyst	Warm water	Cation reaction faster in alkaline electrolyte, higher performance.	Expensive removal of carbon dioxide from fuel and air streams required (carbon dioxide degrades the electrolyte)
	Phosphoric Acid	~35%	150-220 °C	H^+ from anode to cathode	Liquid phosphoric acid soaked in a matrix	Porous carbon coated with Pt catalyst	Hot water	Higher overall efficiency with combine-cycle mode. Increased tolerance to impurities in hydrogen.	Requires expensive platinum catalysts Low current and power Large size/weight

(continued)

TABLE 7 *(continued)*

		Efficiency	Operating temperature	Ion movement	Electrolyte	Electrodes/ Catalyst	Thermal output	Advantages	Disadvantages
	Direct Methanol	25-40%	50-120 °C	H^+ from anode to cathode	PEM (Solid polymer)	Anode: Pt/Ru; Cathode: Pt	Warm water	Liquid methanol is used directly Not reforming required	Methanol crossover from anode to cathode Presence of methanol in the cathode part reduce electricity output and waste methanol.
High-Temperature	Molten Carbonate	~55%	600-700 °C	CO_3^{-2} from cathode to anode	Ceramic matrix containing a molten carbonate	Catalyst: Ni Anode: Ni or NiCr alloy Cathode : NiO doped with Li	Steam	High efficiency Fuel flexibility Can use a variety of catalyst Suitable for combined-cycle mode.	High temperature speeds corrosion and breakdown of cell components Complex electrolyte management Slow start-up
	Solid Oxide	45-50%	650-1000 °C	O^{-2} from cathode to anode	Matrix of yttria-stabilised zirconia; or ceria-gadolinium oxides	Perovskita: Material with similar structure as $CaTiO_3$. General formula is ABO_3.	Steam	High efficiency Fuel flexibility Can use a variety of catalyst Suitable for combined-cycle mode Solid electrolyte reduces electrolyte management problems	High temperature enhances corrosion and breakdown of cell components. Slow start-up Brittleness of ceramic electrolyte with thermal cycling

7.2.1.5 Ion-Conducting Membrane/Electrolyte

The electrolyte in the fuel cell plays one of the most important roles for the proper performance of the device. The membrane allows the crossing of selected ions generated from the chemical reaction and avoids the transfer of electrons through it. So, electrons are passed through the electric circuit, generating an electrical current that is ready to be used. The electrolyte might be an appropriate membrane, a ceramic containing metallic compounds or even a liquid.

Figure 12 shows how the direct methanol fuel cell can be divided into different scales for multiscale modelling. Based only on the performance of the fuel cell, this device might be divided and described in the meso-scale and micro-scale. The anode- and cathode-channel compartments are classified at the meso-scale. Here, the measurement of the rate consumption of raw materials at the different compartments and also the mass transfer phenomena are very important phenomena to take into account. The cathode catalyst layer, anode catalyst layer and proton exchange membrane or electrolyte (all together commonly known as membrane-electrode-assembly MEA) are considered to be in the micro-scale, while the rate of consumption of the raw material, mass transfer resistances, proton exchange rate and charge balances are the important issues to consider.

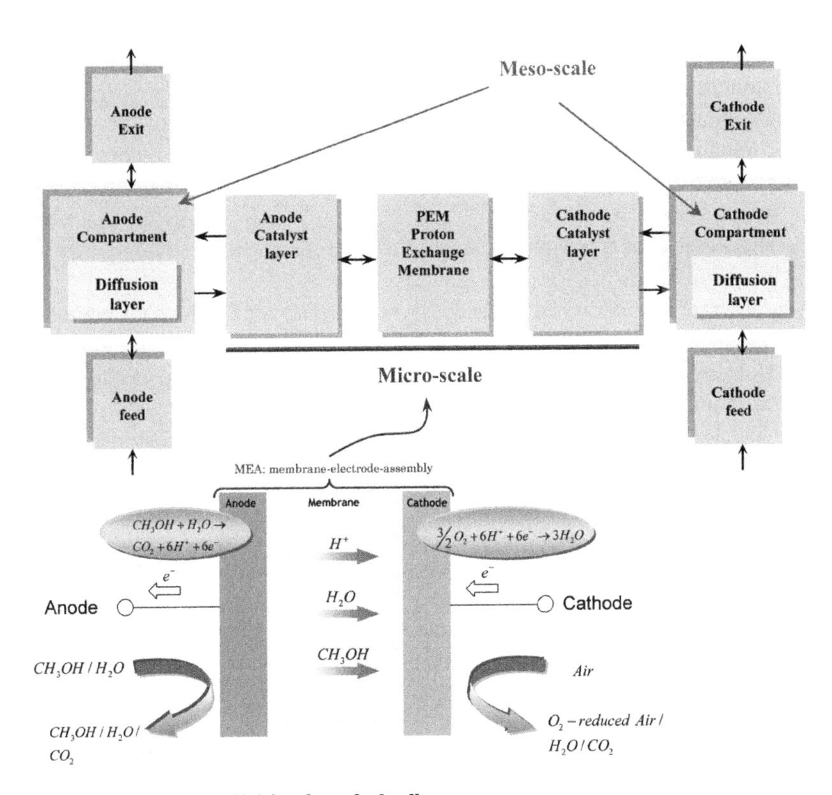

FIGURE 12 Multiscale division for a fuel cell.

7.2.2. Modelling objective

The purpose is to develop a model for the direct methanol fuel cell (DMFC) so that the performance of the DMFC can be studied. A multiscale modelling approach is to be employed.

7.2.3. Model development

The development of the model is described in this section. It is based on the models proposed by Xu et al. (2005) and Sundmacher et al. (2001). First, the explanation of the model equations describing the different sections of the fuel cell is given. The different sections of the fuel cell are also classified in terms of a multiscale modelling framework. The anode-channel compartments are in the meso-scale, whilst the ion conducting membrane and the anode catalyst layer are in the micro-scale.

For the derivation of the mathematical model, the following assumptions also employed by Xu et al. (2005) and Sundmacher et al. (2001) have been considered:

A_1: The anode compartment is treated as a continuous stirred tank reactor (CSTR).

A_2: The ohmic drops in current collectors and electric connections are negligible.

A_3: Isothermal operation in the fuel cell.

A_4: Oxygen is fed in excess. That is, the oxygen conversion in the cathode compartment is negligible, and therefore oxygen mass balance is not required.

A_5: Oxygen and carbon dioxide do not diffuse into the proton electrolyte membrane (PEM).

A_6: The water concentration is assumed to be constant (excess component in liquid mixture).

A_7: Mass transport resistances in the catalyst layers are negligible as these are thin (10 μm) in comparison to the diffusion layers (100 μm) and PEM (200 μm).

A_8: Mass transfer coefficients of methanol and carbon dioxide in the anode diffusion layer are equal.

A_9: The mixtures in the anode compartment and in the anode diffusion layer are treated as a pure liquid phase mixture; i.e., gas-phase formation, by release of carbon dioxide bubbles, is not taken into account.

A_{10}: For other gas components (CO_2, O_2, N_2) pure diffusion is assumed inside the membrane for reasons of simplicity. The mass transport in the membrane is source- and sink-free.

A_{11}: The rate-determining step in the electrochemical methanol oxidation is the dissociative chemisorption of methanol on active platinum catalyst sides, so the formation of hydroxylic groups on the active ruthenium catalyst sites, surface reaction of the absorbed molecules and surface reaction leading to the release of the reaction product carbon dioxide are assumed to be at equilibrium, and are used to determine the surface fraction covered by platinum.

A_{12}: All methanol permeating through the PEM is totally oxidised.

7.2.4. Model analysis

7.2.4.1 List of equations

TABLE 8 List of Equations for Direct Methanol Fuel Cell Model

	Variables			No. of Eq.
Meso-scale	Mass balances for methanol and carbon dioxide for the anode-channel compartment	$\frac{dc^{A}_{CH_3OH}}{dt} = \frac{1}{\tau}\left(c^{FA}_{CH_3OH} - c^{A}_{CH_3OH}\right) - \frac{k^{LSA}A^{SA}}{V_a}\left(c^{A}_{CH_3OH} - c^{CLA}_{CH_3OH}\right)$	(1)	1
	Mass balance for carbon dioxide generated during the chemical reaction	$\frac{dc^{A}_{CO_2}}{dt} = \frac{1}{\tau}\left(c^{FA}_{CO_2} - c^{A}_{CO_2}\right) - \frac{k^{LSA}A^{SA}}{V_a}\left(c^{A}_{CO_2} - c^{CLA}_{CO_2}\right)$	(2)	1
micro-scale	Mass balance for methanol in the catalyst layer with respect to time	$\frac{dc^{CLA}_{CH_3OH}}{dt} = \frac{k^{LSA}A^{SA}}{V^{CL}_a}\left(c^{A}_{CH_3OH} - c^{CLA}_{CH_3OH}\right) - \frac{A^{SA}}{V^{CL}_a} n^{M}_{CH_3OH} - \frac{A^{SA}}{V^{CL}_a} r_1$	(3)	1
	Mass balance of carbon dioxide concentration with respect to time	$\frac{dc^{CLA}_{CO_2}}{dt} = \frac{k^{LSA}A^{SA}}{V^{CL}_a}\left(c^{A}_{CO_2} - c^{CLA}_{CO_2}\right) + \frac{A^{SA}}{V^{CL}_a} r_1$	(4)	1
	Charge balance for the anode catalyst layer	$\frac{d\eta_a}{dt} = \frac{1}{C_a}(i_{cell} - 6Fr_1)$	(5)	1
	Charge balance for the cathode catalyst layer	$\frac{d\eta_c}{dt} = \frac{1}{C_c}\left(-i_{cell} - 6F\left(r_5 + n^{M}_{CH_3OH}\right)\right)$	(6)	1
	Flow velocity in the membrane	$v = -\frac{k_\phi}{\mu} c^{M}_{H^+} F \frac{d\phi}{dz} - \frac{k_p dp}{\mu\, dz} = \frac{k_\phi}{\mu} \frac{i_{cell}/F + c^{M}_{H^+} k_p/\mu((p_c - p_a)/d^M)}{D^{M}_{H^+}/RT + c^{M}_{H^+} k_\phi/\mu} - \frac{k_p}{\mu} \frac{p_c - p_a}{d^M}$	(7)	1
	Methanol mass flux density in the membrane (steady-state)1	$n^{M}_{CH_3OH} = \frac{D^M}{d^M} c^{CLA}_{CH_3OH} \frac{Pe\exp(Pe)}{\exp(Pe)-1}$	(8)	1
	Peclet number	$Pe = \frac{vd^M}{D^M}$	(9)	1

(continued)

TABLE 8 *(continued)*

Variables			No. of Eq.
(R1): First (and rate-determining) step is the dissociative chemisorption of methanol on active platinum catalyst sites	$3Pt + CH_3OH \leftrightarrow Pt_3 - COH + 3H^+ + 3e^-$ $r_1 = k_1 \exp\left(\frac{\alpha_1 F}{RT}\eta_a\right)\theta_{Pt}^3\left\{c_{CH_3OH}^{CLA} - \frac{c_{CO_2}^{CLA}}{K\tilde{K}_2^3 K_3 K_4}\exp\left(-\frac{\alpha_1 F}{RT}\eta_a\right)\right\};$ $\tilde{K}_2 = K_2 \exp\left(\frac{\alpha_1 F}{RT}\eta_a\right)$	(10)	1
(R2): Rate of formation of hydroxylic groups on the active ruthenium catalyst sites	$3Ru + 3H_2O \leftrightarrow 3Ru - OH + 3H^+ + 3e^-$ $r_2 = k_2 \exp\left(\frac{\alpha_2 F}{RT}\eta_a\right)\left\{\theta_{Ru} - \frac{1}{K_2}\exp\left(-\frac{F}{RT}\eta_a\right)\theta_{Ru-OH}\right\}$	(11)	1
(R3): Rate in the surface reaction of the absorbed molecules formed in the previous steps	$Pt_3 - COH + 2Ru - OH \leftrightarrow Pt - COOH + H_2O + 2Pt + 2Ru$ $r_3 = k_3\left\{\theta_{Pt_3-COH}\theta_{Ru-OH}^2 - \frac{1}{K_3}\theta_{Pt-COOH}\theta_{Pt}^2\theta_{Ru}^2\right\}$	(12)	1
(R4): Rate of the surface reaction leading to the release of the reaction product carbon dioxide	$Pt - COOH + Ru - OH \leftrightarrow CO_2 + H_2O + Pt + Ru$ $r_4 = k_4\left\{\theta_{Pt-COOH}\theta_{Ru-OH} - \frac{1}{K_4}c_{CO_2}^{CLA}\theta_{Pt}\theta_{Ru}\right\}$	(13)	1
(R5): Oxygen reduction occurring on the cathode side	$3/2O_2 + 6H^+ + 6e^- \leftrightarrow 3H_2O$ $r_5 = k_5 \exp\left(\frac{\alpha_5 F}{RT}\eta_c\right)\left\{1 - \exp\left(-\frac{F}{RT}\eta_c\right)\left(\frac{p_{O_2}}{p^0}\right)^{3/2}\right\}$	(14)	1
Since the reaction steps (R2)-(R4) are assumed not to be rate determining, their driving forces nearly vanish. Then, the combination of these equations is giving:	$\theta_{Pt}^3 c_{CO_2}^{CLA} + \left(K_2K_4 + c_{CO_2}^{CLA}\right)\left(K_2 \exp\left(\frac{\alpha_1 F}{RT}\eta_a\right)^2\right)^4 K_3 c_{CO_2}^{CLA}\theta_{Pt}$ $-\left(K_2 \exp\left(\frac{\alpha_1 F}{RT}\eta_a\right)^2\right)^3 K_3K_4 = 0$	(15)	1
p_{O_2} is defined	$p_{O_2} = p_c\%_{Oxigeninair}$	(16)	1
Cell voltage of the direct methanol fuel cell	$U_{cell} = U_{cell}^{std} - \eta_a + \eta_c - \frac{i_{cell}d^M}{\kappa^M}$	(17)	1
Mass transfer coefficients determined from measured limiting current densities i_{lim}.	$k^{LSA} = \frac{V^F i_{lim}}{6FV^F c_{CH_3OH}^{FA} - A^{SA} i_{lim}}$	(18)	1

7.2.4.2 List of Variables

TABLE 9 List of Variables Employed for the Direct Methanol Fuel Cell Model

Variable		Units
$c^{A}_{CH_3OH}$	Concentration of methanol in the anode compartment	mol/m^3
$c^{FA}_{CH_3OH}$	Feed concentration of methanol in the anode-channel compartment	mol/m^3
$c^{CLA}_{CH_3OH}$	Concentration of methanol in the anode catalyst layer	mol/m^3
k^{LSA}	Mass transfer coefficient in the anode section	m/s
A^{SA}	Cross-sectional electrode area in the anode section	m^2
V_a	Anode compartment volume	m^3
τ	Resident mean time	S
$c^{A}_{CO_2}$	Concentration of carbon dioxide in the anode compartment	mol/m^3
$c^{FA}_{CO_2}$	Feed concentration of carbon dioxide in the anode-channel compartment	mol/m^3
$c^{CLA}_{CO_2}$	Concentration of carbon dioxide in the anode catalyst layer	mol/m^3
V^{CL}_a	Anode compartment volume in the catalyst layer	m^3
$n^{M}_{CH_3OH}$	Mass flux density of methanol in the membrane	$mol/(m^2 s)$
r_1	Rate of reaction of the anode section	$mol/(m^2 s)$
η_a	Electrode overpotential in the anode	V
i_{cell}	Cell current density	A/m^2
C_a	Double layer capacity in the anode	F/m^2
F	Faraday constant (=96485)	C/mol
η_c	Electrode overpotential in the cathode	V
C_c	Double layer capacity in the cathode	F/m^2
r_5	Rate of reaction of the cathode section	$mol/(m^2 s)$
v	Flow velocity in membrane	m/s
k_ϕ	Electrokinetic permeability of membrane	m^2
μ	Pore fluid viscosity in membrane	$Kg/(ms)$
$c^{M}_{H^+}$	Proton concentration in membrane	mol/m^3
$D^{M}_{H^+}$	Diffusion coefficient of protons in membrane	m^2/s
k_p	Hydraulic permeability of membrane	m^2
d^M	Thickness of membrane	m
Pe	Peclet number	–
D^M	Diffusion coefficient of methanol in membrane	m^2/s
p_c	Pressure in cathode compartment	Pa
p_a	Pressure in anode compartment	Pa
R	Universal gas constant	$J/(mol\ K)$
T	Cell temperature	K
α	Charge transfer coefficient	–
K_j	Equilibrium constant of reaction j	–
k_j	Rate constant of reaction i	$mol/(m^2 s)$
θ_{Pt}	Surface fraction covered by Pt	–
θ_{Pt_3-COH}	Surface fraction covered by Pt_3-COH	–
θ_{Ru}	Surface fraction covered by Ru	–
θ_{Ru-OH}	Surface fraction covered by Ru-OH	–
$\theta_{Pt-COOH}$	Surface fraction covered by Pt-COOH	–
p_{O_2}	Partial pressure of oxygen	Pa
p^0	Pressure at standard conditions	Pa
U_{cell}	Total cell voltage	V

(continued)

TABLE 9 *(continued)*

Variable		Units
U_{cell}^{std}	Standard cell voltage	V
κ^M	Conductivity of membrane	$\Omega^{-1}m^{-1}$
V^F	Anode feed flow rate	ml/min
$i_{\lim}$	Limit cell current density	A/m^2

7.2.4.3 Degrees of freedom for controlled release of active ingredients

- State variables (states)

  $c_{CH_3OH}^A, c_{CH_3OH}^{CLA}, c_{CO_2}^A, c_{CO_2}^{CLA}, \eta_a, \eta_c$
- Parameters

  $A^{SA}, V_a, \tau, V_a^{CL}, i_{cell}, C_a, C_c,$
  $k_\phi, \mu, c_{H+}^M, D_{H+}^M, k_p, d^M, D^M, \alpha,$
  $K_1, K_2, K_3, K_4, k_1, k_5, U_{cell}^{std}, \kappa^M, i_{lim}$
- Constants

  p^0, R, F
- Algebraic variables (algebraics)

  $k^{LSA}, n_{CH_3OH}^M, r_1, r_5, v, Pe, \theta_{Pt}, p_{O_2}, U_{cell}$
  $c_{CH_3OH}^{FA}, c_{CO_2}^{FA}, T, p_c, p_a, V^F$

$N_U = 45$ (states + parameters + algebraics)
$N_E = 15$ (equations)
$N_{DF} = 45–15 = 30$

Note that the constants are not counted in the above degrees of freedom analysis as they are assumed to be already known. The choice for the variables to specify consists of the rate and equilibrium constants of the reactions, inlet conditions, diffusion coefficients, in addition to the parameters involved in the differential algebraic set of equations. Thus, the specifications are:

Parameters	$A^{SA}, V_a, \tau, V_a^{CL}, i_{cell}, C_a, C_c,$
	$k_\phi, \mu, c_{H+}^M, D_{H+}^M, k_p, d^M, D^M, \alpha,$
	$K_1, K_2, K_3, K_4, k_1, k_5, U_{cell}^{std}, \kappa^M, i_{\lim}$
Algebraic	$c_{CH_3OH}^{FA}, c_{CO_2}^{FA}, T, p_c, p_a, V^F$

7.2.4.4 Basic process unit: Incidence matrix

In the analysis of the incidence matrix (see Table 10), it can be noted that a lower tridiagonal structure is obtained for this model. The mathematical equations

TABLE 10 Incidence Matrix for Direct Methanol Fuel Cell Mathematical Model

	Unknown variables														
Equations	$c^{A}_{CH_3OH}$	$c^{A}_{CO_2}$	$c^{CLA}_{CH_3OH}$	$c^{CLA}_{CO_2}$	η_a	η_c	v	Pe	$n^{M}_{CH_3OH}$	k^{LSA}	p_{O_2}	θ_{pt}	r_1	r_5	U_{cell}
(7)							*								
(9)							*	*							
(8)			*					*	*						
(18)										*					
(16)											*				
(15)				*	*							*			
(10)			*	*	*							*	*		
(14)						*					*			*	
(17)					*	*									*
(1)	*		*							*					
(2)		*		*						*					
(3)	*		*						*	*			*		
(4)		*		*						*			*		
(5)					*								*		
(6)						*			*					*	

consists of six differential equations and nine algebraic equations, where it is necessary to specify the initial values for the six dependent differential variables (states) associated with the six ordinary differential equations. In addition, to the values of the variables and parameters given in Table 11.

7.2.4.5 Work flow and data flow

The anode-channel compartment belongs to the meso-scale, whilst the membrane-electrode-assembly is considered to be in the micro-scale. The information that is transferred from the meso-scale to the micro-scale is the concentration of methanol and carbon dioxide. The information transferred from the micro-scale to the meso-scale is methanol and carbon dioxide concentrations, as well as their mass transfer coefficients that are coming from the anode catalyst layer.

The model equations at the two scales could be solved separately and classified in four different ways as proposed by Ingram et al. 2004. For example, to solve the problem only at meso-scale, the information from micro-scale could be "known" data. In this case, the multiscale model is classified as serial: simplification (micro-scale) integration framework. Similarly, solving the model at the micro-scale, information coming from the meso-scale might be considered as "known" data. The problem classification in this case serial: simplification (meso-scale) integration framework. Another interesting feature to highlight for this DMFC model is that the model equation calculations at a smaller scale, for example, adding a model for calculating the diffusion coefficient of methanol through the membrane medium, will require a further

TABLE 11 Data for the Direct Methanol Fuel Cell Model

	Property	Value
Initial values	$c^{A}_{CH_3OH}$	500 mol/m^3
	$c^{A}_{CO_2}$	0 mol/m^3
	$c^{CLA}_{CH_3OH,initial}$	0 mol/m^3
	$c^{CLA}_{CO_2,initial}$	0 mol/m^3
	$\eta_{a,initial}$	0 V
	$\eta_{c,initial}$	0 V
Algebraic	$c^{FA}_{CH_3OH}$	500 mol/m^3
	$c^{FA}_{CO_2}$	0 mol/m^3
	T	343.15 K
	p_c	300000 Pa
	p_a	100000 Pa
	V^F	1.36 ml/min
Parameters	A^{SA}	0.0009 m^2
	V_a	$1.15e^{-05}$ m^3
	τ	500 s
	V^{CL}_a	$4.0e^{-07}$ m^3
	i_{cell}	40 A/m^2
	C_a	40000 F/m^2
	C_c	100000 F/m^2
	k_ϕ	$1.13e^{-19}$ m^2
	μ	0.0003353 $kg/(ms)$
	$c^{m}_{H^+}$	1200 mol/m^3
	$D^{M}_{H^+}$	$5.4e^{-09}m^2/s$
	k_p	$1.57e^{-18}$ m^2
	d^M	0.0002 μm
	D^M	$2.9e^{-10}$ m^2/s
	α	0.5
	K_1	1
	K_2	0.5
	K_3	1
	K_4	0.7
	k_1	$4.9e^{-07} mol/m^2s$
	k_5	$1.8e^{-07} mol/m^2s$
	U^{std}_{cell}	1.2 V
	κ^M	$17(\Omega m)^{-1}$
	i_{lim}	1150 A/m^2
Constants	p^0	101325 Pa
	R	8.314 $J/mol\ K$
	F	96485 C/mol

smaller scale than the 'micro'. That is, the calculation of the diffusion coefficients is performed via dynamic simulation, and the problem classification becomes serial: one-way coupling integration. When experimental values of the diffusion coefficients are regressed through a correlation and values for them are taken from it, this integration could be classified as serial: simplification (nano-scale) structure due to the order of magnitude of the calculation.

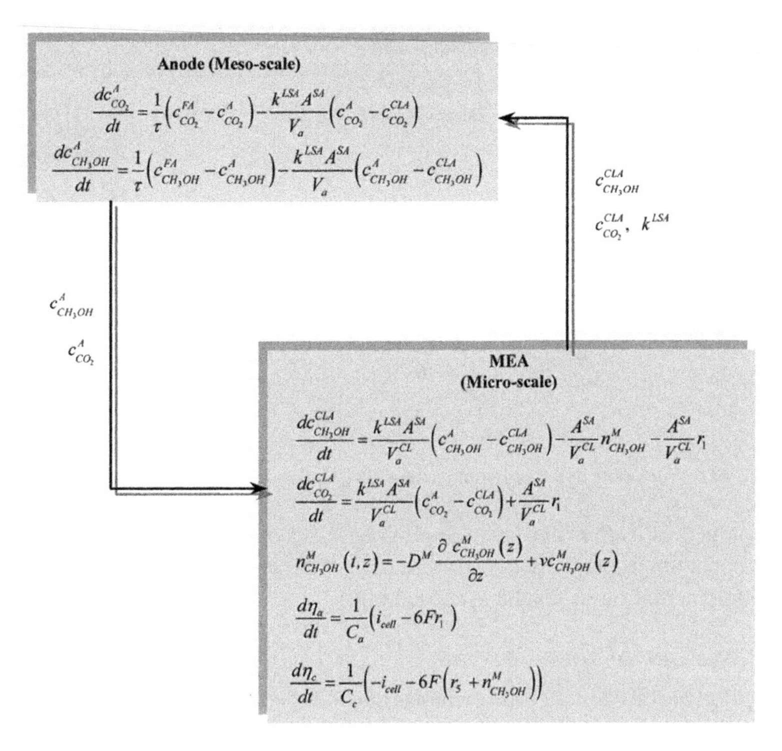

FIGURE 13 **Data flow for the direct methanol fuel cell between the different scales.**

Data flow between the different scales involved in the multiscale model for direct methanol fuel cell is highlighted in Figure 13. In accordance with Ingram et al. 2004, this multiscale model is employing the multidomain framework.

7.2.5. Model Solution

7.2.5.1 Solution strategy

For the DMFC model given in Table 8, the employed the multiscale modelling framework is presented here together with the systematic approach to solve the resulting model equations. The fuel cell is fed with methanol solution in water through the anode compartment. Atmospheric oxygen is fed through the cathode, and a solid polymer membrane is used as an electrolyte.

Solution of the model equations (as listed in Table 8) is done using ICAS-MoT (Sales-Cruz, 2006) and the options used are highlighted in Figure 14.

The work flow for the solution of the DMFC model equations is shown in Figures 13–14, and starts with the introduction of the mathematical model to ICAS-MoT, which performs a translation, analysis and verification of the model. Subsequently, a solution administrator is called and an external software

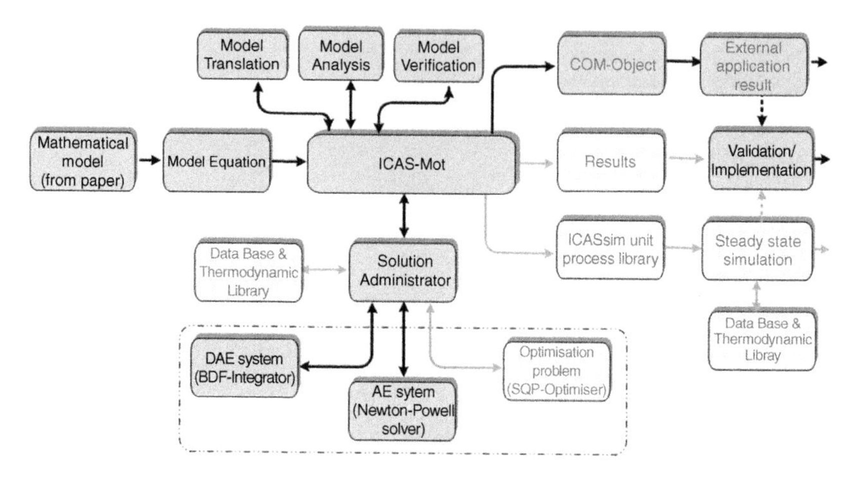

FIGURE 14 **Work flow and tools employed for the solution of the direct methanol fuel cell.**

is employed as a user interface (employing ICAS-MoT as modelling engine). This connection is performed by the use of COM-objects. Finally, the validation and implementation of results is carried out.

7.2.5.2 Values of specified variables and parameters

Table 11 gives the values of the variables that are specified for the solution of the DMFC model equations.

7.2.5.3 Solution of the direct methanol fuel cell

Two different scenarios have been considered for the study of the DMFC process with the model listed in Table 8.

1. Multiscale modelling framework (MS): The set of equations listed in Table 8 is solved (multidomain integration structure). That is, the meso-scale and micro-scale are taken into account simultaneously. The variables that both scales are sharing during the calculations are listed in Table 12.
2. Single-scale modelling framework (SS): This scenario concerns only the solution of the meso-scale equations listed in Table 8. For the micro-scale equations, values of the dependent and explicit variables are specified. In

TABLE 12 Variables Shared Between Meso-scale and Micro-scale

Property	Value
Methanol bulk concentration	$c^{A}_{CH_3OH}$
Concentration in the catalyst layer of methanol	$c^{A}_{CO_2}$
Carbon dioxide bulk concentration	$c^{CLA}_{CH_3OH}$
Concentration in the catalyst layer of carbon dioxide	$c^{CLA}_{CO_2}$

this case study, the values for these variables are taken from the multiscale steady-state simulation. This calculation can be termed as serial: simplification (micro-scale) structure (also called single-scale calculation).

The transient responses of the methanol concentrations and the carbon dioxide concentrations obtained from the single scale and the multiscale scenarios are highlighted in Figures 15–16, respectively. It can be seen that

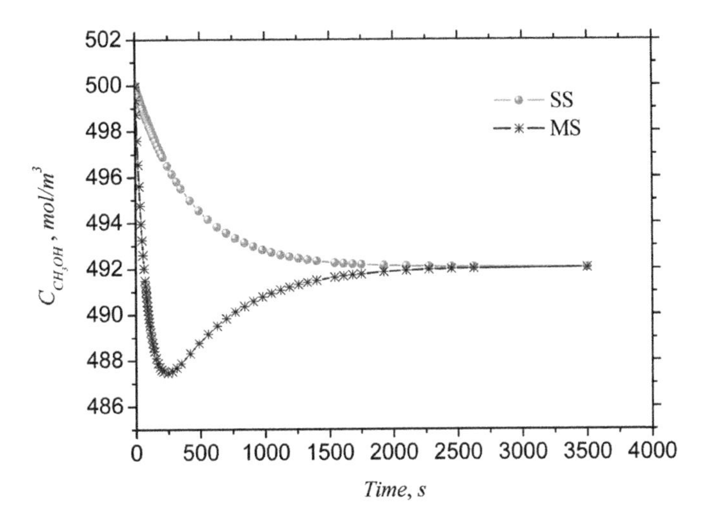

FIGURE 15 Methanol concentration with multiscale modelling and single-scale modelling.

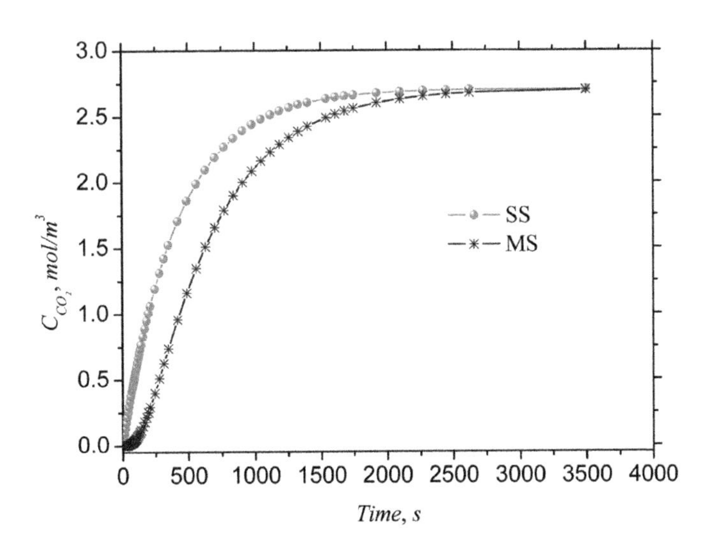

FIGURE 16 Carbon dioxide concentration with multiscale modelling and single-scale modelling.

TABLE 13 Final Results for the Different Scenarios Multiscale (MS) and Single scale (SS)

Variable	MS	SS
$c^{A}_{CH_3OH}$	492.03 mol/m^3	492.03 mol/m^3
$c^{A}_{CO_2}$	2.69 mol/m^3	2.70 mol/m^3
$c^{CLA}_{CH_3OH}$	449.65 mol/m^3	449.65 mol/m^3
$c^{CLA}_{CO_2}$	17.09 mol/m^3	17.09 mol/m^3
θ_{Pt}	0.384 V	0.384 V
U_{cell}	0.6403 V	0.6403 V

while in both cases, the final state states reached are the same, the path taken to reach them are different - first order versus higher order.

Table 13 gives the simulated steady state values for the two scenarios under multiscale modelling approach.

7.2.6. Discussion

Through appropriate models, it is possible to study the behaviour of a product (in this case, the fuel cell) under different application scenarios. This helps to better understand the product, and therefore, to better design the product. The implementation of the model used in this case study is given in Appendix 1 (see chapter-7-2-fuel-cell-dmfc.mot files). See also models for solid oxide fuel cells (Sundmacher et al. 2001; Xu et al. 2005).

7.3. MULTISCALE FLUIDISED BED REACTOR

The schematic diagram of a fluidised bed reactor is shown in Figure 17. The reactant stream enters at a high velocity through a distributor at the bottom of the

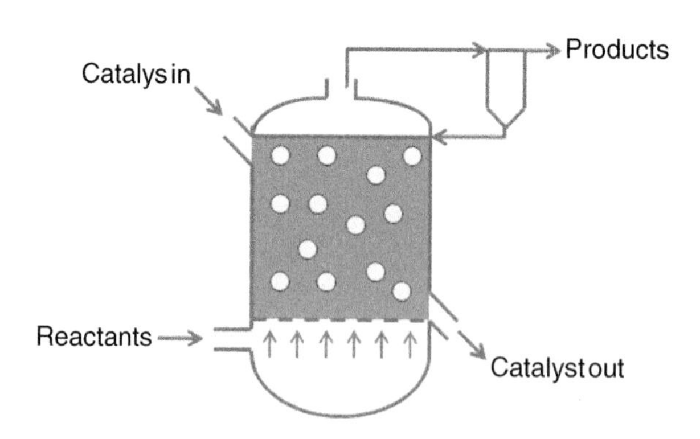

FIGURE 17 **Schematic diagram of a fluidised bed reactor.**

reactor. This generates turbulence. Moreover, the distributor retains the solid catalyst particles in the catalyst bed area. The turbulent reactant flow stirs and fluidises the catalyst particles inside the reactor. The result is uniform particle mixing and temperature distribution. Contact between fluid and catalyst particles promotes the catalyzed reaction. The product stream exits reactor at the top. In order to remove solid particles the product steam is passed through a cyclone.

7.3.1. Modelling Objective

a) Develop a model for a fluidised bed reactor in order to predict the reactant concentrations with respect to time.

b) Analyse the resulting multiscale model to evaluate opportunities for model reduction.

7.3.2. Model Description

Two models are generated – a multiscale model in two sizes considering the balance for the reactor and the particles and a reduced model considering only one balance volume, that is, assuming that the properties remain constant within one of the balance volumes one dimension.

7.3.2.1 Multiscale Model in Two Sizes

Model assumptions: The following assumptions are made.

- Reactants in the gas phase are completely mixed throughout the fluidised bed inside the reactor.
- Changes in the void fraction volume of the bed due to reaction are neglected.
- The particles are assumed to be small enough to consider that heat and mass transfer resistances can be lumped at the particle surface.
- Reactions take place in the porous volume of the catalyst.
- All particles have the same size and are spherical.
- All particles have the same temperature.
- All particles have the same partial pressure of the reactants.
- One irreversible reaction, $A \rightarrow B$, is taking place.
- The reaction rate and heat of reaction are represented by applying the Arrhenius approach.
- The amount of particles is constant during the whole operation.
- Cyclone operation at the top of the reactor is not taken into account in the mathematical model.
- Product of molecular weight and total pressure is constant throughout the whole bed.

Phenomena considered: Mass transfer between steam and particle (mass transfer coefficient); heat transfer between gas and fluidised catalyst bed (heat transfer coefficient); reaction in particle pores, and heat of reaction (Arrhenius); heat transfer between walls and gas-phase.

7.3.2.2 Model Description

Model equation -Macro-scale (bulk): The model equations, i.e. the conservation equations for mass and energy for the balance volume (see Figure 17) the steam phase, are taken from Luss and Amundson (1968) and are given below:

$$\frac{dp}{dt} = p_e - p + H_g \cdot (p_p - p) \tag{1}$$

$$\frac{dT}{dt} = T_e - T + H_w \cdot (T_w - T) + H_t \cdot (T_p - T) \tag{2}$$

Here p and p_p are the partial pressures of the reactant in the steam and inside the catalyst particle, respectively; T and Tp are the corresponding temperatures; p_e and T_e are the pressure and the temperature at the reactor entrance; T_w is the temperature of the reactor wall; H_g, H_w and, H_t are dimensionless numbers defined below:

$$H_g = \frac{a_v \cdot k_g \cdot M \cdot P \cdot V}{q} \tag{3}$$

$$H_w = \frac{2 \cdot h_w \cdot V}{r \cdot c_g \cdot q} \tag{4}$$

$$H_t = \frac{a_v \cdot h_g \cdot V}{q \cdot c_g} \tag{5}$$

where a_v is the interfacial area between the catalyst and the steam per unit volume, k_g is the mass transfer coefficient between the steam and the particles, M is the molecular weight, P is the total pressure, V is the bed volume, q is the gas mass flow rate, h_w is the heat transfer coefficient between the reactor wall and the gas, c_g is the heat capacity of the gas and h_g is the heat transfer coefficient between the gas and the catalyst particles.

Model equations – Micro-scale (catalyst particle): The mass and energy balance for each catalyst pellet is given by,

$$K_k = 0.0006 \cdot exp\left\{\frac{20.7 - 150\,00}{T_p}\right\} \tag{6}$$

$$\frac{dp_p}{dt} = \frac{H_g}{A} \cdot (p - p_p) - \frac{H_g \cdot K_k \cdot p_p}{A} \tag{7}$$

$$\frac{dT_p}{dt} = \frac{H_t}{C} \cdot (T - T_p) + \frac{H_t \cdot F \cdot K_k \cdot p_p}{C} \tag{8}$$

where

$$A = \frac{\alpha \cdot v_p \cdot a_v}{s \cdot s_p} \tag{9}$$

$$C = \frac{a_v \cdot c_s \cdot v_p \cdot \rho_s}{s \cdot s_p \cdot c_g \cdot \rho_g} \tag{10}$$

$$F = \frac{(-\Delta H) \cdot k_g}{h_g} \tag{11}$$

here α is the void fraction of the particles, v_p is the particle volume, a_v is the interfacial area between the catalyst and the steam per unit volume, ε is the void fraction of the bed, s_p is the particle surface, c_s is the heat capacity of the particles, v_p is the volume of a catalyst particle, ρ_s is the particle density, ρ_g is the gas density and ΔH is the heat of reaction.

7.3.2.3 Reduced Model

The multiscale model given above is simplified by neglecting the mass and heat transfer resistences between the particles and the gas phase. The reduced model equations are derived from the multiscale model by setting, $p_p = p$ and $T_p = T$. First, assumptions for the scenario to be simulated are made, as listed below.

Assumptions for scenario to be simulated:

- **i.** Reactants in the gas phase are completely mixed throughout the whole bed inside the reactor. In general ideal mixing in the fluidised bed leads to T and component concentrations being constant.
- **ii.** Changes in the void fraction volume of the bed due to reaction are neglected.
- **iii.** Heat and mass transfer resistance between the particles and gases are neglected.
- **iv.** Reaction A→B takes place in the porous volume of the catalyst.
- **v.** All particles have the same shape and size (this does not matter in this case (only macro-scale) as long as the overall a_v value is known).
- **vi.** All particles have the same temperature (follows from assumptions i and iii).
- **vii.** One irreversible reaction is considered $A \rightarrow B$. Reaction rate and heat can be represented applying the Arrhenius approach.
- **viii.** The amount of particles is constant during the whole operation (a_v = constant).
- **ix.** Cyclone operation found in the top of the reactor is not taken into account in the mathematical model.
- **x.** The product of molecular weight and total pressure is constant throughout the whole bed.

Considered phenomena: Reaction in particle pores, and heat of reaction (Arrhenius); heat transfer between walls and gas phase.

Reduced Model Equations:

$$K_k = 0.0006 \cdot exp\left\{\frac{20.7 - 15\,000}{T}\right\} \tag{12}$$

$$\frac{dp}{dt} = (p_e - p - H_g \cdot K_k \cdot p)/(1 + A) \tag{13}$$

$$\frac{dT}{dt} = (T_e - T + H_w(T_w - T) + H_T \cdot F \cdot K_k \cdot p)/(1 + C) \tag{14}$$

7.3.3. Model Analysis

7.3.3.1 Multiscale Model

Eigenvalue analysis: The steady states of the dynamic model are analysed. In order to do so, the steady-state model needs to be derived first.

Macro-scale (steady state): The steady-state form of the model is obtained by setting the left-hand sides of Eqs. 13–14 to zero.

$$0 = p_e - p + H_g \cdot (p_p - p) \tag{15}$$

$$0 = T_e - T + H_w \cdot (T_w - T) + H_t \cdot (T_p - T) \tag{16}$$

$$K_k = 0.0006 \cdot exp\left\{\frac{20.7 - 150\,00}{T_p}\right\} \tag{17}$$

Micro-scale (steady state):

$$0 = \frac{H_g}{A} \cdot (p - p_p) - \frac{H_g \cdot K_k \cdot p_p}{A} \tag{18}$$

$$0 = \frac{H_t}{C} \cdot (T - T_p) - \frac{H_t \cdot F \cdot K_k \cdot p_p}{C} \tag{19}$$

The above steady-state model equations are solved to determine whether multiple solutions exists. This can be done by analyzing the asymptotic stability of the steady states of the model through an analysis of the eigenvalues of the Jacobian matrix of the steady-state model. Table 14 gives the solutions for the three steady states found together with their corresponding eigenvalues.

For steady states 1 and 3 all the eigenvalues have negative real parts. This means that these steady states are asymptotically stable. The second steady state

TABLE 14 **Steady States and their Corresponding Eigenvalues for Multiscale Model**

	First steady state	Second steady state	Third steady state
p	0.09353	0.06704	0.006822
p_p	0.09351	0.06694	0.006531
T	690.445	758.346	912.764
T_p	690.607	759.170	915.094
λ_1	-0.00632	0.00613	-0.06803
λ_2	-0.91232	-1.26657	-13.4706
λ_3	-270.55	-270.55	-269.20
λ_4	-2187.25	-2189.37	-2261.52

has one eigenvalue with a positive real part, and consequently, it is unstable. All imaginary parts are zero.

Furthermore, the eigenvalue analysis reveals that the system is stiff because the ratio of the absolute value of the real parts of the largest eigenvalue and the smallest eigenvalue is high (>1000). This means that the system consists of variables that reach the steady state much faster than others. This offers the potential for model simplification (reduction) by assuming the fast (dependent differential) variables to be at the steady state.

Multiscale model solution strategy: Note that the multiscale model considers the overall reactor on the macro-scale and the single catalyst pellet on the micro-scale. For both scales a system of ODEs need to be solved. The incidence matrix for the multiscale model is given in Table 15.

The incidence matrix reveals that the macro-scale and the micro-scale equations are coupled because the values for p_p, T_p, p and T need to be exchanged for each time step. Two alternative options to solve the model equations are considered below (see also Figure 18):

1. Solve all the equations together at the time scale of the slowest mode.
2. Solve the micro- and macro-scales separately at their individual time scales and find a strategy for exchanging the required variable values between the scales.

TABLE 15 **Incidence Matrix for Multiscale Model**

	p	T	K_k	p_p	T_p	
Eq. (1)	*			*		macro-scale
Eq. (2)		*			*	
Eq. (6)			*		*	micro-scale
Eq. (7)	*		*	*		
Eq. (8)		*	*	*	*	

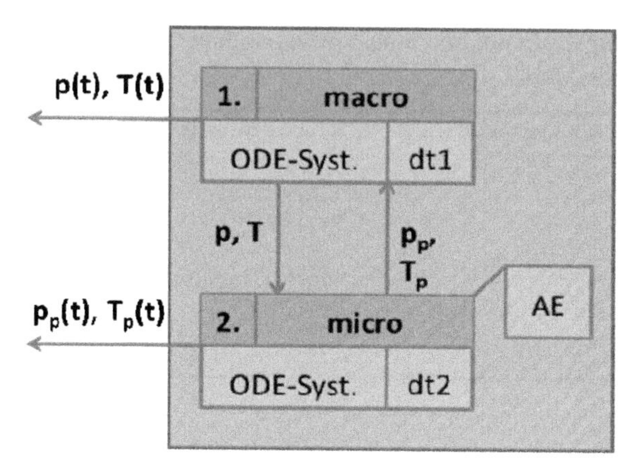

FIGURE 18 Linking scheme of multiscale model.

The linking scheme illustrated in Figure 18 shows the different scales, their output variables, time scales, equation types and the data flow between the scales. The red (coloured) outer-box in Figure 18 symbolises that the scales need to be solved simultaneously.

7.3.3.2 Reduced Model

To perform the reduced model (eigenvalue) analysis, first the steady-state model is considered.

$$K_k = 0.0006 \cdot exp\left\{\frac{20.7 - 15\,000}{T}\right\} \tag{20}$$

$$0 = (p_e - p - H_g \cdot K_k \cdot p)/(1 + A) \tag{21}$$

$$0 = (T_e - T + H_w(T_w - T) + H_T \cdot F \cdot K_k \cdot p)/(1 + C) \tag{22}$$

Table 16 gives the calculated values of p and T obtained by solving Eqs. 20–22. It can be noted that three sets of steady-state solutions are obtained. The corresponding eigenvalues for each solution are also listed in Table 16.

The results of Table 16 show that after the undertaken model reduction only the slow modes are left. The fast modes are represented by mass transfer, which is now assumed to be infinitely fast and thus is not considered in the analysis of the dynamic behaviour. The eigenvalues further reveal that the asymptotic stability of the steady states is also found in the reduced model, indicating thereby that the first and third steady states are asymptotically stable whereas the second steady state is unstable.

<u>Reduced model solution strategy</u>: The reduced (dynamic) model (Eqs. 12–14) is of single scale. The incidence matrix is given in Table 17. It can be

TABLE 16 Calculated Steady States and their Corresponding Eigenvalues for the Reduced Model

	First steady state	Second steady state	Third steady state
p	0.0936	0.0659	0.0069
T	690.22	761.14	911.96
λ_1	-0.00651264	0.00620247	-0.0088858
λ_2	-0.911608	-1.28696	-12.3175

TABLE 17 Incidence Matrix for Reduced Model

	K_k	p	T	
(12)	*		*	AE-part
(13)	*	*		ODE-part
(14)	*		*	

noted that the differential-algebraic equation (DAE) system has index 1, meaning that the DAE system can be solved simultaneously, or, converted to a set of ordinary differential equations (ODE) by inserting Eq. 12 into Eqs. 13–14.

7.3.4. Numerical Solution

The transient responses of the partial pressure and the temperature are obtained by solving the equations for the multiscale model and the reduced dynamic model (see Tables 15 & 17) from an initial condition. In Table 18 (for the multiscale model) and Table 19 (for the reduced model) the input data are given and they correspond to the first steady state. The transient responses are shown in Figures 19–20. The BDF solver available in the ICAS-MoT (Sales-Cruz, 2006) has been used for the multiscale solution strategy given in Figure 18. For the reduced model, a DAE system of equations are solved, also through the BDF solver in ICAS-MoT.

From Figures 19–20 it can be seen that the transient responses of temperatures and partial pressures from both models are the same. With both models, the partial pressures and temperatures reach steady state at about the same time of 600 hours, indicating that at least for this system, model reduction is feasible since the path to reach them are also similar.

The model equations as implemented in ICAS-MoT are given in Appendix 1 (see chapter-7-3-fuidised-bed-multiscale-dyn.mot; chapter-7-3-fuidised-bed-multiscale-ss.mot; chapter-7-3-fuidised-bed-reduced.mot files).

TABLE 18 Initial Conditions and Known Variable Values for Multiscale Model

Dependent	p	0.1 [atm]
	T	600 [°R]
	p_p	0.1 [atm]
	T_p	690.607 [°R]
Known	p_e	0.1 [atm]
	T_e	600 [°R]
	T_w	720 [°R]
	H_g	320 [-]
	H_w	1.6 [-]
	H_t	266 [-]
	F	8000 [-]
	S	0.17142 [-]
	C	205.74 [-]

TABLE 19 Initial Conditions and Known Variable Values for Reduced Model

Dependent	p	0.1 [atm]
	T	600 [°R]
Known	p_e	0.1 [atm]
	T_e	600 [°R]
	T_w	720 [°R]
	H_g	320 [-]
	H_w	1.6 [-]
	H_t	266.667 [-]
	F	8000 [-]
	A	0.17142 [-]
	C	205.74 [-]

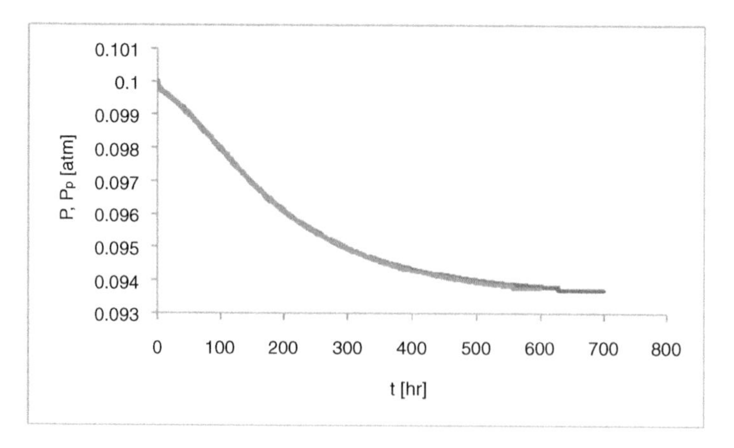

FIGURE 19 Transient behaviour of partial pressures to first steady state for multiscale and reduced model.

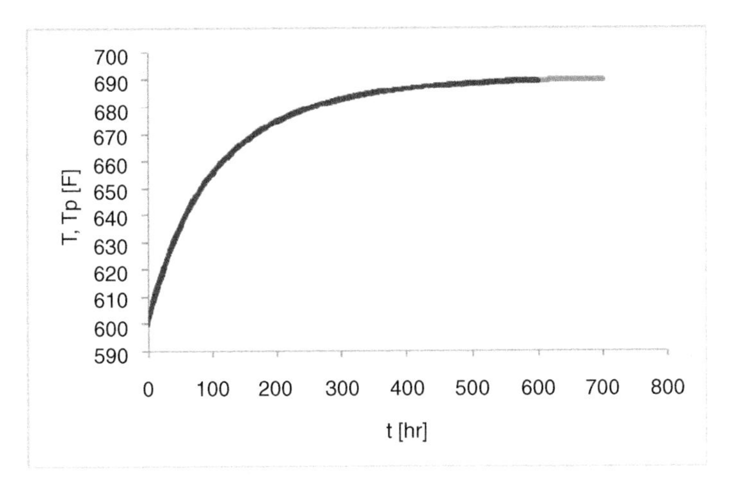

FIGURE 20 Transient behaviour of temperatures to first steady state for multiscale and reduced model.

Discussion: Try to obtain all the three steady states and evaluate their stability through calculating the eigenvalues. Try other schemes for multiscale simulation strategy (Ingram et al. 2004), for example, how would the transient responses look like for another reduction scheme?

7.4. DYNAMIC CHEMICAL REACTOR

Consider the following reactions in series (Fogler, 2005):

$$2A \underset{(1)}{\xrightarrow{k_{1a}}} B \underset{(2)}{\xrightarrow{k_{2b}}} 3C \tag{23}$$

The above reactions are catalysed by H_2SO_4. All reactions are first order with respect to reactant concentratins. The reactions are carried out in a semi-batch reactor (see Figure 21) that has a heat exchanger inside with $UA = 35{,}000.0$ [cal/h K] and ambient temperature of 298 [K]. Pure A enters at a concentration of 4 [mol/dm^3], a volumetric flow rate of 240 [dm^3/hr] and a temperature of 305 [K]. Initially there is a total of 100 [dm^3] in the reactor, which contains 1.0 [mol/dm^3] of A and 1.0 [mol/dm^3] of the catalyst H_2SO_4. The reaction rate is independent of the catalyst concentration. The initial temperature of the reactor is 290 [K].

7.4.1. Modelling Objective

The objective of this modelling exercise is to model, analyse and solve the model equations given as example 9.5 of chapter 9 by Fogler (2005).

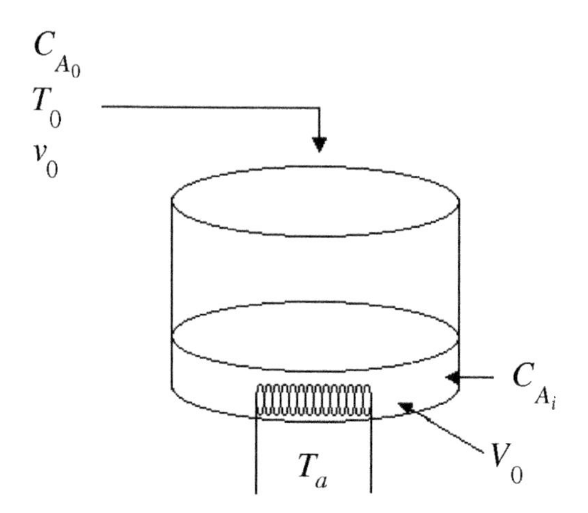

FIGURE 21 **Semi batch reactor scheme.**

7.4.2. Model Description

The component balance equations are written for each compound,

$$\frac{dC_A}{dt} = r_A + \frac{(C_{A_0} - C_A)}{V} V_0 \tag{24}$$

$$\frac{dC_B}{dt} = r_B + \frac{C_B}{V} V_0 \tag{25}$$

$$\frac{dC_C}{dt} = r_C - \frac{C_C}{V} V_0 \tag{26}$$

where the reaction rates are given by

$$-r_{1A} = k_{1A} C_A \tag{27}$$

$$-r_{2B} = k_{2B} C_B \tag{28}$$

and the kinetic constants are of the Arrenhius-type

$$k_{1a} = k_{1A} \exp\left\{ \left(\frac{E_{1A}}{R} \right) \left[\frac{1}{T_{1A}} - \frac{1}{T} \right] \right\} \tag{29}$$

$$k_{2b} = k_{2B} \exp\left\{ \left(\frac{E_{2B}}{R} \right) \left[\frac{1}{T_{2B}} - \frac{1}{T} \right] \right\} \tag{30}$$

The relative rates are obtained by using the stoichiometric coefficients for the reaction series [see Eq. 23]:

$$-r_{2B} = \frac{r_{2C}}{3} \tag{31}$$

$$r_{2C} = -3r_{2B} \tag{32}$$

So that the net reaction rates become:

$$r_A = r_{1A} = -k_{1A}C_A \tag{33}$$

$$r_B = r_{1B} + r_{2B} = \frac{-r_{1A}}{2} + r_{2B} = \frac{k_{1A}C_A}{2} - k_{2B}C_B \tag{34}$$

$$r_C = 3k_{2B}C_B \tag{35}$$

$$N_i = C_iV \tag{36}$$

$$V = V_0 + v_o t \tag{37}$$

$$N_{H_2SO_4} = \left(C_{H_2SO_4,0}\right)V_0 = \frac{1\ mol}{dm^3} \times 100\ dm^3 = 100\ mol \tag{38}$$

$$F_{A_0} = C_{AO}v_O = \frac{4\ mol}{dm^3} \times 240\frac{dm^3}{h} = 960\ \frac{mol}{h} \tag{39}$$

The overall energy balance equations are given by,

$$\frac{dT}{dt} = \frac{UA(T_a - T) - F_{A0}Cp_A(T - T_0) + \left[\left(\Delta H_{Rx_{1A}}\right)(r_{1A}) + \left(\Delta H_{Rx_{2B}}\right)(r_{2B})\right]V}{\left[C_{pmix}\right]V + N_{H_2SO_4}Cp_{H_2SO_4}} \tag{40}$$

where,

$$C_{pmix} = C_AC_{PA} + C_BC_{PB} + C_CC_{PC} \tag{41}$$

7.4.3. Model Analysis

This process model consists of: four ODEs [Eqs. 24–26, and Eq. 40] and 10 Algebraic Equations [Eqs. 27–30, 33–35, 37, 39, 41]. This DAE system is solved simultaneously using any standard DAE or ODE solver (since the AEs are explicit AEs and can be arranged in the lower tridiagonal form).

7.4.4. Numerical Solution

The numerical solution of the model equations is highlighted through the use of the modelling toolbox, ICAS-MoT (Sales-Cruz, 2006). Within ICAS-MoT, the model is first constructed (step-1: write the derived model equations), then translated (step-2: convert the text-form to solver-form), then analysed (step-3: degrees of freedom, incidence matrix, etc.) and finally solved (step-4: using an appropriate numerical solver).

The model developer does not need to write any programming code to enter the model equations. Models are entered (imported) as text files or XML files, which are then internally translated. The step-by-step procedure is highlighted below.

<u>Step 1</u>: Type the model equations in MoT, as shown in Figure 22.

```
#******************************************
#*Non-isothermal Multiple Reaction        *
#*                                        *
#*CAPEC, Department of Chemical Engineering*
#*Technical University of Denmark         *
#                                         *
#******************************************

#The series reactions:
#            k1A      K2B
#       2A -----> B -----> 3C
#            (1)      (2)

#*************
#Solution    *
#*************

#Kinetic
  k1A = k1A0*exp((E1A/R)*(1/T1A0 - 1/T))
  k2B = k2B0*exp((E2B/R)*(1/T2B0 - 1/T))

#Rate Laws
  r1A = -k1A*CA
  r2B = -k2B*CB
  rA = r1A
  rB = k1A*CA/2-k2B*CB
  rC = 3*k2B*CB

#Reactor volume
  V = V0 + vo*time

  FA0 = CA0*vo
  Cpmix = CA*CpA + CB*CpB + CC*CpC

#Mol balances
  dCA = rA + (CA0 - CA)*vo/V
  dCB = rB - CB*vo/V
  dCC = rC - CC*vo/V

#Energy Balance
  dT = (UA*(Ta-T) - FA0*CpA*(T-T0) + (DHRx1A*r1A +
                    DHRx2B*r2B)*V)/(Cpmix*V + CH2SO40*V0*CpH2SO4)
```

FIGURE 22 Typed model equations in MoT.

Step 2: Translate the model equations – MoT employs RPN for this purpose.
Step 3: Model analysis – here, the variables are classified as parameters, known, unknown, dependent and explicit. The dependent variables are related to the differential equations. The unknown and explicit variables are related to the implicit and explicit algebraic equations. The parameters refer to the model parameters, and the default independent variable is the time. In this model, the dependent variables are *T, CA, CB* and *CC*. Since all the algebraic equations are explicit, the corresponding variables are also classified as explicit unknown variables. They are - *k1A, k2B, r1A, r2B, rA, rB, rC, V, FA0, Cpmix*. The remaining variables define the degrees of freedom and have to be specified. These specified variables are classified as parameters (*k1A0, E1A, R, T1A0, k2B0, E2B, T2B0, CpA, CpB, CpC, CpH2SO40, CpH2SO4, DHRx1A, DHRx2B*) and known (*T, vo, UA, Ta, To*). The incidence matrix for the dynamic model is shown in Table 20, where the unshaded cells indicate the lower tridiagonal form for the algebraic equations. Note that at any time t, values for *CA, CB, CC* and *T* are known.
Step 4: Numerical solution – first values for all the parameters and known variables need to be specified. Table 21 provides a list of these variables and their corresponding specified values. The calculated values for the differential variables obtained through the BDF-solver in ICAS-MoT are given in Figures 23–26.

The model equations as implemented in ICAS-MoT are listed in Appendix 1 (see chapter-7-4-chem-reactor.mot file).

TABLE 20 Incidence Matrix

	Variable													
Equations	k1A	k2B	r1A	r2B	rA	rB	rC	V	FA0	Cpmix	CA	CB	CC	T
29	*													+
30		*												+
27	*		*								+			
28		*		*								+		
22			*		*									
34	*	*				*					+	+		
35		*					*					+		
37								*						
39									*					
41										*	+	+	+	
24					*			*			*			
25						*		*				*		
26							*	*					*	
40			*	*				*	*					*

TABLE 21 List of specified variables and their given values (chemical reactor)

Variable	Specified value	Variable	Specified value
T	298	UA	35000
C_A	1	T_a	298
C_B	0	T_O	305
C_C	0	D_{HRx1A}	−6500
R	1.987	D_{HRx2B}	8000
T_{1A0}	320	C_{H2SO40}	1
T_{2B0}	300	C_{pH2SO4}	35
V_O	100	k_{1A0}	1.25
v_o	240	E_{1A}	9500
C_{A0}	4	k_{2B0}	0.08
C_{pA}	30	E_{2B}	7000
C_{pB}	60	t	0
C_{pC}	20		

Discussion: Rewrite the model in a more compact form using the vector and matrix notation. Change the reaction rate parameters to study their effect on the product composition. Evaluate the effect of temperature and/or reactor volume on the reaction residence time as well as the sensitivity of the kinetic model parameters. Also, consider ways to reuse and adapt the model for other chemical and biochemical reactor modelling.

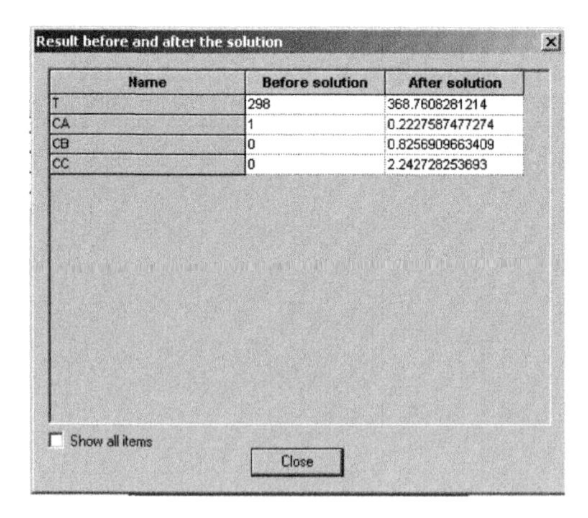

FIGURE 23 Values of dependent variable.

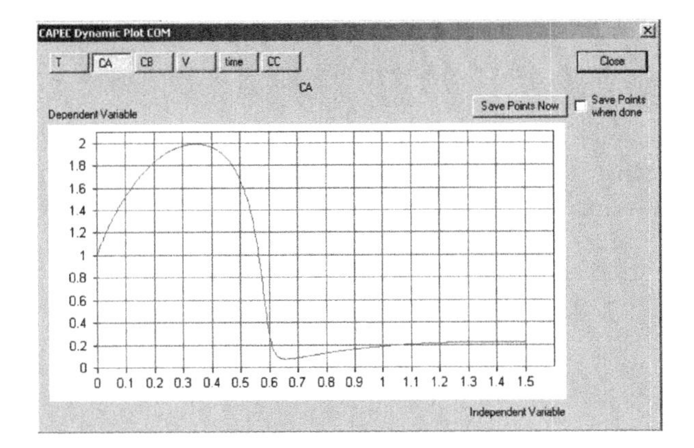

FIGURE 24 Component A dynamic concentration profile.

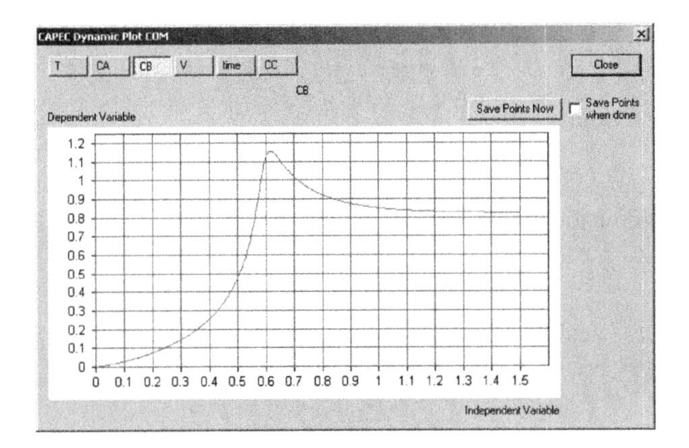

FIGURE 25 Component B dynamic concentration profile.

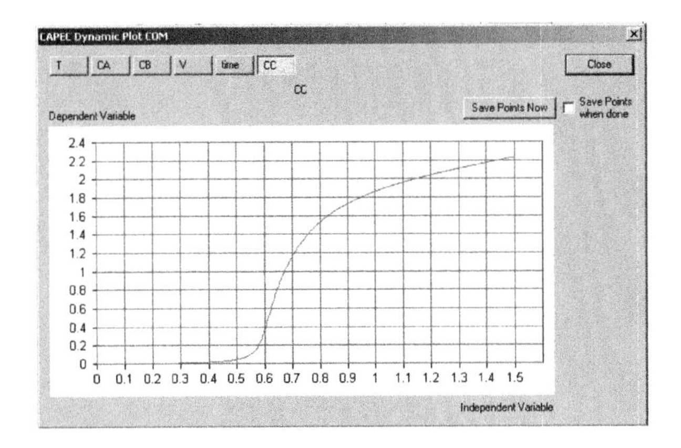

FIGURE 26 Component C dynamic concentration motion.

7.5. POLYMERISATION REACTOR

The methyl methacrylate (MMA) polymerisation takes place in a CSTR (see Figure 27), using azo-iso-butyronitrile (AIBN) as initiator and toluene as solvent. The reaction is exothermic, and a cooling jacket is used to remove the heat of reaction (Silva et al. 2001).

The reaction mechanism of MMA free radical polymerisation consists of the following steps:

Initiation:

$$I \xrightarrow{k_O} 2R$$
$$R + M \xrightarrow{k_I} P_1$$

Propagation:

$$P_i + M \xrightarrow{k_p} P_{i+1}$$

Monomer transfer:

$$P_i + M \xrightarrow{k_{fm}} P_i + D_i$$

Addition termination:

$$P_i + P_j \xrightarrow{k_{tc}} D_{i+j}$$

Disproportionation termination:

$$P_i + P_j \xrightarrow{k_{td}} D_i + D_j$$

where I, P, M, R and D refer to initiator, polymer, monomer, radicals and dead polymer respectively.

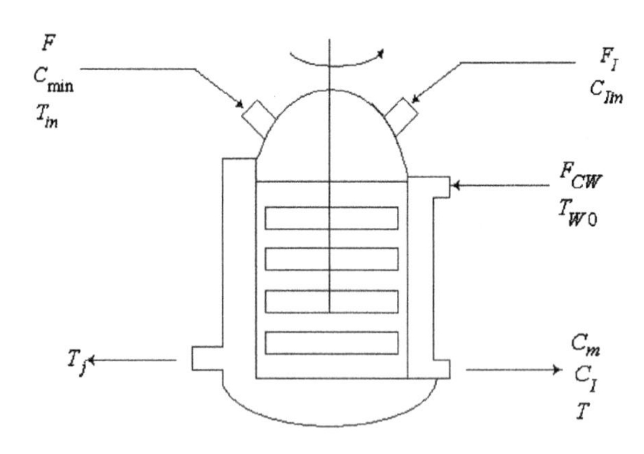

FIGURE 27 Polymerisation reactor.

7.5.1. Modelling Objective

The objective here is to develop a dynamic model so that the transient responses of the compositions and the temperatures during the operation of the polymerisation reactor can be studied; existence of multiple solutions can be identified; and closed-loop simulations performed by adding an appropriate control model.

7.5.2. Model Description

7.5.2.1 Model Assumptions

- The contents of the reactor are perfectly mixed.
- Constant density and heat capacity of the reaction mixture.
- Density and heat capacity of the cooling fluid stay constant.
- Uniform cooling of fluid temperature.
- The reactions only happen inside the reactor.
- There is no gel effect (the conversion of monomer is low, and the proportion of solvent in the reaction mixture is very high).
- Constant volume of the reactor.
- Polymerisation reactions occur according to the free radical mechanism.

7.5.2.2 Model Description

Based on the above assumptions, the mathematical model of the MMA polymerisation is given below.

Mass balance:

$$\frac{dC_m}{dt} = -(k_p + k_{fm})C_m P_0 + \frac{F(C_{min} - C_m)}{V} \tag{42}$$

$$\frac{dC_I}{dt} = -k_I C_I + \frac{(F_I C_{I_{in}} - F C_I)}{V} \tag{43}$$

$$\frac{dD_0}{dt} = -(0.5k_{tc} + k_{td})P_0^2 + k_{fm} C_m P_0 - \frac{F D_0}{V} \tag{44}$$

$$\frac{dD_1}{dt} = M_m (k_p + k_{fm}) C_m P_0 - \frac{F D_1}{V} \tag{45}$$

Energy balance:

$$\frac{dT}{dt} = \frac{(-\Delta H) k_p C_m}{\rho C_P} P_0 - \frac{UA}{\rho C_P V}(T - T_j) + \frac{F(T_{in} - T)}{V} \tag{46}$$

$$\frac{dT_j}{dt} = \frac{F_{CW}(T_{W0} - T_j)}{V_0} + \frac{UA}{\rho_W C_{PW} V_0}(T - T_j) \tag{47}$$

Defined (constitutive) relations:

$$P_0 = \sqrt{\frac{2f^* C_I k_I}{k_{td} + k_{tc}}} \tag{48}$$

$$k_r = A_r e^{-E_r/RT}, \qquad r = p, f_m, I, td, tc \tag{49}$$

Defined polymer average molecular weight (Pm):

$$Pm = D_1/D_0 \tag{50}$$

The percentage monomer conversion (X):

$$X = (C_{min} - C_m)/C_{min} * 100 \tag{51}$$

7.5.3. Model Analysis

The above mathematical model consists of a total of 14 equations out of which 6 are ODEs and 8 are algebraic equations (AEs). The 6 state (dependent) variables for the 6 ODEs are: C_m, C_I, T, D_0, D_1, T_j. The 8 algebraic (unknown) variables for the 8 AEs are: k, P_0, Pm, X (note that k is a vector of 5 variables). The rest of the variables must be specified. Table 22 lists the variables that need to be specified together with their specified values. Table 23 shows the incidence matrix, which is not ordered in a lower tridiagonal form but it can easily be seen that the algebraic equations have a lower tridiagonal form.

Based on the incidence matrix given in Table 23, the calculation procedure is as follows: at any time, t, the values of the state (dependent) variables are known

TABLE 22 List of Specified Variables and their Given Values

$F = 1.0$ m^3/h	$M_m = 100.12$ kg/kgmol
$F^I = 0.0032$ m^3/h	$f^* = 0.58$
$F_{CW} = 0.1588$ m^3/h	$R = 8.314$ kJ/kgmol K)
$C_{min} = 6.4678$ kgmol/m^3	$-\Delta H = 57800$ kJ/kgmol
$C_{Iin} = 8.0$ kgmol/m^3	$E_p = 1.8283 \times 10^4$ kJ/kgmol
$T_{in} = 350$ K	$E_I = 1.2877 \times 10^5$ kJ/kgmol
$T_{W0} = 293.2$ K	$E_{fm} = 7.4478 \times 10^4$ kJ/kgmol
$U = 720$ kJ/(h·k·m^2)	$E_{tc} = 2.9442 \times 10^3$ kJ/kgmol
$A = 2.0$ m^2	$E_{td} = 2.9442 \times 10^3$ kJ/kgmol
$V = 0.1$ m^3	$A_p = 1.77 \times 10^9$ m^3/(kgmo·h)
$V_0 = 0.02$ m^3	$A_I = 3.792 \times 10^{18}$ 1/h
$\rho = 866$ kg/m^3	$A_{fm} = 1.0067 \times 10^{15}$ m^3/(kgmo·h)
$\rho_W = 1000$ kg/m^3	$A_{tc} = 3.8283 \times 10^{10}$ m^3/(kgmo·h)
$C_p = 2.0$ kJ/(kg·K)	$A_{td} = 3.1457 \times 10^{11}$ m^3/(kgmo·h)
$C_{pW} = 4.2$ kJ/(kg·K)	

TABLE 23 Incidence Matrix (note that Eq 49 is actually five equations for the vector k). The shaded Cells Indicate the ODEs and their Corresponding Dependent Variables

Eq\Variable	k	P_0	C_m	C_I	D_0	D_I	T	T_j	Pm	X
Eq. 49	*						*			
Eq. 48	*	*		*						
Eq. 42	*	*	*							
Eq. 43	*			*						
Eq. 44	*	*	*		*					
Eq. 45	*	*	*			*				
Eq. 46	*	*	*				*	*		
Eq. 47							*	*		
Eq. 50					*	*			*	
Eq. 51			*							*

and so the AEs (Eq. 49 and then 48) are solved. Using these variables and the known variables, the right-hand sides (RHSs) of ODES (Eqs. 42–47) are determined, and based on these, new values of the dependent variables are obtained. Equations 50-51 are used to calculate the polymer average molecular weight and the percentage conversion at any time.

7.5.4. Numerical Solution

The model equations have been solved through ICAS-MoT (Sales-Cruz, 2006). The calculated steady-state values for the state variables are given in Table 24. It can be noted that this reactor has multiple steady states, which can be obtained

TABLE 24 Calculated Steady-State Values for the MMA Polymerisation Reactor

	State 1	State 2	State 3
C_m, kgmol/m^3	5.9651	5.8897	2.3636
C_I, kgmol/m^3	0.0249	0.0247	1.7661×10^{-04}
T, K	351.41	353.40	436.20
D_0, kgmol/m^3	0.0020	0.0025	0.4213
D_1, kgmol/m^3	50.329	57.881	410.91
T_j, K	332.99	334.34	390.93
X, %	7.8	8.9	63.5
PM, kg/kgmol	25000	23000	975

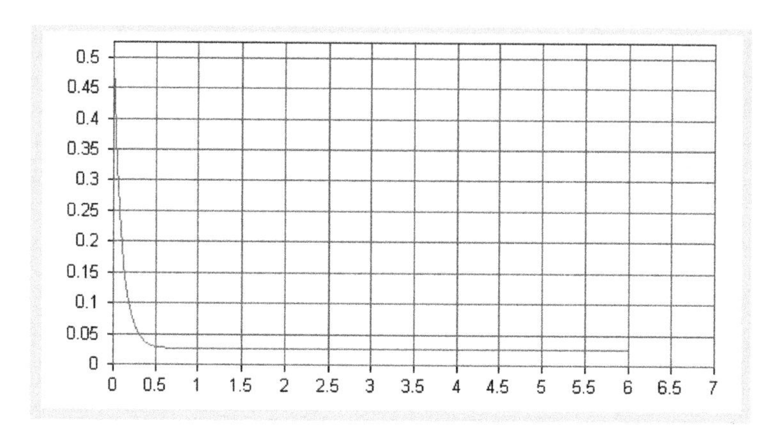

FIGURE 28a **Transient response of C_I (y-axis) versus time (x-axis).**

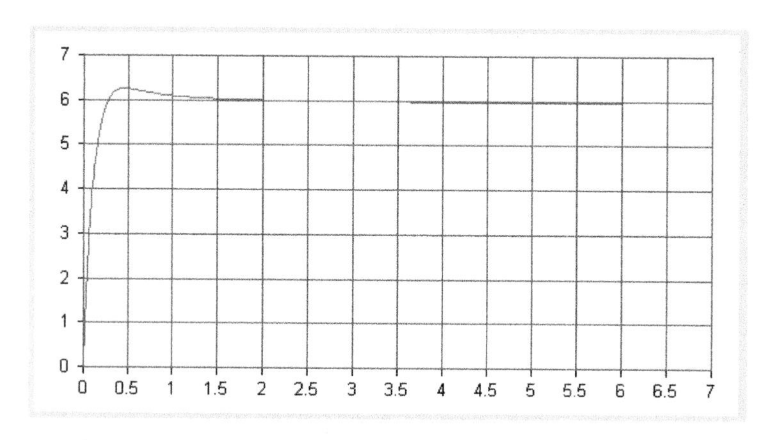

FIGURE 28b **Transient response of C_m (y-axis) versus time (x-axis).**

by starting the dynamic simulation runs from different initial conditions but keeping all the known variables at the listed values of Table 22. This is an example of output multiplicity. The transient responses for the 6 state variables are shown in Figures 28a-f. These results are obtained for the following initial conditions of the state variables: $C_m = 0$; $C_I = 0.5$;$T = 200$; $D_O = 0$; $D_I = 0$; $T_j = 0$. This corresponds to steady state 1 in Table 24.

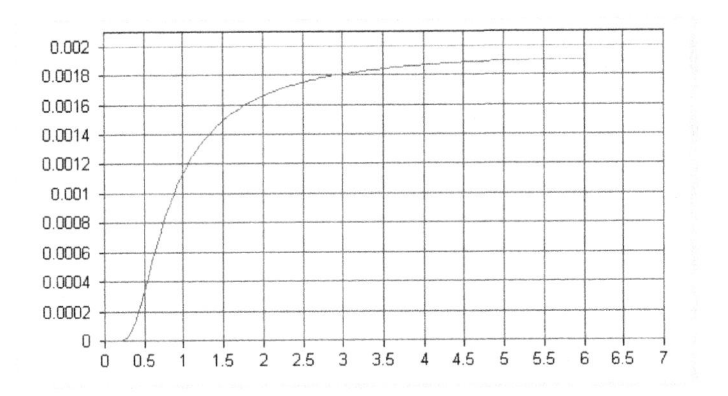

FIGURE 28c **Transient response of D_0 (y-axis) versus time (x-axis).**

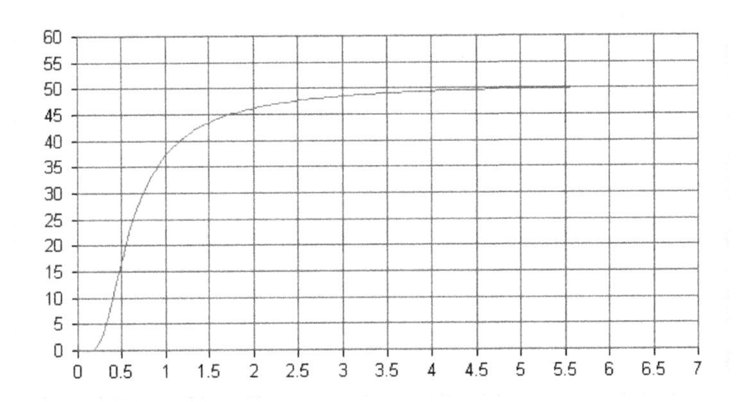

FIGURE 28d **Transient response of D_1 (y-axis) versus time (x-axis).**

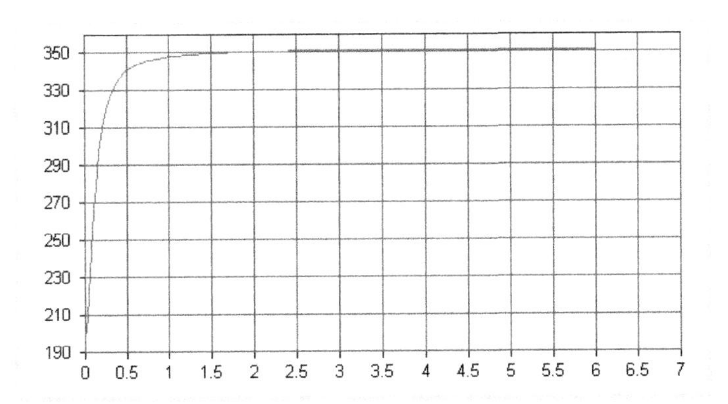

FIGURE 28e **Transient response of T (y-axis) versus time (x-axis).**

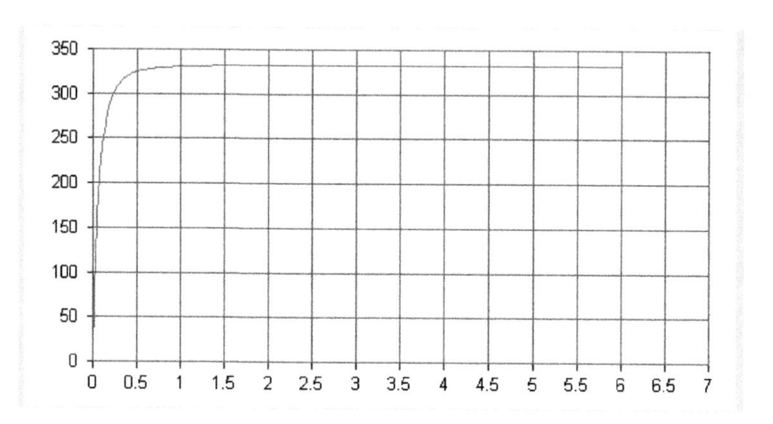

FIGURE 28f Transient response of T_j (y-axis) versus time (x-axis).

The model equations as implemented in ICAS-MoT are given in Appendix 1 (see chapter-7-5-poly-reactor.mot file).

Discussion: Try to obtain all the three steady states and evaluate their stability through calculating the eigenvalues. Add the gel-effect to evaluate the changes in the transient responses. A more detailed polymerisation reactor model is given by López-Arenas et al. (2006) try to implement this model.

REFERENCES

Cooper, H., 2007. Fuel cells, the hydrogen economy and you. Chemical Engineering Progress., 34–43.

Fogler, S. H., 2005. Elements of Chemical Reaction Engineering, 4th Edition, Prentice Hall, USA.

Ingram, C.D., Cameron, I.T., Hangos, K.M., 2004. Chemical Engineering Science 59, 2171–2187.

López-Arenas, T., Sales-Cruz, M., Gani, R., 2006. Chemical Engineering Research and Design 84, 911–931.

Luss, D. and Amundson, N. R., 1968, Stability of Batch Catalytic Fluidized Beds, AIChE Journal, 14 (2), 211–221.

Morales-Rodriguez, R., 2009. Computer-Aided Multi-scale Modelling for Chemical Product-Process Design, PhD-thesis. Lyngby, Denmark: Technical University of Denmark.

Sales-Cruz, M., 2006. Development of a computer aided modelling system for bio and chemical process and product design, PhD-thesis. Lyngby, Denmark: Technical University of Denmark.

Sundmacher, K., Schultz, T., Zhou, S., Scott, K., Ginkel, M., Gilles, G., 2001. Chemical Engineering Science 56, 333–341.

Xu, C., Follmann, P.M., Biegler, L.T., Jhon, M.S., 2005. Computers and Chemical Engineering 29, 1849–1860.

Chapter 8

Distributed Parameter Modelling Applications*

Many processing operations and product descriptions lead to the use of distributed system modelling due to the importance of the spatial variations in system states. This chapter shows some industrial applications of modelling where distributed concepts are vital for gaining insights into the design and operations of these systems.

In this chapter we consider the modelling of a retorting device for the processing of oil shale. This was part of a demonstration plant to test out the performance of such devices. The second example is drawn from the fertiliser industry and involves the modelling of a commercial granulator for producing industrial-grade phosphate fertilisers. A final case study is the steady state behaviour of a short-path evaporator often used in producing concentrates of products such as juices. All these applications have important spatial variations.

8.1. OIL SHALE PRE-HEAT AND COOLING UNIT

8.1.1. Context of the modelling

Shale oil has a long history as an alternate source of hydrocarbon-based fuels. The plants have operated in many countries, either as demonstration plants or for commercial production. Countries such as China, Estonia and Brazil currently produce shale oil. Several demonstration plants operated in the USA, mainly in Colorado. In Australia, oil shale was processed in the early 1840s to produce 'kerosene' from the kerogen content of the shale. In the 1930s shale oil was produced in crude, batch retorts which pyrolised shale that had been suitably mined and crushed. It was an expensive and inefficient process but provided alternative fuels in a time when normal petroleum fuels were in short supply. Much research and development work was done around the work during the 1970s to 1990s to enhance processing and recoverability of products as well as understand the potential environmental impacts.

Processing of oil shale into hydrocarbon products involves a combination of mining technologies, minerals processing and conventional hydrocarbon

*Written in conjunction with Prof Mauricio Sales-Cruz

processing. One of the most important processes is the oil shale preparation and retorting. The preheat-cooler section of a continuous Alberta-Taciuk Processor (ATP) is the focus of this modelling application.

8.1.2. System description

Large-diameter rotating vessels with complex internals can be used to process shale for the production of fuel products. The internal design of the preheat section where the combusted shale is used to transfer heat to the fresh, crushed and dried shale feed is an important process. Figure 1 shows a vessel schematic.

The processor in this case has a large outside drum and a smaller internal drum. The internal drum has a series of tubes or porcupine sections. The inner and outer drums rotate in a synchronised manner. Fresh shale is fed into the inner drum and falls in and out of the heat transfer tubes as it rotates.

On the outer drum, flights lift the bed solids up and cascade them down over the outside surfaces of the tubes. Simultaneously, the hot combustion gas enters around the retort section and carries solids through the section. These gases carry fines out to the stack via cyclones and scrubbers, whilst the larger particles in the bed leave to be cooled, moistened and returned to the mine site.

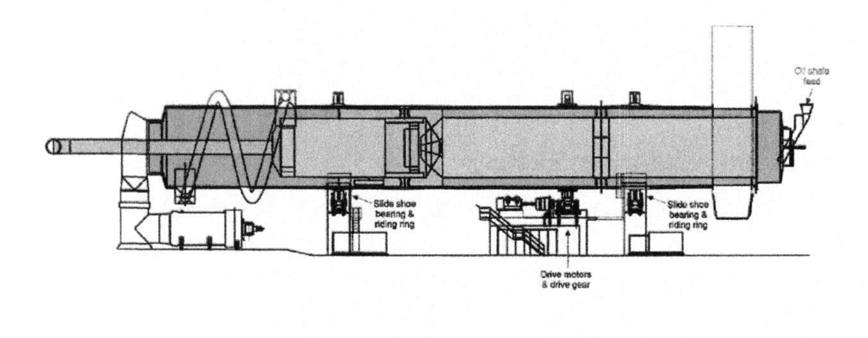

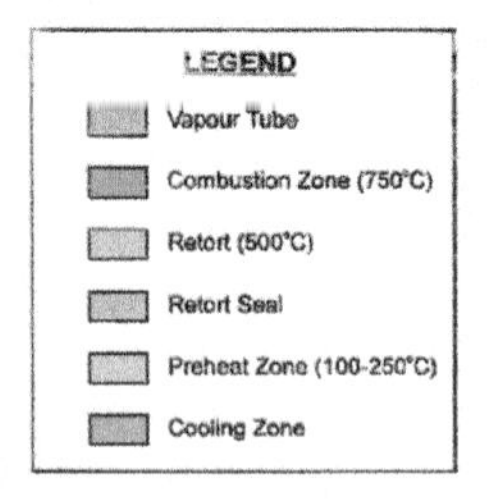

FIGURE 1 **Schematic diagram of an oil shale processor (Courtesy SPPD).**

8.1.3. Purpose of the model

For the purpose of obtaining construction estimates, the required length of the pre-heat-cooling zone is very important given the cost per metre of such a large vessel. The model is required to give designers an indication of the effect of various geometric and operational parameters on the heat transfer capacities under various input disturbances on the performance of the system.

8.1.4. Conceptualisation of the model and balance volume development

In this section the conceptualisation for modelling is discussed and the resultant balance volumes for model development are defined.

8.1.4.1 Flow zones

To enable the modelling tasks to be tractable, a number of simplifications were taken with the system in terms of the major flow zones. Four principal flow zones have been defined based on the cross-section of the ATP and a consideration of the system physics. Figure 2 illustrates the zones in cross-section. These are:

Zone A: Flight zone

Here hot gases can flow in the region between the outer wall and the tip of the flights. Gases through this zone have little resistance to flow and exchange

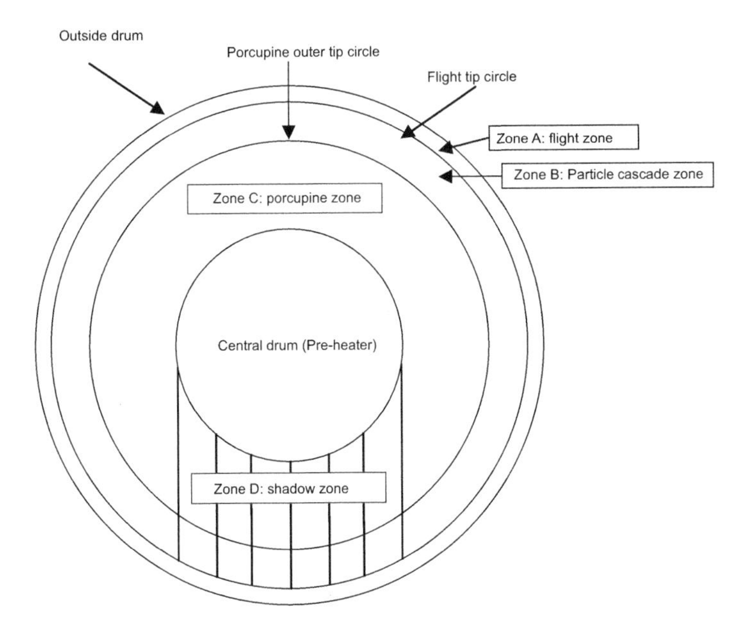

FIGURE 2 **Flow zone definitions for pre-heat/cooler.**

energy with the environment through the outer wall of the ATP section. This can be considered analogous to the flow of gas within a pipe of certain surface roughness. The roughness is determined by the solids surface, flight surface and gusset plates.

Zone B: Particle cascade zone

This zone is defined by the annular region between the flight tips and the outer extent of the porcupine. It is a region where hot gases interact with solids cascading from the rotating flights. Pressure drop of gas in this zone depends on particle curtain geometry a difficult problem to deal with.

Zone C: Heat transfer tubes zone

This is the region where gas–solid mixtures interact with the heat transfer surfaces of the heat transfer tubes. Here solids cascade into this zone from Zone B and act as a solid suspension with the hot gases which enter from the combustion zone. Pressure drop is a complex interaction of solids suspension density and tube geometry.

Zone D: Shadow zone

This is the zone which is directly beneath the internal shale preheat drum. Here, it is assumed that few solids exist, and that hot gases pass through the tubes exchanging heat in the absence of solids. This is analogous to hot gas flows through staggered tube banks. Significant work is available on both pressure drop predictions and estimation of external tube surface heat transfer coefficients.

8.1.4.2 Solids bed zone

This is the region at the bottom of the processor in which the dense phase solids reside and are transported by tumbling motion.

8.1.4.3 Solids–gas interaction

A key issue is the interaction between gas and solids in the cooling section of the ATP. To enable an initial model to be developed the following issues were considered.

In Zone B, gas and solids interact in each axial segment so that the gas and solids are in thermal equilibrium. This is a good assumption in relation to solids <100–$200\ \mu m$ in diameter becoming less valid as solid particle diameters are greater than $1000\ \mu m$.

In Zone C, again gas and solids are assumed to reach thermal equilibrium as well as exchange heat with the incoming shale via the heat transfer tubes.

The transport of solids in the device is a key controlling factor for performance. In this model the following issues were considered:

- Dense bed flow occurs through the effect of the rotation and dynamic angle of repose as set out by Das Gupta et al. (1991).
- Solids uptake in the air phase is extremely complex due to the complex internals within the ATP. At this stage of model development it is assumed that a fraction of a size range cascading from the flights ends up in the gas stream. This can be fixed for each of the 10 size ranges by the user. Independent models suggest that particles >500 μm do not travel axially any significant distance but essential drop back to the dense phase. The presence of the heat transfer tubes in Zone C adds further complications to the solid transport problem.
- Pressure drop induced by gas–solid flows in Zones B and C have initially been considered equivalent to pneumatic conveying of solids as discussed by Zenz and Othmer (1960). This is an area of significant uncertainty in the model as very little work has been published on gas pressure drop through the falling curtains of solids. Some work by Hirosue and Shinohara (1978) is relevant to this issue; however, the sub-model requires pilot plant data to estimate adjustable parameters.

8.1.5. Model development

The following sections outline the development of the model, giving details of assumed balance volumes and the main convective and diffusive flows in the system. The model was developed on conceptual grounds with the intent of looking at relative effects on heat transfer due to changes in geometric design parameters as well as certain heat transfer coefficients.

8.1.5.1 Solids transport model

The mass balance through the cooling section outer drum has been considered and the following conceptualisation used in developing the principal balances. Figure 3 shows the overall concepts. Key assumptions are:

A_1: Dense solids phase includes bed solids plus solids in the flights.
A_2: Airborne solids are those carried by the airflow through the drum and subsequently discharged to cyclones.
A_3: No solids are assumed to be in the inlet gas from the combustion zone, although it is possible to define such flows.
A_4: All solids enter the ATP-cooling zone at the base of the drum.
A_5: It is assumed that the inlet particle distribution can be adequately represented by 10 size fractions, each containing 5 wt % solids.
A_6: It is assumed that air inlet conditions (P, T) are known except for the flow rate.

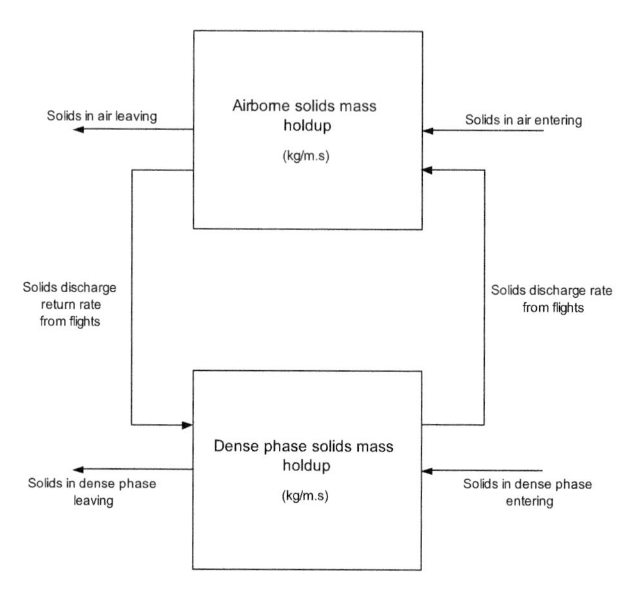

FIGURE 3 **Conceptual solids holdups for cooling zone.**

A_7: It is assumed that solid flow rate to the cooling zone is known as is the temperature from the combustion zone.
A_8: No leakage air into the cooling zone is considered in the model, although it could be introduced if necessary.
A_9: It is assumed that no attrition or breakage of solids occurs in the drum.
A_{10}: It is assumed that overall pressure drop can be defined for the cooling zone.

Figure 4 shows the representative balance volume that has been used to develop the underlying mass conservation equations. The key aspects of this are:

A_{11}: Solids move through the drum under the action of the air flow and through direct dense phase motion in the base of the drum.
A_{12}: It is assumed that the flights are designed as equi angular discharge (E.A.D.) flights.
A_{13}: Solids cascading from the flights fall into Zone B of the ATP. Some solids are separated into the gas flow with the rest returning to the dense bed via Zone C.
A_{14}: Solids leaving Zone B are assumed to be in thermal equilibrium with the gas leaving the zone segment.
A_{15}: Solids leaving Zone C are assumed to be in thermal equilibrium with the gas leaving that zone segment.

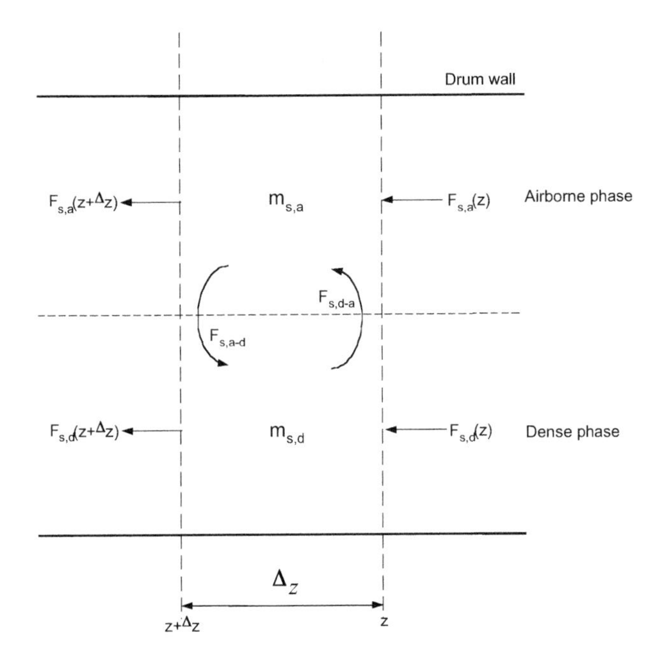

FIGURE 4 Conceptual balance volumes for cooling zone.

8.1.5.2 Solid transport mass balances

The overall mass balances (units: kg.m^{-1}.s^{-1}) for each phase based on Figure 4 schematic are given by:

Dense phase region:

$$\frac{\partial m_{s,d}}{\partial t} = -\frac{\partial F_{s,d}}{\partial z} - F_{net} \tag{1}$$

where:

$m_{s,d}$ = dense phase solid mass (kg.m^{-1})
$F_{s,d(z)}$ = flow of dense phase solids (kg.s^{-1})
$F_{s,d-a(z)}$ = flow of solids from dense phase to air-borne phase (kg.m^{-1}.s^{-1})
$F_{s,a-d(z)}$ = flow of solids from air-borne phase back to dense phase (kg.m^{-1}.s^{-1})
F_{net} = difference between $F_{s,d-a}$ and $F_{s,a-d}$

Airborne region

$$\frac{\partial m_{s,a}}{\partial t} = -\frac{\partial F_{s,a}}{\partial z} + F_{net} \tag{2}$$

with

$$F_{net} = F_{s,d-a} - F_{s,a-d}$$

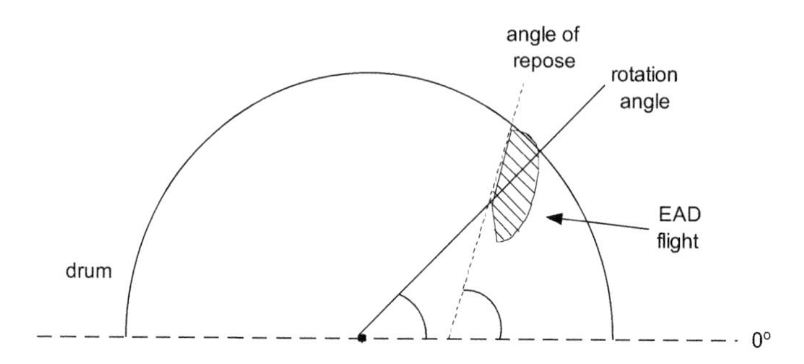

FIGURE 5 Flight geometry and angle.

We neglect the holdup $m_{s,a}$ in the airborne region because of the small amount compared with the dense bed. The term F_{net} incorporates the difference between what is cascaded from the flights at position z and what is deposited into the dense phase bed from further upstream. The two terms which determine the overall flow of solids through the system need careful definition. These two terms are now discussed:

Source term $F_{s,d\text{-}a}$:

$F_{s,d\text{-}a}$ represents the flow (kg.m^{-1}.s^{-1}) from flights containing dense phase material into the airborne phase. This is determined by the flight geometry, rotation rate, solids properties and drum loading. The general expression for the rate of solids discharge R_{Di} (kg.m^{-1}.s^{-1}) from a flight is given by (Wang et al. (1995)):

$$R_{D_i} = -\rho_b \omega \frac{dS_i}{d\theta_i} \tag{3}$$

where:

S_i = cross-sectional area of material in the flight (m^2)
θ_i = angular rotational position of the flight from the horizontal (rads)
ω = angular velocity of the flight (rads/s)
ρ_b = bulk density of the solids (kg/m^3)

The general geometry is shown in Figure 5.

Now if the area S is a function of θ and dynamic angle of repose $\varphi(\theta)$, we can write:

$$R_{D_i} = -\rho_b \omega \frac{dS_i}{d\theta_i} = -\rho_b \omega \left[\frac{\partial S_i}{\partial \theta_i} + \frac{\partial S_i}{\partial \varphi_i} \frac{\partial \varphi_i}{\partial \theta_i} \right] \tag{4}$$

It is useful to check the variation of the dynamic angle of repose against the rotation angle, since if this is close to zero, then the last combination term in the above equation can be eliminated.

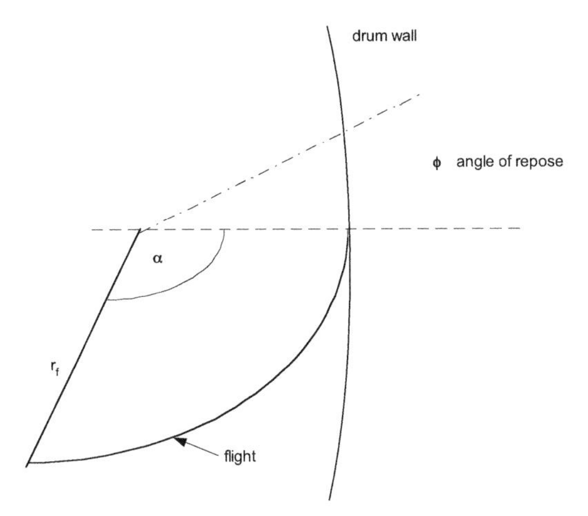

FIGURE 6 Flight design geometry.

It is well established (Schofield and Glikin (1962)) that the dynamic angle of repose is a function of the coefficient of static friction μ rotation rate ω and the radius of the drum R. This is given by:

$$\varphi = \tan^{-1}\left[\frac{\mu - \xi(\mu \sin\theta - \cos\theta)}{1 - \xi(\mu \cos\theta + \sin\theta)}\right] \tag{5}$$

with

$$\xi = \frac{\omega^2 R}{g} \tag{6}$$

To test this, the variation was programmed and computations showed that the variation in φ with θ is indeed close to zero and hence can lead to significant simplification of the expression describing the discharge rate off the flights.

Assuming that the flight geometry is such that it is an equi-angular distribution (EAD) design, we can write that:

$$S(\theta) = S^*\left(1 - \frac{\theta}{\pi}\right) \tag{7}$$

where S* is the design loaded flight area where solids only begin to discharge at a flight rotation angle of zero which is defined as the horizontal point. Figure 6 shows the assumed geometry of this type of flight. This is given by:

$$S^* = r_f^2(\alpha + \varphi)/2 \tag{8}$$

where the angles are given in radians and are shown in Figure 6. The flight radius is r_{f}. This then leads to a discharge rate ($\mathrm{kg.m^{-1}.s^{-1}}$) from a flight i of:

$$R_{D_i} = \rho_b \omega r_{\mathrm{f}}^2 \frac{1}{2\pi}(2.3562 + \varphi) \tag{9}$$

and total discharge rate from n_{f} flights of:

$$F_{\mathrm{s,d-a}} = R_D = \sum_{i=1}^{n_f/2} R_{D_i} = \frac{n_f}{2}\left[\rho_b \omega r_f \frac{1}{2\pi}(2.3562 + \varphi)\right], \tag{10}$$

assuming only half of the flights can discharge (between 0 and π rads). This approach still needs to consider several issues including the need to account for situations where the drum is underloaded as well as overloaded. This will affect the initial discharge point of solids on the flight. In turn it will affect the amount of solids in the gas phase and its redistribution back to the dense phase.

Source term $F_{\mathrm{s,a\text{-}d}}$:

This term represents the redistribution of solids in the air back into the dense phase or remaining in the air phase, to be transported out of the ATP. This is a complex problem due to the following issues:

I_1: The interaction of the solids in the gas flow, especially the effective gas velocity applied to the particles compared with the average gas velocity in the zone.
I_2: The solid curtain voidage, which is extremely hard to estimate.
I_3: The particle size distribution (PSD) effect in creating a distribution of particle trajectories.
I_4: The role played by various zones inside the cooler.
I_5: The probability of collision with the porcupine tubes and the probability that solids will be 'knocked out' of the gas stream and return to the dense bed.
I_6: The voidage of the porcupine tubes, thus changing radial gas velocities.

Some key issues have been considered for this source term:

Particle velocity and travel

We have considered the trajectory of an unhindered particle after discharge from a flight. It has essentially an axial component (z direction) and a vertical component (y direction). The tangential direction perpendicular to the drum axis has been ignored.

Horizontal motion:

A momentum balance on a particle of diameter D_{p} at velocity u_{z} in a gas stream of velocity u_{g} can be performed to give:

$$\frac{dM_{\mathrm{p,z}}}{dt} = \left(\frac{\pi D_{\mathrm{p}}^3}{6}\rho_{\mathrm{p}}\right)\frac{du_{\mathrm{z}}}{dt} = C_{\mathrm{D}}\pi D_{\mathrm{p}}^2 \rho_{\mathrm{g}}\left(\frac{(u_{\mathrm{g}} - u_{\mathrm{z}})^2}{8}\right) \tag{11}$$

where:

M_p = momentum for particle of diameter, D_p (kg m/s)
C_D = drag coefficient (–)
u_g = gas velocity (m/s)
ρ_g = gas density (kg/m^3)

Integrating this equation gives the velocity at any position z from the release point as well as the distance travelled. These are:

$$u_z(t) = \frac{Ku_g^2 t}{(Ku_g t + 1)} \tag{12}$$

$$z(t) = u_g t - \frac{1}{K}\ln(Ku_g t + 1) \tag{13}$$

$$K = \frac{3}{4}\left(\frac{C_D}{D_p}\right)\left(\frac{\rho_g}{\rho_p}\right) \tag{14}$$

with C_D as a function of Reynolds number,

$$C_D = f(Re) \tag{15}$$

$$Re = \frac{D_p \rho_g (u_g - u_z)}{\mu_g} \tag{16}$$

where:

μ_g = gas viscosity (Pa.s)
ρ_g = particle density (kg/m^3)

The drag coefficient C_D is defined for a spherical particle as:

$$C_D = \frac{24}{\text{Re}} \qquad \text{Re} < 0.1 \tag{17}$$

$$C_D = \frac{24}{\text{Re}}\left(1 + 0.14\,\text{Re}^{0.7}\right) \qquad 0.1 < \text{Re} < 1000 \tag{18}$$

$$C_D = 0.445 \qquad 1000 < \text{Re} < 350{,}000 \tag{19}$$

Vertical motion:

For the vertical motion of a particle falling through a gas stream the momentum balance is given by:

$$\frac{dM_{p,y}}{dt} = m_p g - C_D\left(\frac{\pi D_p^2}{4}\right)\left(\rho_g \frac{(u_g - u_z)^2}{2}\right) \tag{20}$$

where:

$g = 9.81$ m/s^2.

Integrating this equation gives velocity and distance travelled at a specific time after release. These are given by:

$$u_y(t) = \sqrt{\frac{g}{K}}\frac{(e^{2\sqrt{g}t} - 1)}{(1 + e^{2\sqrt{g}t})} \tag{21}$$

$$y(t) = \sqrt{\frac{g}{K}}\left[\frac{1}{\sqrt{g}}\ln\left(\frac{1 + e^{2\sqrt{g}t}}{2}\right) - t\right] \tag{22}$$

These allow prediction of the vertical fall distance, and when combined with the horizontal component, it aids in deciding where a particle of average diameter for a given size range might fall.

However due to many unknowns, especially with regard to the behaviour of solids within the porcupine region a simple fractional split on the size ranges has been adopted. This is used in the next section.

8.1.5.3 Mass balances in Zones B and C

In this zone, we assume a pseudo-steady state as the airborne holdup is very small compared with the dense phase. From Figure 4 we obtain the steady-state mass balance in the airborne phase as:

$$F_{\mathrm{s,a}}(z + \Delta z) = F_{\mathrm{s,a}}(z) + \sum_k F_{\mathrm{s,d-a}}(k) frac(k).\Delta z \tag{23}$$

where:

$frac(k)$ = fraction of size range k which goes to the gas phase
$F_{\mathrm{s,d-a}}(k)$ = flow of solids in size range k from flights (kg/m.s)

8.1.5.4 Mass balance on shale and moisture

Dynamics of the fresh (wet) shale feed to the preheater were modelled, based on the balance volume diagram shown in Figure 7.

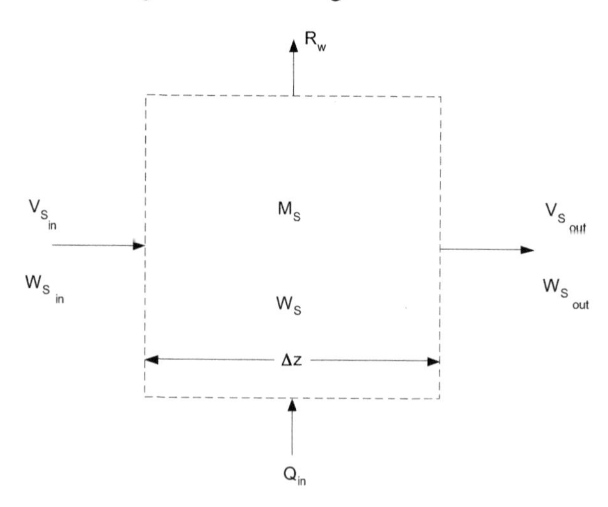

FIGURE 7 Balance Volume diagram for Fresh Shale.

where:

V_S = flow of solids (kg/s)
W_S = moisture of solids (kgH_2O/kg solids)
R_W = drying rate (kgH_2O/s.m)
Q_{in} = energy transfer from spent shale (kW/m)
M_S = mass of shale in segment (kg/m)

The total mass balance is:

$$\frac{\partial}{\partial t}(M_S(1+W_S)) = -\frac{\partial}{\partial z}(V_S(1+W_S)) - R_W \tag{24}$$

The balance on moisture is given by:

$$\frac{\partial}{\partial t}(M_S W_S) = -V_S.\frac{\partial W_S}{\partial z} - R_W \tag{25}$$

since V_S is constant for the shale side.

8.1.5.5 *Energy balances*

Dense phase energy balance

The energy balance for the dense phase Δz can be developed using Figure 4 to give:

$$\frac{\partial E}{\partial t} = -\frac{\partial}{\partial z}(F_{s,d}.c_{ps}.T_s) - F_{s,d-a}.T_s.c_{ps} + F_{s,a-d}.T_{s_{Co}}.c_{ps} \tag{26}$$

where:

C_{ps} = heat capacity of the solids (kJ/kgK)
E = total energy holdup (kJ/m)
Q = heat loss to wall (kW/m)
$T_{S_{Co}}$ = solids return temperature from Zone C
T_S = solid temperature

In the above it is assumed that the specific energy of the solid streams can be approximated by C_pT.

Zone A energy balance

The energy balance in Zone A considers the change of gas temperature in the flight zone as gas exchanges heat through wall losses, which is given conceptually in Figure 8:

The energy balance gives:

$$\frac{\partial E_A}{\partial t} = -\frac{\partial}{\partial z}(T_{gA}.c_{pg}(u\rho A)) - Q_{gw} \tag{27}$$

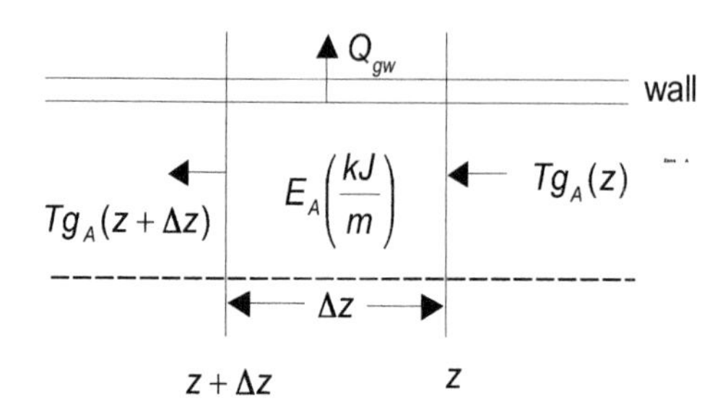

FIGURE 8 **Balance volume diagram for Zone A.**

where:

$u\rho A$ = mass flow rate of gas at the position z (kg/s)
T_{gA} = gas temperature (K)
C_{pg} = gas heat capacity (kJ/kgK)
Q_{gw} = heat loss to wall (kW/m)

We can neglect the holdup of energy in the gas given by $E_A(z, t)$ to get the pseudo-steady-state model

$$0 = -\frac{\partial}{\partial z}(T_{gA}.c_{pg}(u\rho A)) - Q \quad (28)$$

Zone B energy balance

Here the gas and solids interact. Conceptually we have the situation in Figure 9

Here it is assumed that the solids mix intimately with the gas phase and reach a thermal equilibrium in the segment. All solids from the flight Zone A pass into this region. The balance gives:

$$\frac{\partial E_B}{\partial t} = \frac{\partial}{\partial z}(T_{gB}.c_{pg}.(u\rho A)) + F_{s,d-a}(i)c_{ps}T_s(z)_i - F_{s,d-a}(0).c_{ps}T_{sB_0}(z)_0 \quad (29)$$

If we impose thermal equilibrium on the exit flows and assume a pseudo-steady state for the segment we obtain:

$$0 = -\frac{\partial}{\partial z}(T_{gB}.c_{pg}.(u\rho A)) + F_{s,d-a}(i).c_{ps}.T_s(z)_i - F_{s,d-a}(0).c_{ps}T_{gB} \quad (30)$$

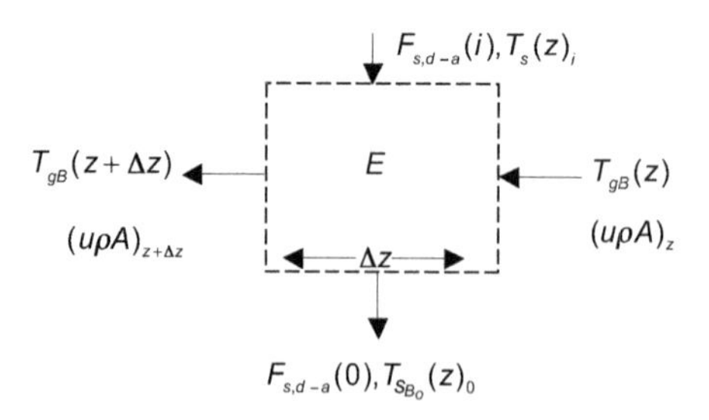

FIGURE 9 **Balance volume diagram for Zone B.**

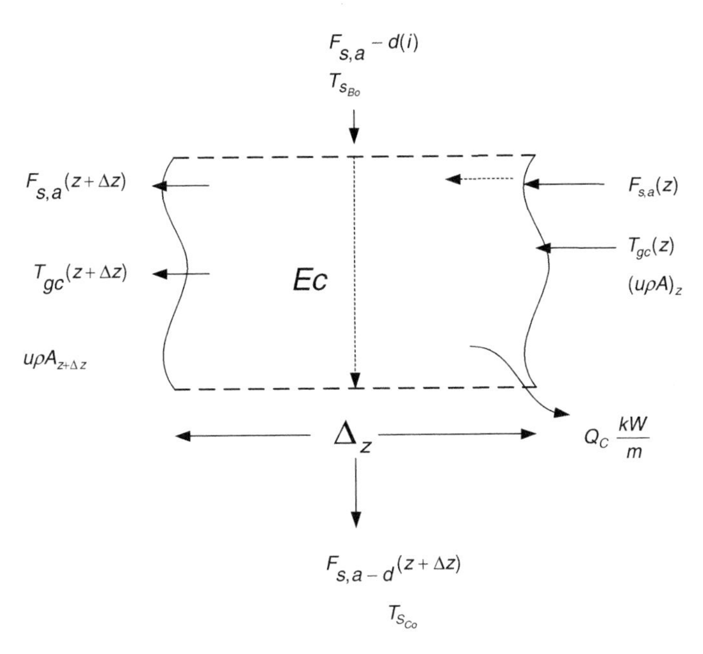

FIGURE 10 **Balance volume diagram for Zone C.**

where:

C_{pg} = heat capacity of gas (kJ/kgK)
C_{ps} = heat capacity of solids (kJ/kgK)
$T_s(z)_i$ = solids temperature entering Zone B (k)
$T_{S_{Bo}}$ = solids temperature leaving Zone B (k)
$u\rho A$= mass flow of gas (kg/s)

Zone C energy balance

The Zone C energy balance includes the heat transfer from the gas–solids flow to the porcupine. Conceptually we have in Figure 10

Again it is assumed that thermal equilibrium is reached between gas and solids flows. The energy balance then becomes:

$$\begin{aligned} 0 \;=\; & -\frac{\partial}{\partial z}(F_{s,a}.c_{ps}T_g) - \frac{\partial}{\partial z}(u\rho A).c_{pg}T_g) \\ & + \; F_{s,a-d}(i).c_{ps}T_{s\text{Bo}} - F_{s,a-d}(0).c_{ps}.T_{s\text{Co}} \\ & - \; Q \end{aligned} \tag{31}$$

where:

$$T_g = T_{s\text{Co}}$$

assuming thermal equilibrium is reached between solids and gas. Here,

E_c = energy content of gas (kJ/m)

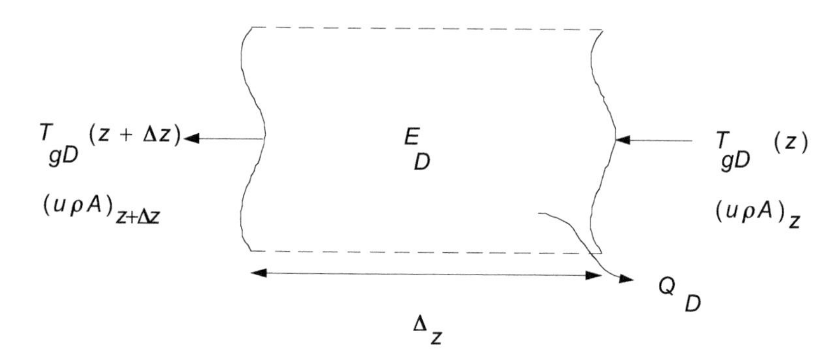

FIGURE 11 **Balance volume diagram for Zone D.**

Q_c = energy transferred to porcupine (kW/m)
$u\rho A$ = mass flow of gas (kg/s)

In this section the solids stream from the flight zone is split, depending on the size range split fraction. In reality some of the solids flow is attributed to Zone B with the rest passing through Zone C. Only large particles < 200 μm return directly to the bed.

Zone D energy balance

In Zone D, which represents the shadow zone beneath the preheater tube, gas from the combustion zone flows through the staggered porcupine tubes, similar to flow across a tube tank. Conceptually the balance volume is given in Figure 11.

The energy balance for this segment is given by:

$$\frac{\partial E_D}{\partial t} = -\frac{\partial}{\partial z}\left((u\rho A).C_{pg}T_{gD}\right) - Q_D \tag{32}$$

and assuming a pseudo-steady-state condition:

$$0 = -\frac{\partial}{\partial z}\left((u\rho A).C_{pg}T_{gD}\right) - Q_D \tag{33}$$

where:

Q_D = energy transfer to heating tubes (kW/m)

Energy Balance on Fresh Shale

The energy balance on the shale side is given by:

$$\frac{\partial}{\partial t}(M_S(\hat{H}_S + W_S\hat{H}_W)) = -\frac{\partial}{\partial z}(V_S(\hat{H}_S + W_S\hat{H}_W)) - \lambda R_W + Q_{in} \tag{34}$$

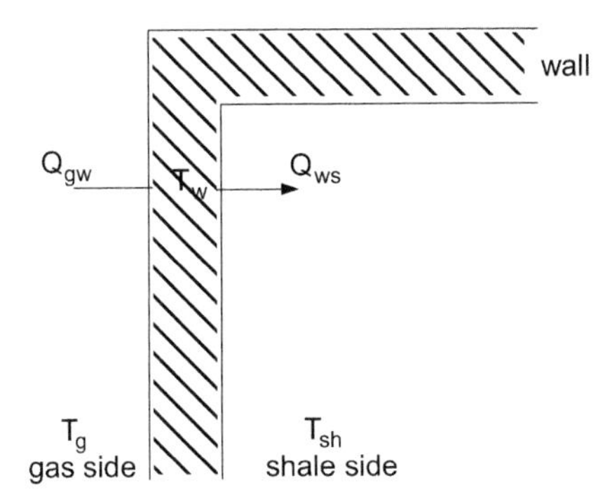

FIGURE 12 Heat transfer in heating tubes.

with

$\hat{H}_S$ = specific enthalpy of solid (kJ/kg)
$\hat{H}_W$ = specific enthalpy of moisture (kJ/kg)
λ = heat of vaporisation of water (kJ/kg)

The specific enthalpies can be approximated by:

$$\hat{H}_S = C_{p_{sh}} T_S \qquad \hat{H}_W = C_{p_W} T_S$$

where C_p is the respective heat capacity of wet shale and water.

Heat tube wall energy balance

It was necessary to consider the energy balance over the porcupine due to differences in heat transfer coefficients on the inside and outside of the porcupine.

The conceptual heat transfer issues are seen in Figure 12.

There are three principal temperatures:

T_g = outside gas–solid mixture temperature
T_w = wall temperature of the heating tubes
T_{sh} = shale temperature on pre-heat side

The overall, lumped parameter energy balance is:

$$\frac{dE_w}{dt} = Q_{gw} - Q_{ws} \tag{35}$$

where:

Q_{gw} = energy from gas to wall (kW/m)
Q_{ws} = energy from wall to shale (kW/m)
E_w = total energy content of wall (kJ) $\equiv M_{porc}.C_{p_{st}}.T_w$
M_{porc} = mass of steel of porcupine (kg/m)
$C_{p_{st}}$ = heat capacity of steel (0.5 kJ/kg.K)

The energy flows are defined by:

$$Q_{gw} = h_o A_o (T_g - T_w) \quad (36)$$

$$Q_{ws} = h_i A_i (T_w - T_{sh}) \quad (37)$$

In the model, there are two principal zones (C, D) which contribute to the overall fluxes Q_{gw}, Q_{ws}. These are considered in the model equations by proportioning the full 360° into 280° and 80°, respectively, for the zonal regions.

Typically the heat transfer coefficients in Zone C are:

$h_o \simeq 100 - 150(W/m^2K)$
$h_i \simeq 60 - 70(W/m^2K)$

These can be varied by the user.
For Zone D, $h_o \simeq 20(W/m^2K)$.

There are, however, several correlations in the literature which attempt to predict heat transfer coefficients in gas–solid flows. Most come from the fluid bed combustion literature. See Murray et al. (1996), Chung and Welty (1989), Wu et al. (1987), Sterritt and Murray (1992), Shah et al. (1981) and Hirosue (1989) as examples.

These show the following facts:

- Heat transfer is enhanced in gas–solid flows compared with gas-only flows by up to an order of magnitude.
- Heat transfer is a function of particle size, with heat transfer increasing for decreasing particle size.
- Heat transfer depends strongly on the gas–solid velocity in the system which provides surface renewal.
- Heat transfer depends on the solid–gas loading ($kg/m^2.s$), generally increasing with increased loading.

The difficulty in using such correlations at the moment is the widespread of heat transfer coefficient predictions and the uncertainties in the gas–solid loading in Zone C of the cooling section. Hence, in the present model heat transfer coefficients can be set by the user. With further plant trials, suitable predictive correlations could be used within the model.

8.1.5.6 Constitutive relations

Besides the mass and energy conservation equations there are a significant number of constitutive equations which help define terms on the right hand side of the conservation balances. These are described in the following sub-sections.

Physical property predictions

Property predictions for gas at various temperatures are needed within the balance equations. Of importance are:

- Thermal conductivity
- Gas density

- Gas viscosity
- Gas heat capacity

The following correlations were developed from data in Perry (2-208/209) for a temperature range of 300–1200 K. The correlations are:

Density:

$$\rho_G = 349.22051T^{-1.0004283}(kg/m^3)$$

Viscosity:

$$\mu_G = -1.0308151 \times 10^{-5} + 1.6669592 \times 10^{-6}T^{1/2}\ (Pa.s)$$

Heat capacity:

$$C_{pg} = \left[0.94796984 + 1.0769646 \times 10^{-5}T^{1.5}\right]^{1/2}\ (kJ/kgK)$$

Thermal conductivity:

$$k_g = -0.0085638959 + 0.00035211183T^{1/2}\ln T\left(\frac{W}{mK}\right)$$

Axial transport of bed solids

The transport of solids in the dense phase was estimated using the work of Das Gupta et al. (1991). This provides an estimate of the axial velocity of the solids as a function of bed slope and angle of repose.

The velocity is given by:

$$\vartheta_{OG} = -\frac{4\pi n\rho_R^2}{3Rf(h)}\left(\frac{dh}{dz}\right)\cot(\beta) \qquad (38)$$

where:

$$f(h) = \left(1 - \frac{(R-h)^2}{R\rho_R}\ln\left(\frac{R+\rho_R}{R-h}\right)\right) \qquad (39)$$

and

n = rotation velocity (s^{-1})
ρ_R = ½ transverse width of bed surface (m)
h = bed height (m)
R = drum radius (m)
β = angle of repose (radians)

In practice the function $f(h)$ has a profound and sensitive effect on axial velocity values and hence on the estimated bed heights. It could be used to help fit experimental data from pilot-scale studies.

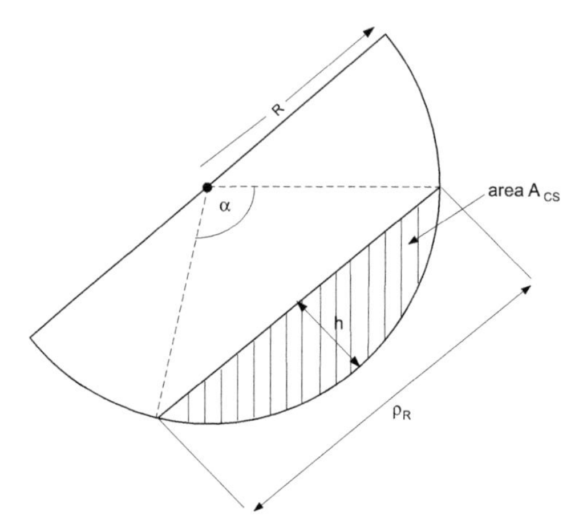

FIGURE 13 Solids bed geometry.

In order to estimate the height of the bed, the total holdup in the bed at any length is used to obtain the cross-sectional area as

$$A_{CS} = \frac{H_s}{\rho_B}(m^2) \tag{40}$$

H_S = solids holdup (kg/m)
ρ_B = solids bulk density (kg/m^3)

A relationship can be developed which relates the area ratio of A_{CS} to the half circle area of the drum, to the chord angle α of the bed as seen in Figure 13.

This relationship is given by:

$$y = a + b.x^c \tag{41}$$

where:

$y = \left(\frac{\alpha}{\pi}\right)$ angle ratio

$x = \left(\frac{A_{CS}}{A_{1/2C}}\right)$ area ratio

$a = 0.013814226$
$b = 0.96386123$
$c = 0.39564869$

from the estimated chord angle (α), chord length (ρ_R) and bed height (h) can be calculated from the system geometry as:

$$\rho_R = 2R\sin\left(\frac{\alpha}{2}\right)$$

$$h = \frac{\rho_R}{2}.\tan\left(\frac{\alpha}{4}\right)$$

This allows the estimate of h at adjacent points in the bed and hence the axial flow rate of solids through the bed.

The factor $f(h)$ is undefined when the bed is empty so it must be set to 1.0 when the model is used for start-up purposes until a bed height is established along the cooling zone. Use of this relation in the model typically accelerates solids flow predictions by a factor of 3 at the inlet and up to 25 at the outlet, compared with a value of $f(h) = 1$. This would be an issue for further investigation at pilot plant trials.

Drying rate of wet shale

The drying rate of the fresh shale feed R_W can be estimated by:

$$R_W = K_W W_S T_S^3 \tag{42}$$

where:

K_W = drying rate constant (kg solid/m.s.C^3)
W_S = moisture (kg H_2O/kg solids)
T_S = temperature of solids

This drying rate expression is a common relation for falling rate drying conditions which exist in the case of the preheater. K_W was chosen as 2.5×10^{-9} from trial and error but could be estimated from plant data.

Pressure drop–flow relations for the zones

The pressure drop–gas flow relations are complex due to the presence of solids in Zones B and C. In Zones A and D it has been assumed that no solids are present in the streams. As such the pressure drop relations are relatively straightforward.

In the model the overall pressure is set for the device and gas flows are then induced by the pressure drop through calculation of the relative flow resistances. In reality it could be anticipated that any distance along the cooler the pressure profile is not uniform across the diameter. Computational fluid dynamics (CFD) could be used to address this issue. Each zone is now addressed.

Zone A pressure drop

In this zone a linear pressure drop with distance is imposed and the properties of the gas and the velocity then vary along the length of the flight zone. This zone is considered to be equivalent to a duct through which the gas flows. The equivalent diameter is based on the wetted perimeter and flow area. The standard pressure drop relation gives

$$\Delta P_f = 4\phi\left(\frac{L}{De}\right)\rho_g u_g^2 \tag{43}$$

so that for a given ΔP_f the velocity is:

$$u_g = \sqrt{\frac{\Delta P_f}{4\rho_g\phi}\left(\frac{De}{L}\right)}\ (m/s) \tag{44}$$

where:

ΔP_f = pressure drop (Pa)
ρ_g = gas density (kg/m^3)
De = equivalent diameter (m)
L = flow length (m)
ϕ = f/2 = friction factor (1/2 of Fanning factor)

Zone B pressure drop

This zone is difficult to assess for ΔP-flow characteristics, since the resistance to flow is due to the complex interaction of the gas with the cascading solids. A number of approaches can be taken including:

- using pneumatic solids transport relations (Zenz and Othmer, 1960);
- using pressure drop correlations from flighted rotary dryer operations (Hirosue and Shinohara (1978)).

The first approach sees the pressure drop being due to horizontal transport of solids through the flow area and pressure loss due to frictional losses due to gas flow and solids flow.

A version of Hinkle's model (Zenz and Othmer (1960)) retaining the key pressure drop contributors gives:

$$\Delta P_T = W_B u_{g_B} + \frac{2f\rho_{g_B}u_B^2 L}{De}\left[1 + \left(\frac{f_p u_p}{f u_{g_B}}\right)\cdot\frac{W_B}{u_{g_B}\rho_{g_B}}\right] \tag{45}$$

where:

W_B = solids flux in flow zone (kg/m^2.ms)
u_{g_B} = gas velocity in Zone B (m/s)
ρ_{g_B} = gas density (kg/m^3)
f = Fanning friction factor (-)
L = flow length (m)
De = equivalent diameter (m)
$\left(\frac{f_p u_p}{f u_{g_B}}\right)$ = ratio of particle to fluid friction (≤ 1)

The pressure drop relation can be solved for u_{g_B} given the overall pressure drop P_T. It should be emphasised that this is a first-pass approximation to the actual situation and is one of the key uncertainties in the model for Zone B. It should also be mentioned that work on rotary dryer pressure drops requires certain model parameters to be estimated from operating data. Again, this approach could be used if pilot plant data were available.

Zone C pressure drop

This region includes the heating tubes, gas flows and accompanying solids flows. Again, this represents an extremely complex flow system.

Pressure drop through staggered tube banks has been well studied. Grimison (1937) developed a correlation (Perry 6-36ff) for staggered tube banks giving

$$\Delta P = \frac{4 f_{\mathrm{tb}} N_{\mathrm{r}} \rho_{\mathrm{g}} u_{\max}^2}{2} \tag{46}$$

where:

f_{tb} = tube bank friction coefficient (-)
N_{r} = number of tube rows
ρ_{g} = gas density (kg/m^3)
$u_{\max}$ = gas velocity through the minimum flow area (m/s)

An analysis of the Zone C situation suggests that $f_{\mathrm{tb}} \varepsilon (0.08 - 0.10)$ for Reynolds numbers between 8,000 and 20,000. Hence the gas velocity can be estimated as:

$$u_{\mathrm{gc}} = \sqrt{\frac{1/2 \Delta P_c}{f N_r \rho_{\mathrm{g}}}} \tag{47}$$

This applies to the gas flow without solids loading. To account for the effect of solids in the gas phase the overall pressure drop is estimated as:

$$\Delta P = \frac{4 f_{\mathrm{tb}} N_r \rho_{\mathrm{g}} u_{\max}^2}{2} \left(1 + \left(\frac{f_p u_p}{f u_{\mathrm{g}}} \right) \frac{\hat{W}}{U_{\mathrm{g}} \sigma_{\mathrm{g}}} \right) \tag{48}$$

This allows a correction to be made to the solids-free flow to account for the presence of falling solids in this zone. In this instance the value $\hat{W}$ is some fraction of the Zone B solids loading W. In the simulation it is currently 0.5 but can be reset based on experimental data.

Again, the overall ΔP equation can be used to estimate the gas flow in Zone C.

Zone D pressure drop

This zone is considered as a flow of gas over a staggered tube bank. As in Zone C, the correlation of Grimison (1937) is used:

$$\Delta P_{\mathrm{D}} = \frac{4 f_{\mathrm{tb}} N_r \rho_{\mathrm{g}} u_{\max}^2}{2} \tag{49}$$

giving an estimated velocity of:

$$u_{\mathrm{gB}} = \sqrt{\frac{\frac{1}{2} \Delta P_{\mathrm{D}}}{f N_r \rho_{\mathrm{g}}}} \tag{50}$$

Parameters in the model are similar to Zone C.

8.1.6. Model analysis and solution

The full model was developed in Daesim Dynamics (Daesim 2005), using a structured text language. Following discretisation of the partial differential system, it led to a large differential-algebraic equation (DAE) set. Model solution was carried out with a proprietary sparse DAE solver based on high-order diagonally implicit Runge-Kutta (DIRK) methods that can effectively handle discontinuous systems as well as index 2 models.

8.1.7. Model simulation and results

This section gives some results of typical runs for the model under different heat transfer coefficient values. Many other options are clearly available to investigate changes in feed materials and system parameters. The scenario considered was

Feed rate of combusted solids	= 143 kg/s
Temperature of gas and solids	= 750 °C
Feed rate of fresh shale	= 200 kg/s
Shale moisture	= 5%
Shale feed temperature	= 100 °C
ATP pressure drop	= 100 Pa
HTC outside	= 150 W/m^2K
HTC inside	= 60 W/m^2K
Fraction of solids to air vector	= [0.5, 0.4, 0.3, 0.2, 0.05, 0.02, 0, 0, 0, 0]

The steady-state values from this simulation are given in Table 1

Consideration was then given to a simulation where the external tube heat transfer coefficient was increased from 150 W/m^2K to 250 W/m^2K at time 1000 s and then dropped to 50 $W/m^2/K$ at time 3000s. Note that the heat transfer coefficient for Zone D remained essentially unchanged. Figure 14 shows the exit temperature profiles with time for four variables:

Zone B gas–solids flow
Zone C gas–solids flow
Zone D gas flow
Bed solids

The temperature range on the vertical scale is between 250 °C and 500 °C.

In Figure 14, gas temperatures from Zones B and C decrease with an increase in heat transfer coefficient on the external wall of the porcupine. Bed solids follow the same trend. The gas temperature in the shadow area (Zone D) shows the opposite trend. This is due to a rise in the wall temperature, thus reducing the energy driving force.

TABLE 1 Steady-state Values for Simulation Study

Property	Value
Outlet gas temperatures:	
Zone A	655°C
Zone B	348°C
Zone C	328°C
Zone D	347°C
Bed solids out of cooler	63.8 kg/s
Bed solids outlet temperature	333°C
Shale out of preheat	275°C
Porcupine wall temperatures	591°C to 220°C
Solids in gas out	79.2 kg/s
Bed height variation along length	0.75 to 0.10 m
Total mass flow of gas	145 kg/s (330 m^3/s)
Internal heat transfer (Zones C&D)	78.7 MW
External heat losses	4.4 MW

Figure 15 shows the wall temperature profile at the inlet, mid-point and exit of the ATP. These show a rise in wall temperatures for an increase in heat transfer coefficient and vice versa. The y-axis range is 150–750 °C.

Figure 16 shows the temperature profiles with time of the bed solids at grid points 1, 2, 5, 10, 15 and exit as the heat transfer coefficient was changed. The trends show an expected behaviour.

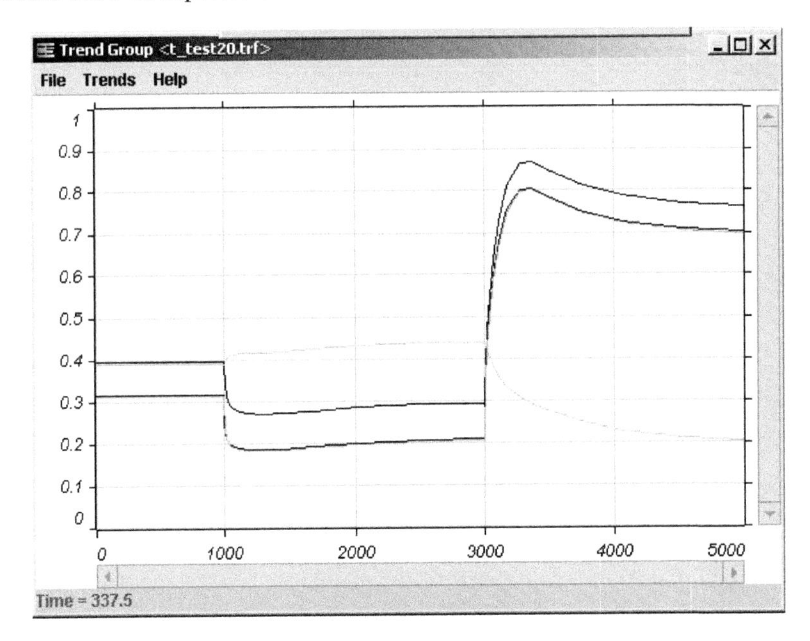

FIGURE 14 Temperature changes vs heat transfer coefficient variations.

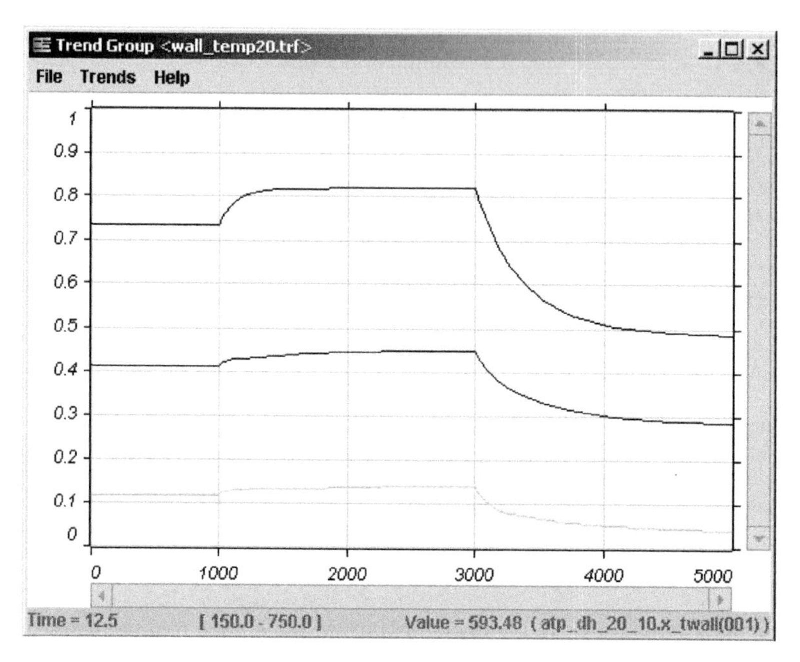

FIGURE 15 **Heating tube wall temperature profiles (entry, mid, exit points).**

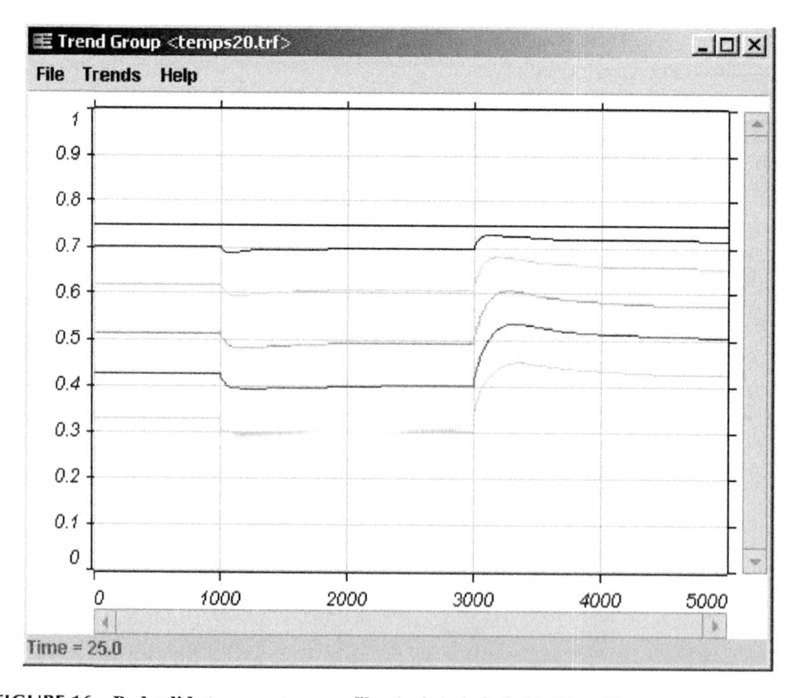

FIGURE 16 **Bed solids temperature profiles (points 1, 2, 5, 10, 15, exit).**

These simulations indicate some of the many options available to the user in order to investigate the dynamic and steady-state behaviour of the model, and particularly parameter sensitivities.

8.2. MODELLING THE DYNAMIC PERFORMANCE OF A COMPLEX GRANULATION SYSTEM FOR MAP–DAP PRODUCTION

8.2.1. Introduction and context of the modelling

The industrial production of fertiliser still remains a very important activity. However, much of the design of these plants is highly empirical. Recent trends have led to approaches that ellucidate the underlying mechanisms in order to allow predictive models to be built. This section sets out the modelling of the complex reaction and crystallisation mechanisms between monoammonium phosphate (MAP) and ammonia (NH_3) to form diammonium phosphate (DAP) during granulation. It is then coupled to a population balance model where the reaction and crystallisation model takes into consideration the dynamics of the liquid phase and the solid phase in the granulator drum. This model plays a central role in the granulator dynamics, since it is now possible to track the change in the amount and composition of the liquid phase in time and space and therefore the moisture content during the granulation process. The moisture content has a significant effect on the granule growth and knowledge of this can lead to improved operational and control strategies based on model simulations. The reaction-crystallisation and growth model was validated by using plant data from a major fertiliser company (Balliu and Cameron 2007).

8.2.1.1 Basic chemistry

In the case of phosphate fertilisers the granulation processes involves the following main reactions:

1. Partial preneutralisation of the phosphoric acid in a preneutraliser:

$$NH_3 + H_3PO_4 \rightarrow (NH_4)H_2PO_4 + \text{Heat} \tag{51}$$

2. Completion of reaction in the drum granulator: the slurry is sprayed onto a rolling bed of recycled dry material and further ammonia is added to form diammonium phosphate (DAP). When the N:P mole ratio is 2, all MAP is converted to DAP.

$$NH_3 + (NH_4)H_2PO_4 \rightarrow (NH_4)_2HPO_4 + \text{Heat} \tag{52}$$

Figure 17 shows a typical process.

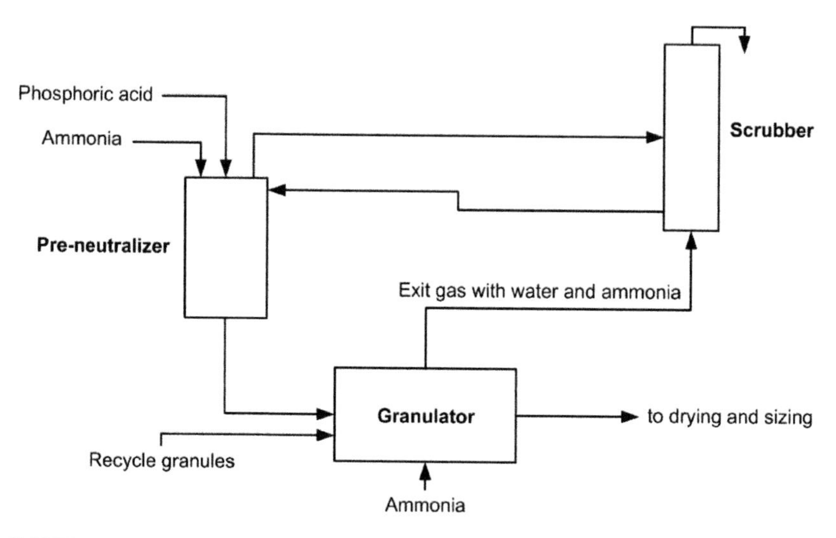

FIGURE 17 **Schematic diagram of typical reaction-granulation process.**

In modelling the granulation process, the focus has always been on granulation mechanisms such as nucleation, layering, agglomeration and breakage. However, in order to obtain granular fertiliser product of required composition and size the following factors related to the reaction in the drum should be taken into consideration:

1. The liquid phase
   The liquid phase consists of free water and the solution of MAP and DAP. The amount of liquid phase is increased by temperature because the constituent compounds are more soluble at high temperature. Hence, the moisture requirements for granulation will depend upon the granulation temperature and the amount of MAP and DAP present in solution.
2. MAP and DAP change in solubility during the process
   Knowledge of solubility in the system is important in order to track the liquid phase composition along the granulator. As seen in Figure 18, when MAP and DAP are mixed to a N:P mole ratio of about 1.45, the solubility of the combined salts is greater than for either of the compounds at the same temperature. At this point, addition of ammonia lowers the salt's solubility, and causes crystallisation of DAP and a decrease in the amount of liquid phase that is present (Wesenberg 1986). This relationship is valid for any other temperature (Davis and Lee 1967). The crystallisation of DAP out of solution acts as a binder to bond particles together in a firm pellet (Nielsson 1986).

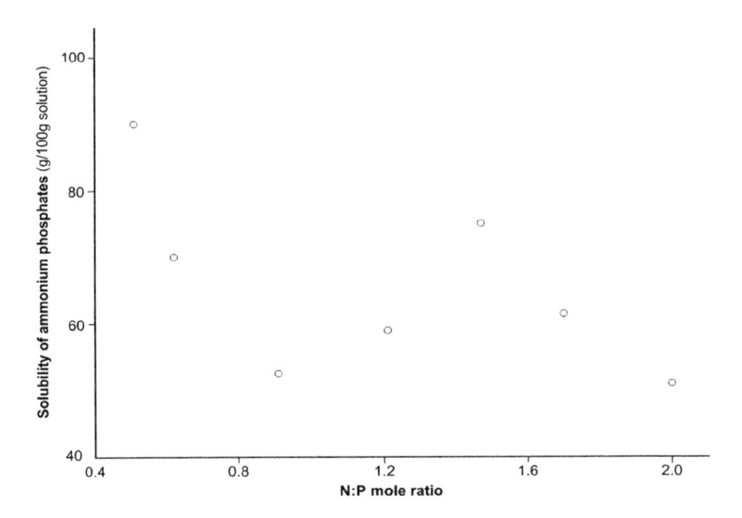

FIGURE 18 **Effect of N: P mole ratio on the solubility of ammonium phosphates at 75°C.**

The phase diagram for the MAP, DAP and H_2O system at equilibrium is well established. In Figure 19, data by Seidell (1941) describing the $H_3PO_4 - NH_3 - H_2O$ solubility system at 75°C was imposed over data describing the MAP, DAP and H_2O system (Stokka 1985). These data are consistent and were used to estimate the change in solubility by adding ammonia to $H_3PO_4 - NH_3 - H_2O$ system.

3. The energy balance during granulation
   It is important to know the amount of heat produced by the reaction as well as the heat required for granulation. The reaction between MAP and ammonia in the drum is exothermic and the amount of heat produced is used to:
   - bring the recycle solids up to the granulation temperature;
   - evaporate the water from the particles surface;
   - heat the gaseous feed (ammonia).

Some of the heat of reaction is also lost through the equipment. If the heat generated by the chemical reaction is not enough to satisfy all these requirements, steam can be added to the process (Nielsson 1986).

Therefore, even though the reaction is not the main mechanism in the granulator, its influence is significant because it affects the granule moisture, which has a great impact on the other granulation mechanisms. Unfortunately, apart from the data presented above, there is no information available in the literature which can allow us to quantify the moisture, heat and solubility during the reaction-crystallisation process.

Zhang (1989) and Zang et al. (2001) were the first to include a very simplistic reaction mechanism into the development of the fertiliser granulation model. Although Zhang's model has helped in understanding the contribution of the reaction mechanism on the granulation process, it didn't consider adequately the

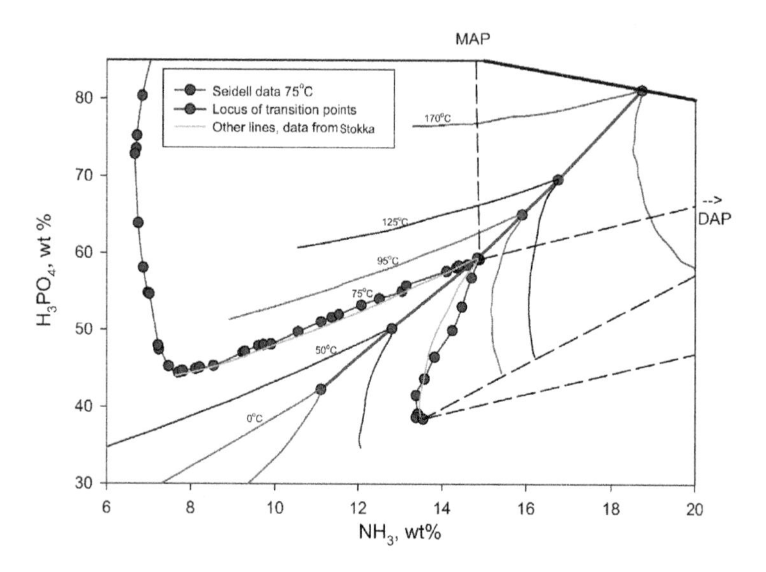

FIGURE 19 **Phase diagram for H_3PO_4 NH_3 H_2O.**

energy balance, phase changes and the liquid phase composition. Improvements are also required for more accurate predictions along with changes in the modelling assumptions:

- The heat of reaction should not be fully used to only evaporate the water from the surface of the granules. From plant observations, depending on the amount of ammonia injected into the drum, sometimes the heat of reaction might be far from sufficient for water evaporation.
- The final product should not be considered as pure DAP, but a mixture of MAP and DAP, otherwise the N:P mole ratio at the granulator exit cannot be satisfied.

8.2.2. Background physico-chemical principles

8.2.2.1 Phase diagram for DAP granulator

Since the granulator is near atmospheric pressure, the heat of reaction of ammonia with the liquid on the granules will lead to a near instantaneous evolution of water as steam, so the temperature will be near the boiling point of the MAP–DAP solution. For illustrative purposes this temperature is taken as 95°C in Figure 20 as this is the recorded data closest to the estimated boiling point.

Figure 20 shows the 95°C equilibrium line plotted in terms of DAP, MAP and H_2O. The lines for the phase regions below the solution transition point S have

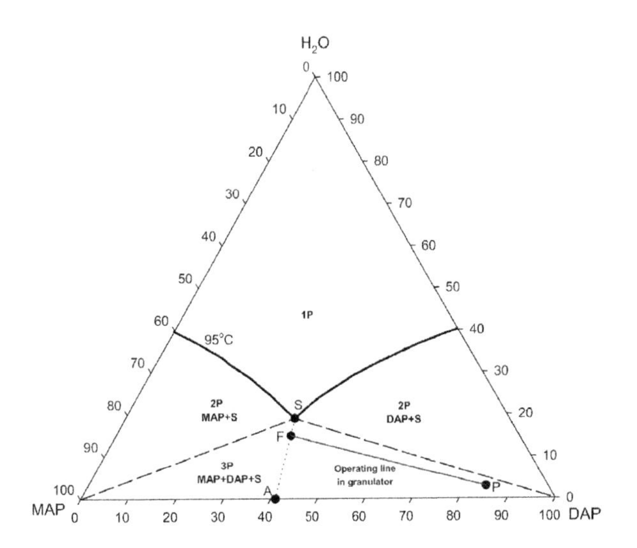

FIGURE 20 **Solubility diagram for MAP-DAP-H_2O at 95°C.**

been added on the diagram. Point F corresponds to a possible feed to the granulator (15% water and mole ratio of N/P = 1.4). This can be interpreted as a mixture of saturated solution (at S) and a mixture of MAP and DAP solids at Point A. Point A is a mixture of DAP (41%) and MAP. Point P corresponds to a final product (3% water, N/P = 1.85).

In the granulator, ammonia is added for reaction. For a small amount of reacting ammonia, one could draw a line linking F to the ammonia point and move along it, depending on the amount of ammonia reacting. The reaction will generate heat which will evaporate water, so this new point should be joined to the water point and a point taken on the extension of this line, since water is being removed. This FP line will represent an operating line for conditions through the granulator. Therefore, in the model the basic crystallisation mechanism considers the dissolution of MAP into solution and the subsequent crystallisation of DAP from solution. That is, the process is solution mediated. The feed slurry, containing appreciable fine MAP, is sprayed onto a larger amount of recycled 'dry' solids, so if the solids are non-porous, this slurry will sit on the surface of the particles or be a capillarity liquid at the contact area between particles.

8.2.2.2 The saturation index

A saturation index depending on the liquid phase temperature (T_l) and mass fractions of MAP and DAP was defined based on the solubility diagram. MAP and DAP compositions on the 75°C, 95°C and 125°C isotherms were plotted and a correlation describing the saturation surface was then generated. This is the concentration surface above which DAP will crystallise out of the liquid phase, and it is presented in Figure 21. The limits of the independent variables are: $0 < \text{MAP} < 50$ (wt%) and $70 < T_{\text{I}} < 130$ (°C).

$$\text{DAP} = a + b(\text{MAP}) + c(T_{\text{I}}) + d(T_{\text{I}})^2 \tag{53}$$

By using this correlation we are able to track the DAP concentration as a function of MAP concentration and temperature of the liquid phase. A saturation index has been defined such that:

$$S_{\text{I}} = \frac{\text{DAP}_{\text{liq}}}{\text{DAP}_{\text{sat}}} \tag{54}$$

where DAP_{liq} represents the DAP composition in the liquid phase and DAP_{sat} the DAP concentration at saturation conditions.

- If $S_{\text{I}} = 1$ then the DAP level is saturated and crystallisation is incipient.
- If $S_{\text{I}} > 1$ then the DAP level is supersaturated and will crystallise at a certain rate.
- If $S_{\text{I}} < 1$ then the DAP is undersaturated and no crystallisation occurs.

Hence by using four we can calculate S_{I} at each time step and see whether crystallisation will occur.

8.2.3. MAP–DAP crystallisation model

The following provides a complete description, justification and development of the reaction-crystallisation model for MAP–DAP. The model was developed

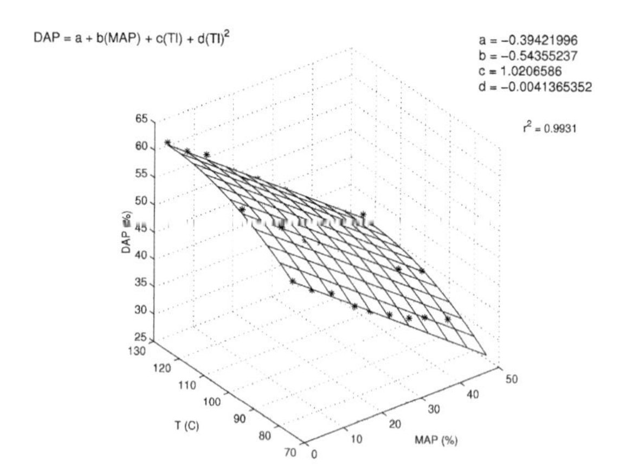

FIGURE 21 **Saturation surface determined with table curve 3d and plotted in matlab.**

taking into consideration the dynamics of the liquid phase and the solid phase in the granulator. The whole granulator drum can be divided into sections. In some of the sections the reaction was considered the main mechanism whilst in others the slurry and ammonia flow can be 'switched off' and only agglomeration/breakage mechanisms will be considered. A generalised balance volume diagram (BVD) representing a section in the drum with both solid and liquid phases entering and leaving the section is presented below in Figure 22:

The slurry flow enters the granulator as MAP, DAP and solution as well as MAP $\left(F_{SL}^{MAP,sol}\right)$ and DAP $\left(F_{SL}^{DAP,sol}\right)$ solids formed in the preneutraliser. A part of $H_2O\left(\varphi F_{SL}^{H_2O}\right)$, as it exits the spray nozzles, flashes out into the vapour phase. The rest $\left((1-\varphi)F_{SL}^{H_2O}\right)$ remains in the liquid phase. The ammonia flow enters the granulator and a part of it is absorbed into the liquid phase $\left(f_{NH_3}\right)$. The rest goes in the vapour space $\left(F_{NH_3}^{vap}\right)$.

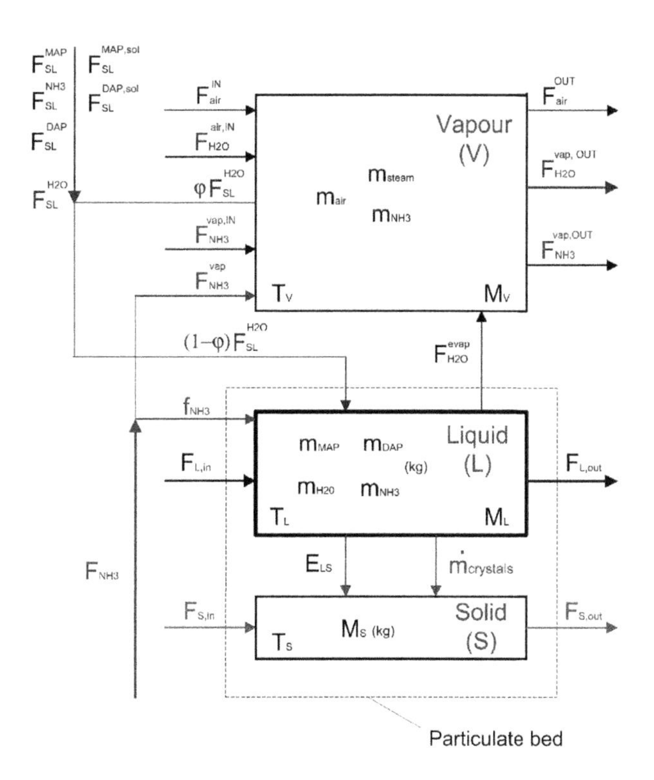

FIGURE 22 Schematic representation of the reaction model.

8.2.3.1 Controlling mechanisms

M1: flow of the solids from the recycle (F_s, in);
M2: flow of the slurry on top of the recycled granules (F_{SL});
M3: flow of sparged NH_3 (F_{NH3});
M4: instantaneous reaction of NH_3(g) with MAP in the solution;
M5: reaction rate;
M6: DAP crystallisation from the solution;
M7: evaporation of the water due to heat of reaction;
M8: heat transfer between the liquid phase and the solid phase;
M9: heat generated by reaction.

8.2.3.2 Assumptions

The following assumptions have been made for the MAP–DAP model:

A1: The liquid phase is considered a single phase unrelated to the particle size distribution.
A2: The dynamics of the liquid phase are important.
A3: The dynamics of the solid phase are important only for the population balance equation.
A4: No physical material losses from the system.
A5: Kinetic and potential terms in the energy balance are neglected.
A6: The heat loss from the granulator vessel is negligible.
A7: All physical properties are constant for the operating range considered.
A8: The mass of MAP and DAP solids carried with the slurry does not react or dissolve and is distributed onto the recycled solids proportional to the surface area of these solids.
A9: The mass of DAP crystals formed is distributed proportionally to the surface area of the recycle solids.
A10: The liquid phase is well mixed.
A11: No accumulation of ammonia, air and steam in the vapour phase.
A12: Ammonia is rapidly taken into the liquid phase.

8.2.3.3 The model equations

Component mass balances – liquid phase

The change of MAP, DAP and H_2O mass over time is described by the following equations:

$$\frac{dm_{\text{MAP}}}{dt} = F_{\text{L,in}}^{\text{MAP}} + F_{\text{SL}}^{\text{MAP}} - F_{\text{L,out}}^{\text{MAP}} - \frac{MW_{\text{MAP}}}{MW_{\text{NH}_3}} r_{\text{MAP/DAP}} \tag{55}$$

$$\frac{dm_{\text{DAP}}}{dt} = F_{\text{L,in}}^{\text{DAP}} + F_{\text{SL}}^{\text{DAP}} - F_{\text{L,out}}^{\text{DAP}} - \dot{m}_{\text{crystals}} + \frac{MW_{\text{DAP}}}{MW_{\text{NH}_3}} r_{\text{MAP/DAP}} \tag{56}$$

$$\frac{dm_{\mathrm{H_2O}}}{dt} = F_{\mathrm{L,in}}^{\mathrm{H_2O}} + (1-\varphi)F_{\mathrm{SL}}^{\mathrm{H_2O}} - F_{\mathrm{L,out}}^{\mathrm{H_2O}} - F_{\mathrm{H_2O}}^{\mathrm{evap}} \tag{57}$$

Overall mass balances

The change of liquid phase mass over time is given by:

$$\frac{dM_{\mathrm{L}}}{dt} = F_{\mathrm{L,in}} + F_{\mathrm{SL}} + f_{\mathrm{NH_3}} - F_{\mathrm{H_2O}}^{\mathrm{evap}} - F_{\mathrm{L,out}} - \dot{m}_{\mathrm{crystals}} \tag{58}$$

The change of solid phase mass over time is:

$$\frac{dM_{\mathrm{S}}(i)}{dt} = F_{\mathrm{S,in}}(i) + F_{\mathrm{SL}}^{\mathrm{MAP,sol}}(i) + F_{\mathrm{SL}}^{\mathrm{DAP,sol}}(i) - F_{\mathrm{S,out}}(i) + \dot{m}_{\mathrm{crystals}}(i) \tag{59}$$

where:

$m_{\mathrm{MAP}}, m_{\mathrm{DAP}}, M_{\mathrm{H2O}}$ = mass of MAP, DAP and H_2O in the liquid phase (kg);
M_{L} = mass holdup of liquid phase (kg);
M_{S} (*i*) = mass holdup of the solids in each particle size interval (*i*) (kg);
$F_{\mathrm{L,in}}^{\mathrm{MAP}}, F_{\mathrm{L,in}}^{\mathrm{DAP}}, F_{\mathrm{L,in}}^{\mathrm{H_2O}}$ = flow of MAP, DAP and H_2O into the liquid phase (kg.s^{-1});
$F_{\mathrm{L,out}}^{\mathrm{MAP}}, F_{\mathrm{L,out}}^{\mathrm{DAP}}, F_{\mathrm{L,out}}^{\mathrm{H_2O}}$ = flow of MAP, DAP and H_2O with the liquid phase out of the drum section (kg.s^{-1});
$F_{\mathrm{L,in}}, F_{\mathrm{L,out}}$ = flow of liquid phase in and out of the drum section (kg.s^{-1});
$F_{\mathrm{S,in}}(i), F_{\mathrm{S,out}}(i)$ = flow of solids in each particle size interval (*i*) flowing in and out of each section (kg.s^{-1});
$F_{\mathrm{H_2O}}^{\mathrm{evap}}$ = flow of water evaporated (kg.s^{-1}). If is constant then this can be computed from the energy balance directly;T_{L};
$F_{\mathrm{SL}}^{\mathrm{MAP}}, F_{\mathrm{SL}}^{\mathrm{DAP}}, F_{\mathrm{SL}}^{\mathrm{H_2O}}$ = flow of MAP, DAP and H_2O solution with the slurry stream (kg.s^{-1});
$F_{\mathrm{SL}}^{\mathrm{MAP,sol}}(i), F_{\mathrm{SL}}^{\mathrm{DAP,sol}}(i)$ = flow of MAP and DAP crystals from the slurry stream deposited onto each size range (*i*) of the recycle stream (kg.s^{-1});
F_{SL} = flow of slurry into the liquid phase section (kg.s^{-1});
$f_{\mathrm{NH_3}}$ = flow of ammonia taken up into the liquid phase (kg.s^{-1});
$r_{\mathrm{MAP/DAP}}$ = reaction rate (kg.s^{-1});
$\dot{m}_{\mathrm{crystals}}$ = mass rate of crystallisation onto existing solid phase (kg.s^{-1});
$\dot{m}_{\mathrm{crystals}}(i)$ = mass rate of crystallisation into each size range onto existing solid phase (kg.s^{-1});
φ = flash fraction of water from slurry flow as it exits the spray nozzle (–).

The energy balance in the liquid phase

The change of the liquid phase energy content over time is described by the following equation:

$$\begin{aligned}\frac{dE_L}{dt} &= F_{L,in}c_{p,L}(T_{L,in} - T_{ref}) + F_{SL}c_{p,SL}(T_{SL} - T_{ref}) \\ &+ f_{NH_3}(c_{p,NH_3}(T_{NH_3} - T_{ref}) + \Delta hp) + \frac{MW_{DAP}}{MW_{NH_3}}\Delta H_{rxn} r_{MAP/DAP} \\ &- F^{evap}_{H_2O}(c_{p,wl}(100 - T_{ref}) + \lambda_{H_2O} + c_{p,wg}(T_V - 100)) \\ &- F_{L,out}c_{p,L}(T_L - T_{ref}) - E_{LS} - \dot{m}_{crystals}\,\Delta H_{crys}\end{aligned} \tag{60}$$

where:

E_L = the energy content in the liquid phase (kJ);
ΔH_{rxn} = heat of reaction (kJ.kg^{-1});
ΔH_{crys} = heat of crystallisation (kJ.kg^{-1});
λ = latent heat of vaporisation (kJ.kg^{-1});
$c_{p,L}$, $c_{p,SL}$ = heat capacity of the liquid phase and slurry (kJ.kg^{-1}.K^{-1});
$c_{p,wl}$, $c_{p,wg}$ = heat capacity of the water liquid and water vapour (kJ.kg^{-1}.K^{-1});
E_{LS} = energy transferred between the liquid phase and the solid phase (kJ.s^{-1});
T_L, T_{SL}, T_{NH3} = temperature in the liquid phase, slurry and ammonia (K);
T_{ref} = reference temperature for the enthalpy calculation (298.15K);
T_v = vapour phase temperature. The energy requirements for water/stream combination includes the heat required to raise from T_{ref} to T_v100 °C plus the latent heat of evaporation plus the amount of heat to raise the steam from 100 °C to the final temperature;
Δhp = correction for the pressure difference in the enthalpy calculation (kJ.kg^{-1}).

Consideration of all potentially significant factors led to the inclusion of the pressure correction factor in this equation. Inclusion of the factor within the simulation allows for any changes in ammonia pressure to be accounted for.

The energy balance in the solid phase

The change of the liquid phase energy content over time is described by the following equation:

$$\begin{aligned}\frac{dE_S}{dt} &= F_{S,in}c_{p,S}(T_{S,in} - T_{ref}) + F^{MAP,sol}_{SL}c_{p,MAP}(T_{SL} - T_{ref}) \\ &+ F^{DAP,sol}_{SL}c_{p,DAP}(T_{SL} - T_{ref}) - F_{S,out}c_{p,S}(T_S - T_{ref}) \\ &+ E_{LS} + \dot{m}_{crystals}\, c_{p,DAP}(T_L - T_{ref})\end{aligned} \tag{61}$$

where:

$F_{S,in}$, $F_{S,out}$ = total flow of solids in and out of each section (kg.s^{-1});
$F^{MAP,sol}_{SL}$, $F^{DAP,sol}_{SL}$ = flow of MAP and DAP crystals with the slurry stream (kg.s^{-1});
E_S = the energy content in the solids (kJ);

$c_{p,S}$ = heat capacity of the solids (kJ.kg^{-1}.K^{-1});
$c_{p,MAP}$, $c_{p,DAP}$ = heat capacity of the MAP and DAP solids (kJ.kg^{-1}.K^{-1});
T_S = solid phase temperature (K).

Constitutive equations

The total flows of solids: The total flow of solids entering/exiting the granulator is a sum of the solids flow in each particle size range from one to maximum size (n size):

$$F_{S,in} = \sum_{i=1}^{nsize} F_{S,in}(i) \tag{62}$$

$$F_{S,out} = \sum_{i=1}^{nsize} F_{S,out}(i) \tag{63}$$

The total mass holdup of the solids: The total solids holdup as a sum of each solids holdup in each particle size range from one to maximum size (n size):

$$M_S = \sum_{i=1}^{nsize} M_S(i) \tag{64}$$

Liquid and solid phase temperature definitions:

$$T_L = \frac{E_L}{M_L c_{p,L}} + T_{ref} \tag{65}$$

$$T_S = \frac{E_S}{M_S c_{p,S}} + T_{ref} \tag{66}$$

Ammonia is rapidly taken into the liquid phase, and the reaction between ammonia and MAP is instantaneous. Therefore, we can relate the reaction rate between ammonia and MAP to the amount of ammonia uptaken into the liquid phase:

$$r_{MAP/DAP} = f_{NH_3} \tag{67}$$

The flow of ammonia taken up by the liquid phase can be written as:

$$f_{NH_3} = k_{uptake} X_{H_2O} \xi S_{part} \tilde{P}_{NH_3} \tag{68}$$

The ammonia flow is assumed to be dependent on the amount of water (mass fraction) available in the liquid layer (X_{H_2O}), the partial pressure of ammonia in the bed voids ($\tilde{P}_{NH_3}$) and the efficiency of contact with the particles surface area (S_{part}) covered by the liquid phase. It can't be assumed that the ammonia flow will access all the liquid phase available in the drum, and hence the term ξ

accounts for the efficiency of ammonia contact with the liquid layer. k_{uptake} represents the mass transfer coefficient $(kg.(m^2)^{-1}.s^{-1}.bar^{-1})$.s

$\tilde{P}_{NH_3}$ is assumed to reach atmospheric pressure very quickly. By choosing different values for the ammonia inlet flow between zero and the maximum acceptable (F_{NH3}) operating value, the following relationship is used to relate $\tilde{P}_{NH_3}$ to the ammonia feed:

$$\tilde{P}_{\mathrm{NH_3}} = 1 - \exp(1 - kF_{\mathrm{NH_3}}) \tag{69}$$

Based on approximations, the value of the coefficient k has been chosen as six. The bigger the k value is, the more rapidly $\tilde{P}_{\mathrm{NH_3}}$ will reach atmospheric pressure.

The amount of water evaporated is also assumed to depend on the efficiency of contact with the particles surface area and the difference between the partial pressure of water in the vapour and air phase.

$$F^{\mathrm{evap}}_{\mathrm{H_2O}} = k_{\mathrm{evap}}\xi_{\mathrm{w}}S_{\mathrm{part}}(\tilde{P}_{\mathrm{H_2O,vap}} - \tilde{P}_{\mathrm{H_2O,g}}) \tag{70}$$

In equation (70) the ξ_{w} coefficient accounts for the efficiency of ammonia contact with the liquid layer and k_{evap} is the mass transfer coefficient $(kg.(m^2)^{-1}.s^{-1}.bar^{-1})$. The water concentration in the air phase is considered to be very small and hence we can assume $\tilde{P}_{\mathrm{H_2O,g}} = 0$.

The water vapour pressure $\tilde{P}_{\mathrm{H_2O,vap}}$ is given by the Antoine equation:

$$log\tilde{P}_{\mathrm{H_2O,vap}} = A - \frac{B}{T_{\mathrm{L}} + C} \tag{71}$$

where A, B and C are coefficients in the Antoine equation for water.

The total surface area of the particles was defined in terms of particles mass holdup, density and average particle size:

$$S_{\mathrm{part}} = \sum_{i=1}^{\mathrm{nsize}} \frac{6M_{\mathrm{S}}(i)}{\rho\bar{d}_{\mathrm{p}}(i)} \tag{72}$$

$\tilde{P}_{\mathrm{NH_3}}$ = ammonia partial pressure in the bed voids (bar);
S_{part} = total surface area of the particles (m^2);
$\bar{d}_p(i)$ = average particle diameter in each size range (m);
ρ = particle density ($kg.m^{-3}$).

The flow of ammonia that was not absorbed into the liquid phase goes into the vapour phase and we have:

$$F^{\mathrm{vap}}_{\mathrm{NH_3}} = F_{\mathrm{NH_3}} - f_{\mathrm{NH_3}} \tag{73}$$

where F_{NH3} is the total amount of ammonia feed to the granulator section ($kg.s^{-1}$) and $F^{\mathrm{vap}}_{\mathrm{NH_3}}$ is the amount of ammonia that is not taken up into the liquid phase and goes to the vapour space $kg.s^{-1}$.

The total amount of ammonia flow that exits each of the intermediate granulation section is given by the following equation:

$$F_{NH_3}^{vap,OUT} = F_{NH_3}^{vap,IN} + F_{NH_3}^{vap} \tag{74}$$

in which $F_{NH_3}^{vap,IN}$ and $F_{NH_3}^{vap,OUT}$ represent the amount of ammonia IN and OUT of the vapour phase $(kg.s^{-1})$. In the first section of the granulator $F_{NH_3}^{vap,IN}$ will be zero.

The mass balance for the air entering and exiting the vapour phase is given by:

$$F_{air}^{OUT} = F_{air}^{IN} \tag{75}$$

where F_{air}^{IN} and F_{air}^{OUT} represent the flow of air IN and OUT of the vapour space $(kg.s^{-1})$;

The amount of water entering each vapour space of the granulator section is the sum of the water coming with the slurry that flushes out into vapour, the water coming with the inlet air and the water evaporated from the reaction:

$$F_{H_2O}^{vap,OUT} = \varphi F_{SL}^{H_2O} + F_{H_2O}^{air,IN} + F_{H_2O}^{evap} \tag{76}$$

The flash fraction is defined by the standard equation:

$$\varphi = 1 - \exp\left(-\frac{c_{p,wl}}{\lambda}(T_{SL} - 100)\right) \tag{77}$$

The rate of crystallisation is given by a first-order relationship between supersaturation and growth:

$$\dot{m}_{crystals} = k_{crys} S_{part} (S_I - 1) \tag{78}$$

where k_{crys} is the mass transfer coefficient $(kg.m^{-2}.s^{-1})$ and S_I is the saturation index. The saturation index was determined based on the phase diagram for MAP DAP H_2O and given by the following equation:

$$S_I = \frac{X_{DAP}}{X_{DAP,sat}} \tag{79}$$

The saturation index depends on liquid phase temperature T_L, MAP and DAP mass fractions such as:

$$X_{DAP,sat} = a + bX_{MAP} + cT_L + d(T_L^2) \tag{80}$$

a, b, c, d = coefficients determined using the phase diagram and the saturation DAP surface.

The components mass fractions are given by the following equations:

$$X_{\mathrm{MAP}} = \frac{m_{\mathrm{MAP}}}{M_{\mathrm{L}}} \tag{81}$$

$$X_{\mathrm{DAP}} = \frac{m_{\mathrm{DAP}}}{M_{\mathrm{L}}} \tag{82}$$

$$X_{\mathrm{H_2O}} = \frac{m_{\mathrm{H_2O}}}{M_{\mathrm{L}}} \tag{83}$$

$X_{\mathrm{MAP}}, X_{\mathrm{DAP}}$ and $X_{\mathrm{H_2O}}$ represents the mass fraction of MAP, DAP and H_2O respectively.

The total liquid phase flow into the granulator section is given by:

$$F_{\mathrm{L,in}} = F_{\mathrm{L,in}}^{\mathrm{MAP}} + F_{\mathrm{L,in}}^{\mathrm{DAP}} + F_{\mathrm{L,in}}^{\mathrm{H_2O}} + F_{\mathrm{L,in}}^{\mathrm{NH_3}} \tag{84}$$

In the first section of the granulator $F_{\mathrm{L,in}}$ will be zero.

The ammonia flow in the liquid phase coming out from the granulator section:

$$F_{\mathrm{L,out}}^{\mathrm{NH_3}} = F_{\mathrm{L,out}} - (F_{\mathrm{L,out}}^{\mathrm{MAP}} + F_{\mathrm{L,out}}^{\mathrm{DAP}} + F_{\mathrm{L,out}}^{\mathrm{H_2O}}) \tag{85}$$

The total flow of slurry into the granulator section is:

$$F_{\mathrm{SL}} = F_{\mathrm{SL}}^{\mathrm{MAP}} + F_{\mathrm{SL}}^{\mathrm{DAP}} + (1-\varphi)F_{\mathrm{SL}}^{\mathrm{H_2O}} + F_{\mathrm{SL}}^{\mathrm{NH_3}} \tag{86}$$

The energy transferred between the liquid phase and the solid phase:

$$E_{\mathrm{LS}} = k_{\mathrm{cv}}\eta S_{\mathrm{part}}(T_{\mathrm{L}} - T_{\mathrm{S}}) \tag{87}$$

$F_{\mathrm{L,in}}^{\mathrm{NH_3}}$ = flow of NH_3 into the section with the liquid phase $(kg.s^{-1})$;
$F_{\mathrm{SL}}^{\mathrm{NH_3}}$ = flow of NH_3 with the slurry stream $(kg.s^{-1})$;
η = contact factor between liquid phase and solid phase;
k_{cv} = thermal resistance to conductive heat transfer $(kJ.(m^2)^{-1}.s^{-1}.K^{-1})$.

The total holdup of the liquid and solid phase in the granulator is given by:

$$M_{LS} = M_L + M_S \tag{88}$$

The total flow discharge from the granulator is given by the following equation: A simplified version of a typical solids transport and discharge equation that relates the flow rate to the flow height (Boyce 1997, Gupta and Khakkar 1991).

$$F_{\mathrm{total,out}} = k_1 \rho_{\mathrm{bulk}}(H)^n \tag{89}$$

$$F_{\mathrm{L,out}} = \frac{M_{\mathrm{L}}}{M_{\mathrm{LS}}} F_{\mathrm{total,out}} \tag{90}$$

$$F_{\mathrm{S,out}} = \frac{M_{\mathrm{S}}}{M_{\mathrm{LS}}} F_{\mathrm{total,out}} \tag{91}$$

The component flows of the liquid phase coming out of the drum section:

$$F^{\mathrm{MAP}}_{\mathrm{L,out}} = X_{\mathrm{MAP}} F_{\mathrm{L,out}} \tag{92}$$

$$F^{\mathrm{DAP}}_{\mathrm{L,out}} = X_{\mathrm{DAP}} F_{\mathrm{L,out}} \tag{93}$$

$$F^{\mathrm{H_2O}}_{\mathrm{L,out}} = X_{\mathrm{H_2O}} F_{\mathrm{L,out}} \tag{94}$$

$F_{\mathrm{total,\ out}}$ = the total flow of solid and liquid phase out of the drum section ($kg.s^{-1}$);
ρ_{bulk} = bulk density ($kg.m^{-3}$);
H = depth of the solids bed in the drum section (m).

The bed depth can be computed via a simple geometric relationship similar to that shown in Figure 13.

8.2.4. Population balance model

A standard population balance for the drum solids was used to incorporate the various mechanisms of nucleation, growth by layering and coalescence and granule breakage. The basic discretised population balance for N_{i}, the number of particles in size interval i (number per m) is given by:

$$\frac{dN_i}{dt} = N_{\mathrm{in}_i} - N_{e_i} + \left(\frac{dN_i}{dt}\right)_{\mathrm{nucl}} + \left(\frac{dN_i}{dt}\right)_{\mathrm{growth}} + \left(\frac{dN_i}{dt}\right)_{\mathrm{break}} \tag{95}$$

Here N_{in_i} is the rate of particles in size range i flowing into the granulator, whilst N_{e_i} represents the outflow of particles in size range i. Other terms represent the contributions in size range i from mechanisms of nucleation, growth and breakage. The adopted discretisation in this model was that proposed by Hounslow et al. (1988).

Nucleation mechanisms were based on work by Wildeboer et al. (2005), whilst a two-stage kernel was used for the coalescence model (Adetayo et al.

1995). Finally a breakage model due to Hill and Ng (1995) was used to account for breakage of granules in the drum.

8.2.5. Parameter estimation

Parameter estimation from experimental data is commonly referred to as regression. In this work, a linear regression routine was used to provide initial estimates for the model parameters.

In general, the fitting problem involves real data y and predicted data y_p generated by some process model of the general form:

$$y_p = m(u, p)$$

where:

u = process input variables;
p = model parameters;
$m()$ = the model relationships.

The fitting or regression task is to find values for the parameters p such that a least square criterion is minimised:

$$\min_p = \sum_i w_i (y_i - y_{p,i})^2 \tag{96}$$

where w_i are the weights given to individual data sets.

By using a steady-state version of the reaction-crystallisation model and employing the least square criterion, initial values for k_{uptake}, k_{evap} and k_{cv} have been generated. By using the dynamic model these values were manually adjusted to fit the plant results.

The parameter k_{cryst} was estimated by using crystal growth data from Mullin (2001). The crystal growth rate can be expressed as:

$$R_G = \frac{6\alpha}{\beta}\rho v = k_{cryst} \tag{97}$$

where:

R_G = mass deposition rate ($kg.m^{-2}.s^{-1}$);
α, β = surface shape factors and $6\alpha/\beta = 1$ for spheres;
v = mean linear velocity ($m.s^{-1}$).

Typical values of the mean linear velocity and supersaturation for DAP crystals were found in Mullin (2001) but only for temperatures between 20°C and 40°C. To determine corresponding values for the granulator operating temperature the following Arrhenius type relationship between the mean linear velocity and supersaturation Ss was assumed:

$$\ln(v) = \ln(k) + (-E/R)\frac{1}{T} + n\ln(Ss - 1) \tag{98}$$

The k, $(-E/R)$ and n values were determined by using the data available and a least square error routine.

8.2.6. Computer modelling and simulation

Computer simulations have been carried out to investigate the behaviour of the reaction-crystallisation model. The granulator drum was divided into three sections as seen in Figure 23. In each section equal amounts of slurry and ammonia were added into the bed of dry recycled particles.

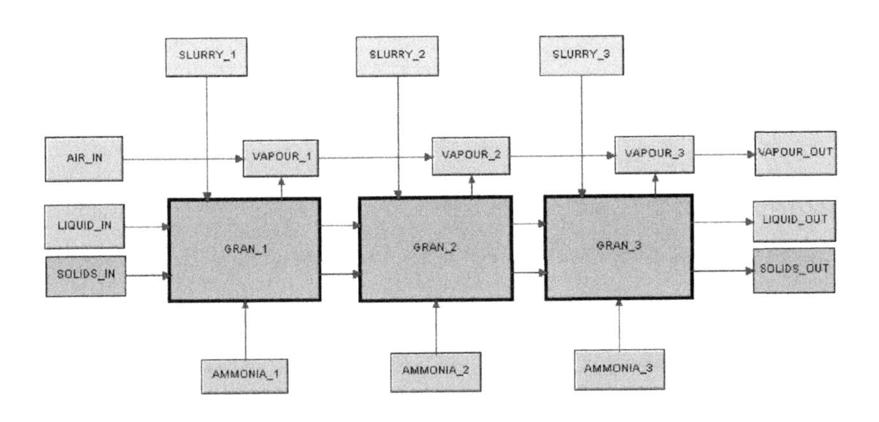

FIGURE 23 The reaction-crystallisation model flowsheet.

The large differential-algebraic equation model was developed within the Daesim dynamics software system and solved for a range of input and parametric changes.

In Figure 24 we can see the change in MAP and DAP compositions together with changes in liquid and solid phase temperatures along the granulator drum. MAP composition drops during the reaction from 0.47 (mass fraction) to 0.14 whilst DAP composition varies from 0.37 (mass fraction) at the inlet of the drum to 0.63 at the exit. Given these compositions in Figure 25 the N:P mole ratio varies from 1.44–1.82 and the moisture in the drum increases from 2.1% to 3.4%. The simulations have been performed by using real plant data and all these results approximate well the measured operating values determined from plant surveys of product mole ratio 1.85 and product moisture 3.13%. The temperature of the product in the case of these data was expected to be 98°C compared with 101°C, the temperature of the solid phase.

Using the same operation conditions we can now change the ammonia flow by ±10% into the first section of the drum. In Figure 26 and 27 we can see the

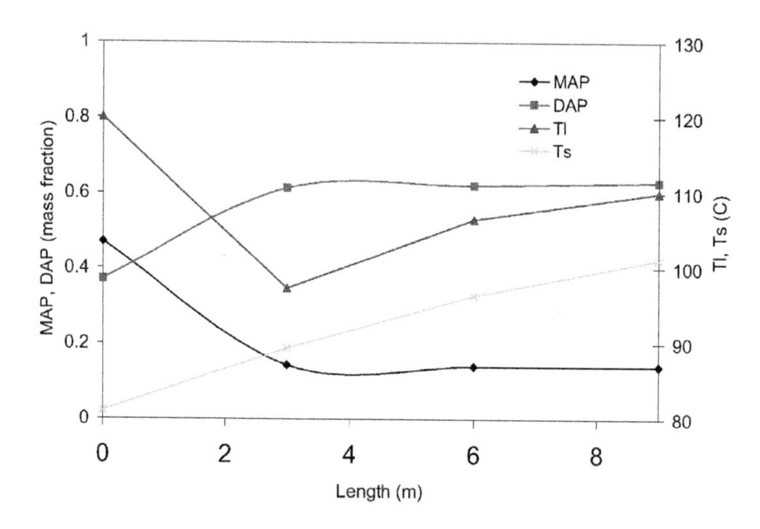

FIGURE 24 Changes in mass fraction and temperatures during reaction-crystallisation.

changes in mole ratio and moisture along the drum. As expected from plant observations, an increase of ammonia flow will generate a decrease of moisture and an increase of mole ratio, especially in the first section. The opposite effect will produce under-ammoniation of the drum bed, leading to an increase in the moisture and a decrease in the mole ratio.

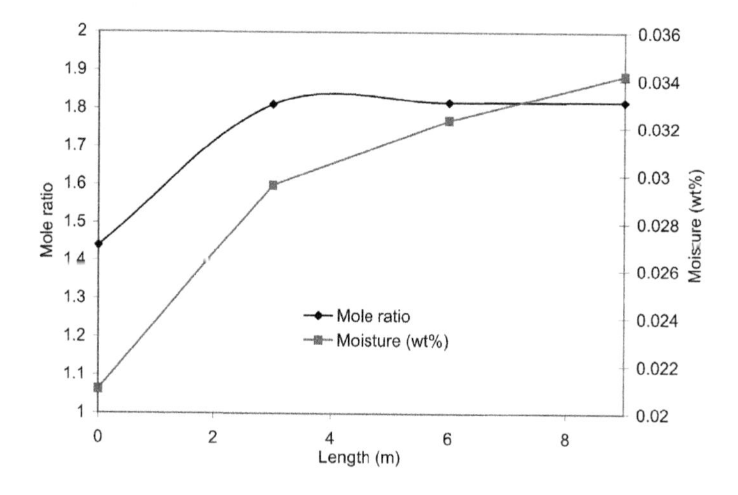

FIGURE 25 Changes in mole ratio and moisture during reaction-crystallisation.

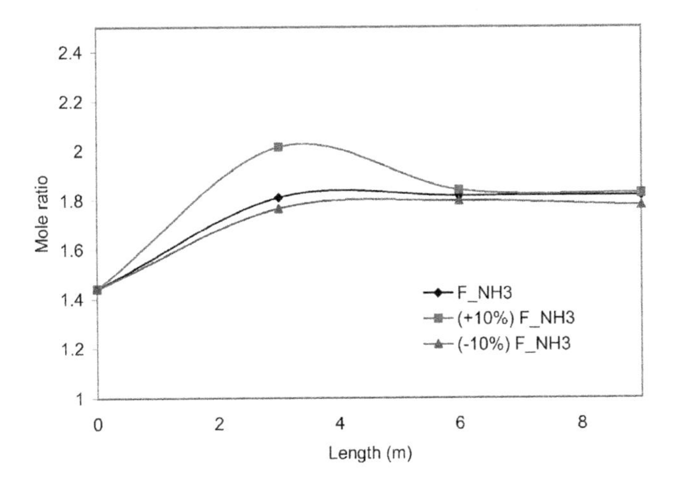

FIGURE 26 **Effect of ammonia flow rate on the mole ratio in the first section of the drum.**

The effect of changing the amount of ammonia into the third or end section of the drum can be seen in Figure 28 and 29. Because the change in the ammonia flow is done at the end of the granulator, we see no variation in moisture and mole ratio in the first two sections (6m) and only a very small change in the last

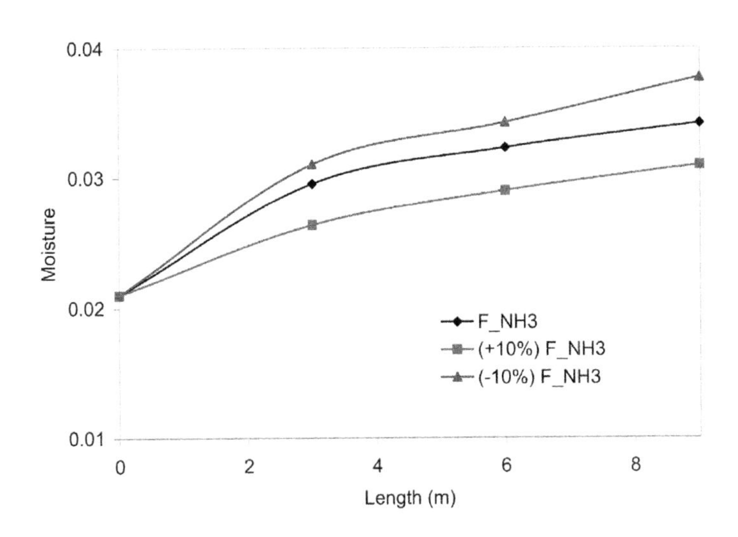

FIGURE 27 **Effect of ammonia flow rate on the moisture in the first section of the drum.**

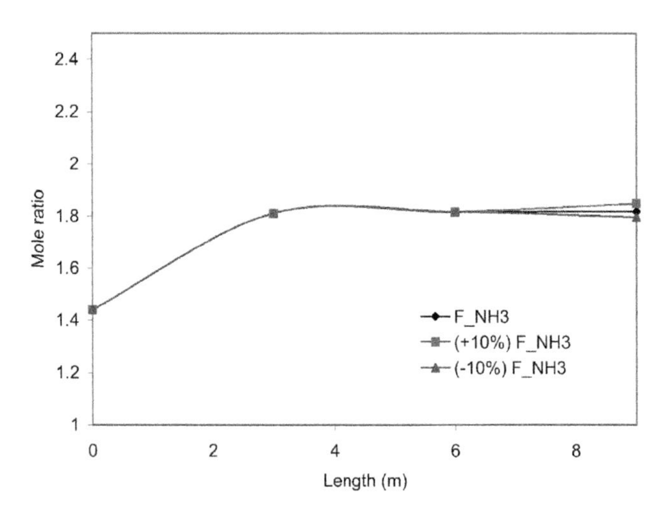

FIGURE 28 **Effect of ammonia flow rate on the mole ratio in the last section of the drum.**

section. Therefore, any change done in the first section of the drum will have a significant impact over the whole granulator whilst little impact occurs for similar changes near the exit of the drum.

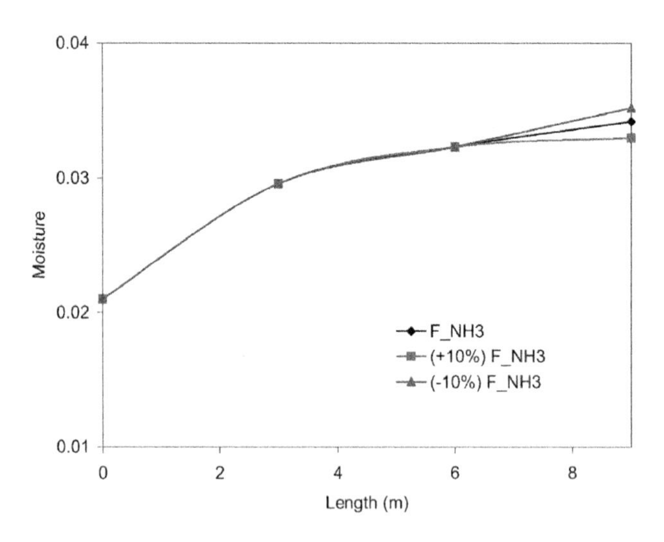

FIGURE 29 **Effect of ammonia flow rate on the moisture in the last section of the drum.**

8.2.7. Summary

A combined reaction and crystallisation model with population balance for the MAP–DAP granulation process has been developed. A saturation index depending on the liquid phase temperature and mass fractions of MAP and DAP was determined based on solubility diagrams. The rate of crystallisation was then defined based on the saturation index and particle surface area. A set of DAEs has been developed and implemented in Daesim dynamics. The model describes the dynamics of the liquid phase and solid phase during the granulation process. The reaction-crystallisation model along with population balance models describing mechanisms such as layering, agglomeration and breakage is now the first comprehensive modelling tool which captures the dynamics of the process.

The reaction-crystallisation model predicts the observed behaviour of the plant quite well and can be used to investigate the effect of changing operating conditions. It was shown that over-ammoniation of the drum bed will generate a decrease of the moisture and an increase of mole ratio whilst under-ammoniation will lead to an increase in the moisture and a decrease in the mole ratio.

Prediction of the DAP formation and crystallisation based on the reaction-crystallisation model is now possible. The model provides a basis for further dertailed work on the better prediction on liquid phase content and its manipulation for improved equipment design and operational efficiency.

8.3. SHORT-PATH EVAPORATOR MODEL

The short-path evaporator consists of an inner cylindrical body surrounded by an outer cylinder. One of the cylinders acts as an evaporation surface whilst the other acts as a condensation surface (see Figure 30). The liquid material to be distilled is fed to the evaporation wall from the top and falls down as a thin film. As the evaporation wall is heated, the lower-boiling compounds vapourise and travel to the condensation wall, which is cooled. The condensed vapour falls down the condensation wall and is collected as the evaporated product. The remaining liquid from the evaporation wall is collected as residue. The evaporation and condensation surfaces are kept at constant temperatures. Due to low pressure inside the separator, a falling film (without boiling) is formed and the concentration and temperature profiles of the most volatile compounds decrease in the axial and radial directions (see Figures 31a and 31b).

8.3.1. Model Objectives

The objective of this modelling exercise is to determine the composition and temperature profiles along the vertical and radial directions of a short-path evaporator operating at a steady state. The profiles are to be generated as a function of the common operational parameters.

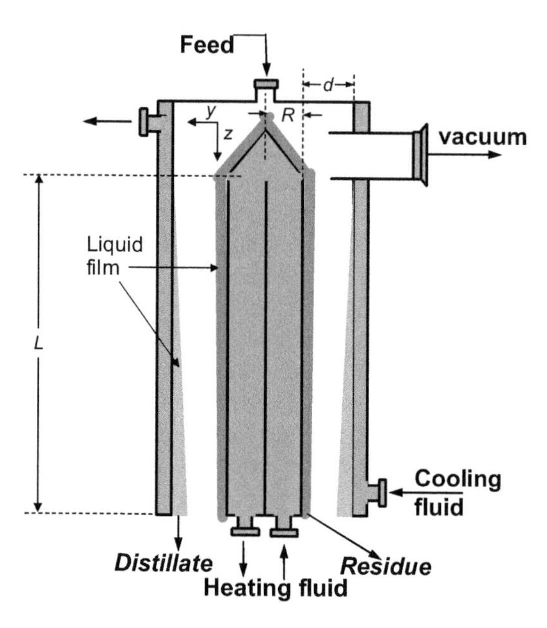

FIGURE 30 Schematic diagram of a short-path evaporator (from Sales-Cruz and Gani, 2006).

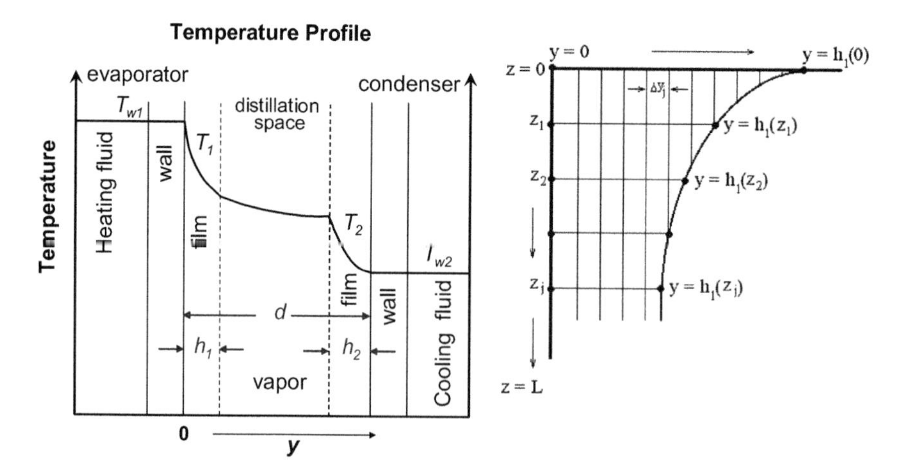

FIGURE 31 Mass and heat transfer directions and discretisation scheme.

8.3.2. Model Description

8.3.2.1 Assumptions

The developed generalized two-dimensional steady state model for short path evaporation considers the following assumptions:

- The process is in steady state.
- The liquid films on the evaporation and condensation walls are much thinner than the corresponding cylinder diameters.
- Rectangular coordinates are used.
- The liquids are Newtonian.
- The flow in the vertical direction is laminar.
- Re-evaporation and splashing phenomena are neglected.
- Operation occurs far from the extremities of the evaporator, that is, for a fully developed flow.
- There is no diffusion in the axial direction and the radial flow is neglected.

8.3.2.2 Model Equations

The model for the short-path evaporator as derived by Sales-Cruz and Gani (2006) is given below:

Momentum balance: In most cases of short-path evaporation, the evaporating liquid is highly viscous and hence the Reynolds numbers are small. The Navier–Stokes equation (at steady state) for the laminar regime describes the velocity profile of falling film

$$v(z)\frac{\partial^2 v(y,z)}{\partial y^2} = -g \tag{99}$$

where, the boundary conditions are given by,

$$v(0,z) = 0, \qquad v(h,z) = v_{max} \tag{100}$$

where v is the velocity, g is the gravity constant, and y and z is the radial and axial coordinates respectively.

Energy balance: The temperature, T, profile in the falling film is given by the equation

$$v(y,z)\frac{\partial T(y,z)}{\partial z} = \frac{\lambda}{\rho Cp}\left[\frac{\partial^2 T(y,z)}{\partial y^2} + \frac{\partial^2 T(y,z)}{\partial z^2}\right] \tag{101}$$

With the boundary conditions

$$T(y,0) = T_F \tag{102a}$$

$$T(0,z) = T_{w1} \tag{102b}$$

$$\lambda \frac{\partial T(y,z)}{\partial y}\bigg|_{y=h_1} = \Delta H^{vap} \cdot k \tag{102c}$$

where l is thermal conductivity; r is density; Cp is thermal capacity; DH^{vap} is heat of evaporation of the multi-component mixture.

Mass balance: The composition, C_i, profiles for each compound is calculated from the diffusion equation.

$$v(y,z)\frac{\partial C_i(y,z)}{\partial z} = D_i\left[\frac{\partial^2 C_i(y,z)}{\partial y^2} + \frac{\partial^2 C_i(y,z)}{\partial z^2}\right], \qquad i = 1,\ldots,N \tag{103}$$

where D_i is the (constant) diffusion coefficient for the i-th component. The boundary conditions for Eq. 103 are

$$C_i(y,0) = C_{i,o} \tag{104a}$$

$$\frac{\partial C_i(0,z)}{\partial y} = 0 \tag{104b}$$

$$D_i \frac{\partial C_i(y,z)}{\partial y}\bigg|_{y=h_1} = I_i(z) \tag{104c}$$

The flow (evaporation) rate I_i for each component is describe by the continuity equation

$$\frac{\partial I_i(z)}{\partial z} = -2\pi \cdot R \cdot k_i, \qquad i = 1,\ldots,N \tag{105}$$

where the effective rate of evaporation, k_i, is calculated through a modified Langmuir–Knudsen equation

$$k_i = \frac{p_i^{vap} T_s(z)}{\sqrt{2\pi R_g M_i T_s(z)}}\left(\frac{P}{P_{ref}}\right)\left\{1 - (1-F)\left[1 - e^{h/(\kappa\beta)}\right]^n\right\}, \quad i = 1,\ldots,N \tag{106}$$

The factor P/P_{ref} is a correction for correcting the vacuum pressure of operation; F is the surface ratio and k is the anisotropy of the vapour phase given by,

$$F = \frac{A_k}{A_k + A_V}, \qquad log\kappa = 0.2F + 1.38(f + 0.1)^4 \tag{107}$$

A_k and A_v are the condensation and evaporation areas, respectively.

The effective rate of evaporation [Eq. 106] is also a function of mixture properties (the vapour pressure p_i^{vap} and molecular weight M_i of each compound) as well as on design parameters (the radius of the evaporator inside cylinder R and the surface temperature T_s).

Another important variable of interest is the thickness film, h_l, along the evaporator height that is calculated as follows:

$$h_1 = \sqrt[3]{\frac{3 \cdot \nu_{mix}}{2\pi g \cdot R_1 \cdot c_{total}} I_{total}} \quad (108)$$

where,

$$I(z) = \sum_{i=1}^{N} I_i(z) \quad (109)$$

$$c = \sum_{i=1}^{N} C_i(z) \quad (110)$$

where n is the kinematic viscosity, I is the total rate of evaporation and c is the total concentration of the multicomponent mixture.

8.3.2.3 Discretisation

The above PDAE system represented by Eqs 99, 100, 101, 102a-102c, 103, 105-110, needs to be discretised in the radial direction and integrated in the axial direction. A generic short-path evaporation model for a multicomponent mixture is described below considering an

N-point discretisation scheme for the radial coordinate y (as shown in Figures 31a-b).

Note that in order to have an easier model analysis, some of the equations given above are repeated below.

Balance equations and boundary conditions:

$$\Delta y = \frac{(h_1 - a_y)}{N} \quad (111)$$

$$y_j = j \cdot \Delta y, \qquad j = 1, ..., N \quad (112)$$

The component flow is computed from the continuity equation for each component i (NC is the number of components in the mixture):

$$\frac{dI_i}{dz} = -2 \cdot \pi \cdot R_1 \cdot k_i \cdot x_i \cdot E_{ref}, \qquad i = NC \quad (113)$$

From a discretisation of the heat balance equation (second order partial differential equation), the temperature profile T_j (meaning the temperature at each y_j) with respect to the position z is calculated as

$$\frac{dT_j}{dz} = \frac{\vartheta}{v_{z,j}} \cdot \left(\frac{T_{j+1} - 2T_j + T_{j-1}}{\Delta y^2} \right), \qquad j = 1, ..., N \tag{114}$$

where,

$$\vartheta = \frac{M_{w,mix} \lambda_{mix}}{C_{P,mix} \cdot \rho_{mix}} \tag{115}$$

Boundary (initial) conditions at $z = 0$ are given by

$$T_0(0) = T_w \tag{116}$$

$$T_j(0) = T_f, \qquad j = 1, ..., N+1 \tag{117}$$

$$T_s(0) = T_{N+1} \tag{118}$$

$$I_i(0) = I_{i,\text{expt}}, \qquad i = 1, ..., NC \tag{119}$$

Boundary condition at $y = 0$

$$\frac{dT_0}{dz} = 0 \tag{120}$$

Boundary condition at $y = h_l$

$$T_{N+1} - T_N + \frac{\Delta y \Delta H_{mix}^{vap} k_{total}}{\lambda_{mix}} = 0 \tag{121}$$

$$T_s = T_{N+1} \tag{122}$$

Constitutive (evaporation rate efficiency) models:

$$E_{ref} = \left[1 - (1 - F)\left(1 - e^{-\frac{d}{\kappa \beta}}\right)^{n_{mc}} \right] \cdot \left(\frac{P}{P_{ref}} \right) \tag{123}$$

where,

$$F = \frac{A_k}{A_k + A_h} \tag{124}$$

$$\kappa = 10^{0.2\,F + 1.38\,(F+0.1)^4} \tag{125}$$

<u>*Constitutive model*</u>: The velocity field is computed from the Nusselt's solution of the Navier–Stokes equation as

$$v_{z,j} = \frac{g \cdot h_1^2}{\nu_{mix}} \left[\frac{y_j}{h_1} - \frac{1}{2} \left(\frac{y_j}{h_1} \right)^2 \right], \qquad j = 1, \ldots, N \tag{126}$$

<u>*Defined surface evaporation rate*</u>:

$$k_i = \frac{\gamma_i \cdot P_i^{vap}}{\sqrt{2\pi R_g M_{w,i} T_s}}, \qquad i = 1, \ldots, NC \tag{127}$$

$$k_{total} = \sum_{i=1}^{NC} (x_i \cdot k_i) \tag{128}$$

<u>*Defined relations*</u>: Volumetric flows, Q_i

$$Q_i = \frac{I_i \cdot M_{w,i}}{\rho_i}, \qquad i = 1, \ldots, NC \tag{129}$$

$$Q_{total} = \sum_{i=1}^{NC} Q_i \tag{130}$$

<u>*Defined relations*</u>: Concentrations, c_i

$$c_i = \frac{I_i}{Q_{total}}, \qquad i = 1, \ldots, NC \tag{131}$$

<u>*Defined relations*</u>: Molar fractions, x_i

$$x_i = \frac{I_i}{I_{total}}, \qquad i = 1, \ldots, NC \tag{132}$$

<u>*Constitutive (property model) equations*</u>: Mixing rules: note that for each of the pure component properties, the corresponding correlation (model) is necessary.

$$I_{total} = \sum_{i=1}^{NC} I_i \tag{133}$$

$$c_{total} = \sum_{i=1}^{NC} c_i \tag{134}$$

$$\rho_{mix} = \sum_{i=1}^{NC} x_i \rho_i \tag{135}$$

$$\lambda_{mix} = \sum_{i=1}^{NC} x_i \lambda_i \tag{136}$$

$$c_{p,mix} = \sum_{i=1}^{NC} x_i c_{p,i} \tag{137}$$

$$M_{w,mix} = \sum_{i=1}^{NC} x_i M_{w,i} \tag{138}$$

$$\eta_{mix} = \sum_{i=1}^{NC} x_i (\eta_i / \rho_i) \tag{139}$$

$$\Delta H_{mix}^{vap} = \sum_{i=1}^{NC} x_i \Delta H_i^{vap} \tag{140}$$

$$\nu_{mix} = \frac{\eta_{mix}}{\rho_{mix}} \tag{141}$$

Defined relations:

$$A_h = PI{*}diameter{*}L \tag{142}$$

$$A_k = PI{*}(diameter - 2.0{*}d){*}L \tag{143}$$

8.3.3. Model Analysis

Consider an evaporation process used in the production of a drug where (after the reaction stage) the active chemicals of the active pharmaceutical ingredient (API) needs to be recovered and purified. As an example, considering the liquid mixture composed of six heat sensitive compounds (called here A, B, C, D, E and F) that needs to be processed. Here, A is the lightest and the most volatile compound whilst F is the compound with the highest boiling point. The role of the short-path evaporator is to separate the active chemicals (mainly C, D and E) together with the inert component F from the chemicals A and B.

The discretised set of equations representing the short-path evaporation model is a DAE system (Eqs. 111–141) consisting of (number of discretisation points, N = 10; number of compounds, NC = 6),

- 118 total equations - 17 ODEs; 1 implicit algebraic and 100 explicit AEs.
- 300 total variables 17 dependent; 1 implicit-unknown; 100 explicit-unknown; 4 known (design-input); 193 parameters.

- Variables to specify 197 variables (193 parameters and 4 design-input variables). The values for the known variables can be found in Sales-Cruz and Gani (2006) and also the supplied ICAS-MoT file (see ch-8-3-short-path-evap.mot file in Appendix-1).

8.3.3.1 Calculation Order

Given: values of the following variables pure component property model parameters ($PVAP_A$, $PVAP_B$, $PVAP_C$, $PVAP_D$ –Vapour Pressure; $HVAP_A$, $HVAP_B$, $HVAP_C$, $HVAP_D$ -Heat of vaporisation; RHO_A, RHO_B, RHO_C, RHO_D Density; CP_A, CP_B, CP_C, CP_D, CP_E -Liquid Heat Capacity; $VISCO_A$, $VISCO_B$, $VISCO_C$, $VISCO_D$, $VISCO_E$ Liquid Viscosity; $THERMAL_A$, $THERMAL_B$, $THERMAL_C$, $THERMAL_D$ -Thermal Conductivity); properties ($\boldsymbol{M_w}$; $\boldsymbol{T_c}$; $\boldsymbol{T_b}$, γ); system variables (P, T_f, $\boldsymbol{I_0}$); evaporator dimensions (L, $diameter$, d, a_y); constants (P_{ref}, R_g, g, π, N).

Calculate: The DAE system of equations are solved to calculate the 17 state variables ($\boldsymbol{T}$, $\boldsymbol{I}$, T_0); 1 implicit unknown variable (T_{10}); and 100 explicit unknown variables (R_1, A_h, A_k, T_s, v_{mix}, ρ_{mix}, λ_{mix}, η_{mix}, $c_{p\text{-}mix}$, ΔH_{mix}^{vap}, $M_{w\text{-}mix}$, c_{total}, I_{total}, Q_{total}, k_{total}, E_{ref}, β, F, κ, θ - scalers; $\boldsymbol{x}$, $\boldsymbol{Q}$, $\boldsymbol{c}$, $\boldsymbol{P}_{vap}$, $\boldsymbol{\rho}$, $\boldsymbol{\lambda}$, $\boldsymbol{\eta}$, $\boldsymbol{c}_p$, $\Delta\boldsymbol{H}^{vap}$, $\boldsymbol{k}$ – vectors of size NC; $\boldsymbol{y}$, $\boldsymbol{v}_z$– vectors of size N).

Calculation order of equations: The order of solving the 118 equations are given in terms of the equation numbers – 142, 143, pure component property models ($\boldsymbol{P}_{vap}$, $\boldsymbol{\rho}$, $\boldsymbol{\lambda}$, $\boldsymbol{\eta}$, $\boldsymbol{c}_p$, $\Delta\boldsymbol{H}^{vap}$ – note that these equations are not given in the above equation set), 129, 130, 131, 133, 134, 132, 135-141, 124, 125, 123, 111, 108, 112, 126, 115, 127, 128, 120, 114, 121, 113.

8.3.4. Numerical Solution

The DAE system is solved through *ICAS-MoT* (Sales-Cruz and Gani, 2003; Sales-Cruz, 2006) using the BDF integration method. The feed flows and calculated exit flows of the distillate and residual are reported in Table 2, where

TABLE 2 Simulated Component Flow rates in the Feed, Residual and Distillate Streams of the Short-path Evaporator

Compound	Feed I_0	Residual I_R	Distillate I_D
A	$6.11*10^{-5}$	0.0	$6.11*10^{-5}$
B	$1.22*10^{-5}$	0.0	$1.22*10^{-5}$
C	$4.72*10^{-2}$	$4.46*10^{-2}$	$2.61*10^{-3}$
D	$1.90*10^{-4}$	$1.87*10^{-4}$	$3.07*10^{-6}$
E	$2.17*10^{-3}$	$2.14*10^{-3}$	$2.46*10^{-5}$
F	$6.66*10^{-4}$	$6.66*10^{-4}$	0.0

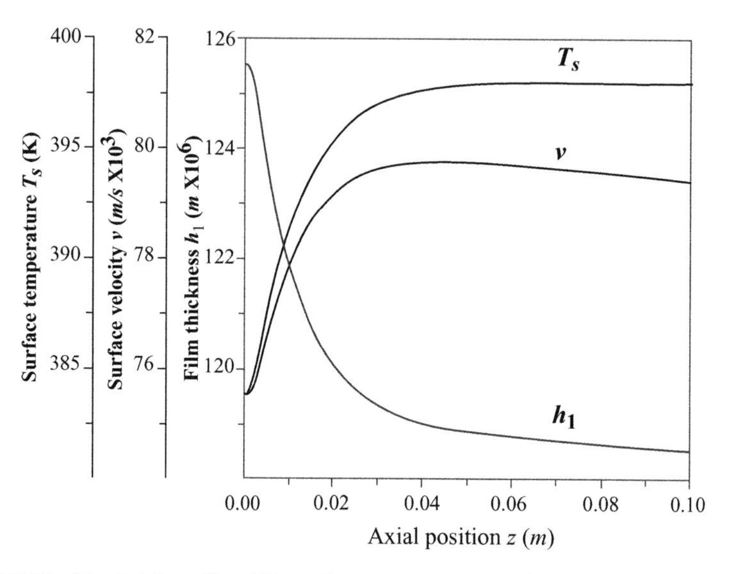

FIGURE 32 **Simulated profiles of the surface temperature, velocity and film thickness in the axial direction.**

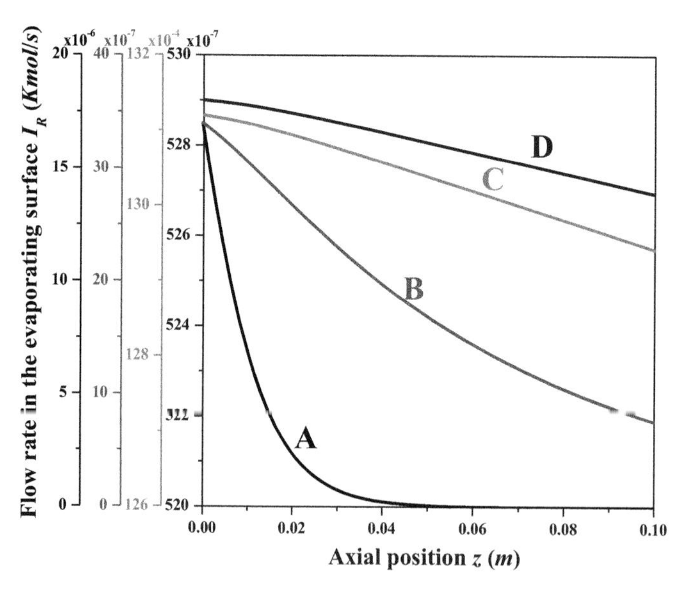

FIGURE 33 **Simulated composition profiles for compounds A, B, C and D in the axial direction.**

it can be seen that the chemical product API (formed by *C*, *D*, *E* and *F*) is obtained as the residual in the short-path evaporator.

The surface velocity, temperature, thickness and flow rates (of components A, B, C and D) for the evaporating film are shown in Figures 32–33. It can be noted that as the lighter compounds evaporate from the film (see Figure 33), the temperature of the film increases in the axial direction whilst the thickness of the film decreases. This shows that qualitatively, the model is correct. To match the exact plant data, the model parameters would need to be tuned. Also, once the model has been validated, then it can be used to study the evaporator operation as a function of the operational (design) parameters.

8.3.5. Discussion

- With the information obtained from the simulation of the short path evaporator, check if it is possible to calculate the heat and mass transfer coefficients?
- Adapt the ICAS-MoT model given in Appendix-1 to study the separation of different binary mixtures.
- Study the effects of changing the values of the feed temperature, feed flows or evaporator dimensions (height, radius, gap between evaporator and condenser) on the mixture separation.

8.3.6. Nomenclature

8.3.6.1 System variables

P	System pressure [Pa]
T	Temperature profile in the liquid film [K]
T_f	Feed temperature [K]
T_s	Film surface temperature [K]
T_{out}	Output temperature [K]
$\boldsymbol{I}$	Partial molar flow rate Input (mol/s)
I_{total}	Total molar flow [mol/s]
$\boldsymbol{k}$	Rate of evaporation (mol/m2.s)
k_{tot}	Total rate of evaporation (mol/m2.s)
$\boldsymbol{v}_z$	Film surface velocity
h_1	Film thickness
c_{tot}	Total concentration mol/m3
$\boldsymbol{x}$	Mole fraction
Gamma	Peripheral liquid load

8.3.6.2 Evaporator Dimensions

L	Length of evaporator [m]
diameter	Diameter of the evaporator [m]
R_1	Evaporator radius [m]

d	Distance evaporator condenser [m]
A_h	Evaporation surface area [m2]
A_k	Condensation surface area [m2]

8.3.6.3 *Evaporation rate efficiency*

nmc	Number of intermolecular collisions before the vapour reaches isotropic state
β	Mean path of the vapour molecule [m]
F	Surface ratio
κ	Degree of anisotropy of the vapour phase in the space between the evaporator and condenser
E_{ref}	Pressure system effect

8.3.6.4 *Constants*

P_{ref}	Reference pressure [Pa]
R_g	Universal gas constant [J/mol K]
g	gravitational constant [m/s2]
π	PI number

8.3.6.5 *Integration parameter*

N	Number of radial integration points
a_y	Initial integration value in y coordinate
b_y	Final integration value in y coordinate (mobile)

8.3.6.6 *Physico-chemical properties*

γ	Activity coefficients
T_c	Critical temperature [K]
T_b	Boiling temperature [K]
M_W	Molecular weight [kg/mol]

8.3.6.7 *Thermo properties*

$\boldsymbol{P}_{vap}$	Pressure of saturated vapour (DIPPR 101) [Pa]
$\Delta\boldsymbol{H}_{vap}$	Heat of vaporisation [kJ/mol]
$\boldsymbol{c}_p$	Liquid heat capacity [J/(mol K)]
ρ	Density [kg/m3]
ς	Liquid viscosity [kg/m s]
λ	Thermal conductivity [W/(m K)]
η	Kinematics viscosity [m2/s]

8.3.6.8 Thermo properties constants

$PVAP_A$, $PVAP_B$, $PVAP_C$, $PVAP_D$	Vapour pressure
$HVAP_A$, $HVAP_B$, $HVAP_C$, $HVAP_D$	Heat of vaporisation
RHO_A, RHO_B, RHO_C, RHO_D	Density
CP_A, CP_B, CP_C, CP_D, CP_E	Liquid heat capacity
$VISCO_A$, $VISCO_B$, $VISCO_C$, $VISCO_D$, $VISCO_E$	Liquid viscosity
$THERMAL_A$, $THERMAL_B$, $THERMAL_C$, $THERMAL_D$	Thermal conductivity

8.4. ACKNOWLEDGEMENTS

Acknowledgement is given to UMATAC Industrial Processes, Alberta, Canada for permission to publish aspects of the ATP processor system. We are thankful to Mr Bill Taciuk for his generosity. We also acknowledge permission to use material from Dr Nicoleta Maynard (nee Balliu) from her research work on fertiliser granulation circuit dynamics.

REFERENCES

Adetayo, A.A., Litster, J.D., Cameron, I.T., 1995. Steady state modelling and simulation of a fertiliser granulation circuit. In Computers & Chemical Engineering. 19 (4), 383–393.

Balliu, N., Cameron, I.T., 2007. Performance assessment and model validation for an industrial granulation circuit. In Powder Technology. 179 (1–2), 12–24.

Boyce, M. P., 1997. Transport and Storage of Fluids. In 'Perry's Chemical Engineers' Handbook' (R. H. Pery and D. W. Green, Eds.), 7th ed. McGraw-Hill.

Chung, T.Y., Welty, J., 1989. Tube array heat transfer in fluidized beds A study of particle size effects. In AIChEJ. 35 (7), 1170–1176.

Daesim, 2005, Daesim Dynamics, htsb://www.daesim.com/.

Das Gupta, S., Khakhar, D.V., Bhatia, S.K., 1991. Axial transport of granular solids in horizontal rotating cylinders. Powder Technology. 67, 145–151.

Davis, C. H., and Lee, R. G., 1967. 'TVA-Publication.' The American Chemical Society Meeting, Atlanta, November.

Gupta, S.D., Khakkar, D.V., 1991. Axial transport of granular solids in horizontal rotating cylinders. Part 1: Theory, Powder Technology. 67, 145–151.

Hill, P.J., Ng, K.M., 1995. A discretized population balance for nucleation, growth and aggregation. AIChEJ. 5 (41), 1204–1216.

Hirosue, H., 1989. Influence of particles falling from flights on volumetric heat transfer coefficient in rotary dryers and coolers. Powder Technology. 59, 125–128.

Hirosue, H. and H. Shinohara, 1978, "Volumetric heat transfer coefficient and pressure drop in rotary dryers and coolers," Proceedings. 1st International Symposium. Drying, 152–159, Science Press, Princeton.

Hounslow, M.J., Ryall, R.L., Marshall, V.R., 1988. A discretized population balance for nucleation, growth and aggregation. AIChEJ. 34 (11), 1821–1832.

Mullin, J.W., 2001. 'Crystallization'. New York: Butterworth-Heinemann.

Murray, D.B., O'Connor, P., Gilroy, P., 1996. Cross-flow heat transfer in the freeboard region of a fluid bed. IMechE. 210, 245–253.

Nielsson, F.T., 1986, Granulation, Volume 5 of Fertilizer Science and Technology, Manual of Fertilizer Processing, Chapter 9, 159–226, Marcel-Dekker.

Perry, R.H., 1997. Chemical Engineers Handbook, 7th Edition. McGraw-Hill Publishers.

Schofield, F.R., Glikin, P.G., 1962. Rotary driers and coolers for granular fertilizers. Transactions of the Institution of Chemical Engineers. 40, 183–190.

Seidell, A., 1941. Solubility of inorganic and organic compounds: A compilation of quantitative solubility data from the periodical literature. New York: Van Nostrand.

Sales-Cruz, M., Gani, R., 2003. A Modelling Tool for Different Stages of the Process Life. In: S. P. Asprey., S. Macchietto (Eds.), *'Computer-Aided Chemical Engineering, Vol. 16, Dynamic Model Development'*. Amsterdam: Elsevier; pp. 209–249.

Sales-Cruz, M., 2006. 'Development of a computer aided modelling system for bio and chemical process and product design, PhD-thesis'. Lyngby, Denmark: Technical University of Denmark.

Sales-Cruz, M., Gani, R., 2006. Computer-aided modelling of short-path evaporation for chemical product purification, analysis and design. In Chemical Engineering Research and Design. 84, 583–594.

Sterritt, D.C., Murray, D.B., 1992. Heat transfer mechanisms in an in-line tube bundle subject to a particulate cross-flow. Institution of Mechanical Engineering. 206, 317–326.

Shah, P. J., Upadhyay, S. N., and Saxena, S. 1981. Heat transfer from smooth horizontal tubes immersed in gas fluidized beds. In AIChE Symposium Series, Heat Transfer 351–358 Milwaukee.

Stokka, P., 1985. Ullmann's Encyclopedia of Industrial Chemistry, ref to Norsk Hydro Internal Report, 5th edit. Germany: Weinheim.

Wang, F.Y., Cameron, I.T., Litster, J., Rudolph, V., 1995. A fundamental study on particle transport through rotary dryers for flight design and system optimization. Drying Technology. 13 (5-7), 1261–1278.

Wesenberg, G.H., 1986, Diammonium Phosphate Plants and Processes, Volume 5 of Fertilizer Science and Technology, Manual of Fertilizer Processing, Chapter 10, 227–287, Marcel Dekker.

Wildeboer, W.J., Litster, J.D., Cameron, I.T., 2005. Modelling nucleation in wet granulation. In Chemical Engineering Science. 60, 3751–3761.

Wu, R.L., Lim, C.J., Chaonki, J., Grace, J., 1987. Heat transfer from a circulating fluidized bed to membrane waterfall surfaces. AIChE Journal. 33 (11), 1888–1893.

Zenz, F.A., Othmer, D.F., 1960. Fluidization and Fluid-particle Systems. USA: Reinhold Publishing Corporation.

Zhang, J., 1998. 'Dynamics and control of a fertiliser granulation circuit'. University of Queensland: Masters Thesis.

Zhang, J., Litster, J.D., Wang, F.Y., Cameron, I.T., 2001. Evaluation of control strategies for fertiliser granulation circuits using dynamic simulation. Powder Technology. 108, 122–129.

Chapter 9

Tennessee Eastman Plant-wide Industrial Process Challenge Problem*

The objective of this case study is to analyse and solve the published models for Tennessee Eastman plant-wide industrial process challenge problem (Downs and Vogel, 1993). Two versions of the models are used:

- simplified model by Ricker and Lee (1995);
- full model by Jockenhovel et al. (2004).

For both the cases, the model equations are presented and analysed, a solution strategy is determined and, based on it, numerical solutions are obtained and compared with the corresponding published results.

9.1. PROCESS DESCRIPTION

Figure 1 shows the flowsheet of the Tennessee Eastman (TE) process (Downs and Vogel, 1993). There are four unit operations in this flowsheet: an exothermic two-phase reactor, a flash separator, a reboiler stripper and a recycle compressor.

The TE process produces two products (G and H) and one undesired by-product (F) from four reactants (A, C, D and E), according to the following reduced reaction scheme:

$$A_{(g)} + C_{(g)} + D_{(g)} \rightarrow G_{(l)} \tag{1}$$

$$A_{(g)} + C_{(g)} + E_{(g)} \rightarrow H_{(l)} \tag{2}$$

$$1/3A_{(g)} + D_{(g)} + 1/3E_{(g)} \rightarrow F_{(l)} \tag{3}$$

9.2. SIMPLIFIED MODEL

9.2.1. Process Model

The objective of this simplified version of the TE process model is to capture the essential characteristics of the process assuming perfect control of the

* Written in conjunction with Prof Mauricio Sales-Cruz

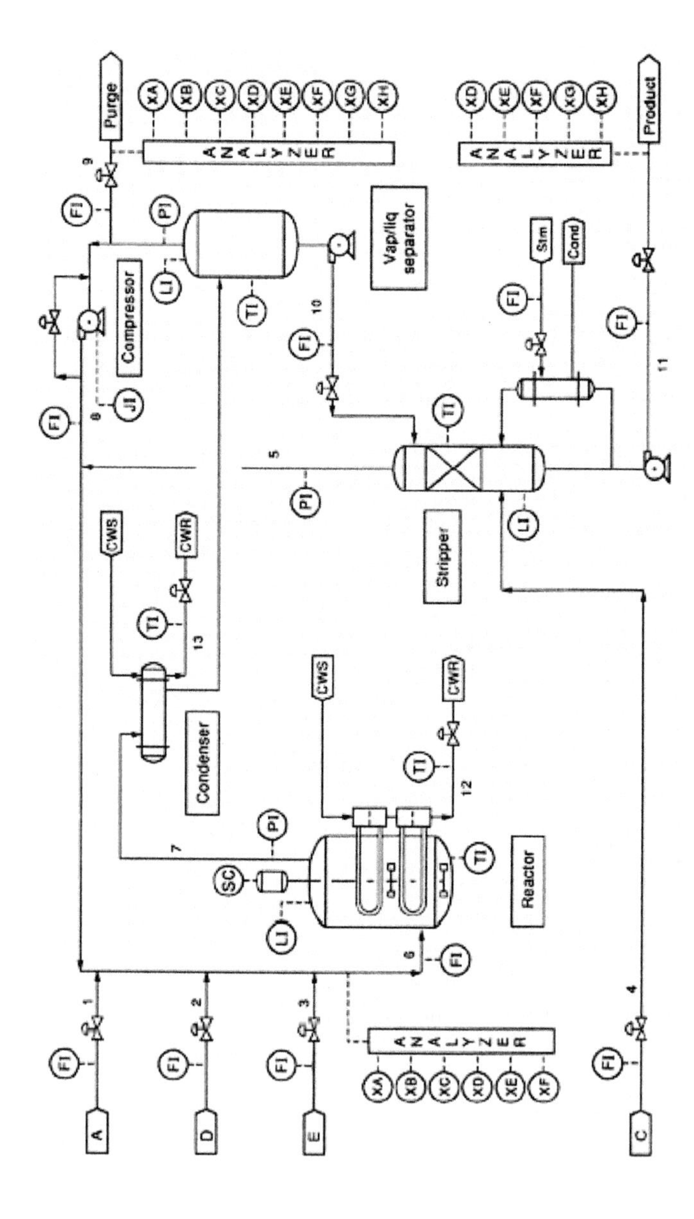

FIGURE 1 The Tennessee Eastman process (Downs and Vogel, 1993).

heating/cooling taking place in the process. That is, the energy balance equations are ignored and only the mass balance–related equations are solved. This means that the dynamics of the compressor and reboiler stripper are ignored but the accumulations of mass in the purge unit and the feed mixer (holding tank) are considered. More details of the model simplification are given in Ricker and Lee (1995).

Mass balance equations (holdup of component i in each unit operation)

$$\frac{dN_{i,r}}{dt} = y_{i,6}F_6 - y_{i,7}F_7 + \sum_{j=1}^{3} \nu_{ij}R_j \quad (i = A, B, \ldots, H) \tag{4}$$

$$\frac{dN_{i,s}}{dt} = y_{i,7}F_7 - y_{i,8}(F_8 + F_9) - x_{i,10}F_{10} \quad (i = A, B, \ldots, H) \tag{5}$$

$$\frac{dN_{i,m}}{dt} = \sum_{j=1}^{3} y_{ij}F_j + y_{i,5}F_5 + y_{i,8}F_8 + F_i^* - y_{i,6}F_6 \quad (i = A, B, \ldots, H) \tag{6}$$

$$\frac{dN_{i,p}}{dt} = (1 - \phi_i)x_{i,10}F_{10} - x_{i,11}F_{11} \quad (i = G, H) \tag{7}$$

Conditional equations (thermodynamic relations) for reactor

$$P_r = \sum_{i=A}^{H} P_{i,r} \tag{8}$$

where

$$P_{i,r} = \frac{N_{i,r}RT_r}{V_{Vr}} \quad (i = A, B, C) \tag{9}$$

$$P_{i,r} = \gamma_{ir}x_{ir}P_i^{sat}(T_r) \quad (i = D, E, \ldots, H) \tag{10}$$

$$V_{Lr} = \sum_{i=D}^{H} \frac{N_{i,r}}{\rho_i} \tag{11}$$

Conditional equations (thermodynamic relations) for separator

$$P_s = \sum_{i=A}^{H} P_{i,s} \tag{12}$$

where

$$P_{i,s} = \frac{N_{i,s}RT_s}{V_{Vs}} \quad (i = A, B, C) \tag{13}$$

$$P_{i,s} = \gamma_{is}x_{i,10}P_i^{sat}(T_s) \quad (i = D, E, \ldots, H) \tag{14}$$

$$V_{Ls} = \sum_{i=D}^{H} \frac{N_{i,s}}{\rho_i} \tag{15}$$

Constitutive (property) models

$$P_i^{sat}(T_s) = A_i - \frac{B_i}{(C_i + T_s)} \quad (i = A, B, C, D, E, F, G, H) \tag{16a}$$

$$P_i^{sat}(T_r) = A_i - \frac{B_i}{(C_i + T_r)} \quad (i = A, B, C, D, E, F, G, H) \tag{16b}$$

Conditional equations (volume) for reactor, separator and purge

$$V_{Vr} = V_r - V_{Lr} \tag{17}$$

$$V_{Vs} = V_s - V_{Ls} \tag{18}$$

$$V_{Lp} = \sum_{i=G}^{H} \frac{N_{i,p}}{\rho_i} \tag{19}$$

Conditional equations (pressure) for mixer

$$P_m = \sum_{i=A}^{H} N_{i,m} \frac{RT_m}{V_m} \tag{20}$$

Defined relations (mole fractions)

$$x_{i,r} = 0 \quad (i = A, B, C) \tag{21}$$

$$x_{i,r} = \frac{N_{i,r}}{\sum_{i=D}^{H} N_{i,r}} \quad (i = D, E, \ldots, H) \tag{22}$$

$$x_{i,10} = 0 \quad (i = A, B, C) \tag{23}$$

$$x_{i,10} = \frac{N_{i,s}}{\sum_{i=D}^{H} N_{i,s}} \quad (i = D, E, \ldots, H) \tag{24}$$

$$x_{i,11} = \chi_{GH} \frac{N_{i,p}}{N_{G,p} + N_{H,p}} \quad (i = G, H) \tag{25}$$

Constitutive (kinetic models) equations

$$R_1 = \alpha_1 V_{Vr} \exp\left[44.06 - \frac{42600}{RT_r}\right] P_{A,r}^{1.08} P_{C,r}^{0.311} P_{D,r}^{0.874} \tag{26}$$

$$R_2 = \alpha_2 V_{Vr} \exp\left[10.27 - \frac{19500}{RT_r}\right] P_{A,r}^{1.15} P_{C,r}^{0.370} P_{E,r}^{1.00} \tag{27}$$

$$R_3 = \alpha_3 V_{Vr} \exp\left[59.50 - \frac{59500}{RT_r}\right] P_{A,r}\left(0.77 P_{D,r} + P_{E,r}\right) \tag{28}$$

Defined relations (vapour mole fractions)

$$y_{i,6} = \frac{N_{i,m}}{\sum_{i=A}^{H} N_{i,m}} \quad (i = A, B, \ldots, H) \tag{29}$$

$$y_{i,8} = y_{i,9} = \frac{P_{i,s}}{P_s} \quad (i = A, B, \ldots, H) \tag{30}$$

$$y_{i,7} = \frac{P_{i,r}}{P_r} \quad (i = A, B, \ldots, H) \tag{31}$$

Vapour (mole fractions) mass balance on the stripper

$$y_{i,5} F_5 = \phi_i z_{i,4} F_4 \quad (i = A, B, C) \tag{12}$$

$$y_{i,5} F_5 = \phi_i x_{i,10} F_{10} \quad (i = D, E, F) \tag{13}$$

$$y_{i,5} F_5 = \phi_i x_{i,10} F_{10} \quad (\phi_G = 0.07, \phi_H = 0.04), (i = G, H) \tag{14}$$

Conditional (flow) relations

$$F_6 = \pm \beta_6 \frac{2413.7}{M_6} \sqrt{||P_m - P_r||} \tag{35}$$

$$F_7 = \pm \beta_7 \frac{5722.0}{M_7} \sqrt{||P_r - P_s||} \tag{36}$$

$$F_{10} = F_{10}^{p} - F_{10}^{*} \tag{37}$$

9.2.2. Model Analysis

The model represented by Eqs. 4–36 can be written in nonlinear state variable form as:

$$\dot{\boldsymbol{x}} = \frac{d\boldsymbol{x}}{dt} = f(\boldsymbol{x}, \boldsymbol{u}, \boldsymbol{d}) \tag{38}$$

$$\boldsymbol{y} = \boldsymbol{g}(\boldsymbol{x}, \boldsymbol{u}, \boldsymbol{d}) \tag{39}$$

where $\dot{\boldsymbol{x}}$ is the state vector, $\boldsymbol{u}$ is a vector of known inputs, $\boldsymbol{d}$ is a vector of unmeasured inputs (disturbances and time-varying parameters) and $\boldsymbol{y}$ is the output vector. There are 26 state variables represented by $\boldsymbol{x}$:

$$\boldsymbol{x}^T = \left[N_{A,r}, N_{B,r}, \ldots, N_{H,r}, N_{A,s}, N_{B,s}, \ldots, N_{H,s}, N_{A,m}, N_{B,m}, \ldots, N_{H,m}, N_{G,p}, N_{H,p}\right] \tag{40}$$

where $N_{i,r}$, $N_{i,s}$ and $N_{i,m}$ are the molar holdups of compound i in the reactor, the separator and the feed mixing zone, respectively; N_{Gp} and N_{Hp} are the molar holdups of compounds G and H in the product reservoir (stripper base), respectively.

The $\boldsymbol{u}$ vector has 10 variables:

$$\boldsymbol{u}^T = \left[F_1, F_2, F_3, F_4, F_8, F_9, F_{10}^p, F_{11}, T_{cr}, T_{cs}\right] \tag{41}$$

where F_j is the molar flow rate of stream j, while T_{cr} and T_{cs} are the reactor and separator temperatures, respectively.

This model consists of 26 ODEs (Eqs. 4–7) and 100 explicit AEs (Eqs. 8–36). The variables are classified as: 26 dependent variables (Eq. 40), 100 explicit unknown algebraic variables (P_r, $\boldsymbol{P}_{i,r}$, V_{Lr}, P_s, $\boldsymbol{P}_{i,s}$, V_{Ls}, V_{Lp}, V_{Vr}, $\boldsymbol{R}$, P_m, $\boldsymbol{x}_{i,r}$, $\boldsymbol{x}_{i,10}$, $\boldsymbol{y}_{i,5}$, $\boldsymbol{y}_{i,7}$, $\boldsymbol{y}_{i,6}$, $\boldsymbol{y}_{i,8}$, $\boldsymbol{y}_{i,9}$, $\boldsymbol{P}^{sat}$, F_6, F_7) and 83 parameters (NC, $\boldsymbol{A}$, $\boldsymbol{B}$, $\boldsymbol{C}$, $\boldsymbol{Mw}$, ρ, $\boldsymbol{v}$, φ, γ_r, γ_s, V_{rT}, V_{sT}, V_{vT}, R_{g1}, R_{g2}, β_6, β_7, α), and 10 known manipulated variables and/or process-design variables (Eq. 41).

9.2.3. Specified Data

The operational scenarios involve four sets of operational conditions specified through the set of manipulated variables vector $\boldsymbol{u}$. This gives the base case as well as three other steady states. The 83 parameters that need to be specified are divided into three groups: (a) pure component property parameters (see Table 1); (b) system parameters $R_{g1} = 8.314$; $R_{g2} = 1.987$; $\beta_6 = 1$; $\beta_7 = 1$; $\alpha_1 =$; $\alpha_2 =$; $\alpha_3 =$; (c) process parameters $NC = 8$; $V_{rT} = 36.8$; $V_{sT} = 99.1$; $V_{vT} = 150.0$.

The initial values of the 26 state variables are given in Table 2 and the known values for the 10 input variables (vector $\boldsymbol{u}$ – see Eq. 41) are given in Table 3.

TABLE 1 Component Physical Properties (at 100 °C)

Compound	Molecular weight	Liquid density (kg/m^3)	γ_r	γ_s	φ	Vapour pressure (Antoine equation)		
						A	*B*	*C*
A	2.0	-	1.0	1.0	1	-	-	-
B	25.4	-	1.0	1.0	1	-	-	-
C	28.0	-	1.0	1.0	1	-	-	-
D	32.0	299	1.0	1.0	1	20.81	-1444	259
E	46.0	365	1.0	1.0	1	21.24	-2114	266
F	48.0	328	1.0	1.0	1	21.24	-2144	266
G	62.0	612	1.0	1.0	0.07	21.32	-2748	233
H	76.0	617	1.0	1.0	0.04	22.10	-3318	250

TABLE 2 Initial values of the state variables

State Variables	Initial values
$N_{A,m}$	51.7765845054
$N_{B,m}$	14.3100440019
$N_{C,m}$	42.4614755860
$N_{D,m}$	11.1168681570
$N_{E,m}$	30.1600733741
$N_{F,m}$	2.66082574139
$N_{G,m}$	5.68007202963
$N_{H,m}$	2.67883480622
$N_{A,r}$	4.71866674478
$N_{B,r}$	1.97858299828
$N_{C,r}$	3.43269304734
$N_{D,r}$	0.18201838615
$N_{E,r}$	10.3059014350
$N_{F,r}$	1.25329368022
$N_{G,r}$	66.0587540475
$N_{H,r}$	67.8869026640
$N_{A,s}$	28.8967272740
$N_{B,s}$	12.1166696783
$N_{C,s}$	21.0215212630
$N_{D,s}$	0.09658565923
$N_{E,s}$	5.90626944116
$N_{F,s}$	0.71825343663
$N_{G,s}$	20.4923456368
$N_{H,s}$	16.1950690323
$N_{F,p}$	21.7412698769
$N_{G,p}$	17.7363533518

TABLE 3 Specified Values of the Manipulated Variables, *u*, at Four Steady-state Conditions

No.	Input	Units	Base case	Mode 1	Mode 2	Mode 3
1	F_1	kmol/h	11.2	11.991	13.818	0.703
2	F_2	kmol/h	114.5	114.314	22.948	161.856
3	F_3	kmol/h	98.0	96.471	174.679	15.216
4	F_4	kmol/h	417.5	413.782	383.109	350.844
5	F_8	kmol/h	1201.5	1441.021	1419.501	880.830
6	F_9	kmol/h	15.1	9.497	16.164	3.895
7	F_{10}	kmol/h	259.5	253.563	243.825	198.512
8	F_{11}	kmol/h	211.3	210.885	194.638	179.021
9	T_{cr}	deg C	120.4	123.074	124.213	121.911
10	T_{cs}	deg C	80.1	92.078	90.259	83.396

9.2.4. Numerical Solution

Open-loop simulation results

The DAE system of equations representing the simplified model [Eqs. 4–36] is solved with the BDF-method in ICAS-MoT (Sales-Cruz 2006) for the data given in Tables 1-3. Simulated values of the state variables and some output variables are given in Tables 4 and 5, respectively. These simulated values show a good match with those reported by Ricker and Lee (1995). Also, screenshots from ICAS-MoT highlighting the steady-state values of some of the state variable, $\boldsymbol{x}$, and the right-hand sides of the ODEs (functions $\boldsymbol{f}$) are seen in Figures 2 and 3. The model as implemented and solved in ICAS-MoT is given in Appendix A (see ch-9-1-te-dynamic-ricker.mot file).

The simulated transient responses for four reactor outputs are shown in Figures 4 and 5 using the steady-state value as the initial condition to verify

TABLE 4 State Variables at Four Steady-state Conditions (all units are kmol)

No.	State	Base case	Mode 1 (50/50)	Mode 2 (10/90)	Mode 3 (90/10)
1	N_{Ar}	4.722	5.4536	6.0899	4.4549
2	N_{Br}	1.9805	3.6096	1.9562	7.2066
3	N_{Cr}	3.4354	2.184	2.4319	1.47350
4	N_{Dr}	0.18231	0.11589	0.01556	0.30814
5	N_{Er}	10.32	7.7163	9.9718	2.1277
6	N_{Fr}	1.2572	2.6844	3.2853	0.98381
7	N_{Gr}	66.06	59.961	10.927	123.15
8	N_{Hr}	67.87	58.48	95.896	14.052
9	N_{As}	28.895	28.565	32.039	25.365
10	N_{BS}	12.119	18.906	10.291	41.032
11	N_{Cs}	21.022	11.44	12.794	8.3895
12	N_{Ds}	0.09656	0.0597	0.00822	0.18371
13	N_{Es}	5.9052	4.2311	5.6803	1.3512
14	N_{Fs}	0.71938	1.4719	1.8714	0.62479
15	N_{Gs}	20.492	21.02	4.0319	41.458
16	N_{Hs}	16.195	16.612	28.679	3.6405
17	N_{Am}	51.776	55.543	59.874	45.817
18	N_{Bm}	14.327	25.607	14.1070	43.208
19	N_{Cm}	42.455	32.324	33.449	27.935
20	N_{Dm}	11.1	10.409	2.206	19.708
21	N_{Em}	30.156	28.351	43.729	5.9845
22	N_{Fm}	2.6668	7.1577	9.6449	2.0160
23	N_{Gm}	5.684	8.5369	1.7076	10.286
24	N_{Hm}	2.6807	4.1394	7.44	0.54295
25	N_{Gp}	21.741	21.741	4.3633	38.999
26	N_{Hp}	17.736	17.736	32.037	3.535

TABLE 5 Output Variables at Four Steady States

No.	Description	Units	Base case	Mode 1 (50/50)	Mode 2 (10/90)	Mode 3 (90/10)
1	Reactor pressure	kPa	2705.0	2800.0	2800.0	2800.0
2	Reactor liquid level	%	75.0	65.0	65.0	65.0
3	Separator pressure	kPa	2633.7	2705.7	2705.4	2764.9
4	Separator liquid level	%	50.0	50.0	50.0	50.0
5	Stripper bottoms level	%	50.0	50.0	50.0	50.0
6	Stripper pressure	kPa	3102.2	3325.7	3327.5	2995.6
7	Reactor feed flow rate	kscmh	42.27	47.27	46.03	32.04
8	A in the reactor feed (stream 6)	mol%	32.19	32.28	34.78	29.47
9	B in the reactor feed	mol%	8.91	14.88	8.19	27.79
10	C in the reactor feed	mol%	26.4	18.79	19.43	17.97
11	D in the reactor feed	mol%	6.9	6.05	1.28	12.68
12	E in the reactor feed	mol%	18.75	16.48	25.4	3.85
13	F in the reactor feed	mol%	1.66	4.16	5.60	1.29
14	A in purge (stream 9)	mol%	32.96	32.82	36.63	27.86
15	B in purge	mol%	13.82	21.72	11.77	45.08
16	C in purge	mol%	23.98	13.14	14.63	9.22
17	D in purge	mol%	1.26	0.88	0.13	2.18
18	E in purge	mol%	18.58	15.93	22.37	3.94
19	F in purge	mol%	2.26	5.54	7.37	1.82
20	G in purge	mol%	4.84	6.70	1.32	9.4
21	H in purge	mol%	2.30	3.27	5.79	0.5
22	G in product (stream 11)	mol%	53.72	53.84	11.66	90.1
23	H in product	mol%	43.83	43.92	85.64	8.17

whether the process moves away from the steady states. It can be noted that the steady state is maintained (during the short time simulated, as shown in Figures 4–5); the small variation (note the scales) is probably due to the precision of the convergence criteria used. However, continuing the simulation for a longer time scale (not shown in the figures) the system moves away from the steady state, indicating an unstable state.

Closed-loop Simulation Results

The model for a simple PI-controller is implemented, considering as measured (controlled) output y, XMEAS_15 = stripper liquid product flow (with a step change of + 15) and as manipulated (actuator) input u, F_{11} = stripper exit feed flow rate. The control law used is as follows (for set point, $Setpt = 65$):

Error and time derivative of the error

$$e = Setpt - X_{meas}(15) \tag{42}$$

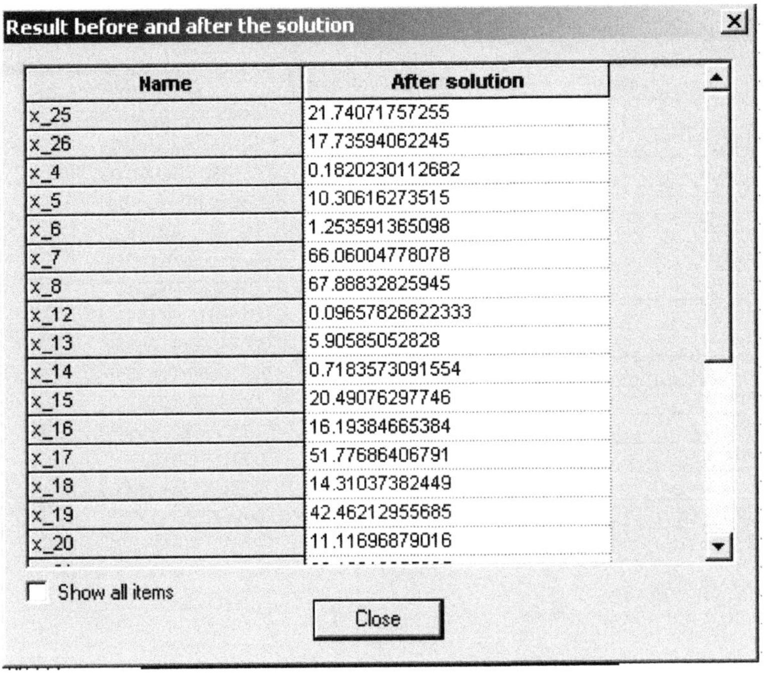

Result before and after the solution

Name	After solution
x_25	21.74071757255
x_26	17.73594062245
x_4	0.1820230112682
x_5	10.30616273515
x_6	1.253591365098
x_7	66.06004778078
x_8	67.88832825945
x_12	0.09657826622333
x_13	5.90585052828
x_14	0.7183573091554
x_15	20.49076297746
x_16	16.19384665384
x_17	51.77686406791
x_18	14.31037382449
x_19	42.46212955685
x_20	11.11696879016

Show all items

Close

FIGURE 2 State variables at the base case steady state (screen shot from ICAS-MoT).

$$\frac{de}{dt} = -22.58\left(\frac{1}{\rho_G}\cdot\frac{dN_{G,p}}{dt} + \frac{1}{\rho_H}\cdot\frac{dN_{H,p}}{dt}\right) \tag{43}$$

Tuning

$$K_\mathrm{p} = \frac{1}{0.3} \tag{45}$$

$$\tau_I = 0.05 \tag{45}$$

Controller: Stripper liquid product flow % (XMV(8))

$$F_{11} = F_{11,reference} - K_\mathrm{p}\cdot\left(e + \int \frac{e}{\tau_I}dt\right) \tag{46}$$

or

$$\frac{dF_{11}}{dt} = -K_\mathrm{p}\left(\frac{de}{dt} + \frac{e}{\tau_I}\right) \tag{47}$$

Adding Eqs. 42–47 to the model allows a closed-loop simulation to be performed. The BDF-method from ICAS-MoT is used as the DAE-solver. The obtained simulation results are shown in Figures 6-8, where it can be seen that

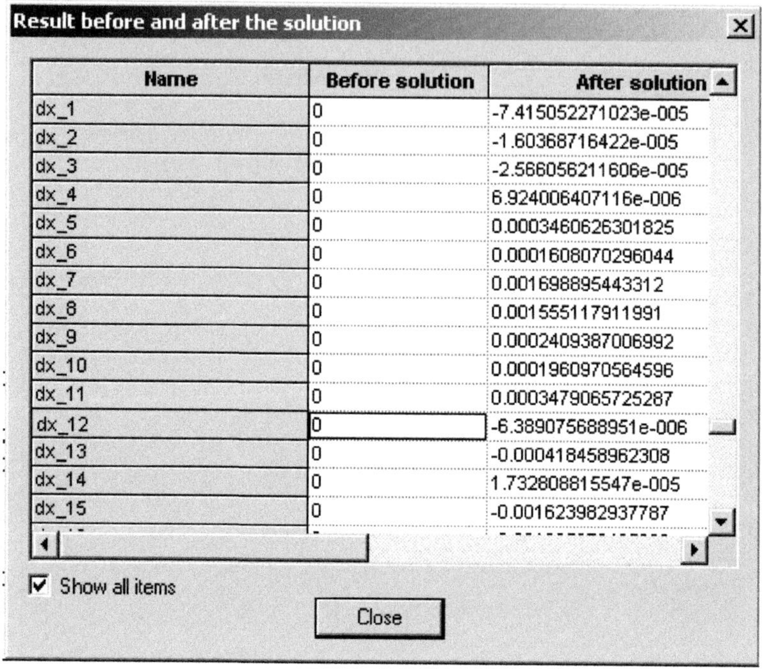

Result before and after the solution

Name	Before solution	After solution
dx_1	0	-7.415052271023e-005
dx_2	0	-1.60368716422e-005
dx_3	0	-2.566056211606e-005
dx_4	0	6.924006407116e-006
dx_5	0	0.0003460626301825
dx_6	0	0.0001608070296044
dx_7	0	0.001698895443312
dx_8	0	0.001555117911991
dx_9	0	0.0002409387006992
dx_10	0	0.0001960970564596
dx_11	0	0.0003479065725287
dx_12	0	-6.389075688951e-006
dx_13	0	-0.000418458962308
dx_14	0	1.732808815547e-005
dx_15	0	-0.001623982937787

Show all items

Close

FIGURE 3 **Function f (right-hand side of Eq 38) values at base case steady state (screenshot from ICAS-MoT).**

the actuator F_{11}, reaches a steady-state value, while the measured (controlled) output XMEAS_15 reaches the given set point = 65. The other outputs (XMEAS_6, XMEAS_8 and XMEAS_12) also reach almost constant values (note the small scale), thereby achieving the desired control objectives.

9.2.5. Discussion

Check the eigenvalues of the Jacobian matrices at the steady states to verify which of the steady states are unstable. Perform a sensitivity analysis to identify the most sensitive actuators (manipulated variables).

Nomenclature (for Section 9.2)

d	unmeasured input vector
F	molar flow rate, kmol/h
F^*	molar 'pseudo-feed', kmol/h (added at the feed mixing point)
$F_{10}{}^{P}$	apparent stream 10 flow rate, as indicated by the valve position, kmol/h
$F_{10}{}^{*}$	bias adjustment for stream 10, kmol/h
N_{ij}	total molar holdup of i-th component in the j-th unit, kmol

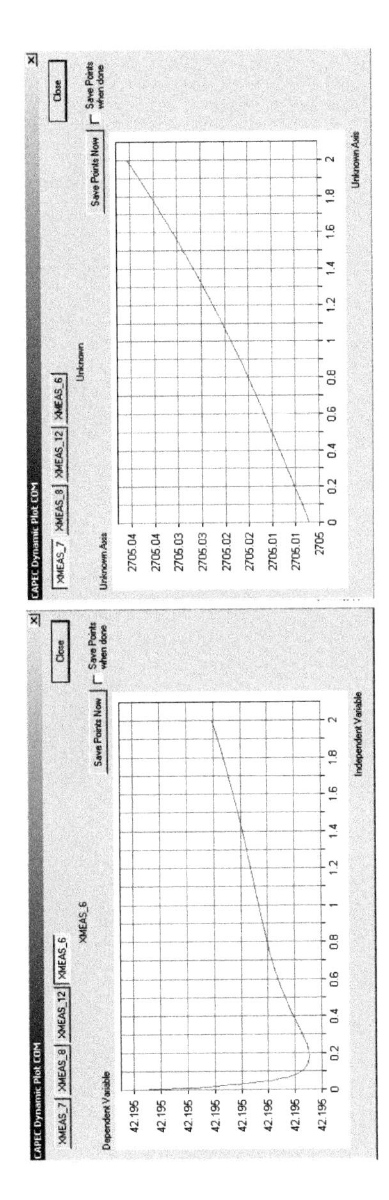

FIGURE 4 Dynamic behaviour of reactor outputs: (a) XMEAS_6 = reactor feed flow rate; (b) XMEAS_7 = reactor pressure.

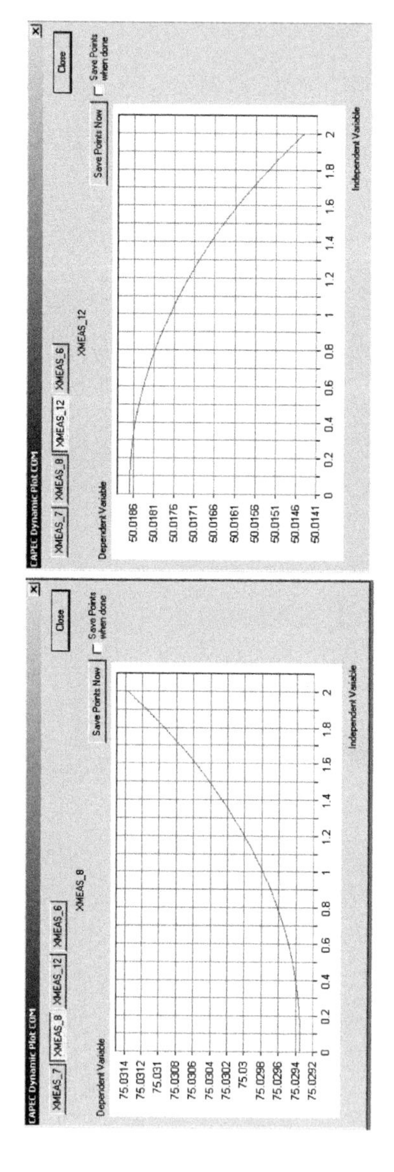

FIGURE 5 **Dynamic behaviour of reactor outputs: (c) XMEAS_8 = reactor liquid level; (d) XMEAS_12 = separator liquid level.**

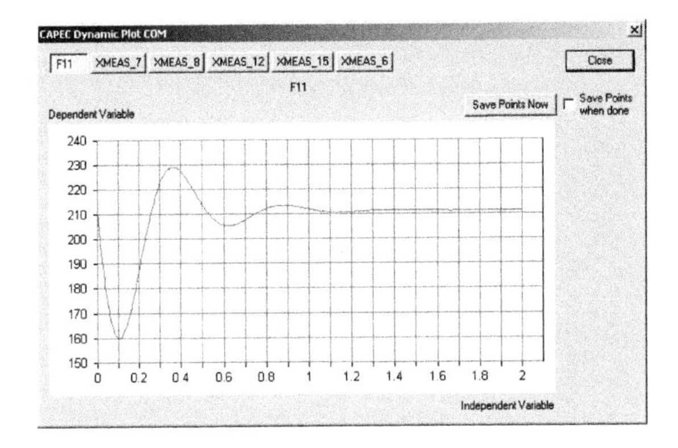

FIGURE 6 Dynamic behaviour of input F11 = stripper exit feed flow-rate.

P^{sat}	vapour pressure
p	partial pressure, kPa
P	total pressure, kPa
$Rnet$	molar rate of reaction, kmol/h
R	gas constant = 1.987 cal/gmol-K in Eqs. 26–28; otherwise it is equal to 8.314 kJ/kmol-K
t	time, h
T	absolute temperature, K
u	known input vector
V	total volume, m^3
$\mathbf{x}$	state vector
x	mole liquid fraction
$\mathbf{y}$	output vector
y	mole vapour fraction
z	mole feed fraction

Subscript

r	reactor
s	separator
m	mixing zone
p	product
L	liquid
V	vapour
i	component
j	stream

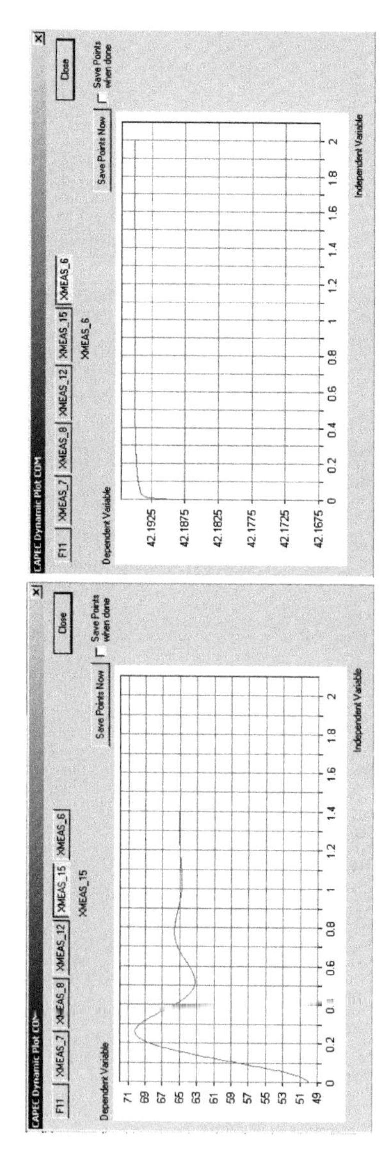

FIGURE 7 **Dynamic behaviour of reactor outputs: (a) XMEAS_15 = stripper level; (b) XMEAS_6 = reactor feed flow-rate.**

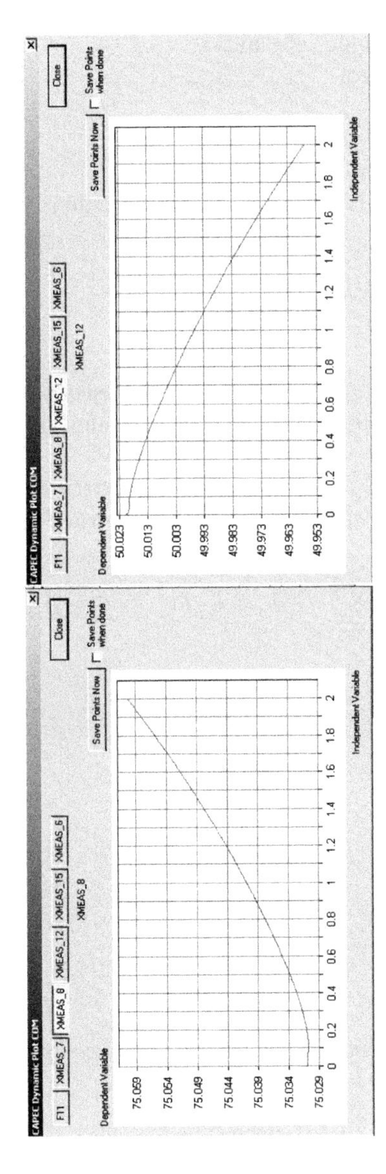

FIGURE 8 **Dynamic behaviour of reactor outputs: (c) XMEAS_8 = reactor liquid level; (d) XMEAS_12 = separator liquid level.**

Greek symbols

α adjustable parameter used in reaction rate equations (Eqs. 26–28); dimensionless with nominal value of unity
β used to adjust flow/pressure drop relation in Eqs. 35–36 for streams 6–7 (nominal values are unity)
γ activity coefficient
φ stripping factor
υ stoichiometric coefficient
ρ molar density, mol/m^3
χ_{GH} purity of G and H in the product (as a fraction)

9.3. FULL (PROCESS) MODEL

9.3.1. Process Model

This model considers the mass as well as the energy balance. The model derivation and assumptions can be found in Jockenhovel et al. (2004).

Mass balances for the mixer, reactor, separator and purge (unit operations operating in the dynamic mode)

$$\frac{dN_{i,m}}{dt} = \sum_{j=1,2,3,5,8} y_{ij}F_j - y_{i,6}F_6 \quad (i = A, B, \ldots, H) \tag{48}$$

$$\frac{dN_{i,r}}{dt} = y_{i,6}F_6 - y_{i,7}F_7 + \sum_{j=1}^{3} \nu_{ij}R_j \quad (i = A, B, \ldots, H) \tag{49}$$

$$\frac{dN_{i,s}}{dt} = y_{i,7}F_7 - y_{i,8}(F_8 + F_9) - x_{i,10}F_{10} \quad (i = A, B, \ldots, H) \tag{50}$$

$$\frac{dN_{i,p}}{dt} = (1 - \phi_i)\left(x_{i,10}F_{10} + y_{i,4}F_4\right) - x_{i,11}F_{11} \quad (i = G, H) \tag{51}$$

Energy balances for the mixer, reactor, separator and purge (unit operations operating in the dynamic mode)

$$\left(\sum_{i=A}^{H} N_{i,m}c_{p,vap,i}\right)\frac{dT_m}{dt} = \sum_{j=1,2,3,5,8} F_j\left(\sum_{i=A}^{H} y_{i,j}c_{p,vap,i}\right)(T_j - T_m) \tag{52}$$

$$\left(\sum_{i=A}^{H} N_{i,r} c_{p,i}\right)\frac{dT_r}{dt} = F_6\left(\sum_{i=A}^{H} y_{i,6} c_{p,vap,i}\right)(T_6 - T_r) - \dot{Q}_r - \sum_{k=1}^{3} \Delta H_{rk} R_k \tag{53}$$

$$\left(\sum_{i=A}^{H} N_{i,s} Cp_i\right)\frac{dT_s}{dt} = F_7\left(\sum_{i=A}^{H} y_{i,7} Cp_i^v\right) (T_r - T_s) - H_0 V_s - \dot{Q}_s \tag{54}$$

$$\left(\sum_{i=G}^{H} N_{i,p} Cp_i\right)\frac{dT_p}{dt} = F_{10}\left(\sum_{i=A}^{H} x_{i,10} Cp_i\right)(T_s - T_p) + F_4\left(\sum_{i=A}^{H} y_{i,4} Cp_i^v\right)(T_4 - T_p) - H_0 V_p + \dot{Q}_p$$

$$F_4\left(\sum_{i=A}^{H} y_{i,4} c_{p,vap,i}\right)(T_4 - T_p) - H_0 V_p + \dot{Q}_p \tag{55}$$

Conditional equations (mixer)

$$P_m = \sum_{i=A}^{H} N_{i,m} \frac{RT_m}{V_m} \tag{56}$$

$$y_{i,6} = \frac{N_{i,m}}{\sum_{i=A}^{H} N_{i,m}} \quad (i = A, B, \ldots, H) \tag{57}$$

Conditional equations (reactor)

$$P_r = \sum_{i=A}^{H} P_{i,r} \tag{58}$$

where

$$P_{i,r} = \frac{N_{i,r} RT_r}{V_{Vr}} \quad (i = A, B, C) \tag{59}$$

$$P_{i,r} = \gamma_{ir} x_{ir} P_i^{sat}(T_r) \quad (i = D, E, \ldots, H) \tag{60}$$

$$P_i^{sat}(T) = 1x10^{-3} \exp\left(A_i + \frac{B_i}{C_i + T_r - T^*}\right) \quad (i = D, E, \ldots, H) \tag{61}$$

Flow rates as a function of pressure drop

$$F_6 = 0.8334 \frac{kmol}{s\sqrt{MPa}} \sqrt{||P_m - P_r||} \tag{62}$$

$$F_7 = 1.5344 \frac{kmol}{s\sqrt{MPa}} \sqrt{||P_r - P_s||} \tag{63}$$

Enthalpy calculation

$$\Delta H_{Rj} = \sum_{i=A}^{H} H_i \nu_{i,j} + H_0 F_j, \quad \text{with} \quad H_i = c_{p,i}(T_r - T^*) \quad (j = 1, 2, 3) \tag{64}$$

Heat exchanger

$$\dot{Q}_r = m_{CW,r} c_{p,CW} (T_{CW,r,out} - T_{CW,r,in}) \tag{65}$$

$$\dot{Q}_r = UA_r \left[\frac{\Delta T_{1,r} - \Delta T_{2,r}}{\ln(\Delta T_{1,r} / \Delta T_{2,r})} \right] \tag{66}$$

where

$$\Delta T_{1,r} = T_r - T_{CW,r,in}; \qquad \Delta T_{2,r} = T_r - T_{CW,r,out} \tag{67}$$

Defined relations (reactor)

$$x_{i,r} = 0 \quad (i = A, B, C) \tag{68}$$

$$x_{i,r} = \frac{N_{i,r}}{\sum_{i=D}^{H} N_{i,r}} \quad (i = D, E, \ldots, H) \tag{69}$$

$$y_{i,7} = \frac{P_{i,r}}{P_r} \quad (i = A, B, \ldots, H) \tag{70}$$

$$V_{Lr} = \sum_{i=D}^{H} \frac{N_{i,r}}{\rho_i} \tag{71}$$

$$V_{Vr} = V_r - V_{Lr} \tag{72}$$

Defined relations (separator)

$$P_s = \sum_{i=A}^{H} P_{i,s} \tag{73}$$

where

$$P_{i,s} = \frac{N_{i,s} R T_s}{V_{Vs}} \quad (i = A, B, C) \tag{74}$$

$$P_{i,s} = \gamma_{is} x_{i,10} P_i^{sat}(T_s) \quad (i = D, E, \ldots, H) \tag{75}$$

Heat transfer

$$T_8 = T_s \left(\frac{P_m}{P_s} \right)^{\frac{1-\kappa}{\kappa}} \tag{76}$$

$$H_0 V_s = \sum_{i=D}^{H} x_{i,10} F_{10} \cdot H_{vap,i} \tag{77}$$

$$\dot{Q}_s = m_{CW,s} c_{p,CW} \left(T_{CW,s,out} - T_{CW,s,in} \right) \tag{78}$$

$$\dot{Q}_s = UA_s \left[\frac{\Delta T_{1,s} - \Delta T_{2,s}}{\ln\left(\Delta T_{1,s} / \Delta T_{2,s} \right)} \right] \tag{79}$$

where

$$\Delta T_{1,s} = T_s - T_{CW,s,in}; \qquad \Delta T_{2,s} = T_s - T_{CW,s,out} \tag{80}$$

$$y_{i,8} = y_{i,9} = \frac{P_{i,s}}{P_s} \quad (i = A, B, \ldots, H) \tag{81}$$

$$x_{i,10} = 0 \quad (i = A, B, C) \tag{82}$$

$$x_{i,10} = \frac{N_{i,s}}{\sum_{i=D}^{H} N_{i,s}} \quad (i = D, E, \ldots, H) \tag{83}$$

$$V_{Ls} = \sum_{i=D}^{H} \frac{N_{i,s}}{\rho_i} \tag{84}$$

$$V_{Vs} = V_s - V_{Ls} \tag{85}$$

$$H_0 V_p = \sum_{i=D}^{H} \left(y_{i,5} F_5 - y_{i,4} F_4 \right) \cdot H_{vap,i} \tag{86}$$

$$\dot{Q}_p = 2258.717 \frac{kJ}{kg} \cdot \dot{m}_{steam} \tag{87}$$

Defined relation (purge)

$$V_{Lp} = \sum_{i=D}^{H} \frac{N_{i,p}}{\rho_i} \tag{88}$$

Other defined relations

$$\phi_i = \sum_{j=0}^{3} a_{i,j}(T_s - 273) \qquad (i = D, E, \ldots, H) \tag{89}$$

$$\phi_i = 1 \qquad (i = A, B, C) \tag{90}$$

$$F_5 = F_{10} + F_4 - F_{11} - \sum_{i=G}^{H} \frac{dN_{i,p}}{dt} \tag{91}$$

$$y_{i,5} = \frac{\phi_i (y_{i,4}F_4 + x_{i,10}F_{10})}{F_5} \qquad (i = A, B, \ldots, H) \tag{92}$$

$$x_{i,11} = \frac{y_{i,4}F_4 + x_{i,10}F_{10} - y_{i,5}F_5}{F_{11}} \qquad (i = D, E, F) \tag{93}$$

$$x_{i,11} = \left(1 - \sum_{j=D}^{F} x_{j,11}\right) \frac{N_{i,p}}{\sum_{j=D}^{H} N_{j,p}} \qquad (i = G, H) \tag{94}$$

Constitutive equations

$$R_1 = \alpha_1 V_{Vr} \exp\left[44.06 - \frac{42600}{RT_r}\right] P_{A,r}^{1.08} P_{C,r}^{0.311} P_{D,r}^{0.874} \tag{95}$$

$$R_2 = \alpha_2 V_{Vr} \exp\left[10.27 - \frac{19500}{RT_r}\right] P_{A,r}^{1.15} P_{C,r}^{0.370} P_{E,r}^{1.00} \tag{96}$$

$$R_3 = \alpha_3 V_{Vr} \exp\left[59.50 - \frac{59500}{RT_r}\right] P_{A,r}(0.77 P_{D,r} + P_{E,r}) \tag{97}$$

9.3.2. Model Analysis

The above full-model can be written, in nonlinear state variable form as,

$$\dot{\boldsymbol{x}} = \frac{d\boldsymbol{x}}{dt} = \boldsymbol{f}(\boldsymbol{x}, \boldsymbol{u}, \boldsymbol{d}) \tag{98}$$

$$\boldsymbol{y} = \boldsymbol{g}(\boldsymbol{x}, \boldsymbol{u}, \boldsymbol{d}) \tag{99}$$

where $\boldsymbol{x}$ is the state vector, $\boldsymbol{u}$ is a vector of known (manipulated) inputs, $\boldsymbol{d}$ is a vector of unmeasured inputs (disturbances and time-varying parameters) and $\boldsymbol{y}$ is the output vector. There are 30 state variables arranged as follows:

$$\mathbf{x}^{\mathrm{T}} = [N_{A,r}, N_{B,r}, \ldots, N_{H,r}, N_{A,s}, N_{B,s}, \ldots, N_{H,s}, N_{A,m}, N_{B,m}, \ldots, N_{H,m}, N_{G,p}, N_{H,p}, T_m, T_r, T_s, T_p] \tag{100}$$

where $N_{i,r}$ is the molar holdup of chemical i in the reactor, and $N_{i,s}$, $N_{i,m}$ and $N_{i,p}$ are in the separator, feed mixing zone and product reservoir (stripper base), respectively. The $\boldsymbol{u}$ vector contains 14 variables

$$\mathbf{u}^{\mathrm{T}} = [F_1, F_2, F_3, F_4, F_8, F_9, F_{10}, F_{11}, T_{CW,r,in}, T_{CW,r,out}, T_{CW,s,in}, T_{CW,s,out}, m_{CW,r}, m_{CW,s}] \tag{101}$$

where F_j is the molar flow rate of stream j, $(T_{CW,s,in}, T_{CW,r,in})$ and $(T_{CW,s,out}, T_{CW,r,out})$ are the cooling water inlet and outlet temperatures in the reactor and separator, respectively, and $m_{CW,s}$ and $m_{CW,r}$ are the cooling water flow rates in the reactor and separator, respectively.

9.3.3. Specified Data

The operational scenarios involve four sets of operational conditions specified through the set of manipulated variables vector $\boldsymbol{u}$. This gives the base case as well as three other steady states. The 127 parameters that need to be specified are divided into three groups (a) pure component property parameters (see Table 6); (b) system parameters $R_{g1} = 8.314$; $R_{g2} = 1.987$; $\beta_6 = 1$; $\beta_7 = 1$; $\alpha_1 = 1$; $\alpha_2 = 1$; $\alpha_3 = 1$; (c) process parameters $NC = 8$; $V_{rT} = 36.8$; $V_{sT} = 99.1$; $V_{vT} = 150.0$.

The initial values of the 30 state variables are given in Table 7 and the known values for the 14 input variables (vector $\boldsymbol{u}$ – see Eq. 101) are given in Table 3 and values of the additional variables are: $T_{CW,r,in} = 308.00$ K; $T_{CW,r,out} = 367.59$ K; $T_{CW,s,in} = 313.00$ K; $T_{CW,s,out} = 350.45$ K; $m_{CW,s} = 93.37$ mol/s; $m_{CW,s} = 49.37$ mol/s. The known measured values for a selection of the output variables are given in Table 8. The simulated values for these 23 variables are compared to these measured values to validate the model.

9.3.4. Numerical Solution

Open-loop Simulation Results

The DAE system of equations representing the simplified model [Eqs. 48–97] is solved with the BDF-method in ICAS-MoT (Sales-Cruz, 2006), for the data

TABLE 6 Component Physical Properties (at 100 °C)

Compound	Molecular weight	Liquid density (kg/m^3)	Liquid heat capacity (kJ/kg·°C)	Vapour heat capacity (kJ/kg·°C)	Heat of vaporisation (kJ/kg)	Vapour pressure (Antoine equation)		
						A	B	C
A	2.0	-	-	14.60	-	-	-	-
B	25.4	-	-	2.04	-	-	-	-
C	28.0	-	-	1.05	-	-	-	-
D	32.0	299	7.66	1.85	202	20.81	-1444	259
E	46.0	365	4.17	1.87	372	21.24	-2114	266
F	48.0	328	4.45	2.02	372	21.24	-2144	266
G	62.0	612	2.55	0.712	523	21.32	-2748	233
H	76.0	617	2.45	0.628	486	22.10	-3318	250

TABLE 7 Initial values of the state variables

State variables	Initial values
$N_{A,m}$	48.83
$N_{B,m}$	13.49
$N_{C,m}$	40.03
$N_{D,m}$	10.44
$N_{E,m}$	28.48
$N_{F,m}$	2.51
$N_{G,m}$	5.40
$N_{H,m}$	2.51
T_m	359.25
$N_{A,r}$	5.20
$N_{B,r}$	2.29
$N_{C,r}$	4.65
$N_{D,r}$	0.12
$N_{E,r}$	7.45
$N_{F,r}$	1.15
$N_{G,r}$	56.10
$N_{H,r}$	59.80
T_r	393.55
$N_{A,s}$	27.50
$N_{B,s}$	12.10
$N_{C,s}$	24.60
$N_{D,s}$	0.0836
$N_{E,s}$	5.86
$N_{F,s}$	0.901
$N_{G,s}$	24.10
$N_{H,s}$	19.80
T_s	353.25
$N_{G,p}$	20.40
$N_{H,p}$	17.30
T_p	338.85

TABLE 8 Elements of the Output Vector (y)

No.	Description	Units	Downs and Vogel (*XMEAS*)
1	Reactor pressure	kPa	7
2	Reactor liquid level	%	8
3	Separator pressure	kPa	13
4	Separator liquid level	%	12
5	Stripper bottoms level	%	15
6	Stripper pressure	kPa	16

(continued)

TABLE 8 *(continued)*

No.	Description	Units	Downs and Vogel (*XMEAS*)
7	Reactor feed flow rate	kscmh	6
8	A in the reactor feed (stream 6)	mol%	23
9	B in the reactor feed	mol%	24
10	C in the reactor feed	mol%	25
11	D in the reactor feed	mol%	26
12	E in the reactor feed	mol%	27
13	F in the reactor feed	mol%	28
14	A in purge (stream 9)	mol%	29
15	B in purge	mol%	30
16	C in purge	mol%	31
17	D in purge	mol%	32
18	E in purge	mol%	33
19	F in purge	mol%	34
20	G in purge	mol%	35
21	H in purge	mol%	36
22	G in product (stream 11)	mol%	40
23	H in product	mol%	41

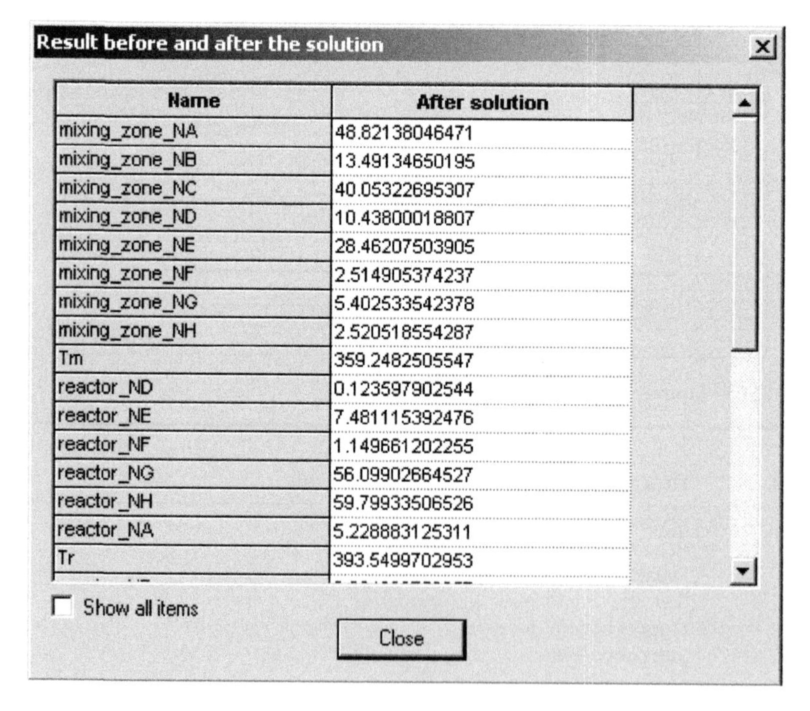

Name	After solution
mixing_zone_NA	48.82138046471
mixing_zone_NB	13.49134650195
mixing_zone_NC	40.05322695307
mixing_zone_ND	10.43800018807
mixing_zone_NE	28.46207503905
mixing_zone_NF	2.514905374237
mixing_zone_NG	5.402533542378
mixing_zone_NH	2.520518554287
Tm	359.2482505547
reactor_ND	0.123597902544
reactor_NE	7.481115392476
reactor_NF	1.149661202255
reactor_NG	56.09902664527
reactor_NH	59.79933506526
reactor_NA	5.228883125311
Tr	393.5499702953

FIGURE 9 **Simulated steady-state values of some state variables (screenshot from ICAS-MoT).**

Result before and after the solution

Name	After solution
dmixing_zone_NA	-0.007675629743988
dmixing_zone_NB	-0.0007147477006369
dmixing_zone_NC	0.01195266583843
dmixing_zone_ND	-0.001795902222394
dmixing_zone_NE	-0.01271804260914
dmixing_zone_NF	0.0004223940780941
dmixing_zone_NG	-0.0001642416008592
dmixing_zone_NH	0.001709599173787
dTm	-0.0008747937840887
dreactor_NA	0.01067841205696
dreactor_NB	0.001088583713205
dreactor_NC	-0.008268228498371
dreactor_ND	0.00311223543362
dreactor_NE	0.01386207056754
dreactor_NF	-0.0003610358693501
dreactor_NG	-0.0005443463092032
dreactor_NH	-0.000870332695363
dTr	-8.405803771109e-006
dseparator_NA	-0.0005493505592055
dseparator_NB	-0.0003725228026639
dseparator_NC	-0.001259539221939
dseparator_ND	0.0001959659570496
dseparator_NE	-0.0003562682920671
dseparator_NF	-7.824647573075e-005
dseparator_NG	-0.001043976294797
dseparator_NH	-0.001062695032336
dTs	0.01402384641581
dstripper_NG	9.298669763263e-005
dstripper_NH	0.0002596694921887
dTstr	-0.002389101436754

FIGURE 10 **Function f [right-hand side of Eq 98] values at steady state (screenshot from ICAS-MoT).**

given in Tables 6-8. The simulated results as screenshots from ICAS-MoT highlighting the steady-state values of some of the state variable, $\boldsymbol{x}$, values and the right-hand sides of the ODEs (functions f Eq. 98) are seen in Figures 9 and 10. The model as implemented and solved in ICAS-MoT is given in Appendix 1 (see ch-9—2-te-dynamic-complete.mot file).

The dynamic behaviour for four reactor outputs is seen in Figures 11 and 12, where the simulation results were generated using the steady-state value as

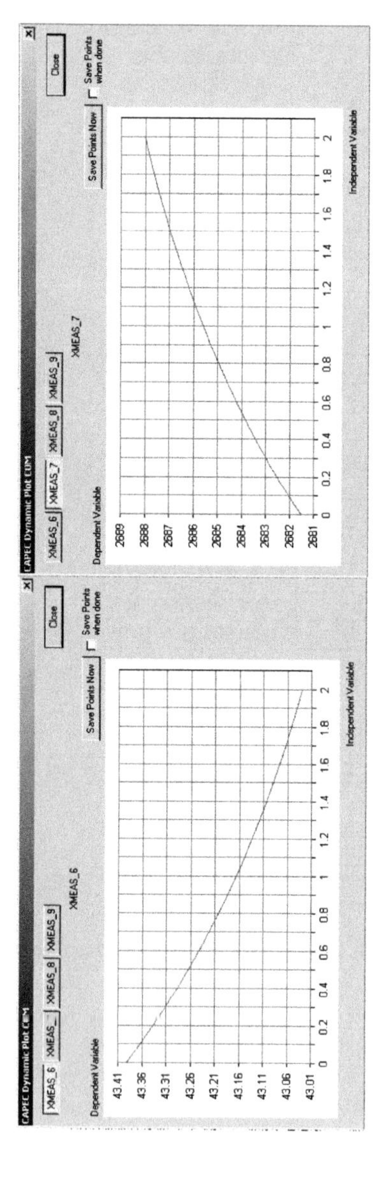

FIGURE 11 Dynamic behaviour of reactor outputs: (a) XMEAS_6 = reactor feed flow rate; (b) XMEAS_7 = reactor pressure.

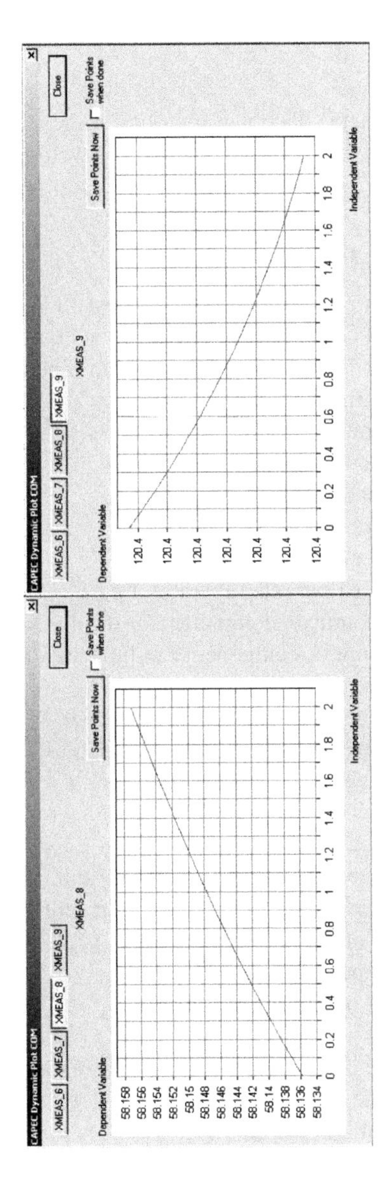

FIGURE 12 Dynamic behaviour of reactor outputs: (c) XMEAS_8 = reactor liquid level; (d) XMEAS_9 = reactor temperature.

initial condition. As can be seen from these figures the steady state almost remains constant (for a short time); the small variation (note the scales on the y-axis) is probably due to numerical accuracy of the computer.

9.3.5. Discussion

Check the eigenvalues of the Jacobian matrices at the steady states to verify whether steady state is stable or unstable. Try to develop a control scheme to keep the operation stable.

Nomenclature (for 9.3)

$C_{p,CW}$	specific heat capacity of cooling water, kJ kg^{-1} K^{-1}
$C_{p,i}$	specific heat capacity of component i in liquid phase, kJ kg^{-1} K^{-1}
$C_{p,vap,i}$	specific heat capacity of component i in vapour phase, kJ kg^{-1} K^{-1}
$\mathbf{d}$	unmeasured input vector
F_j	molar flow rate of stream j, kmol h^{-1}
H_i	enthalpy of component i, kJ
H_0	reference enthalpy, kJ
ΔH_{Rj}	exothermic heat, kJ kmol^{-1}
$m_{CW,r}$	cooling water flow rate reactor, kg h^{-1}
$m_{CW,s}$	cooling water flow rate separator, kg h^{-1}
N_{ik}	total molar holdup of component i in the unit k ($k = m, r, s, p$), kmol
$P_{i,j}$	partial pressure of component i in the unit k ($k = m, r, s, p$), kPa
P_i^{sat}	saturation pressure of component i, kPa
P_k	total pressure in the unit k ($k = m, r, s, p$), kPa
Q_k	energy removed from the unit k ($k = r, s, p$), kW
R_i	reaction conversion component i, kmol h^{-1}
R	gas constant
t	time, h
$T_{cw,k,in}$	cooling water inlet temperature in the unit k ($k = r, s$), K
$T_{cw,k,out}$	cooling water outlet temperature in the unit k ($k = r, s$), K
T_k	temperature of the unit k ($k = m, r, s, p$), K
T^*	absolute temperature, K
UA	specific heat transfer rate, kW K^{-1}
$\mathbf{u}$	known input vector
V_k	total volume of the unit k ($k = m, r, s, p$), m^3
$V_{L,k}$	liquid volume in the unit k ($k = m, r, s, p$), m^3
$V_{V,k}$	vapour volume in the unit k ($k = m, r, s, p$), m^3
$x_{i,k}$	mole liquid fraction of component i in the unit k ($k = m, r, s, p$)
$\mathbf{x}$	state vector
$y_{i,k}$	mole vapour fraction of component i in the unit k ($k = m, r, s, p$)
$\mathbf{y}$	output vector

Subscript

r reactor
s separator
m mixing zone
p product (stripper)
L liquid
V vapour
i component
j stream

Greek symbols

α adjustable parameter used in reaction rate equations.
γ activity coefficient
φ stripping factor
υ stoichiometric coefficient
ρ molar density, mol m^{-3}

REFERENCES

Downs, J.J., Vogel, E.F., 1993. A Plant-Wide Industrial Process Control Problem. Computers and Chemical Engineering. 17, 245–255.

Jockenhovel, T., Biegler, L.T., Wachter, A., 2004. Dynamic Optimization of the Tennessee Eastman Process Using OptControlCentre. Computers and Chemical Engineering. 27, 1513–1531.

Ricker, N.L., Lee, J.H., 1995. Nonlinear Modelling and State Estimation for the Tennessee Eastman Challenge Process. Computers and Chemical Engineering. 19, 983–1005.

Sales-Cruz, M., 2006. Development of a computer aided modelling system for bio and chemical process and product design, PhD-thesis. Technical University of Denmark: Lyngby, Denmark.

Chapter 10

Modelling of Batch Process Operations*

10.1. MODELLING OF A BATCH CRYSTALLISATION PROCESS

Crystallisation is a separation process where mass is transferred from a solute dissolved in a liquid phase to a solid (pure crystal) phase. The crystallised solid particles form a homogeneous phase. At the industrial level, crystal products must have high purity, a specific Crystal Size Distribution (CSD) and a desired crystal shape. Crystals with random CSD are not desired (Quintana-Hernandez et al., 2004). The principal driving force for crystallisation from solution is supersaturation, which is the difference between the solution concentration and the saturation concentration (Myerson, 2001).

The supersaturation profile obtained during the crystallisation operation determines the size, shape and solid-state phase of the product crystals. This supersaturation is usually achieved by cooling, evaporation or addition of anti-solvent. For example, in a cooling crystallisation process, the kinetics of nucleation and crystal growth require supersaturation, which is obtained by a change in temperature. If supersaturation is kept at an optimum value, the growth rate is high and sufficient coarse crystals are formed. This scenario is usually applied for the continuous operational mode. For a batch operation, however, this condition is difficult to achieve because the supersaturation state changes from the beginning to the end of the batch processing step. One way to address this problem is to maintain a cooling profile that keeps supersaturation almost constant, thereby preventing high nucleation rates and limiting the number of nuclei formed. Also, since agitation exerts an additional influence on CSD in terms of low values of induced agglomeration, aggregation and flocculation of particles, whilst high values produce attrition and breakage, it is worthwhile to have an agitation profile that increases mass transfer but reduces crystal destruction. The right combination of temperature and agitation profiles leads to the optimal batch operation crystallisers.

In this chapter, the modelling of a batch cooling crystallisation process (see Figure 1) using the population balance approach is highlighted through the use of a systematic model generation procedure developed by Abdul Samad et al. (2011).

Abdul Samad et al. (2011) has developed a generic model-based framework through which models for specific batch cooling crystallisation operations are

* Written in conjunction with Mr. Noor Asma Fazli Abdul Samad

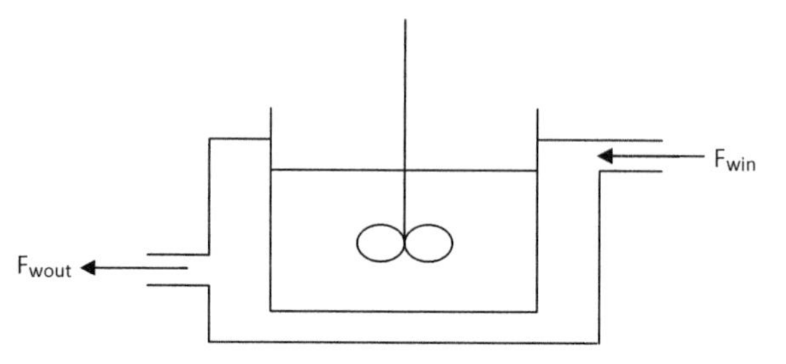

FIGURE 1 Batch cooling crystalliser.

generated for various modelling objectives. The main idea here is that through a sequence of steps, the necessary model equations of different types, forms and complexity are retrieved from a library of models available in the framework. The generic nature of the available models allows different forms of models to be generated according to the needs of the specific model. Once a model is generated, it is linked to a batch operational scenario, and based on this, the model equations are solved and the results of the simulation evaluated with respect to design/analysis of the batch operation.

The batch crystallisation operation is divided into four operational scenarios:

1. Initial cooling scenario: In this operational scenario, we have a solution at a temperature higher than the saturation temperature. If we cool the solution to the solubility line then the solution is saturated. If we continue to cool the solution past the solubility line, the solute may start to crystallise A solution in which the solute concentration exceeds the saturation line at a given temperature is known as a supersaturated solution as shown in Figure 2.

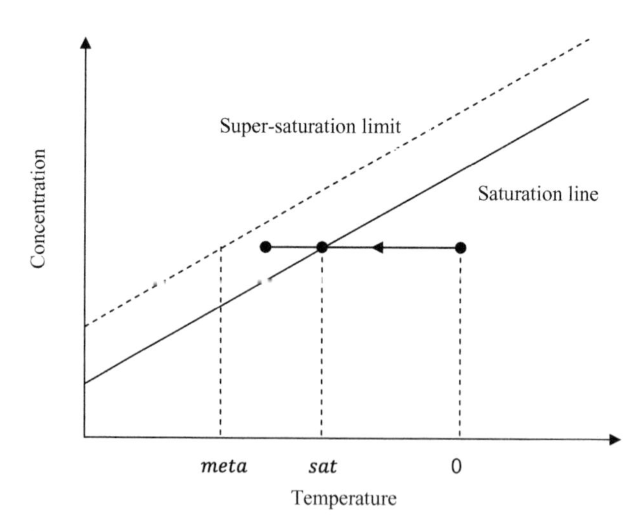

FIGURE 2 Saturation concentrations versus temperature profile for a solid–liquid equilibrium system.

2. Nucleation scenario: Nucleation is the formation of nuclei or minuscule crystals that grow. Once a solute starts to crystallise due to the condition of supersaturation, it will involve the birth of a new crystal and could be regarded as the beginning of the solid separation process. During this operation scenario, the solute molecules form the smallest sized particles possible.
3. Crystal growth scenario: Here, the newly born crystals grow larger by the addition of solute molecules from the supersaturated solution. This is known as crystal growth (Myerson, 2001). Operating along a pre-defined cooling temperature curve allows this scenario to continue until there is no more growth of the crystals. Note that nucleation and crystal growth occur simultaneously.
4. Product removal scenario: At this point we have liquor with solid particles suspended in a liquid. The solid phase is removed from the liquid and processed further to obtain the final dried crystals. The liquid is usually recycled back to the crystalliser.

More details on the batch operational scenarios can be found in Table A in the appendix of this chapter.

10.1.1. Model Description

A brief description of the model generation procedure of Abdul Samad et al. (2011) is given here. For a more detailed description, the reader is referred to the published article.

As shown in the flow diagram (see Figure 3) of the model generation procedure, Abdul Samad et al. proposes a hierarchical stepwise procedure for generating specific batch crystallisation operational models in a systematic and efficient manner. Based on the chemical system description together with information on process types, condition of operation and specific assumptions, the balance, constitutive and conditional equations representing the model for the specified crystallisation operation are obtained. The generated complete set of equations representing the process-operation model are analysed and a solution strategy is determined for the specific process operational scenarios. In the final step, an appropriate numerical solver is employed to solve the model equations.

In a crystallisation process we have crystal particles growing within a liquid solution and usually a population balance approach is employed to model their operations. The solution technique employed to solve the population balance–based model equations is closely related to the crystallisation characteristics (see Figure 4) being investigated. Therefore, choice of the crystallisation characteristics also selects the solution technique for the population balance–based model.

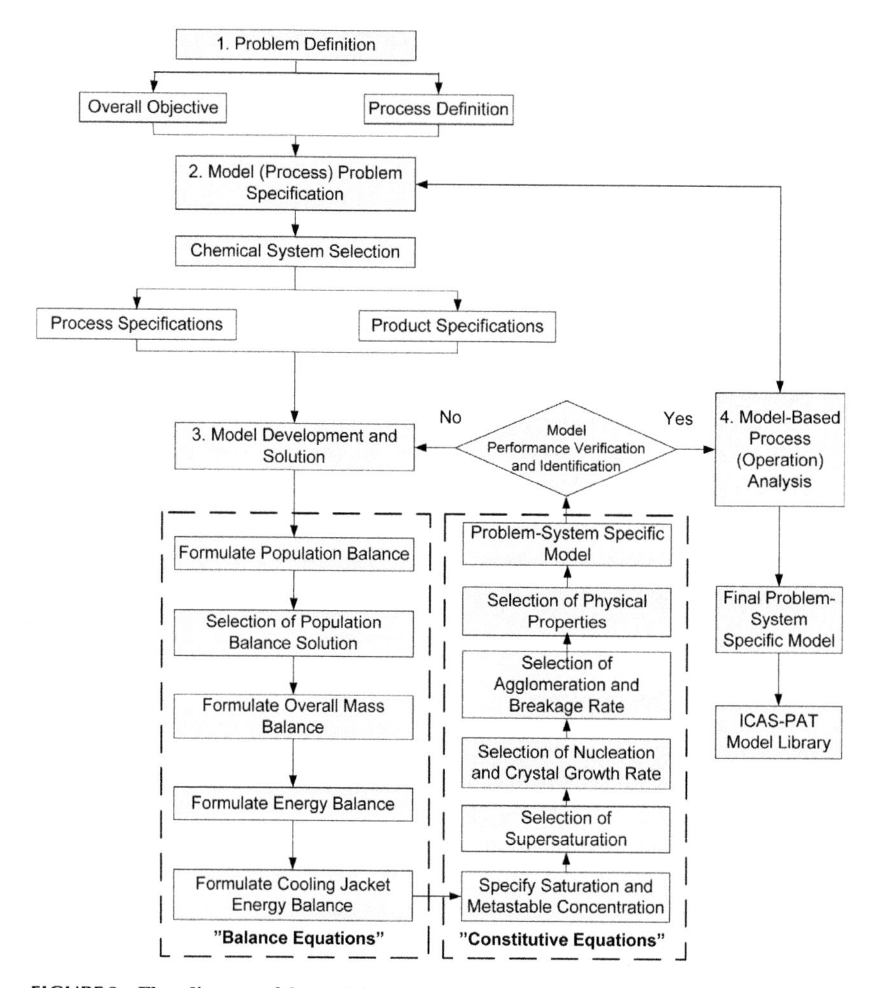

FIGURE 3 **Flow diagram of the model generation procedure (from Abdul Samad et al. (2011).**

10.1.1.1 Process (System) Description

The process is a cooling crystallisation (see Figure 1) operated in the batch mode for the crystallisation of sugar from an aqueous solution.

Assumptions: Two assumptions are considered.

- The solution is well mixed resulting in uniform density and temperature in the crystalliser tank
- The crystals in the tank are well suspended; there is no accumulation of crystals at the bottom of the tank

Modelling Objective: Develop a one-dimensional population balance model under the assumptions listed above for the crystallisation of sugar from an

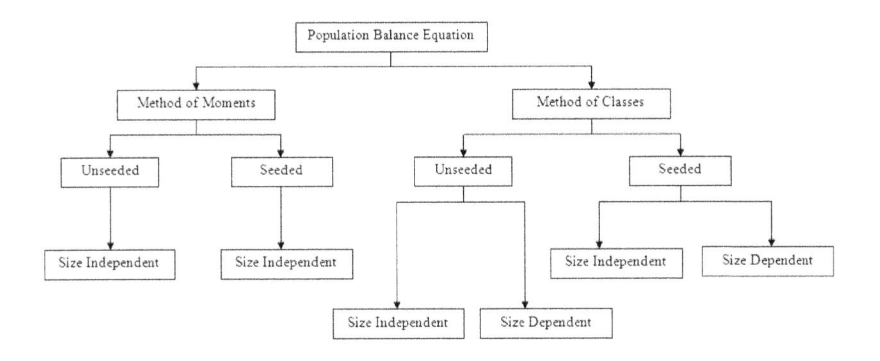

FIGURE 4 PBE solution techniques based on process operation characteristics.

aqueous solution valid for different batch operational scenarios (see Table A1 in the appendix of this chapter).

10.1.1.2 Model Generation

The steps of the model generation procedure, as outlined in Figure 3, are now applied to generate the specific model. Note that steps 2–4 generate the model equations, which are then analysed and solved.

Step 1–Problem definition: Based on the assumptions listed above, the crystallisation process is unseeded.

Step 2–Selection of crystallisation process: The two chemicals involved in this process are sugar and water. The options related to one-dimensional population balance model for unseeded operations are selected. Spherical crystal shape is considered for the particles.

Step 3.1–Formulate the PB: The PBE (Population Balance Equation) is a set of partial differential equations. These are selected from those given in Table 1.

Step 3.2–Selection of PB solution: This is related to the particular form of the solution technique employed. Here, the method of moments is used. Conversion of the PDE model for the PBE into a set of ODEs by the method of moments is highlighted in the appendix of this chapter. The set of ODEs and their representation by the method of moments are given in Table 1.

Step 3.3–Derive overall mass balance equations: In this case the initial size of crystals is assumed to be known. The corresponding equation for the mass balance is listed in Table 2.

Step 3.4–Derive the energy balance equations: For the energy balance, the heat of crystallisation is considered and a temperature difference – based driving is considered for heat exchange. The equations are listed in Table 2.

TABLE 1 PBE and The Method of Moment's Models

[0,1-2]PBE Model	
General	**Problem specific**
This is a general PBE for the x-direction only (0→L) $\frac{\partial n(L,t)}{\partial t} = -\frac{\partial n(L,t)G(L,c,T)}{\partial L} + (B - D)$ where $(B-D) = B_{nuc} + B_{agg} - D_{agg} + B_{br} - D_{br}$	This specific PBE is for the x-direction only (0→L) $\frac{\partial n(L,t)}{\partial t} = -\frac{\partial n(L,t)G(L,c,T)}{\partial L} + B$ where $B = B_{nuc} + \alpha = B^0 + \alpha$
Only size-independent growth has been considered (transformed method of moment's model) $\frac{d\mu_0}{dt} = B_{nuc} + \alpha$ $\frac{d\mu_m}{dt} = mG\mu_{m-1} + B_{nuc}L_0^m; \quad m = 1, \ldots, 4$	$\frac{d\mu_0}{dt} = B^0 + \alpha$ $\frac{d\mu_j}{dt} = jG\mu_{j-1} + B^0 L_0^j,$ $j = 1, 2, \ldots 4$

Step 3.5–Derive cooling jacket energy balance equations: For the energy balance around the cooling jacket, a temperature difference – based driving is considered for the heat exchange. The equations are listed in Table 2.

Step 3.6–Saturation concentration and meta-stable concentration: The parameters for saturation concentration and the heat of crystallisation are selected for the sucrose crystallisation process (see Table 3). Meta-stable concentrations are not used in this case study.

TABLE 2 Overall Mass Balances, Energy Balance and Cooling Jacket Energy Balance Models

General	**Initial size of crystal (L_0) is known (one-dimensional model)**
Overall mass balance: solute concentration $\frac{dc}{dt} = -\frac{\rho_c k_v V}{m_w}\left(3\int_0^\infty Gn(L,t)L^2 dL + B_{nuc}L_o^3\right)$	Overall mass balance: solute concentration $\frac{dC}{dt} = -\rho_c k_v \frac{V}{M_w}[3G\mu_2 + B^0 L_0^3]$
Energy balance $\rho V c_p \frac{dT}{dt} = -\Delta H_c \rho_c k_v V \left(3\int_0^\infty Gn(L,t)L^2 dL + B_{nuc}L_o^3\right) - U_1 A_1 \Delta T$	Energy balance $\rho V c_p \frac{dT}{dt} = -\Delta H_c \rho_c k_v V[3G\mu_2 + B^0 L_0^3] - U_1 A_1 (T - T_w)$
Energy Balance: Cooling Jacket	
$\rho_w V_w c_{pw}\frac{dT_w}{dt} = \rho_w F_{win} c_{pw}(t_{win} - T_w) + U_1 A_1 (T - T_w) + U_2 A_2 (T_{ex} - T_w)$	

TABLE 3 Saturation Concentration Model

General	Parameter coefficient			
Saturation concentration $C^{sat} = a_{i1} + b_{i1}T + c_{i1}T^2 + d_{i1}T^3$	0.64407	7.251×10^{-4}	2.057×10^{-5}	-9.04×10^{-8}
Heat of crystallisation $\Delta H_c = a_{i3} + b_{i3}T + c_{i3}T^2 + d_{i3}T^3$	−12.2115	−0.7937	0	0

Step 3.7–Selection of supersaturation: Relative supersaturation is considered for the nucleation and crystal growth rates. By assuming the solution to be ideal, a relative supersaturation is calculated based on the concentration (see Table 4).

Step 3.8–Selection of nucleation and crystal growth rate: The crystallisation kinetics for nucleation and crystal growth rates are based on relative supersaturation. Secondary nucleation is considered and it is assumed that the crystal growth is independent of size. The effect of agitation on nucleation and crystal growth is also taken into consideration. The nucleation and crystal growth rate model equations and the corresponding model parameters are given in Table 5.

Step 3.9–Select agglomeration and breakage rate: not used in this model.

TABLE 4 Supersaturation Concentration Model

Crystallisation kinetics	Equations
Supersaturation	$\sigma_c = S_c - 1$

TABLE 5 Nucleation and Crystal Growth Rate Models

Crystallisation kinetics	Equations
Nucleation	$B^0 = k_b M_c^j \sigma^b N_{rpm}^p$
Crystal growth rate	$G = k_g \sigma^g N_{rpm}^q$
Birth-death rate (based on agglomeration and breakage)	$\alpha = k_{ab} S^{ab} M_c^k N_{rpm}^r$

TABLE 6 Particle Distribution Model

Type	General
Total no. of particles	$N_c = \mu_0$
Total length	$L_c = k_L \mu_1$
Total area	$A_c = k_A \mu_2$
Total mass	$M_c = \rho_c k_v \mu_3$
Mean size diameter	$D[4,3] = \frac{\mu_4}{\mu_3}$

Step 3.10–Properties from particle distribution: The properties from particle distribution such as total number, total length, total area and total mass as well as mean size diameter can be calculated using the models given in Table 6.

10.1.2. Model Analysis

In this section, the set of equations representing the mathematical model generated in Section 10.1.1.1 is analysed in terms of degrees of freedom, selection of variables for specification and incidence matrix of the model equations and variables. The full set of model equations derived above is given in Table 7.

10.1.2.1 Degree of Freedom Analysis

From Tables 7 and 8, the degree of freedom (DOF) is calculated as follows

$$DOF = N_v - N_e = 56 - 19 = 37$$

where

N_v = number of variables
N_e = number of equations

This means that we have 19 unknown variables and 37 variables that need to be specified. Table 9 shows a classification of the variables into different categories.

10.1.2.2 Incidence Matrix

First differential variables and then unknown algebraic variables are placed in the columns, and regarding the equations, first algebraic equations and then differential equations are placed in the rows (see Table 10). The algebraic equations are ordered to find a lower tridiagonal form.

TABLE 7 List of Complete Model Equations

No.	Equations	Type	No. of Equations
1	$\frac{d\mu_0}{dt} = B_{nuc} + \alpha$	Zeroth moment	1
2	$\frac{d\mu_1}{dt} = G\mu_0$	First moment	1
3	$\frac{d\mu_2}{dt} = 2G\mu_1$	Second moment	1
4	$\frac{d\mu_3}{dt} = 3G\mu_2$	Third moment	1
5	$\frac{d\mu_4}{dt} = 4G\mu_2$	Fourth moment	1
6	$\frac{dc}{dt} = -\frac{\rho_c k_v V}{m_w}\left(3G\mu_2 + B_{nuc}L_0^3\right)$	Overall mass balance	1
7	$\rho V c_p \frac{dT}{dt} = -\Delta H_c \rho_c k_v V\left(3G\mu_2 + B_{nuc}L_0^3\right) - U_1A_1(T - T_w)$	Energy balance	1
8	$\rho_w V_w c_{pw} \frac{dT_w}{dt} = \rho_w F_{win} c_{pw}(T_{win} - T_w) + U_1A_1(T - T_w) + U_2A_2(T_{ex} - T_w)$	Cooling jacket energy balance	1
9	$c^{sat} = a_{i1} + b_{i1}T + c_{i1}T^2 + d_{i1}T^3$	Constitutive	1
10	$\Delta H_c = a_{i3} + b_{i3}T$		1
11	$\sigma = \frac{c - c^{sat}}{c^{sat}}$		1
12	$M_c = \rho_c k_v \mu_3$		1
13	$B_{nuc} = k_b M_c^j \sigma^b N_{rpm}^p$		1
14	$G = k_g \sigma^g N_{rpm}^q$		1
15	$\alpha = k_{ab} S^{ab} M_c^k N_{rpm}^r$		1
16	$N_c = \mu_0$	Algebraic	1
17	$L_c = k_L \mu_1$		1
18	$A_c = k_A \mu_2$		1
19	$D[4,3] = \frac{\mu_4}{\mu_3}$		1
		Total number of equations	19

TABLE 8 List of Variables

Variables Symbol	Number
$\mu_0, \mu_1, \mu_2, \mu_3, \mu_4, B_{nuc}, G, c, \rho_c, k_v, V, m_w, L_0, \rho, c_p, T,$ $\Delta H_c, U_1, A_1, T_w, \rho_w, V_w, c_{pw}, F_{win}, T_{win}, U_2, A_2,$ $T_{ex}, c^{sat}, a_{i1}, b_{i1}, c_{i1}, d_{i1}, a_{i3}, b_{i3}, \sigma, k_b, M_c, j, b,$ $N_{rpm}, p, k_g, g, q, \alpha, k_{ab}, ab, k, r, N_c, L_c, k_L, A_c, k_A, D[4,3]$	56

TABLE 9 Variable Classification

Variable types	Status	Symbol	Number	Total
Known (to be specified)	Fixed by system	$a_{i1}, b_{i1}, c_{i1}, d_{i1}, a_{i3}, b_{i3}, \rho_w, c_{pw}, \rho, c_p, \rho_c$	11	37
	Fixed by model	$k_b, j, b, p, k_g, g, q, k_{ab}, ab, k, r,$	11	
	Fixed by problem	$k_L, k_A, k_v, U_1, A_1, U_2, A_2, T_{ex}, N_{rpm}, V, m_w, L_0, V_w, F_{win}, T_{win}$	15	
	Adjustable parameter	-	-	
Unknown variables (to be calculated)	Algebraic (explicit)	$c^{sat}, \sigma, \Delta H_c, B_{nuc}, G, N_c, L_c, A_c, M_c, \alpha, D[4,3],$	11	19
	Differential (dependent)	$\mu_0, \mu_1, \mu_2, \mu_3, \mu_4, c, T, T_w,$	8	

TABLE 10 Incidence Matrix for Sugar Crystallisation Model

		Differential Variables								Algebraic Variables										
		μ_0	μ_1	μ_2	μ_3	μ_4	c	T	T_w	c_{sat}	ΔH_c	σ	M_c	B_{mc}	G	α	N_c	L_c	A_c	$D[4,3]$
Algebraic Equations	9							*		*										
	10							*			*									
	11									*		*								
	12				*								*							
	13											*	*	*						
	14											*			*					
	15											*	*			*				
	16	*															*			
	17		*															*		
	18			*															*	
	19				*	*														*
Differential Equations	1	*												*						
	2		*												*					
	3			*											*					
	4				*										*					
	5					*									*					
	6			*			*							*	*					
	7							*			*			*	*					
	8							*	*											

10.1.3. Numerical Solution

10.1.3.1 Calculation Procedure

The step by step calculation procedure and a corresponding flow diagram are given in Table 11.

10.1.3.2 Specified data

Table 12 gives the specific process parameters and data used for solving the model equations related to sugar crystallisation (Quantina-Hernandez et al., 2004).

10.1.3.3 Simulation results

The model equations with the specified variable values have been solved under different batch operational scenarios with ICAS-MoT (Sales-Cruz, 2006). These results are given in Table 13. The model as implemented and solved in ICAS-MoT is given in Appendix 1 (see chapter-10-1-sucrose-cryst. mot file).

TABLE 11 Calculation Procedure and Flow diagram

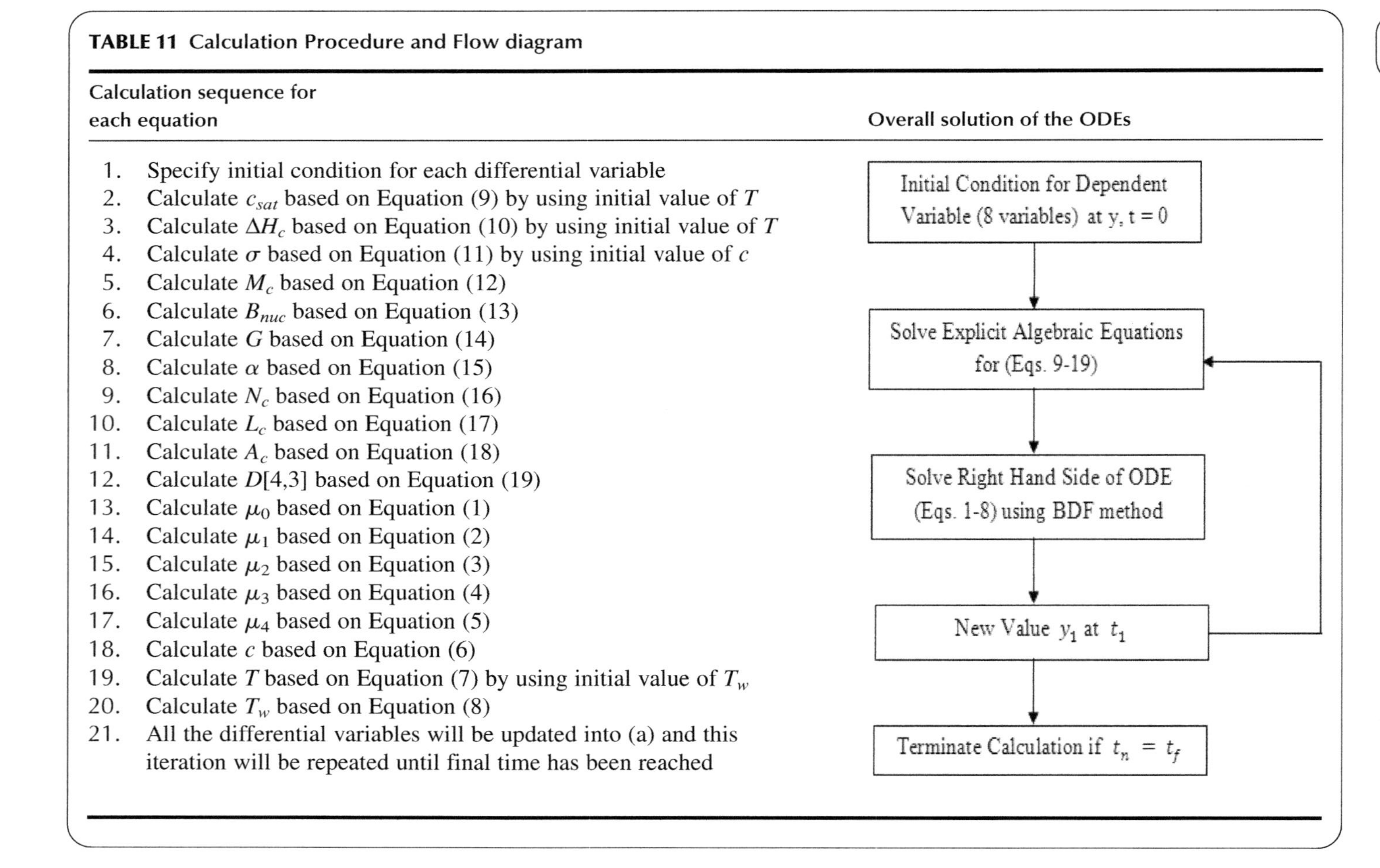

Calculation sequence for each equation	Overall solution of the ODEs
1. Specify initial condition for each differential variable 2. Calculate c_{sat} based on Equation (9) by using initial value of T 3. Calculate ΔH_c based on Equation (10) by using initial value of T 4. Calculate σ based on Equation (11) by using initial value of c 5. Calculate M_c based on Equation (12) 6. Calculate B_{nuc} based on Equation (13) 7. Calculate G based on Equation (14) 8. Calculate α based on Equation (15) 9. Calculate N_c based on Equation (16) 10. Calculate L_c based on Equation (17) 11. Calculate A_c based on Equation (18) 12. Calculate $D[4,3]$ based on Equation (19) 13. Calculate μ_0 based on Equation (1) 14. Calculate μ_1 based on Equation (2) 15. Calculate μ_2 based on Equation (3) 16. Calculate μ_3 based on Equation (4) 17. Calculate μ_4 based on Equation (5) 18. Calculate c based on Equation (6) 19. Calculate T based on Equation (7) by using initial value of T_w 20. Calculate T_w based on Equation (8) 21. All the differential variables will be updated into (a) and this iteration will be repeated until final time has been reached	Initial Condition for Dependent Variable (8 variables) at y, t = 0 → Solve Explicit Algebraic Equations for (Eqs. 9-19) → Solve Right Hand Side of ODE (Eqs. 1-8) using BDF method → New Value y_1 at t_1 → Terminate Calculation if $t_n = t_f$

TABLE 12 Specified data for the sugar crystallization simulation problem

Variable	Value	Units
Cp	2.4687	J/g°C
K_v	$\pi/6$	
U_1	0.8354	J/°C.min.cm^2
A_1	995.32	cm^2
V	2230	cm^3
P_c	1.588	g/cm^3
T_O	70	°C
L_O	15.126	μm
ΔL	20.2525	μm
M_T	3328	G
M_w	800	G
M_s	2528	G
Cp_W	4.18	J/g°C
T_{wi}	70	°C
U_2	0.3236	J/°C.min.cm^2
A_2	1334.88	cm^2
F_{wi}	8100	cm^3/min
T_{ex}	29	°C
P_w	1.0	g/cm^3
V_w	820	cm^3

10.2. MODELLING BATCH DISTILLATION OPERATIONS

10.2.1 Purpose

To set up an operating sequence and validate it through dynamic simulation.

10.2.2 Problem Description

A charge of 45.4 kmol of 25 mole% benzene (B), 50 mole% MonoChloroBenzene (MCB) and 25 mole% of ortho-DiChloroBenzene (DCB) is to be distilled. The objective is to obtain all three products with 99% purity and with a yield of 95%. The operational constraints relationship between the reflux ratio and vapour boil-up rate needs to be checked. The flow diagram is shown in Figure 5. Note that by opening and closing the valves on different streams, different batch operations are generated. For example, the valve in stream 1 is only opened during the charging of the feed; the valve in stream 2 is only opened at the end to discharge the residue; one of the valves in streams 4–6 is opened only when products are to be withdrawn from the distillation column.

TABLE 13 Simulation Results for Sugar Crystallisation Batch Operations

Stage	Time	State	Model
Initial cooling operation	t_0	Prepare a solution A natural cooling profile is imposed	
	$t_1 = t_{sat}$	The solution is initially in the single liquid-phase region at 70°C. Cool the solution to the saturation line (saturation temperature, T_{sat} at t_{sat})	
Nucleation	t_2	Continue to cool the solution past the saturation line ($T < T_{sat}$) In the first 30 minutes of the process, agitation is kept at 600 rpm. This agitation will stir the solution and induce the solution to be saturated and then supersaturated After 30 minutes, the agitation will be decreased linearly until the end of the process, in order to avoid breakage of crystals Once the solution starts to crystallise (due to supersaturation condition), it will promote the formation of a new crystals The molecules start to associate and form aggregates (clusters)	

(continued)

TABLE 13 *(continued)*

Stage	Time	State	Model
		Solute molecules have formed the smallest-size particles possible under the conditions present	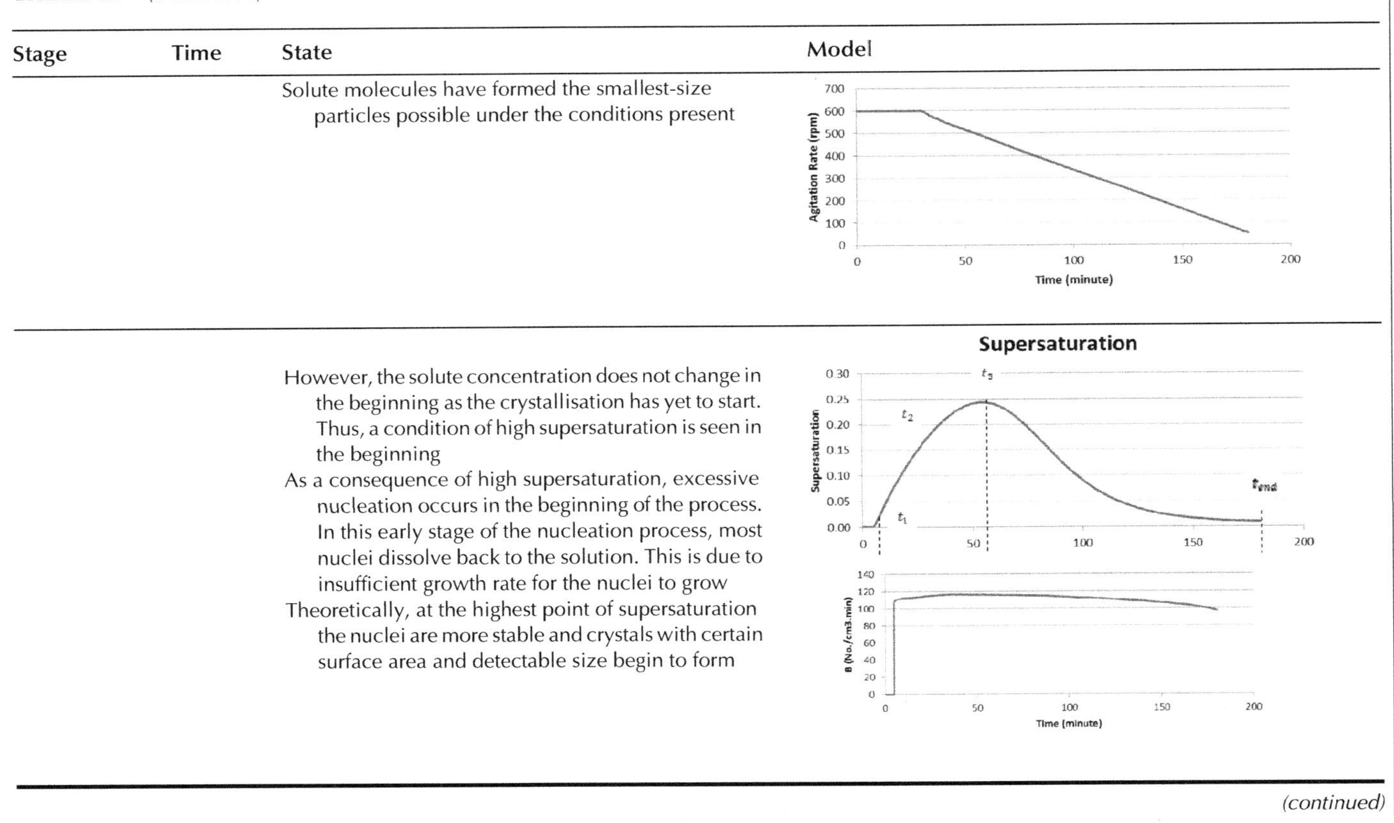
		However, the solute concentration does not change in the beginning as the crystallisation has yet to start. Thus, a condition of high supersaturation is seen in the beginning As a consequence of high supersaturation, excessive nucleation occurs in the beginning of the process. In this early stage of the nucleation process, most nuclei dissolve back to the solution. This is due to insufficient growth rate for the nuclei to grow Theoretically, at the highest point of supersaturation the nuclei are more stable and crystals with certain surface area and detectable size begin to form	

(continued)

TABLE 13 *(continued)*

Stage	Time	State	Model
Crystal growth -a	t_3	During this time, more stable nuclei appear and the growth rate starts to increase due to highest super-saturation. This growth rate is sufficient to grow the nuclei Nuclei start to grow larger by the addition of solute molecules from the supersaturated solution Growth of crystals is linear and described by the change of crystal dimensions with time Assuming the crystal is a sphere, the crystal growth rate is expressed by the increase in diameter with time Low growth rates is obtained near the end of process All the nuclei in the population have grown at the same rate It is important to note that during the crystal growth phase, nucleation also occurs Referring to the nucleation result, the nucleation rate is still high until the end of the process Hence, there are more nuclei formed during the crystal growth phase This phenomena leads to more production of smaller crystals, since the crystal growth rate decreases until end of operation	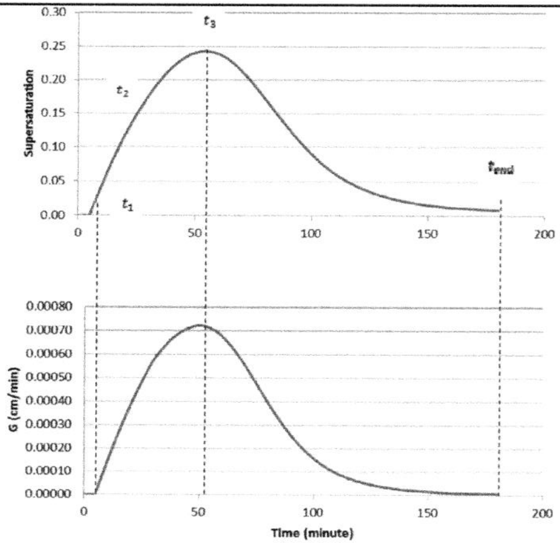

(continued)

TABLE 13 *(continued)*

Stage	Time	State	Model
Crystal growth -b	t_4	In this crystallisation process, the particles are only formed due to nucleation and production-reduction effects The production-reduction effects are generated by agglomeration, aggregation, attrition and breakage of crystals Nuclei are formed at the minimum crystal size assumed Population grows at the same rate resulting into increment in length, area and mass of crystals In the beginning, although the nucleation has occurred, the size of the nuclei initially is neglected. When the nuclei start to grow at maximum supersaturation, the mass of crystals increases steadily Since the nucleation starts early in the process, the mean size of crystals can be measured from the formation of nuclei based on moment model The mean size diameter also increases steadily due to the growth of the crystals Population grows until $t = t_{end}$	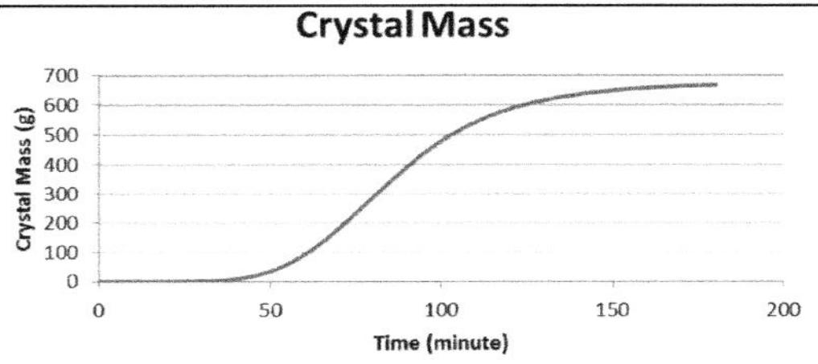 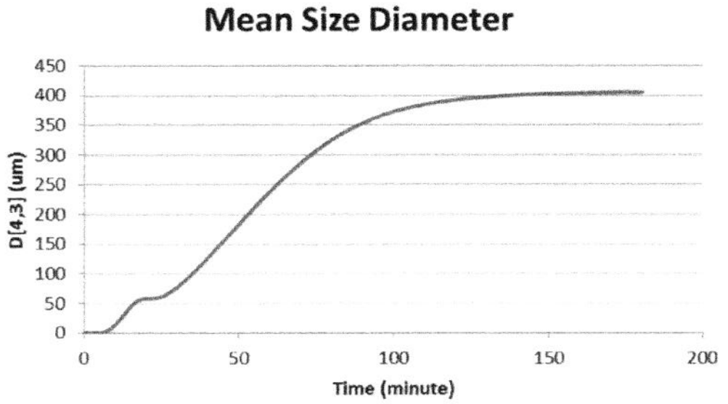 Crystal Mass Mean Size Diameter
Product removal	t_{end}	Discharge product from crystalliser; the product can only be withdrawn from the tank once the operation is completed	Final solute concentration, temperature, cooling temperature, mean size diameter and population density

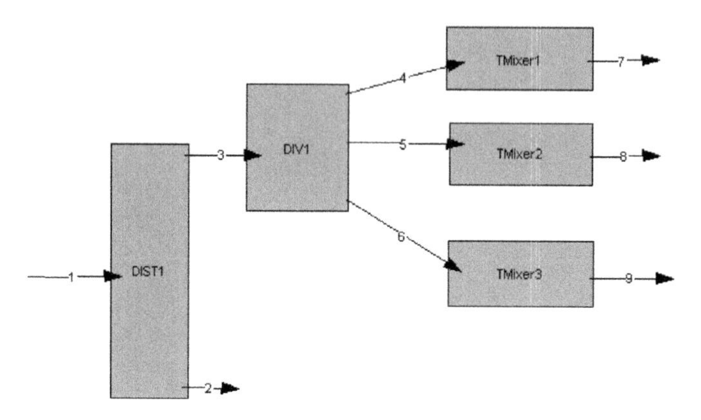

FIGURE 5 Flow diagram for a batch distillation operation.

10.2.3 Modelling Issues

Process model: A dynamic model capable of simulating distillation column operations under different conditions is necessary. In principle, any dynamic model equations allowing multiple runs for changed model parameters should satisfy the modelling requirement. For detailed descriptions of dynamic distillation column models, see Gani et al. (1986) and/or Ruiz et al. (1988). In this example, the dynamic model available in ICAS (Gani et al. 1997) has been used.

Event Modelling: A step-by-step approach (Papaeconomou, 2005) is highlighted to model the batch operation of a distillation column as a function of a specified operating policy.

Step 1: Run a dynamic simulation from ICAS (Dynsim) to generate a reference condition of operation (using the 'Function Evaluation Only' mode). A screenshot from ICAS is shown in Figure 6.

Step 2: Specify the operating steps and conditions for the different batch scenarios. The different operational steps for defining a desired operating sequence are given in Table 14.

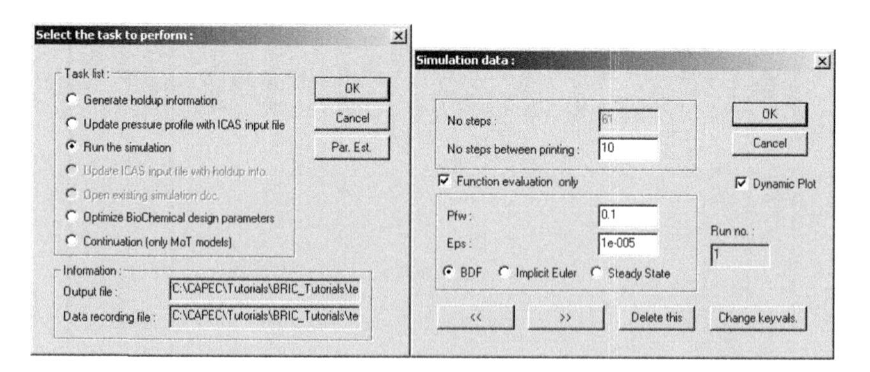

FIGURE 6 Use the dynamic model in ICAS to generate the reference conditions.

TABLE 14 Data for Different Batch Operational Steps

Period number	Reflux ratio	Vapour boil-up rate	Period time (hr)	x^B	Distillate amount	Yield %
1	1.294	100	0.0919	23.102	4.005	35.29
2	1.969	100	0.0441	19.056	1.487	48.39
3	2.65	100	0.0597	14.08	1.635	62.79
4	3.865	100	0.0618	9.761	1.271	73.99
5	6.654	80	0.0993	5.896	1.038	83.14 > 80
6	11.962	80	0.0954	3.55	0.589	88.33

As an example, consider operation 1. Here, we need to set the conditions for the termination of this run (at a specific time) running at a fixed reflux ratio and vapour boil-up. The other conditions being specified are values on the state variables (such as concentration of a specific compound in the residue or the temperature on a specific stage, etc.). A screenshot from ICAS is shown in Figure 7, where the entered data for operation 1 is highlighted.

Repeat this procedure for all operations 1–6 (as shown in Table 14). Note that if the manipulated variables (such as reflux rate or reboiler heat duty) need to be changed at the start of a run, this can be specified as shown in another screenshot from ICAS (see Figure 8).

The procedure is repeated until all the operating steps are defined, the information to be gathered is specified and the appropriate changes in the unit (distillation) parameters are made.

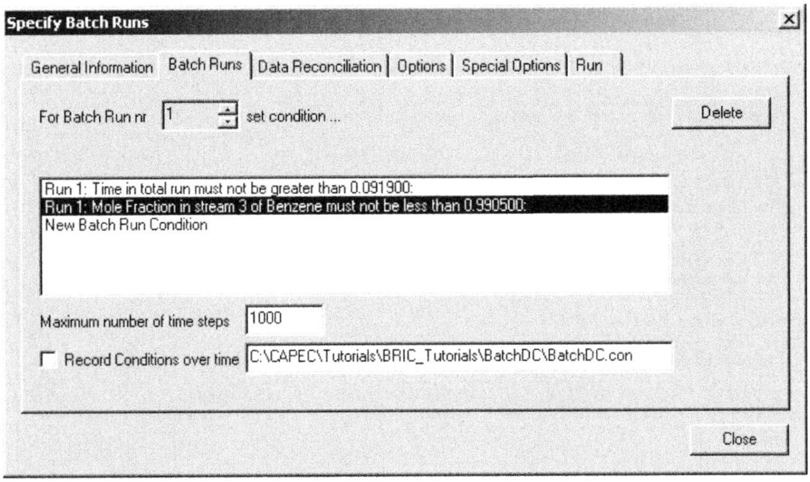

FIGURE 7 **Enter the event data for run 1.**

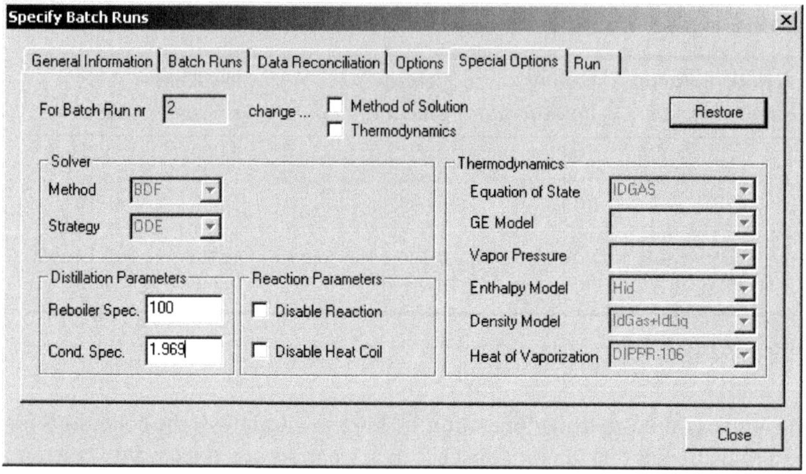

FIGURE 8 Specifying the value of the manipulated variable for a specific run.

Additionally, when the objectives (purity and yield) for product 1 (benzene) are satisfied, we can collect an off-cut product in the second product tank. When all the runs have been specified, the data will look as shown in Figure 9

The final step is to define how the simulation should be performed. The dynamic simulation options need to have the feature for multiple runs. For each run, the simulated data is collected and plotted.

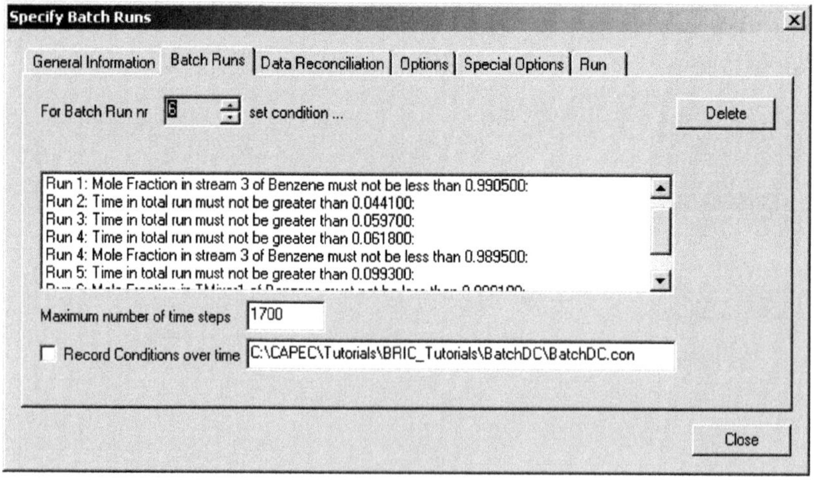

FIGURE 9 Full list of event data entered for each run.

Specify Batch Runs

General Information | Batch Runs | Data Reconciliation | Options | Special Options | Run

Total nr of Runs 7

Run

Pause between each run

Disable Output to Screen during Run Time

Custom Time Scale Value

Enable Normal Output to File during Run Time

Pfw 0.0001

No Steps between Printing 50

Auto Generate Holdup in first Run

All Runs use same Data Reconciliation Settings

Custom Error Value

Eps 1e-005

Reconciliation Options

Merge All Batch Runs to One File

Calculate Time Continuously

Close

FIGURE 10 Simulation run specification dynamic simulation is performed for each of the specified runs.

Step 3: Run the batch simulation (see Figure 10).

10.2.4 Simulation Results

Several dynamic simulation runs are made by ICAS-Dynsim according to the specified event-model. A summary of the results is given in Table 14 where it is pointed out when a product yield of > 80% has been reached. The plots in Figure 11 highlight how the product purity decreases whilst the product yield increases with time. The idea is to operate as long as possible to maintain the condition that gives the desired product. Figure 12 shows how the reflux ratio can be increased step by step to keep the distillate product purity on specification.

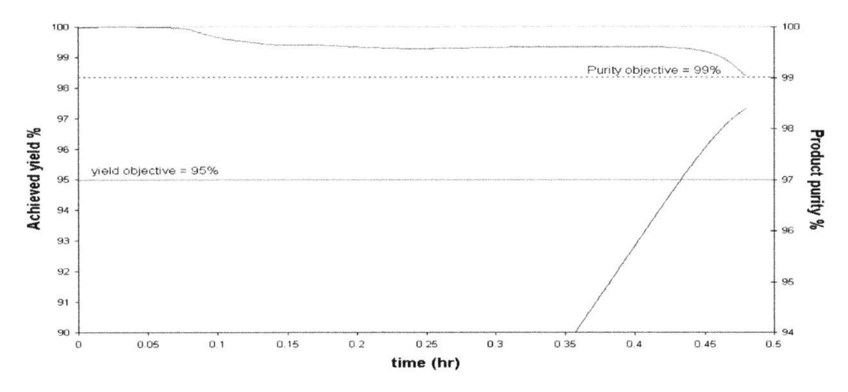

FIGURE 11 Plots of distillation product purity and product yield as a function of time.

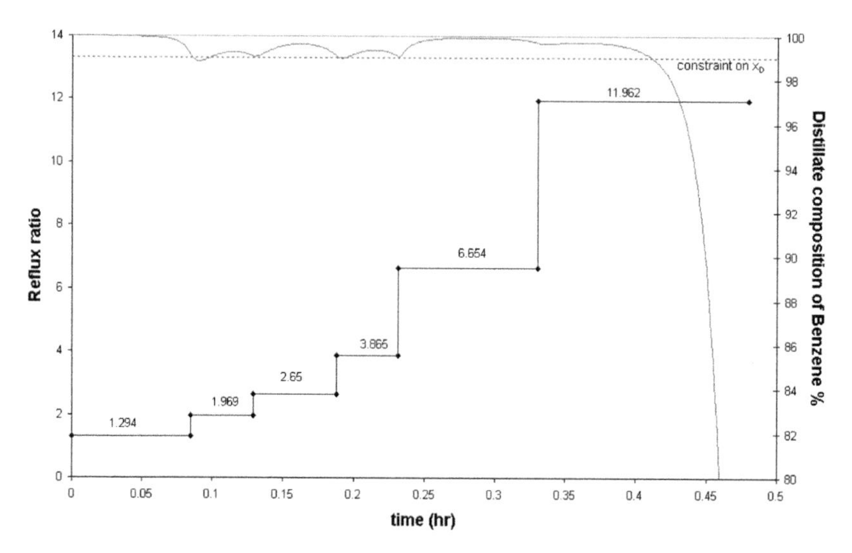

FIGURE 12 **Plot of product purity as a function of reflux ratio.**

In Table 15, the simulated results are shown in terms of cumulative reflux rate and the product purity and product yield.

In a similar manner, the batch operation of a reactor or a crystalliser can also be studied.

Discussion: A number of interesting simulations can be made once a dynamic model for a distillation column is available. In the model library of Appendix 2, a batch distillation model has been included for download by the interested reader. Try to set up multiple runs of the ICAS-MoT model object

TABLE 15 Simulation of Batch Distillation at Constant Reflux Ration During Operation Intervals

Operation	Number of intervals	Constant reflux ratio	Product tank purity %	Yield %	Time (hr)	Total reflux L (kmol)
a)	1	1.294	99.01	44.5	0,1054	6.6
b)	1	1.969	99.01	70.8	0,2104	15.98
c)	1	2.65	99.01	81.5	0,2919	24.76
d)	1	3.865	99.01	89.8	0,4196	39.79
e)	1	6.21	99.01	95	0.805	67.63
f)	1	6.654	99.01	95.6	0.8572	72.92
g)	1	11.962	99.01	98.6	1.4713	135.2
h)	6	[1.294, 11.962]	99.01	97.3	0.48	39.44

from another software environment, for example, EXCEL. Also, try to use the model to generate a distillation residue curve. Investigate what happens if the column is run at total reflux after an initial charge. Change the mixture with the following binary equimolar mixtures: acetone–chloroform; methyl acetate–methanol. What products are obtained as the residue and as the distillate?

NOMENCLATURE (for section 10.1)

Symbols

A_1	Crystalliser's internal area, cm^2
A_2	Crystalliser's external area, cm^2
a_{i3}	Parameter for heat of crystallisation
B^0	Nucleation rate, number of particles/cm^3.min
B	Birth rates
b_{i3}	Parameter for heat of crystallisation
C	Solute concentration, g of sugar/g of water
C^{sat}	Saturation concentration, g of sugar/g of water
C^{met}	Metastable concentration, g of sugar/g of water
c_{i3}	Parameter for heat of crystallisation
c_p	Heat capacity of solution (magma), J/g°C
D	Death rates
$i3$	Parameter for heat of crystallisation
F	Flow rate, cm^3/min
G	Size-independent crystal growth rate, cm/min
k_a	Kinetic coefficient for production reductions, number of particles/cm^3 magma cm^2 min$(g/cm^3)^k$ $(rpm)^r$
k_b	Kinetic coefficient for types of nucleation, number of particles/cm^3 min$(g/cm^3)^j$ $(rpm)^p$
k_g	Kinetic coefficient for growth, cm/min $(rpm)^q$
k_v	Crystal shape factor
L	Crystal length, cm
M	Mass, g
N	Population density (number), number of particles/cm^3slurry cm
N	Population density (number), number of particles/cm
N_{rpm}	Agitation rate, rpm
p	Molecular surface area parameter
R	Ideal gas constant
r	Molecular volume parameter
S	Non-relative supersaturation
T	Time, min
T	Temperature of magma, °C
U_1	Heat transfer coefficient (magma-glass-water), J/°C min cm^2

U_2	Heat transfer coefficient (water-glass-air), J/°C min cm^2
V	Magma volume, cm^3
x	Molar fraction

Greek Letters

$\alpha(L)$	Production-reduction rate, number of particles/cm^3 magma cm min
ΔH_c	Crystallisation enthalpy, J/g
ΔT_{lm}	Log mean temperature
ΔE	Activation energy
ρ	Magma density, g/cm^3
Ø	Estimated parameter vector
σ	Relative supersaturation
μ	Moment indicator
γ	Activity coefficient
φ	Molecular volume fraction
τ	Boltzmann factor
α	Binary interaction parameter
β	Binary interaction parameter

Subscripts and superscripts

A	Production-reduction order
Agg	Aggregation
B	Nucleation order
Br	Breakage
C	Crystal
Ex	Exterior
G	Growth order
I	Size I
In	Into the jacket
J	Mass order at nucleation
K	Mass order at production-reduction
Nuc	Nucleation
P	Agitation order at nucleation
Q	Agitation order at growth
R	Agitation order at production-reduction
W	Water

APPENDIX

A: Batch crystallisation operation scenario versus state, and available data/information

Stage	Time	State	Available data/Information	Model
Initial cooling operation	t_0	Prepare a solution A natural cooling profile is imposed	Initial solute concentration and temperature (T_0, C_0.) Total solution volume Agitation speed for stirring purpose	$\frac{dT}{dt}$ → Specify temperature $\frac{dC}{dt}$ → Specify solute concentration $\frac{dT_w}{dt}$ → Specify cooling temperature
	$t_1 = t_{sat}$	The solution is initially in the single liquid-phase region at 70°C. Cool the solution to the saturation line (saturation temperature, T_{sat} at T_{sat})	$C > C_{sat}$ is known as a supersaturated solution Temperature will be decreased with time if initial solution temperature is outside the solubility temperature Concentration of the solution will be maintained until the solution become supersaturated Agitation speed will be maintained	C^{sat} → Saturation concentration is temperature dependent $\frac{dT}{dt}$ → Temperature will be decreased $\frac{dC}{dt}$ → Concentration will be maintained $\frac{dT_w}{dt}$ → Cooling temperature will be decreased
Nucleation	t_2	Continue to cool the solution past the saturation line ($T < T_{sat}$) In the first 30 minutes of the process, agitation is kept at 600 rpm. This agitation will stir the solution and induce the solution to be saturated and then supersaturated After 30 minutes, the agitation will be decreased linearly until the end of the process, in order to avoid breakage of crystals	$T_{meta} < T < T_{sat}$ Maintain meta-stable limit by manipulating cooling water flow rate Formation of solid crystal from solution; hence, heat of crystallisation is involved Concentration is expected to change due to formation of crystals Solute concentration will be decreased Change of concentration is mainly due to birth of new crystals by nucleation	B^0 → Start to increase S → Supersaturation will be high in the beginning ΔH_c → Starting to affect the temperature profile N_{rpm} → Agitation speed will decrease linearly $\frac{dT}{dt}$ → Temperature will be decreased

(continued)

(continued)

Stage	Time	State	Available data/Information	Model
		Once the solution starts to crystallise (due to supersaturation condition), it will promote the formation of a new crystals The molecules start to associate and form aggregates (clusters) Solute molecules have formed the smallest-size particles possible under the conditions present However, the solute concentration does not change in the beginning as the crystallisation has yet to start. Thus a condition of high supersaturation is seen in the beginning As a consequence of high supersaturation, excessive nucleation occurs in the beginning of the process. In this early stage of the nucleation process, most nuclei dissolve back to the solution. This is due to low growth rate of the nuclei Theoretically, at the highest point of supersaturation the nuclei are more stable and crystals with certain surface area and detectable size begin to form	Influence of agitation rate on nucleation is considered Agitation will be decreasing linearly Increase in total surface area due to nucleation would decrease the supersaturation	$\frac{dC}{dt} \rightarrow$ Concentration will be decreased $\frac{dT_w}{dt} \rightarrow$ Cooling temperature will be decreased
Crystal growth - a	t_3	During this time, more stable nuclei appear and the growth rate starts to increase due to highest supersaturation. This growth rate is sufficient to grow the nuclei Nuclei start to grow larger by the addition of solute molecules from the supersaturated solution	Change of concentration is mainly due to formation of new crystals by growth rate effect Influence of agitation rate on growth rate kinetics Increase in total surface area due to crystal growth would decrease the supersaturation	$G \rightarrow$ Start to increase $M_c \rightarrow$ Start to increase due to crystal growth $S \rightarrow$ Will start to decrease if solute concentration is close to saturation concentration

(continued)

(continued)

Stage	Time	State	Available data/Information	Model
		Growth of crystal is linear and described by the change of crystal dimensions with time Assuming the crystal is a sphere, the crystal growth rate is expressed by the increase in diameter with time Low growth rates are obtained near the end of process All the nuclei in the population have grown at the same rate It is important to note that during the crystal growth phase, nucleation also occurs Referring to the nucleation result, the nucleation rate is still high until the end of the process Hence, there are more nuclei formed during the crystal growth phase This phenomenon leads to more production of smaller crystals since the crystal growth rate decreases until the end of operation	High supersaturation will be generated at the beginning of the run Very low supersaturation will be generated near the end of the run Lengthen the batch cycle time required to achieve the desired crystal size	$\frac{dT}{dt}$ $\frac{dC}{dt}$ $\frac{dT_w}{dt}$
Crystal growth - b	t_4	In this crystallisation process, the particles are only born due to nucleation and production-reduction effects The production-reduction effects is generated by agglomeration, aggregation, attrition and breakage of crystals Nuclei are formed at the minimum crystal size is assumed Population grows at the same rate resulting into increment in length, area and mass of crystals	Number of particles will be increased Number of particles with specific length in the solution Crystal of different sizes have the same crystal habit or shape Total number, length, area and mass of crystals will be neglected in the beginning of the process and start to increase until the end of the process; determine the mean size based on volume/mass	$N_c \rightarrow$ Total number will start to increase $L_c \rightarrow$ Total length will start to increase $A_c \rightarrow$ Total area will start to increase $M_c \rightarrow$ Total mass will start to increase B^0 $\alpha(L)$

(continued)

(continued)

Stage	Time	State	Available data/Information	Model
		In the beginning, although the nucleation has occurred, the size of the nuclei initially is neglected. When the nuclei start to grow at maximum supersaturation, then the mass of crystal increases steadily Since the nucleation starts early in the process, the mean size of crystal can be measured from the birth of nuclei based on moment model The mean size diameter also increases steadily due to the growth of the crystals Population grows until $t = t_{end}$	Required CSD is achieved or not based on volume/mass mean size	$Ø_i$ $n(L_i)$ $D[4,3]$ → Mean size will start to increase $\frac{d\mu_j}{dt}$ $\mathrm{fs}\frac{dN_1}{dt}$
Product removal	t_{end}	Discharge product from crystalliser; the product can only be withdrawn from the tank once the operation is completed		Final solute concentration, temperature, cooling temperature, mean size diameter and population density

B: Derivation of the method of moments equations

Moment Transformation of Population Balance

Consider the general population balance equation

$$\frac{\partial n}{\partial t} = -\frac{\partial nG}{\partial L} + (B - D) \tag{B1}$$

Rearrange Equation (B1) for size-independent growth rate

$$\frac{\partial n}{\partial t} = -G\frac{\partial n}{\partial L} + (B - D) \tag{B2}$$

Define the jth moment of the distribution as

$$\mu_j = \int_0^\infty nL^j dL \quad j = 0, 1, \ldots, 4 \tag{B3}$$

If the population balance in Equation (B1) now be averaged in the L (characteristic length) dimension by multiplying by $L^j dL$ it yields

$$L^j\left[\frac{\partial n}{\partial t} + G\frac{\partial n}{\partial L} + D - B\right]dL = 0 \tag{B4}$$

Integrating Equation (B4) from zero to infinity

$$\int_0^\infty L^j\left[\frac{\partial n}{\partial t} + G\frac{\partial n}{\partial L} + D - B\right]dL = 0 \tag{B5}$$

Expanding Equation (B5)

$$\int_0^\infty L^j\frac{\partial n}{\partial t}dL + \int_0^\infty GL^j\frac{\partial n}{\partial L}dL + \int_0^\infty L^j[D - B]\,dL = 0 \tag{B6}$$

The single internal particle coordinate L can be thought of as a characteristic particle size. Consider the first term in the left-hand side of the Equation (B6). The moment of distribution in the Equation (B3) is related with the first term in the left-hand side of the Equation (B6) by exchanging the order of integration and time differentiation,

$$\frac{d\mu_j}{dt} = \int_0^\infty L^j\frac{\partial n}{\partial t}dL \tag{B7}$$

Substitute Equation (B7) into (B6)

$$\frac{d\mu_j}{dt} + \int_0^\infty GL^j \frac{\partial n}{\partial L} dL + \int_0^\infty L^j [D - B] dL = 0 \tag{B8}$$

The second term in the LHS of the Equation (B8) is integrated by parts and becomes

$$\int_0^\infty GL^j \frac{\partial n}{\partial L} dL = -GnL_0^j - \int_0^\infty jGL^{j-1} n dL \tag{B9}$$

The first term in the right-hand side of Equation (B9) refers to the new nucleated particles, which is nucleation at L_0; therefore, $Gn = B_{nuc}$. Simplify Equation (B9)

$$\int_0^\infty GL^j \frac{\partial n}{\partial L} dL = -B_{nuc} L_0^j - \int_0^\infty jGL^{j-1} n dL \tag{B10}$$

The second term in the right-hand side of Equation (B10) is related to the moment in Equation (4) where

$$\int_0^\infty jGL^{j-1} n dL = jG\mu_{j-1} \tag{B11}$$

Replace Equation (B11) into (B10) becomes

$$\int_0^\infty GL^j \frac{\partial n}{\partial L} dL = -B_{nuc} L_0^j - jG\mu_{j-1} \tag{B12}$$

Substitute Equation (B12) into (B8)

$$\frac{d\mu_j}{dt} - B_{nuc} L_0^j - jG\mu_{j-1} + \int_0^\infty L^j [D - B] dL = 0 \tag{B13}$$

Finally the third term in the right-hand side of Equation (B13) is assumed as mean birth and death rate of crystal particles. This term then is represented as

$$\int_0^\infty L^j [D - B] dL = \bar{D} - \bar{B} \tag{B14}$$

Substitute Equation (B14) into (B13) yields

$$\frac{d\mu_j}{dt} - B_{nuc}L_0^j - jG\mu_{j-1} + \bar{D} - \bar{B} = 0 \tag{B15}$$

Rearrange Equation (B15) becomes

$$\frac{d\mu_j}{dt} = B_{nuc}L_0^j + jG\mu_{j-1} + \bar{B} - \bar{D} \tag{B16}$$

By neglecting the mean birth and death rate, Equation (B16) is simplified as

$$\frac{d\mu_j}{dt} = B_{nuc}L_0^j + jG\mu_{j-1} \tag{B17}$$

Using the Equation (B17), the moment's model consists of zeroth, first, second, third and fourth terms that can be developed by substituting $j = 0,1,2,3,4$.

Moments	Generic moments model
0	$\frac{d\mu_0}{dt} = B_{nuc}$
1	$\frac{d\mu_1}{dt} = B_{nuc}L_0 + G\mu_0$
2	$\frac{d\mu_2}{dt} = B_{nuc}L_0{}^2 + 2G\mu_1$
3	$\frac{d\mu_3}{dt} = B_{nuc}L_0{}^3 + 3G\mu_2$
4	$\frac{d\mu_4}{dt} = B_{nuc}L_0{}^4 + 4G\mu_3$

**Notes that in some works the crystal size at nucleation is considered as being very small ($L_0 \approx 0$), which leads to negligible modelling errors. Therefore, the moment's model becomes:*

Moments	Generic moment's model
0	$\frac{d\mu_0}{dt} = B_{nuc}$
1	$\frac{d\mu_1}{dt} = G\mu_0$
2	$\frac{d\mu_2}{dt} = 2G\mu_1$
3	$\frac{d\mu_3}{dt} = 3G\mu_2$
4	$\frac{d\mu_4}{dt} = 4G\mu_3$

REFERENCES

Abbas, A., Romagnoli, J.A., 2007. Multiscale Modeling, Simulation and Validation of Batch Cooling Crystallization. In Separation and Purification Technology. 53, 153–163.

Abdul Samad, N.A.F., Singh, R., Sin, G., Gernaey, K.V., Gani, R., 2011. A generic multi-dimensional model-based system for batch cooling crystallization processes. Computers and Chemical Engineering. 35 (5), 828–843.

Farrell, R.J., Tsai, Y., 1994. Modeling, Simulation and Kinetic Parameter Estimation in Batch Crystallisation Process. In American Institute of Chemical Engineers Journal. 40 (4), 586–593.

Gani, R., Ruiz, C.A., Cameron, I.T., 1986. Computers and Chemical Engineering. 10, 181–198.

Gani, R., Hytoft, G., Jaksland, C., Jensen, A.K., 1997. Computers and Chemical Engineering. 21, 1135–1146.

Hu, Q., Rohani, S., Jutan, A., 2005. Modeling and Optimization of Seeded Batch Crystallizers. In Computers and Chemical Engineering. 29, 911–918.

Hounslow, M.J., Ryall, R.L., Marshall, V.R., 1988. A Discretized Population Balance for Nucleation, Growth and Aggregation. In American Institute of Chemical Engineers Journal. 34 (11), 1821–1832.

Mullin, J.W., Sohnell, O., 1977. Expression of Supersaturation in Crystallization Studies. In Chemical Engineering Science. 32, 683–686.

Myerson, A.S., 2001. Handbook of Industrial Crystallization, 2nd edit. Butterworth-Heinemann.

Nagy, Z.K., Chew, J.W., Fujiwara, M., Braatz, R.D., 2008. Comparative Performance of Concentration and Temperature Controlled Batch Crystallizations. In Journal of Process Control. 18, 399–407.

Ouiazzane, S., Messnaoui, B., Abderafi, S., Wouters, J., Bounahmidi, T., 2008. Estimation of Sucrose Crystallization Kinetics from Batch Crystallizer Data. In Journal of Crystal Growth. 310, 798–803.

Paengjuntuek, W., Arpornwichanop, A., Kittisupakorn, P., 2008. Product Quality Improvement of Batch Crystallizers by a Batch-to-batch Optimization and Nonlinear Control Approach. In Chemical Engineering Journal. 139, 344–350.

Quintana-Hernandez, P., Bolanos-Reynoso, E., Miranda-Castro, B., Salcedo-Estrada, L., 2004. Mathematical Modeling and Kinetic Parameter Estimation in Batch Crystallization. In American Institute of Chemical Engineers Journal. 50 (7), 1407–1417.

Papaeconomou, F.I., 2005. Integration of Synthesis and Operational Design of Batch Processes, PhD-thesis. Technical University of Denmark: Lyngby Denmark.

Randolph, A.D., Larson, M.A., 1971. Theory of Particulate Processes. New York: Academic Press.

Ruiz, C.A., Cameron, I.T., Gani, R., 1988. Computers and Chemical Engineering. 12, 1–14.

Sales-Cruz, M., 2006. Development of a computer aided modelling system for bio and chemical process and product design, PhD-thesis. Technical University of Denmark: Lyngby, Denmark.

Chapter 11

Parameter Estimation*

Parameter estimation is an important part of the development of a model. There are many ways to estimate the parameters for a model containing PDEs (Partial Differential Equations), ODEs (Ordinary Differential Equations), AEs (Algebraic Equations) or some combination of them. For example, in the case of a model represented by ODEs, the "brute-force" method would be to provide an initial guess for the parameters, then solve the ODEs until the final time t_{end}, then check the simulated data against measured data and, based on this, generate new estimate for the parameters and repeat the steps until convergence is achieved. That is, in this approach, the process model is solved in the inner loop and the optimisation problem (parameter regression) is solved in the outer loop. Another approach would be to convert the ODE system into a system of algebraic equations. Consequently, instead of having to solve the ODE system from an initial to the final value of the independent variable for each (optimisation) iteration step, a large number of algebraic equations are solved. One way to convert the ODEs into algebraic equations is the finite difference approximation. Another method is orthogonal collocation.

Since model parameter estimation is related to matching the model results with measured data, sensitivity of the model parameters, uncertainty of the measured data and uncertainty of model-based predictions all have important roles.

The objective of this chapter is to introduce the reader to the parameter estimation problem through examples taken from reaction kinetic modelling. In Chapter 5, parameter estimation for property models has been highlighted. The sensitivity and uncertainty issues are not discussed in this chapter. The ICAS-MoT (Heitzig et al. 2011) however provides options to study them in detail. In this chapter, simple examples are used to mainly illustrate the concepts.

11.1. MODELLING OF SIMPLE REACTION KINETICS

The objective here is to create a model for use on a stand-alone basis as well as for use in different applications, such as part of other models and/or use of the models to generate chemical (reaction) system information. Start by developing a model representation for a given kinetic mechanism.

* Written in conjunction with Prof Mauricio Sales-Cruz and Miss Martina Heitzig

11.1.1. Model Description

Consider the following first-order reversible series reaction

$$A \underset{k_2}{\overset{k_1}{\rightleftarrows}} B \underset{k_4}{\overset{k_3}{\rightleftarrows}} C \tag{1}$$

The reaction rates for the three compounds (A, B and C) are given by the following ordinary differential equations (ODEs).

$$\begin{aligned} \frac{dz_1}{dt} &= -k_1 z_1 + k_2 z_2 \\ \frac{dz_2}{dt} &= k_1 z_1 - (k_2 + k_3) z_2 + k_4 z_3 \\ \frac{dz_3}{dt} &= -k_3 z_2 - k_4 z_3 \end{aligned} \tag{2}$$

$$z_0[1,0,0] \quad t \in [0,1] \tag{3}$$

where the state vector $\underline{z}$ is defined as $[A, B, C]$. The objective is to calculate the compositions of A, B and C as a function of temperature for given values of the kinetic model parameters $[k_1, k_2, k_3, k_4] = [4, 2, 40, 20]$.

11.1.2. Model Analysis

This is an initial value integration problem. The dependent differential variable is the vector $\underline{z}$. The independent variable is time t. For the given values of the vector $\underline{k}$ (the parameters) and the initial values of the vector $\underline{z}$, it is necessary to integrate with time and calculate the values of the vector $\underline{z}$ at each time step.

Before solving the model equations, it is a good idea to calculate the eigenvalues of the Jacobian matrix ($J = df/dz$, where f is the right-hand side of the ODEs).

11.1.3. Solution Strategy

The solution strategy depends on the eigenvalues of the Jacobian matrix and the right-hand side of the ODEs. In this case

$$J = \begin{matrix} -k_1 & k_2 & 0 \\ k_1 & -(k_2 + k_3) & k_4 \\ 0 & k_3 & -k_4 \end{matrix} \tag{4}$$

For the above Jacobian matrix, the eigenvalues are: $\lambda_1 = -61.44, \lambda_2 = -4.55, \lambda_1 = 0$

11.1.4. Numerical Solution

For the simulation of the reacting mixture compositions as a function of time, the BDF-integration method is used and the calculated results are given in Table 1. The initial condition is given by Eq. 3.

TABLE 1 Calculated Concentration Data for the Reacting System of Eq. 1

Point number	Time	z_1	z_2	z_3
1	0.048	0.83037	0.09175	0.07788
2	0.1	0.68525	0.13454	0.18021
3	0.14	0.59487	0.15981	0.24533
4	0.2	0.48673	0.18993	0.32333
5	0.24	0.42943	0.2059	0.36468
6	0.3	0.36087	0.22499	0.41414
7	0.34	0.32454	0.23511	0.44035
8	0.4	0.28107	0.24722	0.47171
9	0.44	0.25803	0.25363	0.48833
10	0.5	0.23047	0.26131	0.50822
11	0.55	0.21262	0.26628	0.52110
12	0.6	0.19840	0.27024	0.53136
13	0.65	0.18708	0.3734	0.53952
14	0.7	0.17807	0.27590	0.54602
15	0.75	0.17090	0.2779	0.55120
16	1	0.15184	0.28321	0.56495

11.2. KINETIC MODELLING – PARAMETER ESTIMATION (1)

For the model developed above in Section 11.1, we will now use the predicted time-dependent compositions as our experimental data and set up a parameter estimation problem using the following initial estimates for the parameters (see Table 2).

11.2.1. Model Description

The model is the same as before. We need to add an objective function in the outer loop and solve Eqs. 2–3 in the inner loop. Note, however, that Eqs. 2–3 is a set of ODEs; therefore, for every iteration of the outer loop, a dynamic

TABLE 2 Initial Values and Bounds for the Parameters

Optimisation parameter	Initial value	Lower bound	Upper bound
k_1	10	0	10
k_2	10	0	10
k_3	30	10	50
k_4	30	10	50

TABLE 3 Comparison of the Optimum Parameters for Reacting System Eq. 1

Parameter	Real value	Esposito and Floudas[1]	Tjoa and Biegler[2]	Katare et al.[3]	ICAS-MoT
k_1	4	4.000	3.997	3.9013	3.9999
k_2	2	2.001	1.998	1.9759	1.9999
k_3	40	39.80	40.538	39.787	39.9338
k_4	20	19.90	20.264	19.887	19.9699
Objective	-	3.367×10^{-7}	4.125×10^{-5}	8.1313×10^{-5}	1.843×10^{-8}

[1] *Esposito and Floudas (2000);*
[2] *Tjoa and Biegler (1991);*
[3] *Katare et al. (2004).*

simulation needs to be performed to generate the calculated values of the reacting system compositions. The optimisation results are given in Table 3, where the results obtained through the ICAS-MoT (Sales-Cruz *et al.* 2003; Sales-Cruz, 2006) are compared with those reported by others. A copy of the ICAS-MoT file (ch-11-2-kin-fit-ode-1.mot) is given in Appendix 1.

11.3. KINETIC MODELLING – PARAMETER ESTIMATION (2)

This problem describes an overall reaction of catalytic cracking of gas oil (A) to gasoline (Q) and other products (S). This model proposed by Froment and Bischoff (1979) is given by

$$\begin{array}{ccc} A & \xrightarrow{k_1} & Q \\ k_3{\rightarrow} & & {\leftarrow}k_2 \\ & S & \end{array} \tag{5}$$

Only the concentrations of A and Q are measured; therefore, the concentration of S does not appear in the model (see below). This reaction scheme involves nonlinear reaction kinetics rather than simple first-order kinetics in the previous two examples. The ordinary differential equation model takes the form:

$$\begin{aligned} \frac{dz_1}{dt} &= -(k_1 + k_3)z_1^2 \\ \frac{dz_2}{dt} &= k_1 z_1^2 - k_2 z_2 \end{aligned} \tag{6}$$

$$z_0[1, 0] \quad t \in [0, 0.95] \tag{7}$$

where z_1 represents concentration of compound A and z_2 represents concentration of compound Q.

The model defined by Eq. (6) is nonlinear in the states.

11.3.1. Numerical Solution

Like in the above example, first a ODE-based model is developed. With this model and the known parameters (Eq. 7), pseudo-experimental data are generated (see Table 4). Next, for the initial estimates given in Table 5, the parameter regression is performed. Again, since a dynamic (process) model is involved, in

TABLE 4 Generated Concentration Data

Point number	Time	z_1	z_2
1	0.025	0.7408	0.1994
2	0.052	0.5787	0.2845
3	0.07	0.5050	0.3049
4	0.1	0.4167	0.3067
5	0.12	0.3731	0.2958
6	0.16	0.3086	0.2616
7	0.18	0.2841	0.2423
8	0.2	0.2631	0.2230
9	0.22	0.2451	0.2043
10	0.26	0.2155	0.1699
11	0.3	0.1923	0.1403
12	0.36	0.1656	0.1048
13	0.4	0.1515	0.0864
14	0.46	0.1344	0.0650
15	0.5	0.1250	0.0541
16	0.55	0.1149	0.0433
17	0.65	0.0990	0.0287
18	0.7	0.0926	0.0237
19	0.75	0.0870	0.0199
20	1	0.0667	0.0096

TABLE 5 Initial Values and Bounds for Parameter Estimation Problem

Optimisation parameter	Initial value	Lower bound	Upper bound
k_1	10	0	20
k_2	15	0	20
k_3	1	0	20

TABLE 6 Regressed Optimal Parameters

Parameter	Real Value	Esposito and Floudas[1]	Tjoa and Biegler[2]	Katare et al[3]	MoT
k_1	12	12.212	11.998	12.246	11.9996
k_2	8	7.980	7.993	7.9614	7.9996
k_3	2	2.222	2.024	2.2351	2.0004
Objective	-	2.6384×10^{-3}	8.221×10^{-5}	2.6802×10^{-3}	6.090×10^{-10}

[1] *Esposito and Floudas (2000)*;
[2] *Tjoa and Biegler (1991)*;
[3] *Katare et al. (2004).*

the inner loop an initial value integration problem is solved; this provides the calculated values for the pseudo-measured data for the outer-loop optimisation. The regressed parameters from optimisation are given in Table 6. A copy of the ICAS-MoT file (ch-11-3-kin-fit-ode-2.mot) is given in Appendix 1.

11.4. KINETIC MODELLING – PARAMETER ESTIMATION (3)

11.4.1. Problem Statement

Through this case study, aspects of model discrimination and regression of model parameters are highlighted. Consider the reaction $A + B \rightarrow C + D$, taking place in a differential reactor. A feed stream $v_o = 100.0$ cm³/min is fed to a reactor, which contains $W = 0.0001$ kg of catalyst. The reactor is operated in a differential mode and the outlet concentrations P_A, P_B and P_C in [atm.] are measured; the conversion χ of the reaction is calculated and used to determine the rate of reaction from the design equation

$$r = \chi v_o C_{Ao}/W = \chi v_0 Pa_0/RTW \quad [gmol/kgCat/min] \tag{8}$$

The objective is to carry out a nonlinear least-squares analysis for the data given in Table 7, to determine which of the following rate laws best describe the data:

$$\text{Model} - 1 \quad r = k_4 {P_A}^{\alpha} P_B^{\beta} P_c^{\gamma} \tag{9}$$

$$\text{Model} - 2 \quad r = \frac{k_4 K_1 P_A P_B}{1 + K_1 P_A + K_3 P_C} \tag{10}$$

where

$$k_4 = k_{40}\exp(-E_4/RT) \tag{11}$$

TABLE 7 Differential Reactor Data

Run no.	Temp [K]	P_{Ao} [atm]	P_{Bo} [atm]	P_{Co} [atm]	Rate, r [gmol/kg cat/min]	Conv, χ	P_A [atm]	P_B [atm]	P_C [atm]
2	600	0.10	0.10	0.00	0.0395129	0.00854	0.09915	0.0991	0.000854
3	600	2.00	0.10	0.00	0.0741632	0.00011	1.99978	0.0998	0.000222
4	600	0.10	2.00	0.00	0.3449568	0.03959	0.09604	1.9960	0.003959
5	600	2.00	2.00	0.00	1.197896	0.00221	1.99558	1.9956	0.004422
6	600	0.10	0.10	0.20	0.0012132	0.00049	0.09995	0.1000	0.200049
7	600	2.00	0.10	0.20	0.0098456	9.5E-05	1.99981	0.0998	0.200189
8	600	0.10	2.00	0.20	0.0178852	0.00977	0.09902	1.9990	0.200977
11	650	0.10	0.10	0.00	0.0014893	0.00729	0.09927	0.0993	0.000729
12	650	2.00	0.10	0.00	0.1356959	0.00065	1.99869	0.0987	0.001305
13	650	0.10	2.00	0.00	0.326404	0.12557	0.08744	1.9874	0.012557
14	650	2.00	2.00	0.00	3.3834579	0.01284	1.97432	1.9743	0.025685
15	650	0.10	0.10	0.20	0.0058049	0.00258	0.09974	0.0997	0.200258
16	650	2.00	0.10	0.20	0.0906307	0.00056	1.99888	0.0989	0.201122
17	650	0.10	2.00	0.20	0.1200444	0.04913	0.09509	1.9951	0.204913
18	650	2.00	2.00	0.20	1.4929583	0.01107	1.97786	1.9779	0.222144
21	700	2.00	0.10	0.00	0.1853618	0.00232	1.99535	0.0954	0.004649
22	700	0.10	2.00	0.00	0.1910585	0.12880	0.08712	1.9871	0.012880
23	700	2.00	2.00	0.00	3.6766623	0.04408	1.91183	1.9118	0.088167
34	700	0.10	0.10	0.20	0.0107313	0.00533	0.09947	0.0995	0.200533
25	700	2.00	0.10	0.20	0.1492342	0.00207	1.99585	0.0959	0.204147
26	700	0.10	2.00	0.20	0.1948485	0.09651	0.09035	1.9903	0.209651
29	600	0.10	0.10	0.00	0.0640631	0.00208	0.09979	0.0998	0.000208
a	600	2.00	2.00	0.20	0.39614	0.00189	1.99622	1.9962	0.203780
b	650	1.05	1.05	0.10	0.79191	0.01081	1.03865	1.0386	0.111350
c	700	0.10	0.10	0.00	0.01256	0.00737	0.09926	0.0992	0.000740
d	700	2.00	2.00	0.20	2.98675	0.03958	1.92084	1.9208	0.279160

$$K_1 = K_{10}\exp(-\Delta H_1/RT) \quad (12)$$

$$K_2 = K_{20}\exp(-\Delta H_2/RT) \quad (13)$$

$$K_3 = K_{30}\exp(-\Delta H_3/RT) \quad (14)$$

For model 1 (power law kinetics) the model parameters are: $\{k_{40}, E_4, \alpha, \beta, \gamma\}$. For model 2 (Langmuir-Hinshelwood kinetics) the model parameters are:

$$\{k_{40}, E_4, K_{10}, \Delta H_1, K_{20}, \Delta H_2, K_{30}, \Delta H_3\}$$

.

11.4.2. Power Law Kinetics – Model 1

Model Analysis & Solution Strategy

First we will consider the following pseudo-kinetic model as given by Eq. 9. For this model the orders of the reactions α, β and γ as well as pre-exponential factor k_{40} and activation energy E_4 for the Arrhenius relationship need to be estimated. That is, the estimation of the following model parameters: $P = \{\alpha, \beta, \gamma, k_{40}, E_4\}$.

This pseudo-kinetic model is useful to compare the optimisation methods and for guesses or *insights* into the Langmuir-Hinshelwood (L-H) model. First the model parameters are estimated using a linear approach (through a minimisation by least squares). Then based on this solution, the model parameters are regressed using a nonlinear approach. Next, the initial condition and bound values are stated for the nonlinear problem. Next, the Langmuir-Hinshelwood (L–H) model parameters are regressed by the nonlinear approach as well as the maximum likelihood principle method.

Linear Solution (Model 1)

Taking the logarithm of both sides of Eq. (9), we get

$$\ln(r) = \ln(k_4) + \alpha\ln(P_A) + \beta\ln(P_B) + \gamma\ln(P_C) \quad (15)$$

Substituting Eq. (10) in Eq. (15), we get

$$\ln(r) = \ln(k_{40}) + \left(\frac{-E_4}{R}\right)\left(\frac{1}{T}\right) + \alpha\ln(P_A) + \beta\ln(P_B) + \gamma\ln(P_C) \quad (16)$$

Introducing the following definitions

$$Y = \ln(r),\ X_1 = \frac{1}{T},\ X_2 = \ln(P_A),\ X_3 = \ln(P_B),\ X_4 = \ln(P_C) \quad (17)$$

$$a_0 = \ln(k_{40}),\ a_1 = \frac{-E_4}{R},\ a_2 = \alpha,\ a_3 = \beta,\ a_4 = \gamma \quad (18)$$

Then Eq. 16 is written as (after inserting subscript j to indicate data point j),

$$Y_j = a_0 + a_1 X_{1j} + a_2 X_{2j} + a_3 X_{3j} + a_4 X_{4j} \tag{19}$$

where $X_{1j} = 1/T_j, X_{2j} = \ln(P_{A0j}), X_{3j} = \ln(P_{B0j})$ and $X_{4j} = \ln(P_{C0j})$, with (T_j, P_{AO}, P_{BO}, P_{CO}) for data set j.

In linear regression the objective is to minimise the sum of the squares of the differences between the measured data and the calculated data, that is, finding the values of a_0, a_1, a_2, a_3, and a_4 that minimise the sum,

$$f = \sum_{j}^{N} \left(Y_j - Y_{j,\exp}\right)^2 \tag{20}$$

Substituting Eq. 19 in Eq. 20, we get

$$f = \sum_{j=1}^{N} \left(a_0 + a_1 X_{1j} + a_2 X_{2j} + a_3 X_{3j} + a_4 X_{4j} - Y_{j,\exp}\right)^2 \tag{21}$$

At least a local minimum is found when for the model parameters, the first derivative of f with respect to the parameters is zero. That is

$$\frac{\partial f}{\partial a_k} = 0, \qquad k = 0, \ldots, 4 \tag{22}$$

Deriving the first partial derivatives and inserting them in Eq. 22 gives the following five linear equations, where the five unknown variables are the five model parameters:

$$Na_0 + a_1 \sum_{j=1}^{N} X_{1j} + a_2 \sum_{j=1}^{N} X_{2j} + a_3 \sum_{j=1}^{N} X_{3j} + a_4 \sum_{j=1}^{N} X_{4j} - \sum_{j=1}^{N} Y_{j,\exp=0} \tag{23}$$

$$\begin{aligned} &a_0 \sum_{j=1}^{N} X_{1j} + a_1 \sum_{j=1}^{N} (X_{1j})^2 + a_2 \sum_{j=1}^{N} (X_{1j} X_{2j}) + a_3 \sum_{j=1}^{N} (X_{1j} X_{3j}) \\ &+ a_4 \sum_{j=1}^{N} (X_{1j} X_{4j}) - \sum_{j=1}^{N} \left(Y_{j,\exp\, X_{1j}}\right) = 0 \end{aligned} \tag{24}$$

$$\begin{aligned} &a_0 \sum_{j=1}^{N} X_{2j} + a_1 \sum_{j=1}^{N} (X_{1j} X_{2j}) + a_2 \sum_{j=1}^{N} (X_{2j})^2 + a_3 \sum_{j=1}^{N} (X_{2j} X_{3j}) \\ &+ a_4 \sum_{j=1}^{N} (X_{2j} X_{4j}) - \sum_{j=1}^{N} \left(Y_{j,\exp} X_{2j}\right) \\ &= 0 \end{aligned} \tag{25}$$

$$
\begin{aligned}
& a_0 \sum_{j=1}^{N} X_{3j} + a_1 \sum_{j=1}^{N} \left(X_{1j}X_{3j}\right) + a_2 \sum_{j=1}^{N} \left(X_{3j}X_{2j}\right) + a_3 \sum_{j=1}^{N} \left(X_{3j}\right)^2 \\
& + a_4 \sum_{j=1}^{N} \left(X_{3j}X_{4j}\right) - \sum_{j=1}^{N} \left(Y_{j,\exp} X_{3j}\right) \\
& = 0
\end{aligned} \tag{26}
$$

$$
\begin{aligned}
& a_0 \sum_{j=1}^{N} X_{4j} + a_1 \sum_{j=1}^{N} \left(X_{1j}X_{4j}\right) + a_2 \sum_{j=1}^{N} \left(X_{4j}X_{2j}\right) + a_3 \sum_{j=1}^{N} \left(X_{4j}X_{3j}\right) \\
& + a_4 \sum_{j=1}^{N} \left(X_{4j}\right)^2 - \sum_{j=1}^{N} \left(Y_{j,\exp} X_{4j}\right) \\
& = 0
\end{aligned} \tag{27}
$$

This system of equations [23–27] are now solved for the five unknown variables a_0, a_1, a_2, a_3, and a_4 using any linear equations solver. The following solution is obtained (see also Appendix 1 where the ICAS-MoT file for this problem is given: ch-11-4-kin-fit-pl-1.mot): $a_0 = 6.77080$; $a_1 = -5315.23$; $a_2 = 0.82535$; $a_3 = 1.12517$; $a_4 = -0.24615$. Using Eq. 38, the values of k_4, E_4, α, β and γ are obtained: $k_{40} = \exp(a_0) = 872.00522$; $E_4 = -a_1 R = 436115.98$; $\alpha = a_2 = 0.825347$; $\beta = a_3 = 1.125166$; $\gamma = a_4 = -0.246146$.

The corresponding calculated rate values are given in Table 8 (column 2) and some regression statistics are given in Table 9 (row referring to linear model 1).

Nonlinear Solution (Model 1)

The optimal values for the parameters $\{\alpha, \beta, \gamma, k_{40}, E_4\}$ are found by minimising the function,

$$
\min_{\alpha,\beta,\gamma,k_{40},E_4} f = \sum_{j}^{N} \left(r_j - r_{j,\exp}\right)^2 \tag{28}
$$

subject to Eqs. 9 and 11 and using as initial estimates for the parameters as given in Table 10.

Using the same type of optimisation method, the optimal parameters are obtained and listed in Table 8 (column 3). The ICAS-MoT file corresponding to this optimal solution is given in Appendix 1 (see ch-11-4-kin-fit-pl-2.mot). The regression statistics are given in Table 9 (row referring to nonlinear model 1). ICAS-MoT also provides a number of regression statistics and analysis (for more details, see Heitzig *et al.* 2011).

TABLE 8 Experimental and Calculated Rate Values for Model 1 and Model 2

Data	$r_{calc\text{-}1a}$	$r_{calc\text{-}1b}$	$r_{calc\text{-}2}$
2	0.007776	0.004923	0.005799
3	0.130267	0.110573	0.080776
4	0.152223	0.105898	0.108456
5	1.811431	1.686778	1.550592
6	0.002062	0.001286	0.001554
7	0.024406	0.020309	0.027642
8	0.059478	0.041028	0.030692
11	0.016018	0.012698	0.013911
12	0.164341	0.173575	0.154914
13	0.208598	0.179323	0.223512
14	2.274230	2.632659	2.753032
15	0.004059	0.003167	0.005144
16	0.047664	0.049611	0.079042
17	0.112909	0.097067	0.096737
18	1.341914	1.544013	1.491171
21	0.207109	0.263757	0.248612
22	0.370561	0.385874	0.480743
23	2.827598	3.935937	4.026789
24	0.007237	0.006841	0.013661
25	0.082138	0.103476	0.166415
26	0.192517	0.199555	0.243085
29	0.011145	0.007091	0.005928
a	0.706048	0.650178	0.545519
b	0.452849	0.480131	0.647789
c	0.028645	0.027517	0.029031
d	2.148796	2.982841	2.888178

TABLE 9 Statistics Report for Model 1 and Model 2

	Statistics				
Problem	Correlation	R^2	Adjusted R^2	Standard Error	$\Sigma(Y_{j,exp} - Y_j)^2$
Linear (model 1)	0.965515	0.884074	0.879244	0.374390	3.3641
Nonlinear (model 1)	0.979672	0.959691	0.958011	0.220767	1.1687
Nonlinear (model 2)	0.985057	0.970336	0.969100	0.189386	0.8608

TABLE 10 Bounds and Initial Values for Model 1 (Nonlinear Fit)

Parameter	Lower bound	Upper bound	Initial value	Optimal value
k_{40}	0	1×10^6	872.0052	13468.37
E_4	0	1×10^6	436115	579587.2
α	−1	2	0.825347	0.921559
β	−1	2	1.125166	1.159066
γ	−1	2	−0.24615	−0.24907

Non-Linear Solution (Model 2)

Consider the LH kinetic models by Eqs. 10–14. For this model, the kinetic parameters to be estimated are: $P_2 = \{k_{40}, E_4, K_{10}, \Delta H_1, K_{30}, \Delta H_3\}$.

Then, the optimal values of the parameters for this model are found by solving the minimising the objective function (Eq. 28), with initial estimates and bounds given in Table 11. The calculated values for the regressed parameters are given in Table 8 (column 4), and the regression statistics are given in Table 9 (row referring to nonlinear model 2). The ICAS-MoT file corresponding to this optimal solution is given in Appendix 1 (see ch-11-4-kin-fit-lh-1.mot).

11.5. KINETIC MODELLING – PARAMETER ESTIMATION WITH MAXIMUM LIKELIHOOD PRINCIPLE

Here, we consider the same reaction as in Section 11.4. The available data, however, is slightly different (see Table 12). Comparing Tables 8 and 12, one can see that there are new data points in Table 12 (data points 1, 9–10, 19–20, 27–28, 30–32). Also, comparing data points in Table 12, it is seen that the following data points (1, 10, 19, 28), (27, 30) and (20, 32) have different reaction rates even though they are operated under the same conditions. Therefore, a least-squares

TABLE 11 Bounds and Initial Estimates for Model 2

Parameter	Lower bound	Upper bound	Initial value model 2	Optimal value model 2
k_{40}	0	1×10^8	1	202.8362
E_4	0	1×10^8	1×10^4	218319.7693
K_{10}	0	1×10^8	1	297.2284
ΔH_1	-1×10^8	1×10^8	1×10^4	347710.2856
K_{30}	0	1×10^8	1	0.0329
ΔH_3	-1×10^8	1×10^8	1×10^4	−299822.8755

TABLE 12 Differential Reactor Data

Run no.	Temp [K]	P_{Ao} [atm]	P_{Bo} [atm]	P_{Co} [atm]	Rate, r [gmol/kg cat/min]	Conv, χ	P_A [atm]	P_B [atm]	P_C[atm]
1	650	1.05	1.05	0.10	0.7947824	0.01081	1.03865	1.0387	0.111346
2	600	0.10	0.10	0.00	0.0395129	0.00854	0.09915	0.0991	0.000854
3	600	2.00	0.10	0.00	0.0741632	0.00011	1.99978	0.0998	0.000222
4	600	0.10	2.00	0.00	0.3449568	0.03959	0.09604	1.9960	0.003959
5	600	2.00	2.00	0.00	1.1978960	0.00221	1.99558	1.9956	0.004422
6	600	0.10	0.10	0.20	0.0012132	0.00049	0.09995	0.1000	0.200049
7	600	2.00	0.10	0.20	0.0098456	9.5E-05	1.99981	0.0998	0.200189
8	600	0.10	2.00	0.20	0.0178852	0.00977	0.09902	1.999	0.200977
9	600	2.00	2.00	0.20	0.3425765	0.00189	1.99622	1.9962	0.203777
10	650	1.05	1.05	0.10	0.8671039	0.01081	1.03865	1.0387	0.111346
11	650	0.10	0.10	0.00	0.0014893	0.00729	0.09927	0.0993	0.000729
12	650	2.00	0.10	0.00	0.1356959	0.00065	1.99869	0.0987	0.001305
13	650	0.10	2.00	0.00	0.3264040	0.12557	0.08744	1.9874	0.012557
14	650	2.00	2.00	0.00	3.3834579	0.01284	1.97432	1.9743	0.025685
15	650	0.10	0.10	0.20	0.0058049	0.00258	0.09974	0.0997	0.200258
16	650	2.00	0.10	0.20	0.0906307	0.00056	1.99888	0.0989	0.201122
17	650	0.10	2.00	0.20	0.1200444	0.04913	0.09509	1.9951	0.204913
18	650	2.00	2.00	0.20	1.4929583	0.01107	1.97786	1.9779	0.222144
19	650	1.05	1.05	0.10	0.7068661	0.01081	1.03865	1.0387	0.111346
20	700	0.10	0.10	0.00	0.0187180	0.00737	0.09926	0.0993	0.000737
21	700	2.00	0.10	0.00	0.1853618	0.00232	1.99535	0.0954	0.004649
22	700	0.10	2.00	0.00	0.1910585	0.12880	0.08712	1.9871	0.012880
23	700	2.00	2.00	0.00	3.6766623	0.04408	1.91183	1.9118	0.088167

(continued)

TABLE 12 *(continued)*

Run no.	Temp [K]	P_{Ao} [atm]	P_{Bo} [atm]	P_{Co} [atm]	Rate, r [gmol/kg cat/min]	Conv, χ	P_A [atm]	P_B [atm]	P_C[atm]
34	700	0.10	0.10	0.20	0.0107313	0.00533	0.09947	0.0995	0.200533
25	700	2.00	0.10	0.20	0.1492342	0.00207	1.99585	0.0959	0.204147
26	700	0.10	2.00	0.20	0.1948485	0.09651	0.09035	1.9903	0.209651
27	700	2.00	2.00	0.20	3.0341589	0.03958	1.92084	1.9208	0.279156
28	650	1.05	1.05	0.10	0.7988874	0.01081	1.03865	1.0387	0.111346
29	600	0.10	0.10	0.00	0.0640631	0.00208	0.09979	0.0998	0.000208
30	700	2.00	2.00	0.20	2.9393335	0.03958	1.92084	1.9208	0.279156
31	600	2.00	2.00	0.20	0.4496993	0.00189	1.99622	1.9962	0.203777
32	700	0.10	0.10	0.00	0.0064090	0.00737	0.09926	0.0993	0.000737

method is not able to simultaneously fit all these data points. An alternative approach is to use the maximum likelihood principle approach.

11.5.1. Theoretical Background – Maximum Likelihood Principle

The maximum likelihood principle estimates are derived from the probability density function of the measurement errors. Under certain assumptions these estimates coincide with the least squares or weighted least-squares estimates. The idea behind maximum likelihood principle–based parameter estimation is to determine the parameters that maximise the probability (likelihood) of the collected data.

Mathematically, the likelihood is defined as the probability of matching a set of measurements. If we have N observations of different quantities x_j, for instance, then the likelihood is defined as:

$$L = p(x_1,\ x_2,\ x_3,\ \ldots,\ x_N) \tag{29}$$

Note that this is a joint probability distribution of all the measurements (as indicated by the commas in Eq. 29). Often the measurements are independent of one another, in which case the joint probability is simply the product of all the individual probabilities:

$$L = \prod_{j=1}^{N} p(x_j) \tag{30}$$

The assumption of independent probabilities certainly makes the regression step simpler because joint distributions can be extremely difficult to compute. On the other hand, there may be a loss of accuracy if the errors are correlated, so one must be careful when applying this approach.

Sums are easier to deal with than products (and the product of a lot of small numbers may be too small to represent on a computer), so a good idea is to work with the log of the likelihood. The log varies monotonically with its argument (i.e. $ln(x)$ increasing when x increases and decreasing when it decreases), so the log of a function will have its maximum at the same position. Usually one is interested to find the position of the maximum of the likelihood function.

$$LL = \ln(L) = \sum_{j=1}^{N} \ln\left[p(x_j)\right] \tag{31}$$

In fact, to make the analogy with least squares more obvious, often the minus log likelihood is minimized. One way to think about likelihood is that we imagine we have not measured the data yet. We have a model, with various parameters to adjust, and some idea of sources of error and how they would propagate. This allows calculating the probability of any possible set of measurements. Finally, we bring in the actual measurements and see how well they agree with the model.

Let us apply maximum likelihood to the problem where the errors in predicting the observations follow a Gaussian distribution. With a few simple manipulations, maximum likelihood is equivalent to least squares in this case.

First, note the equation for a Gaussian probability distribution.

$$p(x_j) = \frac{1}{\sqrt{2\pi\sigma_j^2}} \exp\left[-\frac{1}{2\sigma_j^2}(x_j - \hat{x}_j)^2\right] \tag{32}$$

where $\hat{x}_j$ is the predicted value of x_j.

In log likelihood, we will need the log of this probability distribution

$$\ln[p(x_j)] = -\frac{1}{2}\ln(2\pi) - \ln(\sigma_j) - \frac{1}{2\sigma_j^2}(x_j - \hat{x}_j)^2 \tag{33}$$

Now, the likelihood is the probability of making the entire set of observations. Here it is assumed that the observations are independent, so that the likelihood is the product of all the individual probabilities and the log likelihood is the sum of the logs. To make the comparison with least-squares minimisation clearer, a minus sign is added as shown in Eq. 34.

$$-LL = \sum_{j=1}^{N}\left[\frac{1}{2}\ln(2\pi) + \ln(\sigma_j) + \frac{1}{2\sigma_j^2}(y - y*_j)^2\right] \tag{34}$$

11.5.2. Parameter Estimation Problem for Models 2

To estimate the model parameters with the likelihood estimation, it is necessary to consider that the data were generated using the following probability distribution:

$$y_i = r_i(1 + r_i^{\gamma^* - 1}\varepsilon) \tag{35}$$

where

$$\varepsilon \approx NID(0, \omega r_{\max}) \tag{36}$$

The parameters $\gamma*$ and ω of the error distribution needs to be estimated together with the parameters for each model.

TABLE 13 Bounds, Initial and Optimal Values, and ±CI95% for Model 2 (Maximum Likelihood)

Parameter	Lower bound	Upper bound	Initial value model 2	Optimal value model 2	±CI95%
k_{40}	0	1×10^8	1	486.9774	174.2776525
E_4	0	1×10^8	-1×10^4	272397.6240	97484.64398
K_{10}	0	1×10^8	1	122.3300	43.77900337
ΔH_1	-1×10^8	1×10^8	-1×10^4	286917.8500	102681.0882
K_{30}	0	1×10^8	1	0.0355	0.012704607
ΔH_3	-1×10^8	1×10^8	1×10^4	−299901.9718	−107327.7972
ω	0	1×10^8	0.1	0.1220	-
γ^*	0	1×10^8	0.1	0.0976	-

Numerical Solution

Consider the L-H kinetic model described by Eqs. 10–14. The maximum likelihood approach is applied and the models are solved using ICAS-MoT (see the corresponding codes in Appendix 1). The bounds and initial values for each model are given in Table 13. The optimal values are reported in Table 13; the corresponding calculated rate values in comparison with the experimental ones are given in Table 14; some regression plots are shown in Figures 1a–1c; and the regression statistics are given in Table 15.

11.5.3. Discussion

Even though the performance of model 1 with the maximum likelihood principle has not been demonstrated, the regression statistics indicate the model 2 is likely to be better than model 1 for the data given in Tables 8 and 12. When there are uncertainties in the data, the least-squares technique is not suitable. Try to regress model 1 parameters with the maximum likelihood principle and model 2 parameters with the nonlinear least squares for the data set given in Table 12 and confirm whether model 2 is really the best. The ICAS-MoT file for model 2 with MLE is given in Appendix 1 (see file ch-11-5-kin-fit-lh-1.mot).

TABLE 14 Experimental and Calculated Rate Values for Model 2 (Maximum Likelihood)

No. Point	r_{exp}	Model 2 r_{calc}
1	0.794782	0.668596
2	0.039513	0.006119
3	0.074163	0.079776
4	0.344957	0.120686
5	1.197896	1.543512
6	0.001213	0.001330
7	0.009846	0.028045
8	0.017885	0.032337
9	0.342577	0.561634
10	0.867104	0.668596
11	0.001489	0.014990
12	0.135696	0.152649
13	0.326404	0.245699
14	3.383458	2.737282
15	0.005805	0.005100
16	0.090631	0.079593
17	0.120044	0.102353
18	1.492958	1.512501
19	0.706866	0.668596
20	0.018718	0.031410
21	0.185362	0.248676
22	0.191058	0.525314
23	3.676662	4.061234
24	0.010731	0.014082
25	0.149234	0.167829
26	0.194848	0.257205
27	3.034159	2.930159
28	0.798887	0.668596
29	0.064063	0.006266
30	2.939333	2.930159
31	0.449699	0.561634
32	0.006409	0.031410

TABLE 15 Statistics Reports for Model 2 (Maximum Likelihood)

	Statistics				
Problem	Correlation	R^2	Adjusted R^2	Standard Error	$\Sigma(Y_{j,exp} - Y_j)^2$
Model 2	0.985430	0.971069	0.970105	0.183845	1.013964

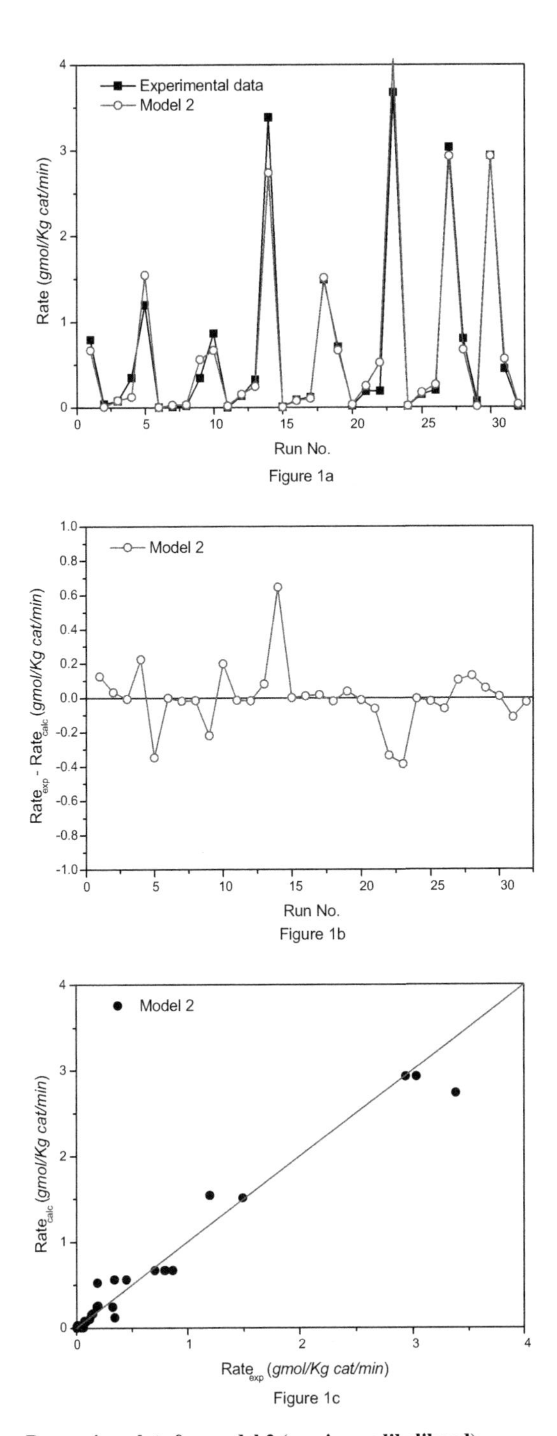

FIGURE 1a–1c Regression plots for model 2 (maximum likelihood)

11.6. PARAMETER REGRESSION - DYNAMIC OPTIMISATION

11.6.1. Problem Formulation

We consider again the first-order reversible chain reaction from Section 11.2. The task is to identify the optimal values of the parameters: k_1, k_2, k_3, k_4 that fit a given set of data. We use again data set given in Table 1 of this chapter.

For this problem, the approach of re-formulating the parameter estimation problem to solving a set of algebraic equations by orthogonal collocation has been applied (Tjoa and Biegler, 1991). In orthogonal collocation the dependent variable is discretised into elements and the solution $y(t)$ of an ODE having the form:

$$\frac{dy}{dt} = f(y, P), \quad y(0) = y_0 \tag{37}$$

is approximated by piecewise polynomials for each element. For each element, i the polynomial has the form:

$$y(t) = \sum_{j=0}^{NCOL} y_{ij}\phi_j(t) \tag{38}$$

In Eq. 38, j is the number of collocation points in which each element is subdivided and y_{ij} are parameters that need to be determined and are equal to the solution of the differential equation at the collocation points. P is the vector of parameters. $\phi_j(t)\ j = 0, NCOL$ are the Lagrange interpolation polynomials which are given by the following equation.

$$\phi_j(t) = \prod_{n=0,\quad n\neq j}^{NCOL} \frac{(t - t_{in})}{(t_{ij} - t_{in})} \tag{39}$$

Here, t refers the absolute locations of the collocation points. They result for each element from the relative locations of the collocation points τ_j from equation 40:

$$t_{ij} = \alpha_i + \tau_j \cdot (\alpha_{i+1} - \alpha_i) \tag{40}$$

Here i is the element index and j the index of the collocation point. Furthermore, α_i is the starting point of the ith element. The relative collocation points are the same for each element so that the Lagrange polynomials, independently from the element, can be written as:

$$\phi_j(t) = \phi_j(\tau) = \prod_{n=0,\quad n\neq j}^{NCOL} \frac{(t - t_n)}{(t_j - t_n)} \tag{41}$$

The collocation points are determined as the roots of the nodal polynomial given in equation 42.

$$P_N^{(\alpha,\beta)}(t) = \sum_{i=0}^{N} (-1)^{N-i} \gamma_i t^i \tag{42}$$

Here, N is the number of collocation points in the element. The γ_i are obtained by exploiting the orthogonality property of the nodal polynomials. They result from equation 43.

$$\gamma_i = \frac{N-i+1}{i} \cdot \frac{N+i+\alpha+\beta}{i+\beta} \gamma_i - 1, \quad \gamma_0 = 1, \quad i = 1, 2, \ldots, N \tag{43}$$

In order to determine the parameters γ_{ij} of the piecewise polynomials these polynomials are introduced into the differential equation (37). The parameters are determined such that the resulting residuals are equal to zero. Inserting equation (38) into the differential equation (Eq. 37)

$$\sum_{n=0}^{NCOL} y_{in} \frac{\phi_{in}(\tau_j)}{\Delta\alpha_i} - f(y_{ij}, P) = 0, \quad for\, j = 1, \ldots, NCOL; \quad i = 1, \ldots, NEL \tag{44}$$

These residuals need to be equal to zero. In this equation y_{ij} is the value of the dependent variable y at the collocation point ij. $\Delta\alpha_i$ is the length of the element i. Furthermore, the continuity between the elements has to be guaranteed, so that equation (45) is satisfied:

$$y_{i0} = \sum_{n=0}^{NCOL} y_{i-1,\,n}\, \phi_n(\tau = 1) \tag{45}$$

With this background the optimisation problem consisting of a model represented by ODEs can be rewritten as a nonlinear programming problem. The objective function is given by equation (46):

$$F_{obj} = \sum \left(\underline{y}(t_k) - \underline{y}^{Exp}(t_k) \right) \tag{46}$$

The residuals in equation (44) and the continuity conditions (45) are added as constraints:

$$\sum_{n=0}^{NCOL} y_{in} \frac{\phi_{in}(\tau_j)}{\Delta\alpha_i} - f(y_{ij}, P) = 0, \quad for\, j = 1, \ldots, NCOL; \quad i = 1, \ldots, NEL$$

$$y_{i0} = \sum_{n=0}^{NCOL} y_{i-1,\,n}\, \phi_n(\tau = 1) = 0, \quad for\, i = 2, \ldots, NEL$$

A further constraint is that the element length must be larger than or equal to zero. The initial conditions for the problem are re-derived so that $y_{10} = y(0)$.

For the given problem, the above equations have to be written for each of the three components A, B and C. To summarise, the parameters to be fitted in this problem are the parameters for the piecewise polynomials as well as the model parameters k_1, k_2, k_3, k_4.

An alternative approach is to apply the determined collocation points for the orthogonal collocation, but instead of fitting the piecewise polynomials, the finite difference approximation is applied using the collocation points as the grid points. Use of this hybrid method to solve the given model identification problem is illustrated below.

There are several ways to approximate the $\frac{dy}{dt}$-term in Eq. 37 by its differential quotient. We have chosen a first-order approximation, which in this case results in a system of implicit algebraic equations. For the concentration of the component A at the collocation point j (time t_j) the ODE becomes:

$$y_{Aj} = y_{A,j-1} + (-k_1 \cdot y_{A,j-1} + k_2 \cdot y_{B,j-1}) \cdot \Delta t_j \tag{47}$$

The equations for components B and C are transformed in the same way. The number of collocation points N is important for the success of the method, and in this example, it is set to 10. The resulting nodal polynomial from which the relative collocation points are obtained as roots is determined from Eqs. 42 and 43. The α and β parameters are obtained such that the residuals given in Eq. 44 are minimised. In this case, we have set these parameters to 1. Consequently, the nodal polynomial becomes:

$$\begin{aligned} P_{10}^{(1,1)} &= 1 - 65t + 1365t^2 - 13650t^3 + 76440t^4 - 259896t^5 \\ &\quad + 556920t^6 - 755820t^7 + 629850t^8 - 293930t^9 + 58785t^{10} \end{aligned} \tag{48}$$

The resulting roots, that is the relative collocation points, are: 0.02755036, 0.09036034, 0.18356192, 0.30023453, 0.43172353, 0.56827647, 0.69976547, 0.81643808, 0.90963966, 0.97244963.

The time interval considered is between 0 and 1. It has been divided into four elements, and in each of these elements, there are 10 collocation points. The values of the concentrations of A, B and C at the available experimental data points are determined by linear interpolation between the collocation points. The problem has been implemented into ICAS-MoT and solved. The equations implemented in ICAS-MoT are given in Appendix 1 (see ch-11-6-finite-element.mot file and ch-11-6-ortho-colloc.mot files). Table 16 gives the initial values of the parameters and the resulting optimised values.

TABLE 16 **Solution of Optimisation Problem Applying Finite Differences and 10 Collocation Points for Each Element**

Parameter:	Initial value	Fitted value
k_1	10.000	3.758
k_2	10.000	1.902
k_3	30.000	39.304
k_4	30.000	19.705

The resulting value of the objective function is 2.2×10^{-6}. As stated in Section 11.2, the experimental data set was generated for parameter values of 4, 2, 40 and 20 for k_1, k_2, k_3 and k_4, respectively. It can be seen that the resulting regressed values are quite close but there are still some deviations (compare these values with those given in Table 3). This means that the number of collocation points needs to be increased. In general the number of collocation points should be increased until the resulting parameter values do not change significantly. Consequently, the number of collocation points has been increased to 20 in each of the four elements. Below is given the nodal polynomial to determine the relative collocation points.

$$
\begin{aligned}
P_{10}^{(1,1)} &= 1 - 230t + 17480t^2 - 655500t^3 + 14486550t^4 - 208606320t^5 \\
&+ 2086063200t^6 - 15123958200t^6 + 81921440250t^7 \\
&- 3,38609{\cdot}10^{11}t^8 + 1,08355{\cdot}10^{12}t^9 - 2,70887{\cdot}10^{12}t^{10} \\
&+ 5,31355{\cdot}10^{12}t^{11} - 8,17469{\cdot}10^{12}t^{12} + 9,80963{\cdot}10^{12}t^{13} \\
&- 9,07391{\cdot}10^{12}t^{14} + 6,33839{\cdot}10^{12}t^{15} - 3,23134{\cdot}10^{12}t^{16} \\
&+ 1,1338{\cdot}10^{12}t^{17} - 2,44663{\cdot}10^{11}t^{18} + 24466267020t^{19} + t^{20}
\end{aligned} \tag{49}
$$

The resulting relative collocation points are: 0.0079, 0.0264, 0.0494, 0.093, 0.1398, 0.1942, 0.2551, 0.3212, 0.3912, 0.4635, 0.5365, 0.6088, 0.6788, 0.7448, 0.8061, 0.8599, 0.9074, 0.9446, 0.9738, 0.9920, and the absolute collocation points are given in Table 17. The results of the optimisations are shown in Table 18.

It can be seen that these results are better than the ones obtained for 10 collocation points in each element. This is also indicated by the improved value of the objective function of 2.28×10^{-7}. However, the resulting parameter values still differ from their true values so that the number of collocation points should be further increased to obtain better results. This is not done here but the procedure is analogous to the previously described solution. Figures 2–4 show the plots of the experimental data and the predictions of the discretised model for 20 collocation points in each element applying the obtained parameter values from the optimisation.

The alternative approach of using orthogonal collocation applying the methodology as described above (equation 46 and the corresponding constraints) has likewise been implemented in MoT.

TABLE 17 Absolute Collocation Points (Time t) for Subdivision into Four Elements With 20 Collocation Points Each

t_{el1}	t_{el2}	t_{el3}	t_{el4}
0.00198095	0.24998095	0.49798095	0.74637713
0.00659897	0.25459897	0.50259897	0.75191876
0.01234725	0.26034725	0.50834725	0.75881671
0.02325	0.27125	0.51925	0.7719
0.03495	0.28295	0.53095	0.78594
0.04855	0.29655	0.54455	0.80226
0.063775	0.311775	0.559775	0.82053
0.0803	0.3283	0.5763	0.84036
0.0978	0.3458	0.5938	0.86136
0.115875	0.363875	0.611875	0.88305
0.134125	0.382125	0.630125	0.90495
0.1522	0.4002	0.6482	0.92664
0.1697	0.4177	0.6657	0.94764
0.1862	0.4342	0.6822	0.96744
0.201525	0.449525	0.697525	0.98583
0.214975	0.462975	0.710975	1.00197
0.22685	0.47485	0.72285	1.01622
0.23615	0.48415	0.73215	1.02738
0.24345	0.49145	0.73945	1.03614
0.248	0.496	0.744	1.0416

TABLE 18 Solution of Optimisation Problem Applying Finite Differences and 20 Collocation Points for Each Element

Parameter:	Initial value	Fitted value
k_1	10.000	3.871
k_2	10.000	1.935
k_3	30.000	39.653
k_4	30.000	19.848

The solution procedure is as follows.

Step 1: Determine the values of the interpolation polynomials $\phi_j(\tau)$ for each relative collocation point j. Note: they do not differ for the different elements.
Step 2: Formulate the residual functions for each of the component concentrations (that is, each of the ODEs) at each collocation point j of each element i. Note, in the case study, $i = 4$; $j = 10$; NC (number of compounds) = 3.

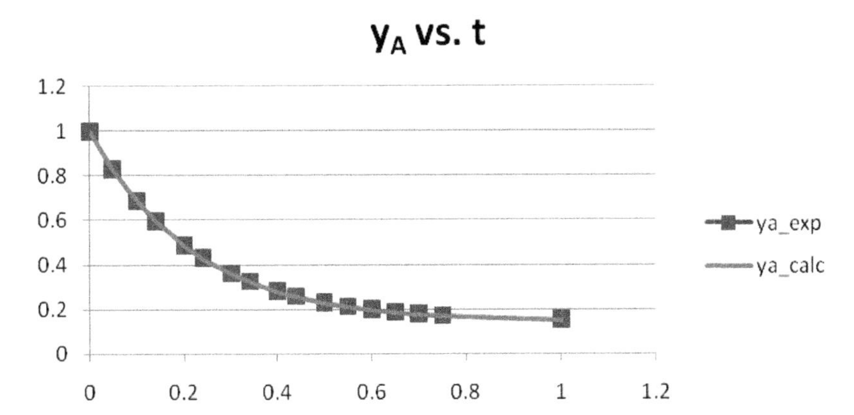

FIGURE 2 Concentration of component A versus time (solid line: prediction of discretised model with 20 collocation points in each of the four elements; dots: experimental measurements)

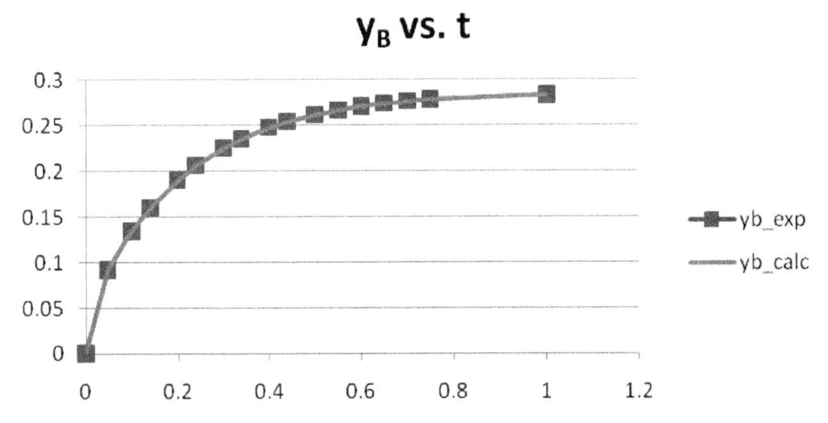

FIGURE 3 Concentration of component B versus time (solid line: prediction of discretised model with 20 collocation points in each of the four elements; dots: experimental measurements)

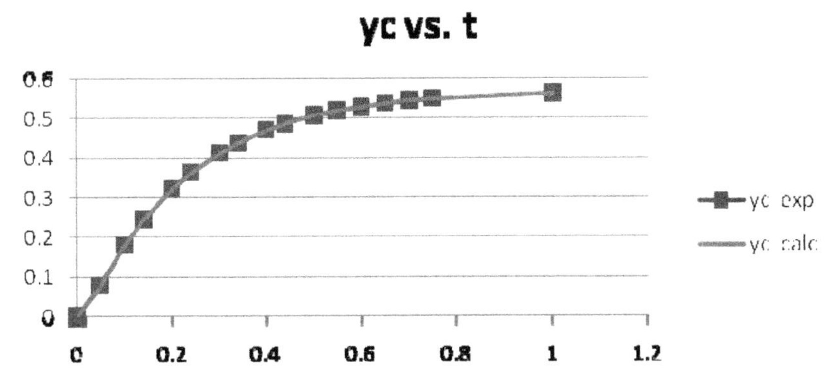

FIGURE 4 Concentration of component C versus time (solid line: prediction of discretizsd model with 20 collocation points in each of the four elements; dots: experimental measurements)

Step 3: Determine the values of $\phi_j(\tau = 1)$ for each relative collocation point and then derive the continuity conditions at the boundary between the elements.
Step 4: Calculate the objective function (Eq. 46).
Step 4.1 Determine the values for $\phi_j(\tau^{exp})$ at all absolute time values of the experimental data.
Step 4.2 Calculate the model outputs at these times $(\phi_j(\tau^{exp}))$ by applying Eq. 38.
Step 4.3 Calculate the objective function (Eq. 46).

Note that the values of the parameters of the polynomials need to be determined in addition to the values of the model parameters, which should be regressed. Initial values for the polynomial parameters can be generated by linearly interpolating between the experimental model output values, since the polynomial parameters give the model outputs at the times of the collocation points. Implementation of this problem in ICAS-MoT is given in Appendix 1 (kfit-ortho.mot).

REFERENCES

Esposito, W.R., Floudas, C.A., 2000. Industrial & Engineering Chemistry Research. 39, 1291–1310.

Froment, G.F., Bischoff, K.B., 1979. In Chemical reactor analysis and design. New York: Wiley.

Katare, S., Bhan, A., Caruthers, J.M., Delagas, E.N., Venkatasubramanian, V., 2004. Computers and Chemical Engineering. 28, 2569–2581.

Heitzig, M., Sin, G., Sales-Cruz, M., Glarborg, P., Gani, R., (2011)., A computer-aided modeling framework for efficient model development, analysis and identification: Combustion and reactor modeling, Industrial & Engineering Chemistry Research,. 50, 5253–5265.

Sales-Cruz, M., Gani, R., 2003. A Modelling Tool for Different Stages of the Process Life, Computer-Aided Chemical Engineering, 16, 209–249.

Sales-Cruz, M., 2006. Development of a computer aided modelling system for bio and chemical process and product design, PhD-thesis. Lyngby, Denmark: Technical University of Denmark.

Tjoa, I., Biegler, L.T., 1991. Industrial & Engineering Chemistry Research. 30, 376–385.

Chapter 12

Modelling for Bio-, Agro- and Pharma-Applications*

12.1. MICROCAPSULE-CONTROLLED RELEASE OF ACTIVE INGREDIENTS

Microcapsules are the products of microencapsulation that basically consists of the wrapping of substances (for example, an active ingredient dissolved in a solvent) within some porous medium (for example, a polymeric shell). Microcapsules usually have a spherical shape on a small size scale, where the wall/shell of the microcapsule envelops the core that consists of the active ingredient and solvent. In some cases the solution of the active ingredient and solvent is dispersed within the body of the microcapsule, where the phase of the active ingredient and/or solvent dispersing out might be liquid or gas.

Controlled release may be a technology that involves a product (for example, a microcapsule device) that is specifically designed to provide the delivery of an active ingredient to a specific target or medium at a desired rate and duration.

Microcapsules for controlled release of the active (pharmaceutical) ingredients are quite common in the pharmaceutical industry. However, the application of microcapsules can also be found in a broad variety of industries, such as cosmetics, agrochemicals, consumer and personal care products, biotechnology, food, bio medicals, and sensors, which have contributed to the research in this field. Microencapsulation of active ingredients is a multidisciplinary task. For instance, the application of this activity in pharmaceuticals might require the scientific knowledge of experts in physics, physiology, biology, pharmacology, and, chemistry, supported by modelling where chemical engineering plays an important role.

In general, the methods of microencapsulation can be broadly classified into two groups: physical methods and chemical methods. For example,

- Physical methods: Pan coating, fluid bed coating, centrifugal head coextrusion, rotary disk atomisation, stationary nozzle coextrusion, spray chilling, spray drying, submerged nozzle coextrusion.

*Written in conjunction with Dr. Ricardo Morales-Rodriguez and Dr. Ravendra Singh

- Chemical methods: interfacial polymerisation, solvent evaporation, solvent extraction, *in situ* polymerisation, complex and simple coacervation, liposome technology, nano-encapsulation, phase separation.

The core of the microcapsule, specifically the active ingredient with or without solvent, is chosen for a specific purpose. The wall or shell of the microcapsules could be synthetic and natural polymers (widely used), proteins, fats, polysaccharides, waxes, resins and starches.

This section of the chapter focusses on the development and solution of a mathematical model of microcapsule based control release of active ingredients. The model can be used to evaluate the performance of the product (microcapsule) with respect to different design parameters, such as, diameter and thickness of the microcapsule, polymer used and solvent used. Here, the mathematical modelling of a polymeric microcapsule, suitable for controlled release of active ingredients is described with emphasis on the modelling and numerical solution issues.

12.1.1. Problem description: Microcapsule-controlled release of pesticide or pharma-products

In the agrochemical industry, the use of controlled release technology for the delivery of pesticides has numerous advantages, ranging from optimised delivery of the active ingredients to a specific site, to the reduction of hazards to humans and environment. That is, the amount of pesticide used on the field can be reduced and also the safety level during its use can be improved. As far as controlled release of drugs or pharma-products is concerned, this technology is mainly aimed at controlling the amount of the drug delivered. The main benefits of controlled release are that it helps to keep an effective level of drug in the body for a specific period of time, and thereby side effects generated by drug overdosing and/or under-dosing may be avoided.

Figure 1 illustrates one of the most common trends in the controlled release of pesticides or pharmaceuticals, as well as the desired lower and upper limits of the delivered active ingredient. The lower limit in Figure 1 represents the minimum non-toxic concentration, whilst the upper limit represents the maximum effective concentration of the active ingredient. The dashed line represents an example of release without a device for controlled release, whilst the continuous line is an example of controlled release through a device within specified limits. The behaviour described above can be generated by employing different models for controlled release of an active ingredient.

System description

Device-based controlled release is possible through a variety of devices for the delivery of a product. Polymeric microcapsules are one of them, and only this type is considered in this chapter.

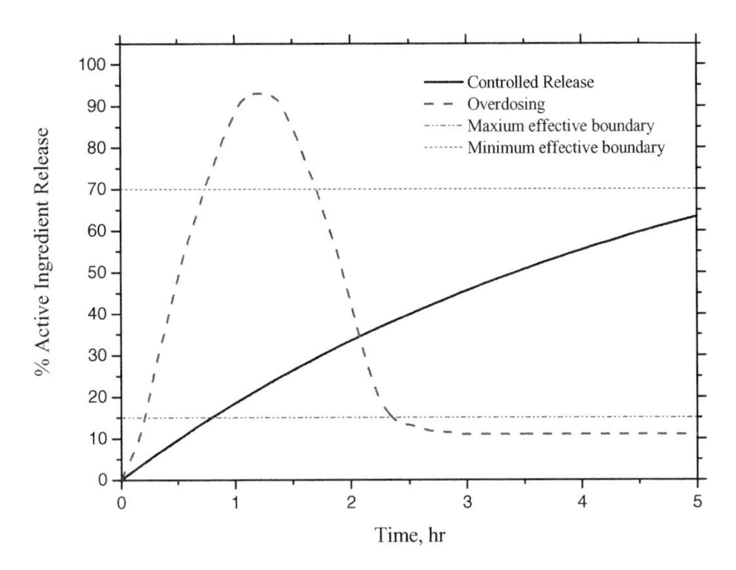

FIGURE 1 Percent of release of active ingredient versus time.

Figure 2 shows the schematic representation of a microcapsule, where the active ingredient (AI, with concentration $C_{d,i}$) is enclosed within the core of the device made of a polymeric membrane of inside radius r_i, and thickness $r_o – r_i$. The AI is delivered from the core of the microcapsule to the release medium with concentration C_r. Prediction and/or measurement of this concentration therefore allows the evaluation of the performance of the microcapsule for controlled release.

Brief descriptions of the main items within the system seen in Figure 2 are given below.

Active Ingredient

This is the chemical substance that is encapsulated within the core of the microcapsule. The identity of the substance differs according to the applications of the microcapsule device. For instance, if the microcapsule is applied for pest control, a pesticide or fungicide is in the core of the microcapsule. For delivery of a pharmaceutical product, the substance is a drug molecule or formulation, such as an antibiotic, antibodies, antioxidants or probiotics, which is placed in the core of the device.

Donor Medium

Usually, the active ingredients are solid substances that need to be dissolved in a solvent (donor medium).

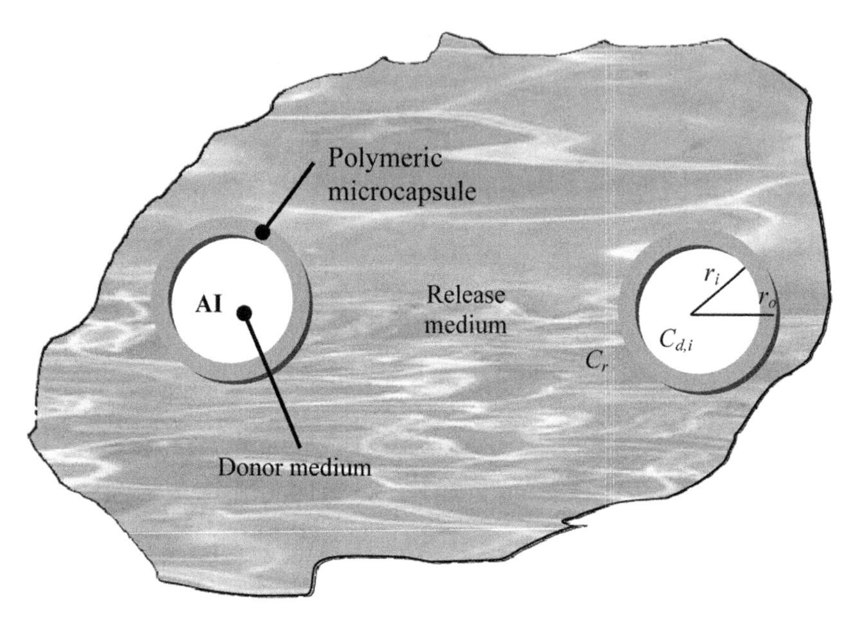

FIGURE 2 Schematic representation of a microcapsule.

Microcapsule wall (polymeric membrane)

The active ingredient, together with the solvent, is encapsulated by a polymeric membrane. One of the most important phenomenon occurring here is the mass transfer (by diffusion) of the active ingredient through this wall.

Release Medium

The active ingredient diffuses out from the microcapsules into the release medium, which depends on the application field. For instance, in the agrochemical field the most common release medium is water, whilst for a pharmaceutical product, the release medium could be gastric acid, blood or some other medium found within the human body.

12.1.2. Objective of the model

The purpose of the model is to study the behaviour of the microcapsule for controlled release of active ingredients. The developed model should be able to predict the concentration of the active ingredient in the release medium as a function of time and subject to an initial population of microcapsules.

12.1.3. Model Development

A mathematical model (Muro-Suñé et al., 2005) representing the controlled release of active ingredients is employed for this case study. Multi-scale modelling issues are considered. Two different scales have been identified: micro-scale (release of active ingredient from the microcapsules) and meso-scale (normal distribution of microcapsules).

Equations at the micro-scale represent the behaviour of a single microcapsule. These equations describe the concentration of the active ingredient in the donor medium as well as in the release (also known as receiver) medium as a function of time. The concentration profile follows the Fick's law of diffusion and has a dependence on time. As far as meso-scale is concerned, it is considered that not all the microcapsules have the same size. Hence, the meso-scale calculations account for the number of microcapsules of different sizes according to a specified (normal) distribution of the microcapsules in terms of their size.

The model considers the following assumptions.

A_1: Diffusion occurs through a film that is thin enough so that the diffusion can be considered one-dimensional.
A_2: Initial concentration of AI is equal for all the microcapsules.
A_3: Diffusion coefficient is independent of concentration.
A_4: Concentration of the AI changes in the donor and release mediums due to the diffusivity of the active ingredient in the polymer, and also, due to the partition coefficients between the wall of the microcapsule (polymeric membrane) and the donor and release mediums.
A_5: Isothermal condition during the controlled release.
A_6: Non-constant activity source.
A_7: The concentration of the AI in solution is below their solubility limit.

In addition, the concentration of the AI is assumed to be subject to the burst and lag time effects during the initial periods of release. These phenomena depend mainly on the diffusivity of the solute (AI) in the polymer, the thickness of the membrane and the storage as well as the usage conditions.

The *burst effect* occurs when, for example, the devices are stored for a period, giving time for the AI to diffuse into the polymer membrane and saturate it. Thus, the initial delivery rate from the microcapsule becomes greater than that of the steady state.

In the *lag effect*, there is no lapse between fabrication and use of the device, the active ingredient does not have time to partition into the membrane and there is a delay before the steady-state gradient is reached.

Mathematical representation for burst and lag effects are also illustrated in this chapter.

12.1.4. Model Analysis

List of equations

TABLE 1 List of Equations for Controlled Release of Active Ingredients

Variables			No. of Eq.
Active ingredient concentration in the donor	$\frac{dC_{d,i}}{dt} = -\frac{DA_i}{hV_{d,i}} K_{m/d} C_{d,\text{initial}} \exp\left(-\frac{DA_i K_{m/d}}{V_r h}\left(\frac{K_{m/r}}{K_{m/d}} + \frac{V_r}{V_{d,i}}\right) t\right)$	(1)	n_f
Active ingredient concentration in the receiver	$\frac{dC_{r,i}}{dt} = \frac{DA_i}{V_r h} K_{m/d} C_{d,\text{initial}} \exp\left(-\frac{DA_i K_{m/d}}{V_r h}\left(\frac{K_{m/r}}{K_{m/d}} + \frac{V_r}{V_{d,i}}\right) t\right)$	(2)	n_f
Microcapsule size calculation	$n_f = \frac{\lvert r_{\max} - r_{min} \rvert}{r_{step}}$	(3)	1
Average radius for one set of microcapsules	$r_i = \frac{r_{min,i} + r_{\max,i}}{2}$	(4)	n_f
Area for each microcapsules	$A_i = 4\pi r_i^2$	(5)	n_f
Volume for each microcapsule (donor volume)	$V_{d,i} = \frac{4}{3}\pi r_i^3$	(6)	n_f
Normal distribution function for microcapsules	$N'_{p,i} = F(r_i; \mu; \sigma) = \int_{-\infty}^{r'} \frac{1}{\sqrt{2\pi}\sigma} \exp\left(-\frac{(r-\mu)^2}{2\sigma^2}\right) dr$	(7)	n_{ff}
Total donor volume	$V_{d,calc} = \sum N'_{p,i} V_{d,i}$	(8)	1
Scaling for the number of microcapsules of radii i	$N_{p,i} = N'_{p,i} \frac{V_d}{V_{d,\text{calc}}}$	(9)	n_f

(continued)

TABLE 1 *(continued)*

Variables			No. of Eq.
Total number of microcapsules.	$N_{T,p} = \sum_i N_{p,i}$	(10)	1
Percent of particles for each size	$\%particles_i = \frac{N_{p,i}}{N_{T,p}}$	(11)	n_f
Mass of the active ingredient in the donor	$\Delta M_{d,i} = C_{d,i} V_{d,i} N_{p,i}$	(12)	n_f
Initial mass of the active ingredient in the donor	$M_{\text{initial}} = \sum_i C_{d,\text{initial}} V_{d,i} N_{p,i} = C_{d,\text{initial}} V_d$	(13)	1
Mass of the active ingredient in the release medium	$\Delta M_{r,i} = C_{r,i} V_r N_{p,i}$	(14)	n_f
Total mass change (to the release medium)	$\Delta M_{\text{total}} = \sum_i \Delta M_{r,i}$	(15)	1
Release percentage for the total set of microcapsules	$\%release = \frac{(M_{\text{initial}} - \Delta M_{\text{total}})}{M_{\text{initial}}} \times 100$	(16)	1
Burst time effect: Mass of the active ingredient in the release medium	$M_{r,i}(t) = \frac{V_r C'_{d,\text{initial}}}{(K_{m/r}/K_{m/d} + V_r/V_{d,i})}(1 - \exp(-\alpha t)) + J_{\max} A_i \frac{2}{\alpha'}\left(1 - \exp\left(-\alpha' t\right)\right)$	(17)	n_f
Lag time effect: Mass of the active ingredient in the release medium	$M_{r,i}(t) = \frac{V_r C'_{d,\text{initial}}}{(K_{m/r}/K_{m/d} + V_r/V_{d,i})}(1 - \exp(-\alpha_i t)) - J_{\max} A_i \frac{2}{\alpha'}\left(1 - \exp\left(-\alpha'_i t\right)\right)$	(18)	n_f

(continued)

TABLE 1 *(continued)*

Variables			No. of Eq.
Burst/lag time term	$\alpha_i = \frac{DA_i}{V_r h} K_{m/d} \left(\frac{K_{m/r}}{K_{m/d}} + \frac{V_r}{V_{d,i}} \right)$	(19)	n_f
Burst/lag time term	$\alpha'_i = \frac{D\pi^2}{h^2} K_{m/d}$	(20)	n_f
Burst/lag time term	$J_{max} = \frac{DC_{d,initial}}{h} K_{m/d}$	(21)	n_f
Initial mass of the active ingredient in the donor for particles with radii i	$M_{d,initial,i} = C_{d,initial} V_{d,i} N_{p,i}$	(22)	n_f
Burst time effect for initial mass	$M_{burst,\infty} = \frac{2}{\alpha'} J_{max} A_i N_{p,i}$	(23)	n_f
Lag time effect for initial mass	$M_{lag,\infty} = -\frac{2}{\alpha'} J_{max} A_i N_{p,i}$	(24)	n_f
Initial mass of for burst and lag time effect	$M'_{d,initial,i} = M_{d,initial,i} - M_{burst/lag,\infty,i}$	(25)	n_f
Initial concentration for burst/lag time effect	$C'_{d,initial,i} = \frac{M'_{d,initial,i}}{V_{d,i}}$	(26)	n_f
Mass for different microcapsule sizes for burst/lag time effect	$\Delta M_{r,i} = M_{r,i} N_{p,i}$	(27)	n_f

List of Variables

TABLE 2 List of Variables Employed for Controlled Release of Active Ingredients

Variable		Units
$C_{d,i}$	Concentration of donor as function of time	g/m^3
$C_{d,\text{initial}}$	Initial concentration of donor as function of time	g/m^3
$K_{m/d}$	Partition coefficient of the AI between the donor and the polymer membrane.	-
$K_{m/r}$	Partition coefficient of the AI between the receiver and the polymer membrane.	-
$V_{d,i}$	Donor volume for radii i	m^3
V_r	Receiver volume for radii i	m^3
A_i	Surface area through which diffusion takes place for radii i	m^2
D	Polymer solvent binary mutual diffusion coefficient	m^2/s
h	Thickness of the microcapsule wall	m
t	Time	s
$r_{\max}$	Maximum particle radii	m
$r_{\min}$	Minimum particle radii	m
r_{step}	Radii increment	m
r_i	Average radius	m
n_f	Number of microcapsule sizes	-
μ	Mean distribution value (mean radius)	m
σ	Standard deviation	m^3
$V_{d,\text{calc}}$	Total donor volume	m^3
$N'_{p,i}$	Number of microcapsule	-
V_d	Total donor volume (specified)	m^3
$N_{T,p}$	Total number of particles	-
$N_{p,i}$	Scaled number of particles of each size	-
$\%\ particles_i$	Percent of particles for each size	-
$C_{d,i}$	Donor concentration in a particle with radii i	g/m^3
$C_{r,i}$	Receiver concentration from a particle with radii i	g/m^3
V_r	Receiver volume (specified)	m^3
$\Delta M_{d,i}$	Mass of the active ingredient in the donor for different particle sizes	g
$M_{r,i}$	Mass concentration of the active ingredient in the receiver as a function of time	g/m^3
$\Delta M_{r,i}$	Mass of the active ingredient in the receiver medium for different particle sizes	g
ΔM_{total}	Total mass change (from donor to release medium)	g

(continued)

TABLE 2 *(continued)*

Variable		Units
M_{initial}	Initial mass of active ingredient in the donor compartment	g
% release	Percent of active ingredient released in the receiver medium	-
α_i	Burst/lag time term	hr
α'_i	Burst/lag time term	hr
$J_{\max}$	Burst/lag time term	hr
$M_{\text{burst},\infty}$	Burst time effect for initial mass	g
$M_{\text{lag},\infty}$	Lag time effect for initial mass	g
$C'_{d,\text{initial},i}$	Initial concentration for burs/lag time effect	g/m^3
$M_{d,\text{initial},i}$	Initial mass of the active ingredient in the donor for particles with radii i	g
$M'_{d,\text{initial},i}$	Initial mass of for burst and lag time effect	g

First order model: Degrees of freedom

The variables are classified in terms of the following types:

- State variables (states)

$$C_{d,i}, C_{r,i}$$

- Parameters

$$K_{m/d}, K_{m/r}, D, \mu, \sigma$$

- Algebraic variables (algebraics)

$$C_{d,\text{initial}}, h, t, r_{\max}, r_{min}, r_{\text{step}}, n_f, V_{d,\text{calc}}, V_d, N_{T,p}, V_r, \Delta M_{\text{total}}, M_{\text{initial}}, \%release, V_{d,i}, A_i, r_i, N'_{p,i}, N_{p,i}, \%particles_i, C_{d,i}, C_{r,i}, \Delta M_{d,i}, \Delta M_{r,i}$$

$N_U = 10n_f + 19$ (states + parameters + algebraics)
$N_E = 10n_f + 6$ (equations)
$N_{DF} = 10n_f + 19 - 10n_f - 6 = 13$

The 13 variables to specify may be selected from the parameters and those representing the initial conditions and microcapsule dimension, so the specifications are:

Parameters	$K_{m/d}, K_{m/r}, D, \mu, \sigma$
Algebraic variables (algebraics)	$C_{d,\text{initial}}, h, t, r_{\max}, r_{min}, r_{step}, V_d, V_r$

First order model: Incidence matrix

Based on the calculated degrees of freedom, for a prediction considering first-order release, equations (1), (2), (4)–(7), (9), (11), (12) and (14) are solved n_f times, once for each microcapsule (that is, this set of equations are solved n_f times). Equations (3), (8), (10), (13), (15) and (16) are solved once, giving $10n_f +$ 6 equations to solve. Table 3 shows the incidence matrix for this model, where it is possible to observe the order for the solution of the ordinary differential algebraic set of equations.

Burst-time/lag-time models: Degrees of freedom

The variables are classified in terms of the following types:

- Parameters

$$K_{m/d}, K_{m/r}, D, \mu, \sigma$$

- Algebraic variables (algebraics)

$C_{d,\text{initial}}, h, t, r_{\max}, r_{min}, r_{\text{step}}, n_f, V_{d,calc}, V_d, V_r, N_{T,p}, \Delta M_{\text{total}}, \%release,$
$V_{d,i}, A_i, r_i, N'_{p,i}, N_{p,i}, \%particles_i, M_{r,i}, \Delta M_{r,i}, \alpha_i, \alpha'_i, J_{\max}, M_{\text{burst},\infty}, C'_{d,\text{initial},i},$
$M_{d,\text{initial},i}, M'_{d,\text{initial},i}$

$N_U = 15n_f + 18$ (parameters + algebraics)
$N_E = 15n_f + 5$ (equations)
$N_{DF} = 15n_f + 18–15n_f + 5 = 13$

The 13 variables to specify may be selected from the parameters and those representing the initial conditions and microcapsule dimension, so, the specifications are:

Parameters $K_{m/d}, K_{m/r}, D, \mu, \sigma$
Algebraic $C_{d,\text{initial}}, h, t, r_{\max}, r_{min}, r_{\text{step}}, V_d, V_r$

Burst-time/lag-time models: Incidence matrix

Table 4 gives the incidence matrix for the model with the burst time effect. This model consists of equations (17), (19)–(26), (7), (4)–(6), (9), (11) and (27), which are solved n_f times, once for each microcapsule, whilst equations (3), (8), (10), (15) and (16) are solved once, giving a total set of $15n_f + 5$ equations to solve. The incidence matrix also shows the order for the solution of the algebraic set of equations; note that the equation for the mass released is given as analytical expression. For the model with a lag-time effect, a similar calculation order is employed, where equation (17) representing the burst effect, is substituted by equation (18). Also, equation (23) needs to be replaced by equation (24) to model controlled release with lag-time effect.

TABLE 3 Incidence Matrix for Controlled Release of Active Ingredients (First-Order Behaviour)

Equations	Unknown variables															
	$c_{d,i}$	$c_{r,i}$	n_f	r_i	A_i	$V_{d,i}$	$F(r;\mu;\sigma) = N'_{p,i}$	$V_{d,cal}$	$N_{p,i}$	$N_{T,p}$	$\%particles_i$	$\Delta M_{d,i}$	$\Delta M_{initial}$	$\Delta M_{r,i}$	ΔM_{total}	$\%release$
3			*													
4				*												
5				*	*											
6				*		*										
7				*			*									
8						*	*	*								
9							*	*	*							
10									*	*						
11									*	*	*					
12	*					*			*			*				
13						*			*				*			
14		*							*					*		
15														*	*	
16													*		*	*
1	*				*	*										
2		*			*	*										

TABLE 4 Incidence Matrix for Controlled Release of Active Ingredients (Burst Time Behaviour)

	Unknown variables																			
Equations	n_f	r_i	A_i	$V_{d,i}$	$F(r;\mu;\sigma) = N'_{p_i}$	$V_{d,cal}$	$N_{p,i}$	$N_{T,p}$	$\% particles_i$	α	α'	J_{max}	$M_{d,initial,i}$	$M_{burst,\infty}$	$M'_{d,initial}$	$c'_{d,initial}$	$M_{r,i}$	$\Delta M_{r,i}$	ΔM_{total}	$\% release$
3	*																			
4		*																		
5		*	*																	
6		*		*																
7		*			*															
8				*	*	*														
9					*	*	*													
10							*	*												
11							*	*	*											
19			*	*						*										
20																				
21												*								
22				*			*						*							
23			*				*				*	*		*						
25														*	*					
26				*											*	*				
17			*	*						*	*	*					*			
27							*										*	*		
15																		*	*	
16													*						*	*

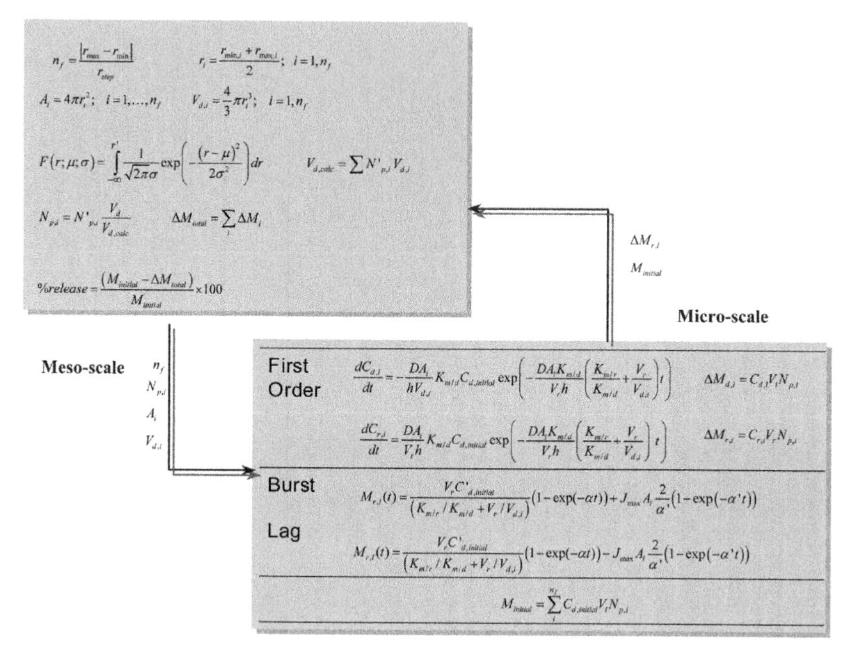

FIGURE 3 Data flow for the microcapsule-controlled release between meso-scale and micro-scale.

Data flow

Data flow for the microcapsule-based controlled release involves the transfer of data between the meso-scale and the micro-scale. Figure 3 highlights the data flow between these scales. Meso-scale calculations provide the number of microcapsule sizes (n_f), the number of particles of radius $i(N_{p,i})$, the surface area of the microcapsules where the diffusion is taking place (A_i) and the microcapsule volume of radius i ($V_{d,i}$) data to the micro-scale. This information is used in the micro-scale to calculate the mass of active ingredient (released) from each microcapsule. The data on the amount of AI released to the receiver medium from the set of microcapsules is transferred to the meso-scale, where the calculation of the total amount released to the receiver medium as well as the percent of the active ingredient released from the microcapsules is performed.

Based on the classification proposed by Ingram et al., 2004, the mathematical model for microcapsule-based controlled release involves the use of different integration frameworks to link the models at different scales. The connection between micro-scale and macro-scale involves the *simultaneous integration strategy*. The connection may also involve a *serial: transformation integration structure*, if the values of diffusion coefficients are estimated through a correlation obtained by fitting measured data.

12.1.5. Model solution

Solution strategy

The solution of the equations representing the model for the microcapsule-based controlled release of active ingredients is performed through the *V*irtual *P*roduct-*P*rocess *D*esign Lab (*VPPD*L) software (Morales-Rodriguez and Gani, 2009). This modelling tool is able to perform various product-process design/analysis tasks for different types of products-processes on the *VPPD*L manual guides the user through the solution of the product-process design problem in a systematic way, allowing easy product-process design/analysis.

Controlled release of a drug

This case study concerns a pharmaceutical product, Codeine [CAS number 76-57-3], which is encapsulated together with a carrier (an ion exchange resin) in microcapsules made of polyurea. The polymer wall is formed by water-promoted polyreaction of the monomer methylene diphenyl diisocyanate (MDI [CAS number 101-68-8]) that can give a cross-linked polymer.

The simulated results are compared against measured values taken from Lukaszczyk and Urbas (1997). The compositions of the AI within the microcapsules and the conditions of the release experiment are reported by Lukaszczyk and Urbas (1997). This information together with the data that are needed by the controlled release model is given in Table 5 for three different

TABLE 5 List of Values of the Input Data Required for the Mathematical Release Model

	Scenario		
Variable	1	2	3
MDI/resinate	0.25	0.5	1.0
Wall thickness $h(m)$	6.72 E-09	12.23 E-09	20.85 E-09
Maximum radius ($r_{\max}$)	*329 E-09*	*329 E-09*	*329 E-09*
Minimum radius (r_{min})	*29 E-09*	*29 E-09*	*29 E-09*
Mean radius (μ)	*129 E-09*	*129 E-09*	*129 E-09*
Standard deviation (σ)	*3 E-08*	*3 E-08*	*3 E-08*
Radius step (r_{step})	*1 E-08*	*1 E-08*	*1 E-08*
$V_r(m^3)$	400 E-06	400 E-06	400 E-06
$t(s)$	12600	12600	12600
$C_{d,\text{initial}}(g/m^3)$	324.44 E+03	280.697 E+03	220.825 E+03
$V_d(m^3)$	0.536 E-06	0.620 E-06	0.788 E-06

TABLE 6 List of Values for Diffusion and Partition Coefficients for Controlled Release Model

Variable	All scenarios	Source
$D(m^2/s)$	1.027 E-19	Estimated
$K_{m/r}$	2.67	Adapted
$K_{m/d}$	0.11	Assumed

simulation scenarios that have been investigated. Each scenario has a different membrane thickness defined through a different monomer-MDI resinate ratio. Due to the lack of data for all variables needed by the model, values for some variables have been assumed and marked in italic in Table 5. For instance, the size distribution data for the capsules were not available. Therefore, based on information of commercial microcapsules, values for these variables have been assumed.

The required properties of the chemical system are summarised in Table 6. The diffusivity (D) has been predicted with the constitutive model reported by Muro-Suñe (2005), whilst the value for the partition coefficient between codeine and polyurea ($K_{m/r}$) is adapted from data of Kubo et al. (2001). The partition coefficient between codeine and solvent ($K_{m/d}$) has been assumed based on values for similar systems. The values of the partition coefficients used for the different modelled scenarios are the same, since the donor and release mediums do not change in these scenarios.

The value of the diffusion coefficient of codeine through polyurea is estimated, as mentioned above, in a completely predictive manner through the extended predictive model developed by Muro-Suñe (2005). In order to perform the calculations with this model (based on the free volume theory), parameters related to the polymer viscosity with respect to temperature are also required. For the polymer of interest, polyurea, these parameters or experimental data are not available. Therefore, the parameters corresponding to polyurethane are employed making the assumption that both behave similarly. The value for the diffusion coefficient is estimated at the temperature for which the release experiments are reported, that is 309.15 K.

First, a distribution of microcapsules of different sizes is generated as seen in Figure 4.

For the three scenarios illustrated in Figure 5, the comparison of experimental data with simulated results from the model show a good agreement. This confirms the predictive power of the mathematical model as well as the applicability to a pharmaceutical product. Note, however, that the employed model

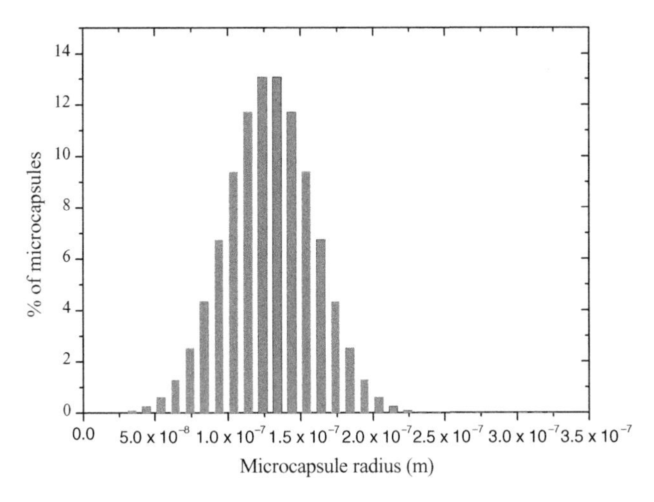

FIGURE 4 Normal distribution of microcapsules for controlled release of codeine.

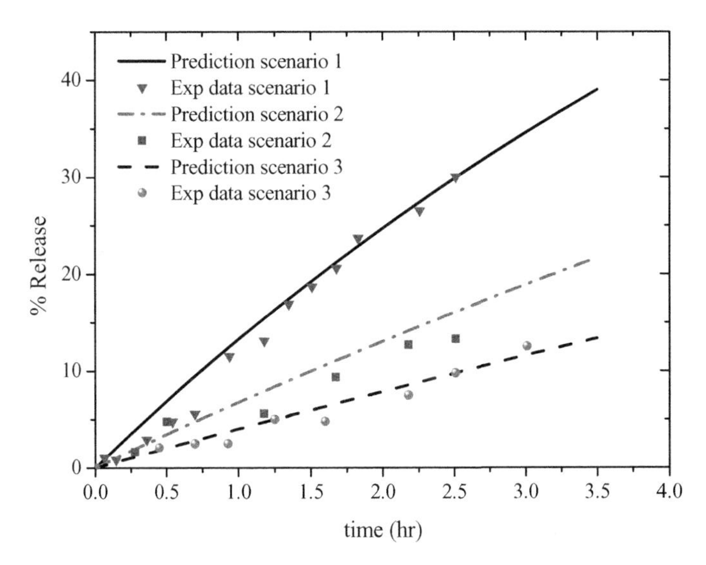

FIGURE 5 Comparison between the experimental values and model predictions of the release of codeine.

corresponds to a first-order system, since no time effects were found in this experimental data.

Table 7 illustrates the values for the release percent for the three different scenarios proposed in Table 5.

TABLE 7 Result of the Mathematical Release Model for the Pharmaceutical Product

	Scenario		
Variable	1	2	3
% release	38.7	21.2	13.5

12.2. FERMENTATION PROCESS MODELLING

Fermentation is a widely used process for production of different biological and pharmaceutical products, for example, to produce insulin. Mathematical modelling plays an important role to increase the process understanding. A fermentation process model is presented here. The model objective is to study and understand the process behaviour under typical operational conditions. This model is analysed systematically to characterise the model equations and model variables. Based on the model analysis, an incidence matrix is prepared to identify the solution strategy, and subsequently the model is implemented into a simulation tool. The detail description of the fermentation process model is available in the literature (Singh et al., 2009).

12.2.1. Process (System) Description

The considered fermentation process is shown in Figure 6 (adapted from Petrides et al., 1995). Fermentation medium is prepared in a mixing tank and sterilised in a continuous heat steriliser. The axial compressor and the absolute filter provide sterile air and ammonia to the fermentor. A jacketed fermentor is used to carry out an aerobic batch fermentation with *E. coli* cells. These cells are used to produce the Trp-LE'-MET-pro-insulin precursor of insulin, which is not excreted in the fermentation broth but retained in the cellular biomass. The fermentation time in the production fermentor is about 18 hours per batch, and the fermentation temperature is 32 oC. Water is used as a coolant to maintain the fermentor temperature. The final concentration of *E. coli* cells in the production fermentor is about 50 g/l (30 g/liter dry cell weight). The chimeric protein Trp-LE'-MET-pro-insulin accumulates intracellularly as insoluble aggregates (inclusion bodies) and this decreases the rate at which the protein is degraded by proteolytic enzymes. In the base case, it was assumed that the inclusion bodies (IB's) constitute 20% of total dry cell mass, similar to Petrides et al. (1995). At the end of the fermentation, the broth is cooled down to 10 °C to minimise cell lysis. As shown in Figure 6, mixing, heat sterilisation and fermentation operations are involved in this process. The mathematical model of the fermentation is described in the following section.

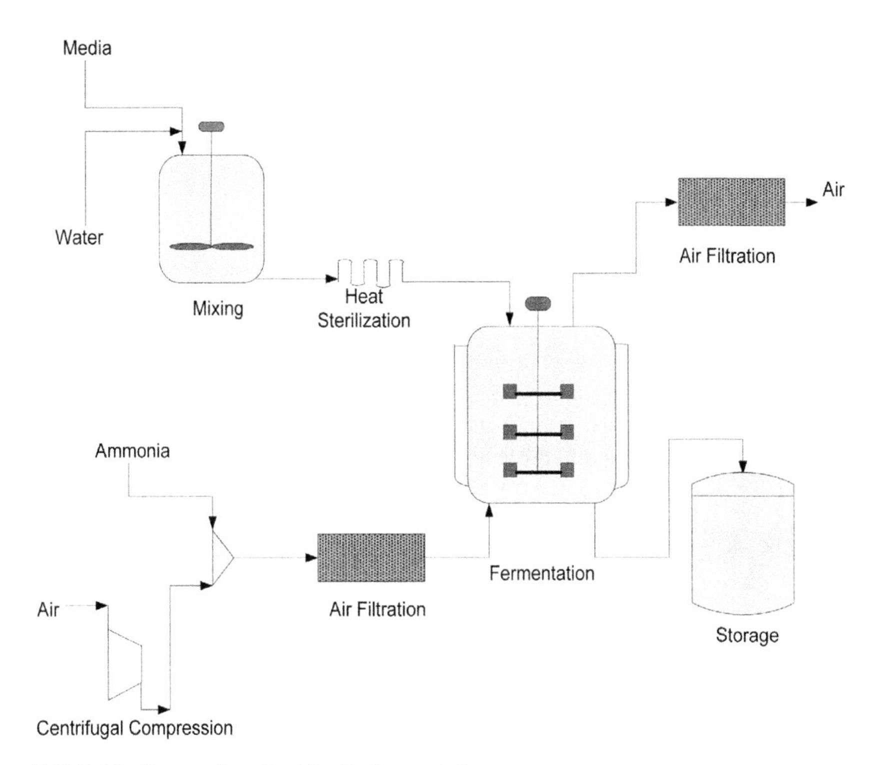

FIGURE 6 **Process flow sheet for the fermentation process.**

12.2.2. Model Description

The fermentation process model is described as follows:

Assumptions

Perfectly mixed and sterilised fermentation medium is fed to the fermentor, and *E. coli* cells are used as the starter culture. Death rate of cells is assumed to be negligible. Air and ammonia are supplied to the fermentor at fixed rate. The stirrer speed is also assumed to be constant. Spatial distribution of concentrations and temperature in the fermentor is not considered. The model is simulated in open-loop conditions.

Balance equations

Accumulation of carbonic acid: The difference between the formation and the dissociation rate of carbonic acid gives the rate of change of the carbonic acid concentration during the fermentation process:

$$\frac{dC_{H_2CO_3}}{dt} = r_{CO_2} - r_{H_2CO_3} \tag{1}$$

Accumulation of bicarbonate ion: The rate of change of the bicarbonate ion concentration can be obtained by subtracting the bicarbonate ion dissociation rate from the carbonic acid dissociation rate. The rate of change of the carbonate ion concentration is equal to the bicarbonate dissociation rate.

$$\frac{dC_{HCO_3^-}}{dt} = r_{H_2CO_3} - r_{HCO_3^-} \quad (2)$$

$$\frac{dC_{CO_3^{--}}}{dt} = r_{HCO_3^-} \quad (3)$$

Accumulation of phosphate ions concentration: The difference of the dissociation rates of phosphoric acid and the dihydrogen phosphate ion gives the accumulation rate of the dihydrogen phosphate ion as follows:

$$\frac{dC_{H_2PO_4^-}}{dt} = r_{H_3PO_4} - r_{H_2PO_4^-} \quad (4)$$

The rate of accumulation of the hydrogen phosphate ion is obtained by subtracting the rate of dissociation of the hydrogen phosphate ion from the dissociation rate of the dihydrogen phosphate ion as follows:

$$\frac{dC_{HPO_4^{--}}}{dt} = r_{H_2PO_4^-} - r_{HPO_4^{--}} \quad (5)$$

The rate of phosphate ion accumulation is the same as the rate of the hydrogen phosphate ion dissociation as follows:

$$\frac{dC_{HPO_4^{--}}}{dt} = r_{H_2PO_4^-} - r_{HPO_4^{--}} \quad (6)$$

Accumulation of ammonium ion concentration: The rate of ammonium ion accumulation is equal to the rate of dissociation of the ammonia as follows:

$$\frac{dC_{NH_4^+}}{dt} = r_{NH_3} \quad (7)$$

Accumulation of sulphate ion concentration: The rate of accumulation of the hydrogen sulphate ion is obtained by subtracting the dissociation rate of the hydrogen sulphate ion from the dissociation rate of sulphuric acid as follows:

$$\frac{dC_{HSO_4^-}}{dt} = r_{H_2SO_4} - r_{HSO_4^-} \quad (8)$$

The rate of accumulation of the sulphate ion is equal to the rate of dissociation of the hydrogen sulphate ion, which can be given as follows:

$$\frac{dC_{SO_4^{--}}}{dt} = r_{HSO_4^-} \quad (9)$$

Rate of change of biomass concentration: The change in the biomass concentration during the fermentation process is governed by the following equation (Doran, 2006):

$$\frac{dC_x}{dt} = \mu C_X \tag{10}$$

Rate of change of substrate concentration: The substrate is mainly consumed for biomass production but some amount of it is also used for maintenance. The consumption rate of substrate can be expressed as follows (Doran, 2006):

$$\frac{dC_S}{dt} = -\left(\frac{\mu}{Y_{XS}} + m_S\right)C_X \tag{11}$$

Rate of change of dissolved oxygenconcentration: The oxygen in gaseous form is supplied to the fermentor. The gaseous oxygen then dissolves in the liquid (rate of dissolution=$kla_{O_2}(C^*_{O_2} - C_{O_2})$). Some quantity of dissolved oxygen($q_{O_2}C_X$) is assumed to be used for cell growth, some quantity for maintenance (mC_X) and some quantity ($b_X C_X$) for biomass decay. The rate of accumulation of dissolved oxygen can be expressed as follows:

$$\frac{dC_{O_2}}{dt} = kla_{O_2}\left(C^*_{O_2} - C_{O_2}\right) - q_{O_2}C_X - m_s C_X - b_x C_X \tag{12}$$

Rate of change of dissolved CO_2concentration: CO_2 is produced through biomass production, maintenance and biomass decay. A fraction of it can be converted into carbonic acid, and another fraction can be removed by aeration. The rate of accumulation of dissolved CO_2 can be expressed as follows:

$$\frac{dC_{CO_2}}{dt} = q_{CO_2}C_X + m_s C_X + b_X C_X + kla_{CO_2}\left(C^*_{CO_2} - C_{CO_2}\right) - r_{CO_2} \tag{13}$$

Rate of change of phosphoric acid concentration: Phosphoric acid is used as a nutrient for the biomass production. Some phosphoric acid is assumed to be regenerated by biomass decay. The rate of accumulation of phosphoric acid can be expressed as follows:

$$\frac{dC_{H_3PO_4}}{dt} = \frac{(bx - \mu)}{Y_{XH_3PO_4}} C_X - r_{H_3PO_4} \tag{14}$$

Rate of change of ammonia concentration: Ammonia is used as a nutrient for the biomass production. Some ammonia is assumed to be regenerated as a consequence of biomass decay, and a considerable amount of ammonia can be present in the form of ammonium ions. An external supply of ammonia is assumed for pH control, where the supplied gaseous form of ammonia is converted to the liquid form. The rate of accumulation of ammonia can be expressed as follows:

$$\frac{dC_{\mathrm{NH_3}}}{dt} = \frac{(bx-\mu)}{Y_{X\mathrm{NH_3}}} C_X + kla\,\mathrm{NH_3}\left(C^*_{\mathrm{NH_3}} - C_{\mathrm{NH_3}}\right) - r\,\mathrm{NH_3} \tag{15}$$

Rate of change of sulphuric acid concentration: Sulphate is consumed as an essential nutrient during biomass production. A fraction of it can be released again as a consequence of biomass decay. The equilibrium between sulphuric acid, hydrogen sulphate and sulphate ions is assumed. The rate of accumulation of sulphate can be expressed as follows:

$$\frac{dC_{\mathrm{H_2SO_4}}}{dt} = \frac{(bx-\mu)}{Y_{X\mathrm{H_2SO_4}}} C_X - r_{\mathrm{H_2SO_4}} \tag{16}$$

Rate of change of hydroxyl ion concentration: The dissociation of water in hydroxyl and hydrogen ions is assumed. The rate of change of the hydroxyl ion concentration is given as follows:

$$\frac{dC_{\mathrm{OH^-}}}{dt} = r_{\mathrm{H_2O}} \tag{17}$$

Rate of change of hydrogen ion (proton) concentration: Protons are formed in the medium, due to the dissociation of salts, salt ions and water. Some fraction of these protons can be combined with ammonia to form the ammonium ion. The rate of change of the hydrogen ion concentration can be expressed as follows:

$$\begin{aligned}\frac{dC_{\mathrm{H^+}}}{dt} =& \left(r_{\mathrm{H_3PO_4}} + r_{\mathrm{H_2PO_4^-}} + r_{\mathrm{HPO_4^{--}}}\right) + \left(r_{\mathrm{H_2SO_4}} + r_{\mathrm{HSO_4^-}}\right) \\ &+ \left(r_{\mathrm{H_2CO_3}} + r_{\mathrm{HCO_3^-}}\right) + r_{\mathrm{H_2O}} - r_{\mathrm{NH_3}}\end{aligned} \tag{18}$$

Fermentor temperature: In the fermentation process heat is generated due to the fermentation process itself and due to agitation. Some heat is lost directly to the surroundings through the fermentor walls, some is lost due to evaporation of water into the air stream and some heat is also lost as a sensible heat loss to the air stream (Cooney et al., 1968). Water is used as a coolant to maintain the fermentor temperature. The rate of change of fermentor temperature can be expressed as follows:

$$\begin{aligned}\rho_b V C_{p_b} \frac{dT_b}{dt} =& -U_1 A_1 (T_b - T_w) \\ &+ V\left(H_{gr} + H_{ag} - H_{surr} - H_{evp} - H_{sen}\right)\end{aligned} \tag{19}$$

Coolant temperature: The coolant is circulated in a coil, and exchanges heat with the fermentation broth. Finally, some heat is lost to the surroundings. A simplified expression for heat exchange between coolant and vessel contents in the jacketed vessel has been adopted from the literature (Quintana-Hernàndez et al., 2004). The rate of change of coolant temperature can be expressed as follows:

$$\rho_w C_{p_w} V_w \frac{dT_w}{dt} = \rho_w f_{w_{in}} C_{p_w} (T_{w_{in}} - T_w) + U_1 A_1 (T_b - T_w) + U_2 A_2 (T_{surr} - T_w) \tag{20}$$

Constitutive equations (explicit)

Power dissipation: Power dissipation due to agitation can be expressed as a function of power number (P_n), impeller speed (n), impeller diameter (d_i) and fluid density (ρ_b) as follows (Angst and Kraume, 2006):

$$Pow = P_n n^3 d_i^5 \rho_b \tag{21}$$

Homogeneity: An intermediate parameter (λ) relating homogeneity with the fermentor dimensions is given as below:

$$\lambda = K.n \left(\frac{d_i}{d_{\text{tank}}} \right)^{\alpha} \left(\frac{d_{\text{tank}}}{T_L} \right)^{\beta} \tag{22}$$

The homogeneityhomogeneity in the fermentor is calculated as follows (Ochieng & Onyango, 2008):

$$HO = 1 - \exp(-\lambda t) \tag{23}$$

Gas liquid mass transfer coefficient : The gas liquid mass transfer coefficient for oxygen can be correlated with the power density (Pow/V) and the superficial gas velocity (u_{O_2}) as follows (Linek et al., 1987):

$$kla_{O_2} = 0.00495 \left(\frac{Pow}{V} \right)^{0.593} u_{O_2}^{0.4} \tag{24}$$

The gas liquid mass transfer coefficient for CO_2 depends on the gas liquid mass transfer coefficient of oxygen:

$$kla_{CO_2} = kla_{O_2} \left(\frac{C_{CO_2}}{C_{O_2}} \right)^{0.5} \tag{25}$$

Similar to the gas liquid mass transfer coefficient of oxygen, the gas liquid mass transfer coefficient for ammonia is also correlated with the power density (Pow/V) and the superficial velocity (u_{NH_3}) (Linek et al., 1987):

$$kla_{NH_3} = 0.00495 \left(\frac{Pow}{V} \right)^{0.593} u_{NH_3}^{0.4} \tag{26}$$

Specific cell growth rate : The dependency of the specific cell growth rate (μ) on the dissolved oxygen concentration (C_{O_2}) and the nutrient concentrations (glucose (C_S), ammonia (C_{NH_3}) and phosphoric acid ($C_{H_3PO_4}$)) was incorporated in the model as follows:

$$\mu = \mu_{max} \frac{C_S}{(k_s + C_S)} \frac{C_{O_2}}{(k_{O_2} + C_{O_2})} \frac{C_{NH_3}}{(k_{NH_3} + C_{NH_3})} \frac{C_{H_3PO_4}}{(k_{H_3PO_4} + C_{H_3PO_4})} \quad (27)$$

Maintenance coefficient: Maintenance coefficientA fraction of the substrate and dissolved oxygen, which are taken up by the *E. coli* cells, is assumed to be consumed for cell maintenance. The dependency of the maintenance coefficient on the substrate (C_S) and dissolved oxygen (C_{O_2}) concentration can be expressed as follows:

$$ms = \frac{ms_1 C_S C_{O_2}}{(k_s + C_S)(k_{O_2} + C_{O_2})} \quad (28)$$

Biomass decay: *Biomass decay* The specific biomass decay can be expressed as a function of the dissolved oxygendissolved oxygen concentration (C_{O_2}), as follows:

$$bx = bx_1 \frac{C_{O_2}}{k_{O_2} + C_{O_2}} \quad (29)$$

Oxygen saturation concentration: The temperature is inversely related to the solubility of dissolved oxygen. As a consequence, the driving force for mass transfer $\left(C^*_{O_2} - C_{O_2}\right)$ is reduced with increasing temperature. Oxygen solubility in pure water as a function of temperature has been included in the model by the following equation (Truesdale et al., 1955):

$$C^*_{O_2} = 14.161 - 0.3943T_b + 0.007714T_b^2 - 0.0000646T_b^3 \quad (30)$$

Specific oxygen uptake rate: The specific oxygen uptake rate (oxygen uptake rate per unit biomass concentration) depends on the specific cell growth rate (μ) and the yield coefficient for oxygen (Y_{XO_2}) as given below:

$$q_{O_2} = \frac{\mu}{Y_{XO_2}} \quad (31)$$

Rate of oxygen consumption for cell growth rate: The oxygen uptake rate per unit volume of broth can be expressed as follows:

$$Q_{O_2} = q_{O_2} C_X \quad (32)$$

Specific CO_2production rate: The specific CO_2 production rate (rate of production per unit biomass concentration) can be expressed as a function of the specific cell growth rate and the yield coefficient of CO_2 as follows:

$$q_{CO_2} = \frac{\mu}{Y_{XCO_2}} \quad (33)$$

Rate of CO_2 production by cells: The rate of carbon dioxide production due to the fermentation process can be expressed as a function of the biomass concentration as follows:

$$Q_{CO_2} = q_{CO_2} C_X \tag{34}$$

pH : pH and all relevant chemical equilibrium were modelled as described in Sin et al. (2008). pH can be expressed as a function of the hydrogen ion concentration present in the broth:

$$\mathrm{pH} = -log\left(C_{\mathrm{H}}^{+}\right) \tag{35}$$

Algebraic equations (explicit)

Carbon dioxide and bicarbonate equilibrium: The dissolved CO_2 can react with the water to form H_2CO_3 (carbonic acid). The carbonic acid can dissociate into a proton and a bicarbonate ion, and the bicarbonate ion can subsequently dissociate into a proton and a carbonate ion by subsequent equilibrium reactions (Descoins et al., 2006).

The rate of CO_2 dissociation into bicarbonate can be expressed as follows:

$$r_{CO_2} = kf_{CO_2}\left(C_{CO_2} - \frac{C_{H_2CO_3}}{K_{CO_2}}\right) \tag{36}$$

Equilibrium between the carbonate and bicarbonate ions is assumed (Descoins et al., 2006), and the rate of carbonic acid dissociation into the bicarbonate ion can be expressed as follows:

$$r_{H_2CO_3} = kf_{H_2CO_3}\left(C_{H_2CO_3} - \frac{C_{HCO_3^-} C_{H^+}}{K_{H_2CO_3}}\right) \tag{37}$$

Equilibrium between bicarbonate ion and carbonate ion is also assumed and the rate of bicarbonate ion dissociation can be expressed as follows:

$$r_{HCO_3^-} = kf_{HCO_3^-}\left(C_{HCO_3^-} - \frac{C_{CO_3^{--}} C_{H^+}}{K_{HCO_3^-}}\right) \tag{38}$$

Phosphate dissociation: Phosphoric acid (H_3PO_4) is used as one of the nutrients for the fermentation process. It is assumed that phosphoric acid can dissociate and form phosphate ions through the following set of equilibrium reactions (Musvoto et al., 2000):

$$H_3PO_4 \Leftrightarrow H_2PO_4^- + H^+ \qquad K_{P1}$$

$$H_2PO_4^- \Leftrightarrow HPO_4^{--} + H^+ \qquad K_{P2}$$

$$HPO_4^{--} \Leftrightarrow PO_4^{3-} + H^+ \qquad K_{P3}$$

The rate of phosphoric acid dissociation can be expressed as follows:

$$r_{H_3PO_4} = kf_{H_3PO_4}\left(C_{H_3PO_4} - \frac{C_{H_2PO_4^-} C_{H^+}}{K_{H_3PO_4}}\right) \tag{39}$$

The rate of dihydrogen phosphate and hydrogen phosphate dissociation can be expressed as follows:

$$r_{H_2PO_4^-} = kf_{H_2PO_4^-}\left(C_{H_2PO_4^-} - \frac{C_{HPO_4^{--}}C_{H^+}}{K_{H_2PO_4^-}}\right) \tag{40}$$

$$r_{HPO_4^{--}} = kf_{HPO_4^{--}}\left(C_{HPO_4^{--}} - \frac{C_{PO_4^{---}}C_{H^+}}{K_{HPO_4^{--}}}\right) \tag{41}$$

Ammonia dissociation: A chemical reaction also occurs when ammonia dissolves in water. In aqueous solution, ammonia acts as a base, acquiring hydrogen ions from H_2O to yield ammonium and hydroxide ions (Shakhashiri, 2007). Equilibrium between ammonia and the ammonium ion is assumed (Musvoto et al., 2000). The rate of ammonia dissociation is equal to the rate of ammonium ion formation, which can be expressed as follows:

$$r_{NH_3} = -kf_{NH_4^+}\left(C_{NH_4^+} - \frac{C_{NH_3}C_{H^+}}{K_{NH_4^+}}\right) \tag{42}$$

Sulphate dissociation: Equilibrium between sulphuric acid (a strong acid), hydrogen sulphate ions and sulphate ions is assumed. The rate of dissociation of sulphuric acid can be expressed as follows:

$$r_{H_2SO_4} = kf_{H_2SO_4}\left(C_{H_2SO_4} - \frac{C_{HSO_4^-}C_{H^+}}{K_{H_2SO_4}}\right) \tag{43}$$

The rate of hydrogen sulphate ion dissociation can be expressed as follows:

$$r_{HSO_4^-} = kf_{HSO_4^-}\left(C_{HSO_4^-} - \frac{C_{SO_4^{--}}C_{H^+}}{K_{HSO_4^-}}\right) \tag{44}$$

Water dissociation: The rate of water dissociation can be expressed as follows:

$$r_{H_2O} = kf_w\left(1 - \frac{C_{H^+}C_{OH^-}}{K_w}\right) \tag{45}$$

12.2.3. Model Analysis

There are 15 constitutive equations and 30 balance equations (20 differential and 10 algebraic) in the fermentation process operation model. The total number of variables involved in the model is found to be 104 (see Table 8). The degree of freedom is calculated to be 59. The detailed classification of the variables is given in Table 8. The incidence matrix of this model has shown a lower triangular form (see Table 9, 25 by 25 square matrix forming the bottom right part of the table), and therefore all the equations can be solved sequentially.

TABLE 8 Classification of Variables of the Fermentation Process Model

Variable	Status	Symbol	Number	Total
To be specified	Constant	C_{p_w}, ρ_w, α, β, K	5	59
	Fixed by problem	$C^*_{CO_2}$, $C^*_{NH_3}$, C_{p_b}, V_w, U_1, U_2, ρ_b, $T_{w_{in}}$, T_{surr}, H_{gr}, H_{ag}, H_{surr}, H_{evp}, H_{sen}, n, u_{O_2}, u_{NH_3}, $F_{w_{in}}$, P_n, μ_{max}, ms_1, bx_1, Y_{XO_2}, Y_{XCO_2}, Y_{XS}, $Y_{XH_3PO_4}$, Y_{XNH_3}, $Y_{XH_2SO_4}$, k_s, k_{O_2}, k_{NH_3}, $k_{H_3PO_4}$, kf_{CO_2}, K_{CO_2}, $kf_{H_2CO_3}$, $K_{H_2CO_3}$, $kf_{HCO_3^-}$, $K_{HCO_3^-}$, $kf_{H_3PO_4}$, $K_{H_3PO_4}$, $kf_{H_2PO_4^-}$, $K_{H_2PO_4^-}$, $kf_{NH_4^+}$, $K_{NH_4^+}$, $kf_{H_2SO_4}$, $K_{H_2SO_4}$, $kf_{HSO_4^-}$, $K_{HSO_4^-}$	48	
	Fixed by system	V, A_1, A_2, d_i, d_{tank}, T_L,	6	
To be predicted	Algebraic (explicit)	Pow, kla_{O_2}, kla_{CO_2}, kla_{NH_3}, ms, bx, $C^*_{O_2}$, q_{O_2}, Q_{O_2}, q_{CO_2}, Q_{CO_2}, pH, r_{CO_2}, $r_{H_2CO_3}$, $r_{HCO_3^-}$, $r_{H_3PO_4}$, $r_{H2PO_4^-}$, $r_{HPO_4^{--}}$, r_{NH_3}, $r_{H_2SO_4}$, $r_{HSO_4^-}$, r_{H_2O}, μ, λ, HO	25	45
	Differential	C_{O_2}, C_{CO_2}, C_S, C_{NH_3}, $C_{H_3PO_4}$, C_X, $C_{H_2CO_3}$, $C_{HCO_3^-}$, C_{H^+}, $C_{CO_3^{--}}$, $C_{H_2PO_4^-}$, $C_{HPO_4^{--}}$, $C_{PO_4^{--}}$, $C_{NH_4^+}$, $C_{H_2SO_4}$, C_{HSO_4}, $C_{SO_4^{--}}$, C_{OH^-}, T_b, T_w,	20	
	Total number of variables involved in the model			104
	Total number of constitutive equations (explicit)			15
	Total number of balance equations (20 differential, 10 algebraic (explicit))			30

TABLE 9 Incidence Matrix of Fermentation Process Model

Differential Equations: Eq. No. 1–20

	Differential Variables																																												
Eq. No.	C_{H2CO3}	C_{HCO3^-}	$C_{CO3^{--}}$	C_{H2PO4^-}	$C_{HPO4^{--}}$	$C_{PO4^{---}}$	C_{NH4^+}	C_{HSO4^-}	$C_{SO4^{--}}$	C_X	C_S	C_{O2}	C_{CO2}	C_{H3PO4}	C_{NH3}	C_{H2SO4}	C_{OH^-}	C_{H^+}	T_b	T_a	Pow	λ	HO	kla_{O2}	kla_{CO2}	kla_{NH3}	μ	ms	bx	C^*_{O2}	q_{O2}	Q_{O2}	q_{CO2}	Q_{CO2}	pH	r_{CO2}	r_{H2CO3}	r_{HCO3^-}	r_{H3PO4}	r_{H2PO4^-}	$r_{HPO4^{--}}$	r_{NH3}	r_{H2SO4}	r_{HSO4^-}	r_{H2O}
1	*																																			*	*								
2		*																																			*	*							
3			*																																										
4				*																																		*	*	*					
5					*																																			*	*				
6						*																																			*				
7							*																																			*			
8								*																																			*	*	
9									*																																			*	
10										*																	*																		
11										*	*																*	*																	
12										*		*																*	*	*	*														
13										*			*											*				*	*				*			*									
14										*				*											*				*										*						
15										*					*											*	*		*													*			
16										*						*											*		*														*		
17																	*																												*
18																		*																			*	*	*	*	*	*	*	*	*
19																			*	*																									
20																			*	*																									
21																					*																								
22																						*																							
23																						*	*																						
24																					*			*																					
25												*	*											*	*																				
26																					*					*																			
27											*	*		*	*												*																		
28											*	*																*																	
29												*							*										*																
30																														*															
31																											*				*														
32										*																					*	*													
33																											*						*												
34										*																							*	*											
35																		*																	*										
36	*												*																							*									
37	*	*																*																			*								
38		*	*															*																				*							
39				*										*				*																				*	*						
40				*	*													*																						*					
41					*	*												*																							*				
42							*								*			*																								*			
43								*								*		*																									*		
44								*	*									*																										*	
45																	*	*																											*

12.2.4. Model Solution

Open-loop simulation of an aerobic batch fermentation process is performed with constant and limited supply of air and ammonia during the fermentation process. The stirrer speed is assumed to be constant at 480 rpm. The simulation tool, ICAS-MoT (Sales-Cruz, 2006) is used for simulation.

Some of the important results of the fermentation process model are described in this section. Figure 7 shows the profile of dissolved oxygen concentration during the fermentation process. The consumption of dissolved oxygen increases during the fermentation process whilst the inlet oxygen flow rate remains constant therefore the dissolved oxygen concentration in the fermentor first increases and then decreases (see Figure 7). The profile of pH is seen in Figure 8, where the pH value first decreases and then increases. It should be noted that the ammonia flow rate is assumed to be constant during the fermentation process. The profile of specific growth rate (mue) is shown in Figure 9. In order to achieve the optimal specific growth rate the dissolved oxygen concentration and pH need to be maintained within a specific limit. However, as seen in Figures 7 and 8 the dissolved oxygen concentration and the pH value are changing during the fermentation process, which leads to the variation in specific growth rate.

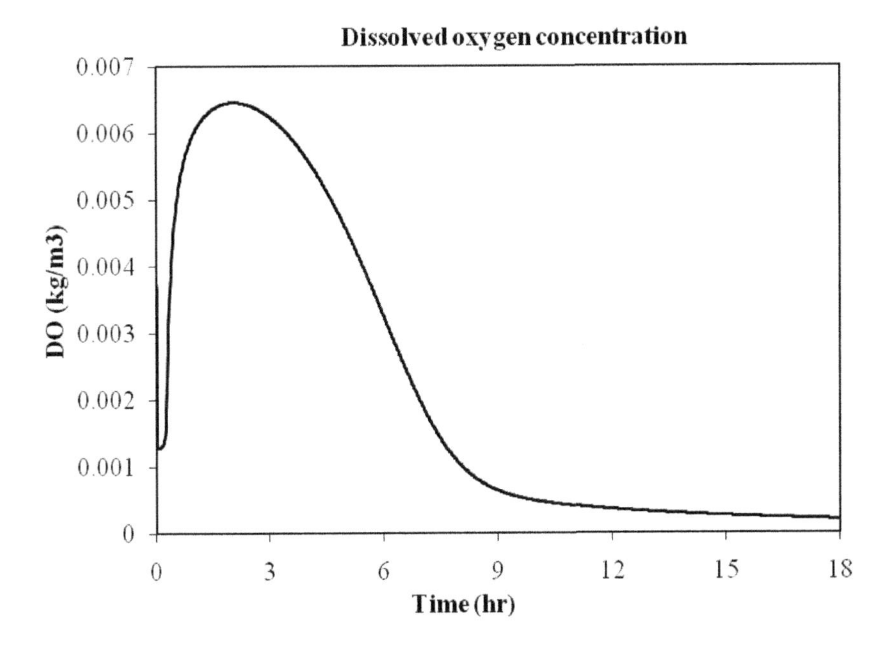

FIGURE 7 **Simulated dynamic response of the dissolved oxygen concentration.**

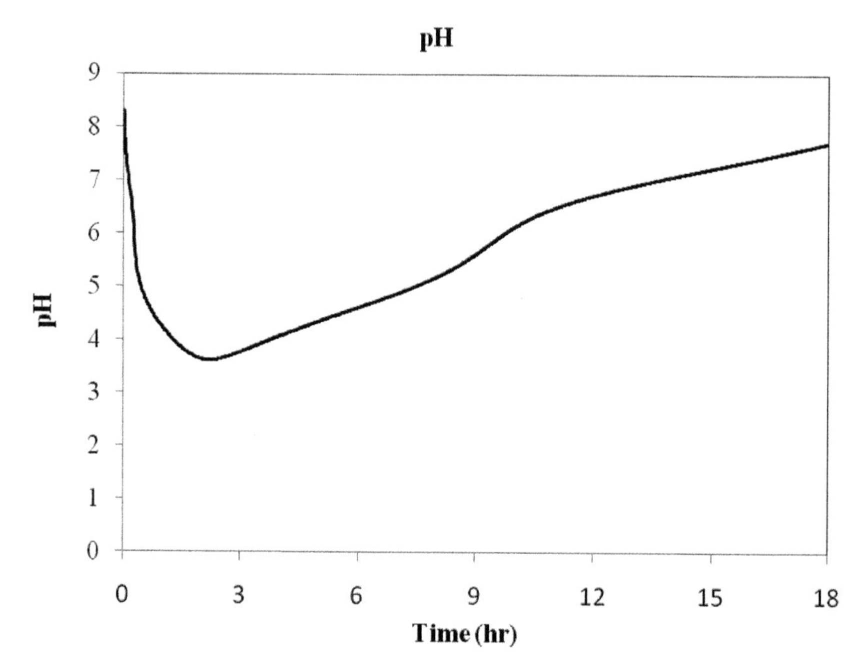

FIGURE 8 Simulated dynamic pH profile.

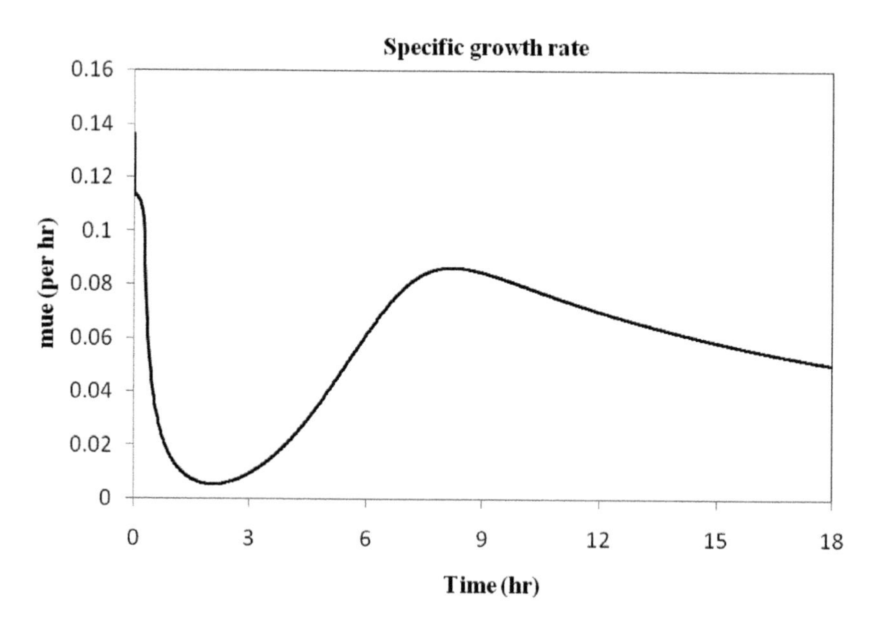

FIGURE 9 Simulated dynamic response of the specific growth rate (mue).

VALUE OF KNOWN VARIABLES AND PARAMETERS

Differential Variables	Symbol	Initial value
Dissolved oxygen concentration (kg/m^3)	C_{O_2}	0.004
Dissolved CO_2 concentration (kg/m^3)	C_{CO_2}	1.000
Substrate concentration (kg/m^3)	C_S	150.000
Dissolved NH_3 concentration (kg/m^3)	C_{NH_3}	2.000
Phosphate concentration (kg/m^3)	$C_{H_3PO_4}$	14.000
Biomass concentration (kg/m^3)	C_X	5.000
Bicarbonate concentration (kg/m^3)	$C_{H_2CO_3}$	1.000
HCO_3^- concentration (kg/m^3)	$C_{HCO_3^-}$	0.100
Hydrogen ion concentration (kg/m^3)	C_{H^+}	1.00E-07
CO_3^- concentration (kg/m^3)	$C_{CO_3^-}$	0.001
$H_2PO_4^-$ concentration (kg/m^3)	$C_{H_2PO_4^-}$	2.000
HPO_4^{--} concentration (kg/m^3)	$C_{HPO_4^{--}}$	1.000
PO_4^{---} concentration (kg/m^3)	$C_{PO_4^{---}}$	0.100
NH_4^+ concentration (kg/m^3)	$C_{NH_4^+}$	1.000
Sulfate concentration (kg/m^3)	$C_{H_2SO_4}$	5.000
HSO_4^- concentration (kg/m^3)	$C_{HSO_4^-}$	1.000
SO_4^{--} concentration (kg/m^3)	$C_{SO_4^{--}}$	1.000
OH^- concentration (kg/m^3)	C_{OH^-}	1.00E-07
Fermentation temperature (K)	T_b	310.000
Coolant temperature (K)	T_w	283.000

Known variables	Symbol	Value
Saturation concentration of CO_2 (kg/m^3)	$C^*_{CO_2}$	1.450
Saturation concentration of NH_3 (kg/m^3)	$C^*_{NH_3}$	500.000
Heat capacity of broth (kJ/(kg K))	C_{P_b}	4.200
Heat capacity of water (kJ/(kg K))	C_{P_w}	4.200
Volume of cooling jacket (m^3)	V_w	4.000
Overall heat transfer coefficient (broth-coolant) (W/m^2.K)	U_1	370.000
Overall heat transfer coefficient (coolant-air) (W/m^2 K)	U_2	11.300
Broth density (kg/m^3)	ρ_b	1200.000
Water density (kg/m^3)	ρ_w	1000.000
Inlet temperature of coolant (K)	T_{w_i}	278.000
Surrounding temperature (K)	T_{surr}	298.000
Heat of fermentation (kJ/m^3 hr)	H_{gr}	4704.000
Heat of agitation (kJ/m^3 hr)	H_{ag}	13944.000
Heat lost to surroundings (kJ/m^3 hr)	H_{surr}	2562.000
Heat of evaporation (kJ/m^3 hr)	H_{evp}	96.000
Sensible heat lost (kJ/m^3 hr)	H_{sen}	21.000
Stirrer speed (rps)	n	8.000
Superficial velocity of air (m/s)	u_{O_2}	0.100
Superficial velocity of ammonia (m/s)	u_{NH_3}	1.000E-005
Coolant flow rate (m^3/hr)	$F_{W_{in}}$	71.000
Power number	P_n	2.000

Parameters	Symbol	Value
Maximum specific cell growth rate (hr^{-1})	μ_{max}	0.150
Maintenance coefficient (hr^{-1})	ms_1	0.020
Decay coefficient (hr^{-1})	bx_1	0.005
Yield coefficient (oxygen)	Y_{XO_2}	0.482
Yield coefficient (CO_2)	Y_{XCO_2}	0.470
Yield coefficient (substrate)	Y_{XS}	0.320
Yield coefficient (phosphate)	$Y_{XH_3PO_4}$	54.348
Yield coefficient (ammonia)	Y_{XNH_3}	5.988
Yield coefficient (sulphate)	$Y_{XH_2SO_4}$	312.500
Monod constant for substrate (kg/m^3)	k_s	0.004
Monod constant for oxygen (kg/m^3)	k_{O_2}	0.0004
Monod constant for ammonia (kg/m^3)	k_{NH_3}	0.0004
Monod constant for phosphate (kg/m^3)	$k_{H_3PO_4}$	0.0004
Forward rate constant (CO_2)	kf_{CO_2}	4.000
Equilibrium constant ($CO_2 - H_2CO_3$)	K_{CO_2}	0.0016
Forward rate constant (H_2CO_3)	$kf_{H_2CO_3}$	4.444
Equilibrium constant ($H_2CO_3 - HCO_3^-$)	$K_{H_2CO_3}$	4.440E-07
Forward rate constant (HCO_3^-)	$kf_{HCO_3^-}$	0.468
Equilibrium constant ($HCO_3^- - CO_3^{--}$)	$k_{HCO_3^-}$	4.680E-11
Forward rate constant (H_3PO_4)	$kf_{H_3PO_4}$	7.089E5
Equilibrium constant ($H_3PO_4 - H_2PO_4^-$)	$K_{H_3PO_4}$	7.089E-3
Forward rate constant ($H_2PO_4^-$)	$kf_{H_2PO_4^-}$	6.303E4
Equilibrium constant ($H_2PO_4^- - HPO_4^{--}$)	$K_{H_2PO_4^-}$	6.300E-08
Forward rate constant (NH_4^+)	$kf_{NH_4^+}$	5.678E2
Equilibrium constant ($NH_4^+ - NH_3$)	$K_{NH_4^+}$	5.680E-10
Forward rate constant (H_2SO_4)	$kf_{H_2SO_4}$	1.000
Equilibrium constant ($H_2SO_4^- - HSO_4^-$)	$K_{H_2SO_4}$	1000.00
Forward rate constant (HSO_4^-)	$kf_{HSO_4^-}$	20.000
Equilibrium constant ($HSO_4^- - SO_4^{--}$)	$K_{HSO_4^-}$	1.200e-002
Fermentor volume (m^3)	V	30.000
Heat exchange area (fermentor-cooling jacket) (m^2)	A_1	50.000
Heat exchange area (cooling jacket-environment) (m^2)	A_2	55.000
Impeller diameter (m)	d_i	0.830

Nomenclature (for section 12.2)

Symbols

A_1	Heat exchange area (fermentor-cooling jacket) (m^2)
A_2	Heat exchange area (cooling jacket-surroundings) (m^2)
bx	Specific biomass decay (hr^{-1})
bx_1	Decay constant (hr^{-1})
C	Concentration (kg/m^3)
C_p	Heat capacity (KJ/(Kg. K))
d_i	Impeller diameter (m)

F_s	Steam flow rate (m^3/hr)
F_w	Coolant flow rate (m^3/hr)
H	Heat transfer rate (KJ/ (m^3.hr))
HO	Homogeneity
k	Monod constant (kg/m^3)
K	Equilibrium constant
kf	Forward rate constant
kla	Mass transfer coefficient (m/s)
ms	Specific maintenance coefficient (hr^{-1})
ms_1	Maintenance constant (hr^{-1})
N	Stirrer speed (revolution per second)
pH	pH
P_n	Power number
Pow	Power dissipation through agitation (W)
q_{CO2}	Specific CO_2 production rate (hr^{-1})
Q_{CO2}	CO_2 production rate (Kg/m^3.hr)
q_{O2}	Specific oxygen uptake rate (hr^{-1})
Q_{O2}	Oxygen uptake rate (Kg/(m^3.hr))
r	Rate of dissociation (Kg/(m^3.hr))
T	Temperature (K)
ts	Stirring duration (s)
$\underline{u}$	Superficial velocity (m/s)
U_1	Over all heat transfer coefficient (broth-coolant) (W/(m^2 K))
U_2	Over all heat transfer coefficient (coolant- surrounding air) (W/(m^2 K))
V	Fermentor volume (m^3)
V_w	Volume of cooling jacket (m^3)
Y_{XZ}	Yield coefficient for compound Z on compound X

Greek letters

μ	Specific growth rate (hr^{-1})
ρ	Density (kg/m^3)

Subscripts

ag	Agitation
b	Broth
CO_2	Carbon dioxide
CO_3^{--}	Carbonate ion
evp	Evaporation
gr	Generation
H^+	Hydrogen ion
H_3PO_4	Phosphoric acid
$H_2PO_4^-$	Dihydrogen phosphate ion

HPO_4^{--}	Hydrogen phosphate ion
H_2CO_3	Carbonic acid
HCO_3^-	Bicarbonate ion
H_2O	Water
H_2SO_4	Sulphuric acid
HSO_4^-	Hydrogen sulphate ion
H_2CO_3	Carbonic acid
max	Maximum
MT	Mixing tank
NH_3	Ammonia
NH_4^+	Ammonium ion
O_2	Oxygen
OH^-	Hydroxyl ion
PO_4^{---}	Phosphate ion
S	Substrate
sen	Sensible
ster	Steriliser
SO_4^{--}	Sulphate ion
surr	Surrounding
W	Water (coolant)
w_{in}	Coolant inlet
X	Biomass

Superscripts

*	Saturation
0	Steady-state value

Abbreviations

DO	Dissolved oxygen
DOF	Degree of freedom
ICAS	Integrated computer aided system
MoT	Modelling test bed
rpm	Revolution per minute

12.3. MILK PASTEURISATION PROCESS MODELLING

Pasteurisation process is commonly used in the food industry involving the milk product such as cheese, butter and curd manufacturing. A mathematical model of milk pasteurisation process is presented here. This pasteurisation system is used in the cheese manufacturing process. The model objective is to study the process behaviour. This model is analysed systematically to characterise the model equations and model variables. Based on the model analysis, an incidence matrix is prepared to identify the solution strategy and subsequently the model is implemented into a simulation tool. The detail description of the pasteurisation process model is available in the literature (Singh et al., 2009).

12.3.1. Process (System) Description

The schematic diagram of a continuous pasteurisation is shown in Figure 10 (adapted from Armenante & Leskowicz, 1990). The whole process can be divided into 4 sections - preheating section, heating section, holding section and cooling section. In the preheating section the heat content in the hot milk is utilised to heat the cold milk. The preheated milk is then heated up to its pasteurisation temperature in the heating section and is kept at this temperature as it flows in the holding section where pasteurisation takes place. The pasteurised milk is then passed through the pre-heater and then cooled to its final temperature in a cooler. This milk is then fed to the cheese vat. Counter current double pipe heat exchangers are used for heat exchange.

12.3.2. Model Description

Assumptions

The milk is fed to the pasteuriser continuously at a fixed rate. The inlet flow rate and the inlet temperature of the heating/cooling fluid are assumed to be constant. Heat capacity and density of milk are also constant (independent of temperature). No loss of heat to the surroundings (perfect insulation). The process is simulated in open loop conditions. The process model equations are given below.

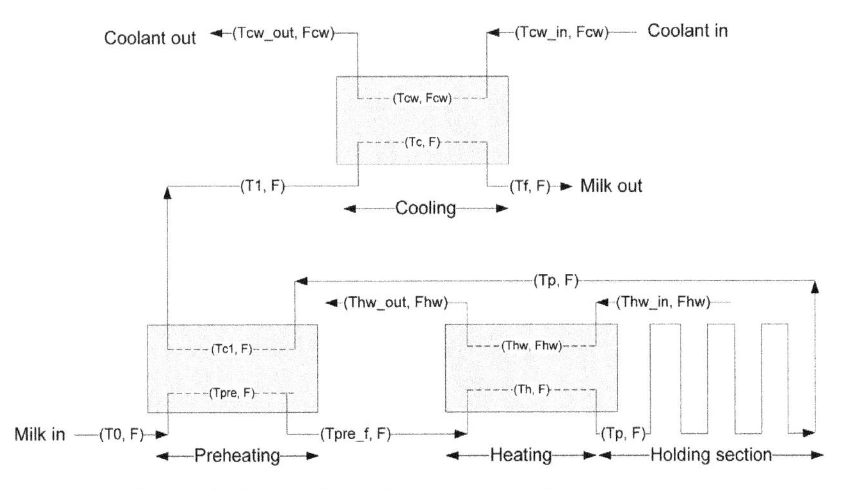

FIGURE 10 Schematic diagram of a continuous pasteurisation system. (adapted from Armenante & Leskowicz, 1990).

Steady state mass and energy balance equations

The mass flow rate of milk to the pasteuriser is calculated as follows:

$$F = \rho_{\text{milk}} f \tag{1}$$

The mass fraction of living microorganisms in the milk is calculated as follows:

$$X_{\text{organism}} = \frac{NM_{0\text{organism}}}{\rho_{\text{milk}}} \tag{2}$$

The mass of living microorganisms in the milk is calculated as follows:

$$F_{\text{organism}} = FX_{\text{organism}} \tag{3}$$

The mass of minerals in the milk is calculated as follows:

$$F_{minerals} = FX_{minerals} \tag{4}$$

The mass of whey protein in the milk is calculated as follows:

$$F_{\text{whey}_p} = FX_{\text{whey}_p} \tag{5}$$

The mass of casein protein in the milk is calculated as follows:

$$F_{\text{case}_p} = FX_{\text{case}_p} \tag{6}$$

The mass of fat in the milk is calculated as follows:

$$F_{\text{fat}} = FX_{\text{fat}} \tag{7}$$

The mass of lactose in the milk is calculated as follows:

$$F_{\text{lactose}} = FX_{\text{lactose}} \tag{8}$$

The mass fraction of water in the milk is calculated as follows:

$$X_w = 1 - X_{\text{fat}} - X_{\text{case}_p} - X_{\text{whey}_p} - X_{\text{minerals}} - X_{\text{organism}} - X_{\text{lactose}} \tag{9}$$

The mass of water in the milk is calculated as follows:

$$F_w = FX_w \tag{10}$$

<u>*Preheating section*</u>: The temperature of the milk in the pre-heater is given as follows:

$$T_{\text{pre}} = T_0 + b_{\text{pre}}(T_1 - T_0)t \tag{11}$$

The expression for the parameter b_{pre} is a function of the system that is studied. The time, t is counted from the instant the fluid element enters the heat exchanger (Armenante & Leskowicz, 1990):

$$b_{\text{pre}} = \frac{A_{pre} U_{\text{pre}}}{S_{\text{pre}} \rho_{\text{milk}} C_{p_{\text{milk}}} l_{pre}}$$

For a double-pipe counter current heat exchanger b_{pre} becomes:

$$b_{\text{pre}} = \frac{2U_{\text{pre}}}{\rho_{\text{milk}} r_{\text{pre}} C_{p_{\text{milk}}}} \tag{12}$$

The outlet temperature of the milk from the pre-heater is calculated as follows:

$$T_{\text{pre\_}f} = T_0 + b_{\text{pre}}(T_1 - T_0)t_{\text{pre}} \tag{13}$$

The outlet temperature of the hot milk is given as follows:

$$T_1 = T_p - (T_{\text{pre\_}f} - T_0) \tag{14}$$

The holding time in the pre-heater is calculated as follows:

$$t_{\text{pre}} = \frac{A_{\text{pre}} l_{\text{pre}}}{f} \tag{15}$$

The flow area in the pre-heater is defined as follows:

$$A_{pre} = \pi r_{pre}^2 \tag{16}$$

Heating section: The temperature of the milk in the heat exchanger is given as follows:

$$T_h = T_{\text{pre\_}f} + \left(\frac{T_{\text{pre\_}f} - T_{hw\_\text{out}}}{a_h - 1}\right)[1 - \exp((a_h - 1)b_h t)] \tag{17}$$

The parameter a_h is defined as follows:

$$a_h = \frac{f \rho_{\text{milk}} C_{p_{\text{milk}}}}{f_{hw} \rho_w C_{p_w}} \tag{18}$$

The expression for the parameter b_h is a function of the specific heat exchanger system that is studied. The time, t is counted from the instant the fluid element enters the heat exchanger (Armenante & Leskowicz, 1990):

$$b_h = \frac{A_h U_h}{S_h \rho_{\text{milk}} C_{p_{\text{milk}}} l_h}$$

For a double-pipe counter current heat exchanger bh becomes:

$$b_h = \frac{2U_h}{\rho_{\text{milk}} r_h C_{p_{\text{milk}}}} \tag{19}$$

The outlet temperature of the milk from the heating section is calculated as follows:

$$T_p = T_{\text{pre}\_f} + \left(\frac{T_{\text{pre}\_f} - T_{hw\_\text{out}}}{a_h - 1}\right)\beta_h \tag{20}$$

The outlet temperature of the heating fluid is calculated as follows

$$T_{hw\_\text{out}} = \frac{c_h T_{\text{pre}\_f} - T_{hw\_\text{in}}}{c_h - 1} \tag{21}$$

where c_h is defined as follows:

$$c_h = \frac{a_h \beta_h}{a_h - 1} \tag{22}$$

$$\beta_h = 1 - \exp((a_h - 1)b_h t_h) \tag{23}$$

The residence time (t_h) of milk in the heater is defined as follows:

$$t_h = \frac{A_h l_h}{f} \tag{24}$$

The flow area in the heater is defined as follows:

$$A_h = \pi r_h^2 \tag{25}$$

Holding section: The temperature of milk in the holding section is assumed to be constant and it will be equal to the outlet temperature from the heater. The holding time is calculated as follows:

$$t_{\text{hold}} = \frac{A_{\text{hold}} l_{\text{hold}}}{f} \tag{26}$$

The flow area of the pipe in the holding section is defined as follows:

$$A_{\text{hold}} = \pi r_{\text{hold}}^2 \tag{27}$$

Cooling section: The temperature of the milk in the heat exchanger is given as follows:

$$T_c = T_1 + \left(\frac{T_1 - T_{cw\_\text{out}}}{a_c - 1}\right)[1 - \exp((a_c - 1)b_c t)] \tag{28}$$

The parameter a_c is defined as follows:

$$a_c = \frac{f \rho_{\text{milk}} C_{p_{\text{milk}}}}{f_{cw} \rho_w C_{p_w}} \tag{29}$$

b_c is the function of the system. The time, t is counted from the instant the fluid element enters the heat exchanger (Armenante & Leskowicz, 1990):

$$b_c = \frac{A_c U_c}{S_c \rho_{\text{milk}} C_{p_{\text{milk}}} l_c}$$

For a double-pipe counter current heat exchanger b_c becomes:

$$b_c = \frac{2U_c}{\rho_{\text{milk}} r_c C_{p_{\text{milk}}}} \tag{30}$$

The outlet temperature of the milk from the cooler is calculated as follows:

$$T_f = T_1 + \left(\frac{T_1 - T_{cw\_\text{out}}}{a_c - 1}\right) \beta_c \tag{31}$$

The outlet temperature of the cooling liquid is calculated as follows

$$T_{cw\_\text{out}} = \frac{c_c T_1 - T_{cw\_\text{in}}}{c_c - 1} \tag{32}$$

where c_c is defined as follows:

$$c_c = \frac{a_c \beta_c}{a_c - 1} \tag{33}$$

$$\beta_c = 1 - \exp((a_c - 1) b_c t_c) \tag{34}$$

The residence time (t_c) of milk in the cooler is defined as follows:

$$t_c = \frac{A_c l_c}{f} \tag{35}$$

The flow area in the cooler is defined as follows:

$$A_c = \pi r_c^2 \tag{36}$$

The temperature of the milk throughout the different stages of the process is defined as follows:

$$T = \begin{cases} T_{\text{pre}};\ \text{if } t \le t_{\text{pre}} \\ T_h;\ \text{if } t_{\text{pre}} < t \le (t_{\text{pre}} + t_h) \\ T_p;\ \text{if } (t_{\text{pre}} + t_h) < t \le (t_{\text{pre}} + t_h + t_{\text{hold}}) \\ T_{c1};\ \text{if } (t_{\text{pre}} + t_h + t_{\text{hold}}) < t \le (t_{\text{pre}} + t_h + t_{\text{hold}} + t_{c1}) \\ T_c;\ \text{if } (t_{\text{pre}} + t_h + t_{\text{hold}} + t_{c1}) < t \le (t_{\text{pre}} + t_h + t_{\text{hold}} + t_{c1} + t_c) \\ T_f;\ \text{if } t > (t_{\text{pre}} + t_h + t_{\text{hold}} + t_{c1} + t_c) \end{cases} \tag{37}$$

The holding time of hot fluid in the pre-heater is defined as follows:

$$t_{c1} = \frac{A_{c1} l_{\text{pre}}}{f} \tag{38}$$

The flow area is defined as follows:

$$A_{c1} = \pi\left(r_{c1}^2 - r_{\text{pre}}^2\right) \tag{39}$$

Constitutive equations

The rate of thermal death for the contaminating organisms is assumed to follow first-order kinetics where the specific reaction kinetic rate, K_{d}, is a function of the Arrhenius constant K_{dO}, the activation energy E_{d}, and the pasteurisation temperature T, according to the following equation (Armenante & Leskowicz, 1990):

$$K_d = K_{do}\exp\left(-\frac{E_d}{R^*T}\right) \tag{40}$$

For a generic pasteurisation process, a balance for the final number of surviving microorganisms per unit volume of fluid, N, can be written as:

$$\ln\left(\frac{N}{N_0}\right) = -\int_0^t K_d dt \tag{41}$$

12.3.3. Model Analysis

The total number of equations in this case is 41. The total number of variables involved in the model is 74 (see Table 10). Based on the total number of variables and the total number of equations the degrees of freedom is calculated to be 33. The detailed classification of the variables is given in Table 10. Based on the classification of variables an incidence matrix between the number of equations and the number of unknown variables is prepared as shown in Table 11. Equations 13, 14, 20, 21, 31 and 32 are implicit because there are crosses on the upper triangular (see shaded equations in incidence matrix), and therefore these equations need to be solved iteratively.

TABLE 10 Classification of Variables of the PasteurisationSterilisation Process Model

Variable types		Variables	No. of variables	Total no
To be specified	Constant	π, R, Cpw, ρ_w	4	33
	Fixed by problem	K_{do}, E_d, N_0, T_0, ρ_{milk}, $C_{p_{milk}}$, f, f_{hw}, f_{cw}, T_{hw_in}, T_{cw_in}, U_{pre}, U_h, U_c, $M_{0\_organism}$, $X_{minerals}$, X_{whey_p}, X_{case_p}, X_{fat}, $X_{lactose}$,	20	
	Fixed by system	l_{pre}, l_h, l_{hold}, l_c, r_{pre}, r_h, r_{hold}, r_{c1}, r_c	9	
To be predicted	Explicit	A_{pre}, t_{pre}, A_h, t_h, A_{hold}, t_{hold}, A_{c1}, t_{c1}, A_c, t_c, b_{pre}, a_h, b_h, β_h, c_h, a_c, b_c, β_c, c_c, T, K_d, N, T_{pre}, T_h, T_c, F, $X_{organism}$, $F_{organism}$, $F_{minerals}$, F_{whey_p}, F_{case_p}, F_{fat}, $F_{lactose}$, F_w, X_w	35	41
	Implicit	T_{pre_f}, T_1, T_p, T_{hw_out}, T_f, T_{cw_out}	6	
Total number of variables involved in the model				74
Total number of constitutive equations (explicit)				2
Total number of balance algebraic equations (explicit: 33, implicit: 6)				39

TABLE 11 Incidence Matrix of PasteurisationSterilisation Process Model

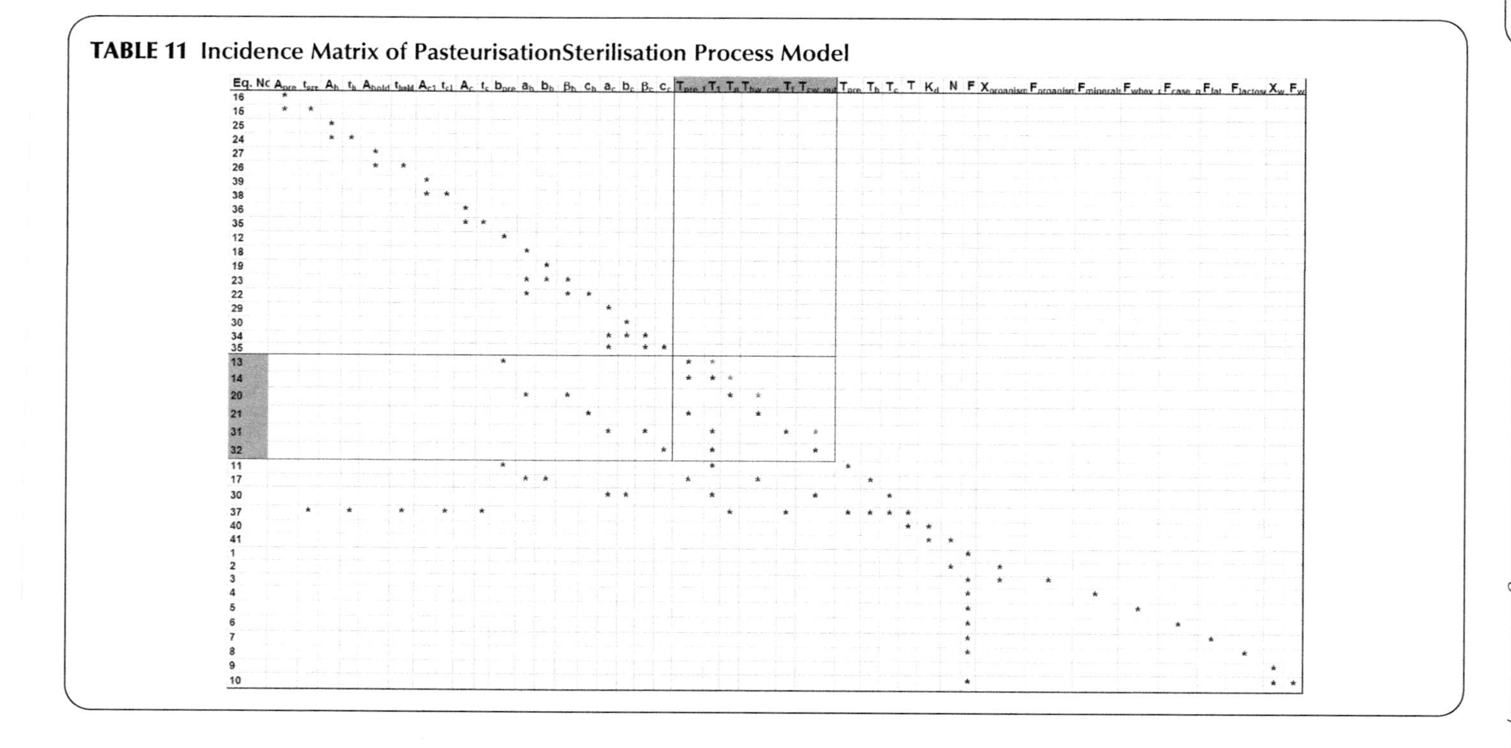

12.3.4. Model solution

The pasteurisation process model is simulated in ICAS-MoT (Sales-Cruz, 2006). The main stream (milk) temperature profile is shown in Figure 11. The temperature profile of the milk in different sections of the pasteurisation process is also shown in this figure. As seen in the figure, the milk is first pre-heated in the preheating section then the milk temperature is raised to the pasteurisation temperature in the main heating section and the milk is hold at pasteurisation temperature in the holding section. The pasteurised milk is then pre-cooled in the pre-cooling section and at the end cooled to a final temperature in the cooling section.

VALUE OF KNOWN VARIABLES AND PARAMETERS

Known variables	Symbol	Value
Length of heat exchanger in preheating section (m)	L_{pre}	2
Length of heat exchanger in cooling section (m)	l_c	2
Length of heat exchanger in heating section (m)	l_h	2
Length of holding pipe (m)	l_{hold}	1
Radius of inner pipe of heat exchanger in preheating section (m)	r_{pre}	0.6
Radius of outer pipe of heat exchanger in preheating section (m)	r_{c1}	1
Radius of inner pipe of heat exchanger in cooling section (m)	r_c	0.6
Radius of inner pipe of heat exchanger in heating section (m)	r_h	0.6
Radius of holding pipe (m)	r_{hold}	0.6
Heat capacity of milk (J/kg K)	$C_{p_{milk}}$	3770
Activation energy (J/mol)	Ed	278924
Milk flow rate (m^3/hr)	f	10
Hot water flow rate (m^3/hr)	f_{hw}	12
Coolant flow rate (m^3/hr)	f_{cw}	11
Arrhenius type constant	K_{d0}	1E44
Initial number of microorganism (no./m^3)	N_0	1E10
Milk temperature in feed (°C)	T_0	4
Hot water inlet temperature (°C)	T_{hw_in}	70
Cold water inlet temperature (°C)	T_{cw_in}	10
Overall heat transfer coefficient in cooler (J/hr m^2 K)	U_c	1.8E07
Overall heat transfer coefficient in cooler (J/hr m^2 K)	U_h	1.8E07
Overall heat transfer coefficient in cooler (J/hr m^2 K)	U_{pre}	1.8E07
Density of milk (kg/m^3)	ρ_{milk}	1030

Parameters	Symbol	Value
Pi	π	3.14
Heat capacity of water (J/kg K)	C_{pw}	4186
Universal gas constant (J/mol K)	R	8.314
Density of water (kg/m^3)	ρ_w	1000

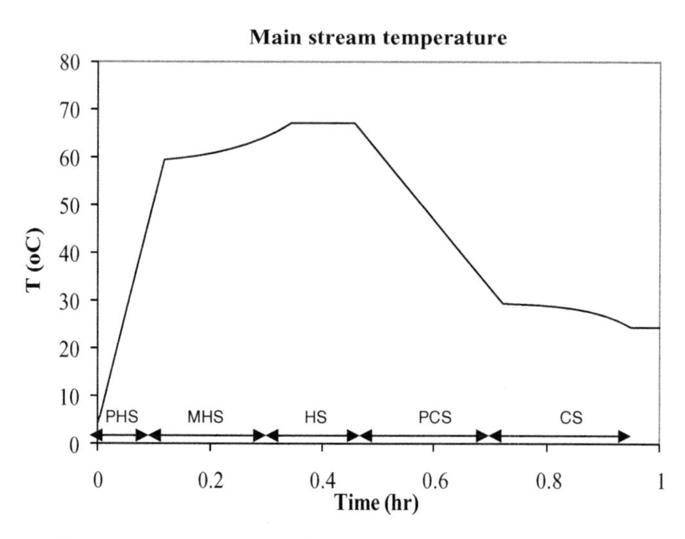

FIGURE 11 Simulated main stream (milk) temperature in open loop (PHS = preheating section; MHS = main heating section; HS = holding section; PCS = pre-cooling section; CS = cooling section).

Nomenclature (for section 12.3)

A	Area
C_p	Heat capacity
E_d	Activation energy
F	Volumetric flow rate
K_d	Rate constant
K_{d0}	Arrhenius type constant
L	Length
N	Number of microorganism
R	Radius
T	Time
U	Overall heat transfer coefficient

Greek letters

ρ	Density

Subscripts

0	Initial
H	Heating
Pre	Preheating
P	Pasteurisation

Hold	Holding
C	Cooling
F	Final
W	Water (coolant)

12.4. MILLING PROCESS MODEL

The pharmaceutical tablet manufacturing process involves many unit operations such as mixing, milling, granulation, tablet storage, tablet pressing and tablet coating. Singh et al. (2010) have proposed mathematical models for these unit operations. One of them, the milling process model is presented here.

12.4.1. Process (System) Description

A batch wet milling machine is shown in Figure 12. The feed is introduced batch wise in the machine and after the milling time the size of the particles is reduced to nanoscale.

Assumptions

- The particle size reduction due to attrition is negligible
- The spatial distribution of particle size is neglected

The model objective is to study the process behaviour under different operational scenarios. This model is analysed systematically to characterise the model equations and model variables. Based on the model analysis, an incidence matrix is generated to identify the appropriate solution strategy and subsequently the model is implemented into a simulation tool.

12.4.2. Model Description

The milling process model is described as follows:

Differential equations

By performing a rate–mass balance on material in each size interval i present at time t in a mill, the mass fraction of particles of size x_i is given

FIGURE 12 Milling process.

as follows (Tangsathitkulchai, 2002):

$$\frac{dw_i}{dt} = -S_i w_i + \sum_{j=1}^{i-1} b_{ij} S_j w_j; \quad \mathrm{n} \geq \mathrm{i} \geq \mathrm{j} \tag{1}$$

Algebraic equations (explicit)

Mass of one particle in the size interval i is given as follows (Hogg, 1999):

$$M_{p\_i} = \rho K_v x_i^3 \tag{2}$$

The mass of solids in the feed is calculated as follows:

$$M_{s\_F} = X_s M_F \tag{3}$$

The total mass of the particles in the size interval i is given as follows:

$$M_i = X_i M_{s\_F} \tag{4}$$

The number of particles in the size interval i is calculated as follows:

$$N_{p\_i} = \frac{M_i}{M_{p\_i}} \tag{5}$$

The total number of particles in the milling machine is given as follows:

$$N_{p\_\mathrm{total}} = \sum_{i=1}^{z} N_{P\_i} \tag{6}$$

The average size of the particles in the milling machine is calculated as follows:

$$L_{\mathrm{avg}} = \frac{\sum_{i=1}^{z} N_{P\_i} x_i}{N_{P\_\mathrm{total}}} \tag{7}$$

The breakage parameter b_{ij} (primary breakage distribution), is defined as below (Hogg, 1999):

$$b_{ij} = 1 - \exp\left[-\left\{ \frac{x_i}{x_{cr}} \left(\frac{x_j - x_{cr}}{x_j - x_i} \right)^{\gamma} \right\} \right] \tag{8}$$

The critical rotational speed is given as follows (McCabe et al., 2005):

$$n_{cr} = \frac{1}{2\pi} \sqrt{\frac{g}{R - r}} \tag{9}$$

The ratio between the operational speed of the mill and the critical speed of the mill (Φ_{cr}) is given as below:

$$\Phi_{cr} = \frac{n}{n_{cr}} \tag{10}$$

The parameter a depends on the rotational speed of the mill as follows (Austin & Brame, 1983):

$$a = \begin{cases} \dfrac{k(\Phi_{cr} - 0.1)}{1 + \exp[15.7(\Phi_{cr} - 0.95)]}; & \Phi_{cr} > 0.1 \\ 0; & \Phi_{cr} \leq 0.1 \end{cases} \tag{11}$$

The specific rate of breakage is calculated as (Austin & Brame, 1983; Hogg, 1999):

$$S_i = a\left(\frac{S\left(\frac{x_i}{x_1}\right)^{\alpha}}{1 + \left(\frac{x_i}{x_m}\right)^{\beta}}\right) \tag{12}$$

12.4.3. Model Analysis

The total number of equations in this case (for z size intervals) is [z (z-1)/2+5z+6]. The total number of variables involved in the model is found to be [z (z-1)/2+6z+22] (see Table 12). The degrees of freedom is found to be (z+16). The detailed classification of the variables is given in Table 12. The incidence matrix of this model has shown a lower triangular form (see Table 13), meaning that the equations of this model can be solved sequentially.

12.4.4. Model Solution

The milling process model is simulated in ICAS-MoT (Sales-Cruz, 2006). In this section selected simulation results are highlighted. During the milling process the bigger particles break down to smaller particles, therefore, the number of particles increases. Figure 13 shows the increase in the total number of the particles during the milling process. Figure 14 shows the weight fractions of the particles in different size intervals. It should be noted that the feed is assume to be the uniform in size (size interval x1).

VALUE OF KNOWN VARIABLES AND PARAMETERS

Differential variables	Symbol	Initial value
Mass fraction of particles of size x_1	X_1	1
Mass fraction of particles of size x_i; i >1	X_i	0

Known variables	Symbol	Value
Weight of feed (kg)	M_F	100
Mill rotational speed (rph)	n	300
Characteristic size (m)	x_m	1E-04
Critical size of particle (m)	x_{cr}	1E-05
Weight fraction of solid in feed	X_s	0.9
Particle density (kg/m^3)	ρ	2650

Parameters	Symbol	Value
Acceleration due to gravity (m/s^2)	g	9.18
Volume shape factor	K_v	0.5
Coefficient	α	0.5
Coefficient	β	0.5
Coefficient	γ	0.3
Coefficient	k	1

TABLE 12 Classification of Milling Process Model Variables (*z* is the Number of Size Interval Considered for the Particle Size Distribution)

Variable types		Variables	No. of variables	Total no
To be specified	Constant	π, g	2	$z + 16$
	Fixed by problem	M, n, x_m, x_c, X_s	5	
		$\underline{x} = [x_1 ... x_i .... x_z]$	z	
	Fixed by system	$\rho, K_v, \alpha, \beta, \gamma, k, S, R, r$	9	
To be predicted	Algebraic (explicit)	$a, \Phi_{cr}, N_{cr}, M_s, N_{p\_total}, L_{avg}$	6	$z(z - 1)/2 + 5z + 6$
		$\underline{b_{ij}} = [b_{11} ... b_{ij} .... b_{zz}]$	$z(z - 1)/2$	
		$\underline{S} = [S_1 ... S_i .... S_z]$	z	
		$\underline{M} = [M_1 .... M_i .... M_z]$	z	
		$\underline{M_P} = [M_{P_1} .... M_{P_i} .... M_{P_z}]$	z	
		$\underline{N_P} = [N_{P_1} .... N_{P_i} .... N_{P_z}]$	z	
	Differential	$\underline{X} = [X_1 ... X_i .... X_z]$	z	
Total number of variables involved in the model				$z(z - 1)/2 + 6z + 22$
Total number of differential equations				z
Total number of algebraic equations				$z(z - 1)/2 + 4z + 6$

TABLE 13 Incidence Matrix of Milling Process Model

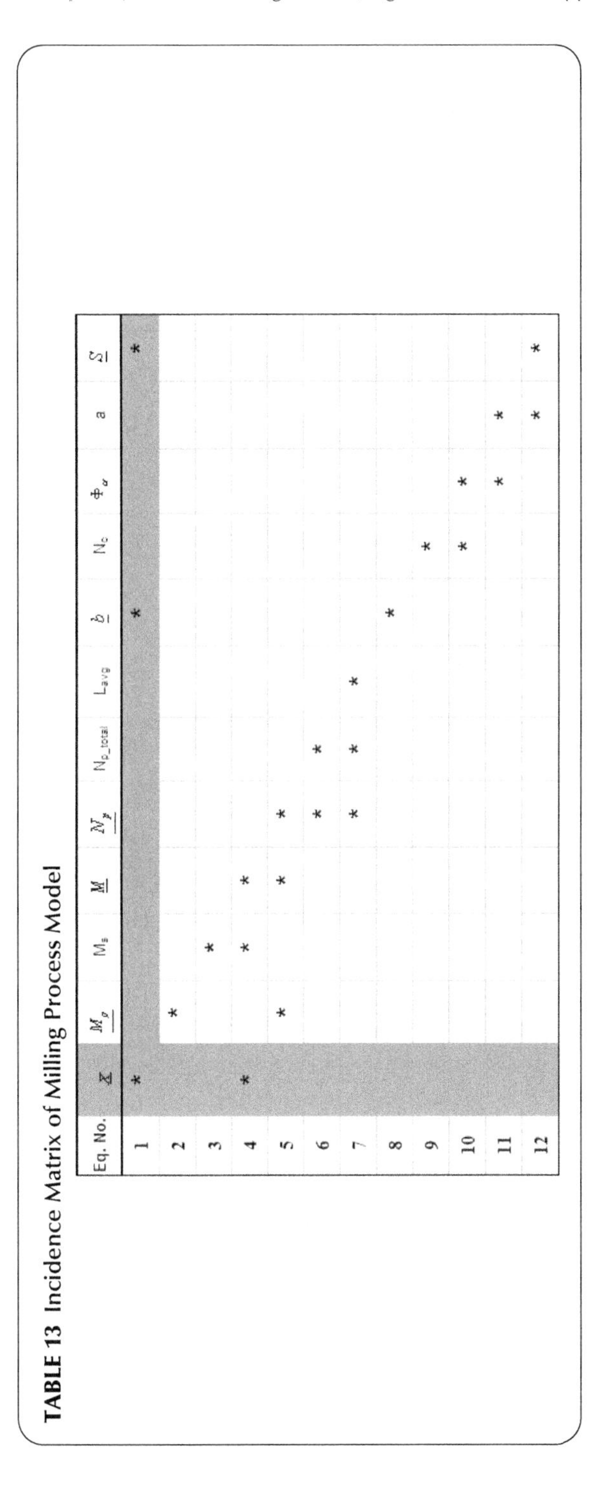

Eq. No.	$\underline{X}$	$\underline{M_g}$	M_s	$\underline{M}$	$\underline{N_p}$	N_{p_total}	L_{avg}	$\underline{b}$	N_c	Φ_α	a	$\underline{S}$
1	*							*				*
2		*										
3			*									
4	*		*	*								
5		*		*	*							
6					*	*						
7					*	*	*					
8								*				
9									*			
10									*	*		
11										*	*	
12											*	*

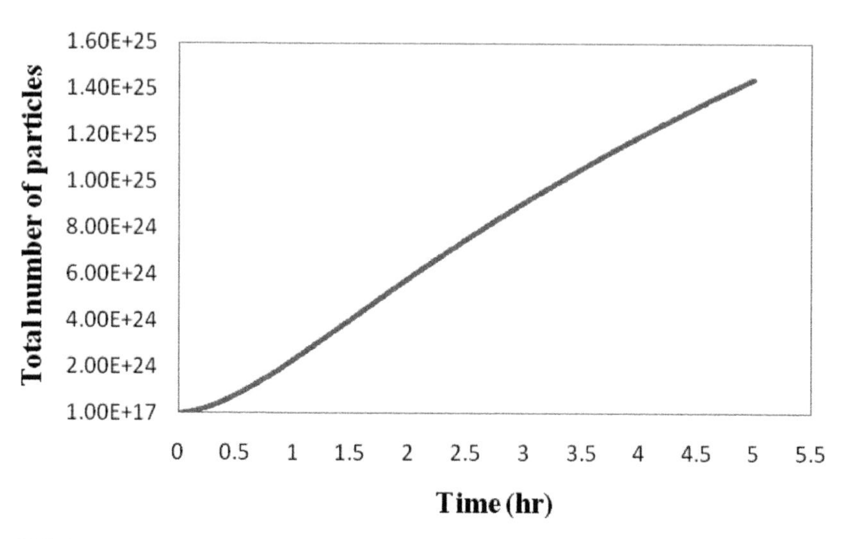

FIGURE 13 Total number of particles as a function of time.

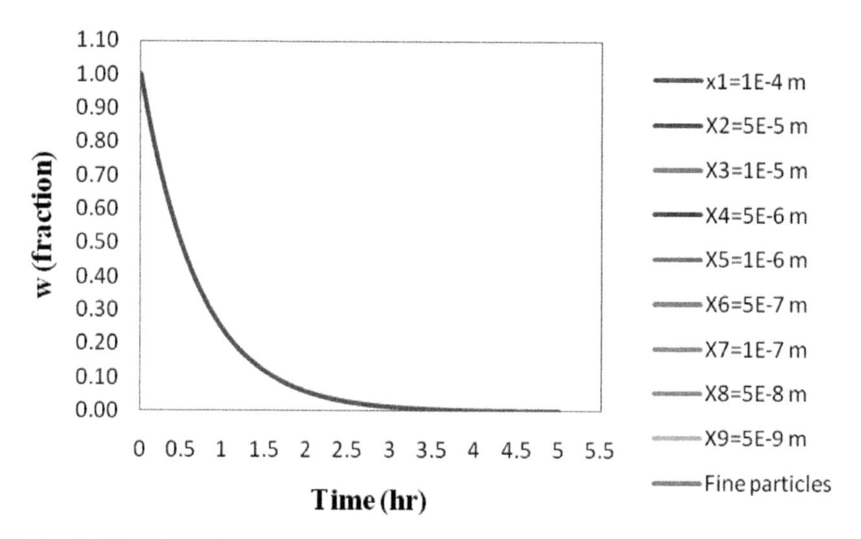

FIGURE 14 Weight fraction of particles in different size intervals.

Nomenclature (for section 12.4)

Symbol	*Name*
a	Milling speed correction parameter
$\underline{b}$	Primary breakage distribution
g	Acceleration due to gravity
K_v	Volume shape factor

k	Fitting parameter
L_{avg}	Average size of the particles
m	Mass of particles of specific size range
$\bar{m}_p$	Mass of one particle in a specific size range
$\bar{N}$	Mill rotational speed
N_c	Critical mill speed
N_p	Number of particles in a specific size range
$\bar{N}_{p\_total}$	Total number of particles in the mill
R	Mill diameter
r	Diameter of grinding media
S	Specific rate of breakage
$\bar{S}$	Characteristic rate
W	Weight of feed
W_s	Weight of solid in mill
w	Weight fraction of solid in a specific size range
$\bar{x}$	Particle size interval
x_c	Critical size of particle
x_m	Characteristic size
y_s	Weight fraction of solid in feed

Greek letters

ρ	Particle density
Φ_c	Critical velocity fraction
α, β, γ	Fitting parameters

Subscript

c	Critical
i,j	Particle size interval

12.5. GRANULATION PROCESS MODEL

The pharmaceutical tablet manufacturing process involves many unit operations such as mixing, milling, granulation, tablet storage, tablet pressing and tablet coating. Singh et al. (2010) have proposed the mathematical models of these unit operations. In this section, a granulation process model is presented.

12.5.1. Process (System) Description

The fluidised bed granulator is shown in Figure 15 (adopted from de Jong (1991)). As shown in this figure, heated air is used as the fluidisation medium. Above the powder bed, there is a nozzle by which the binder solution is atomised

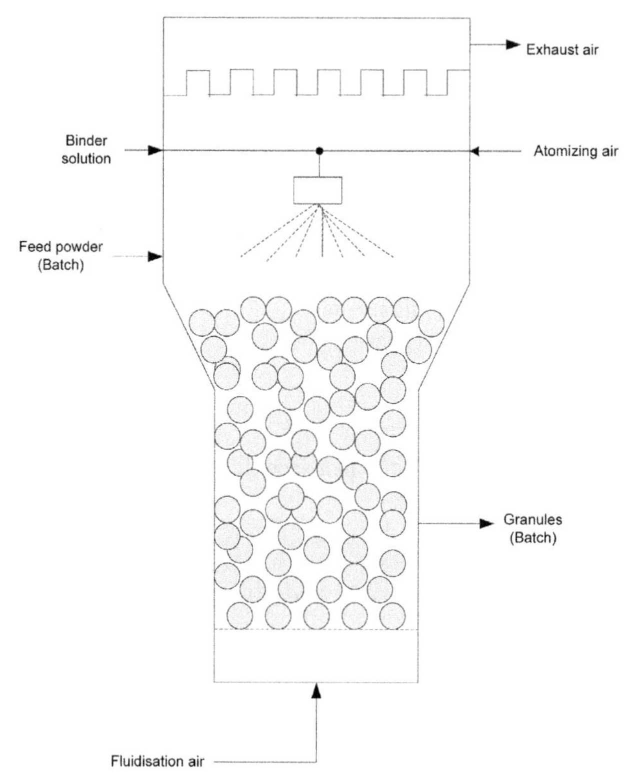

FIGURE 15 Fluidised bed granulation process.
(adapted from de Jong (1991)).

by adding air. A shaking filter is used to prevent the entrainment of the fine particles.

Assumptions

- Collision frequency and collision efficiency factors are constant (independent of agglomerate size, that is, of the extent of agglomeration)
- Uniform particle size of the feed powder
- Uniform size of granules in the granulator (size distribution is not considered)
- Powder particles and granules are spherical in shape
- Temperature is uniform in the granulator
- Constant growth rate of granules

The model objective is to study the process behaviour under different operational scenarios. This model is analysed systematically to characterise the model equations and model variables. Based on the model analysis, an incidence matrix is prepared to identify the solution strategy and subsequently the model is implemented into a simulation tool.

12.5.2. Model Description

The granulation process model is described as follows:

Differential equations

The rate of change of mass of binder spread is given by:

$$\frac{dM_b}{dt} = (1 - X_{bw})F_b \tag{1}$$

The rate of change of liquid mass in the granulator is given by:

$$\frac{dM_w}{dt} = F_{bw} - F_{ev} \tag{2}$$

The rate of change of the total mass of the gas phase is given by:

$$\frac{dM_g}{dt} = F_{a\_\mathrm{in}} - F_{a\_\mathrm{out}} + F_{ev} \tag{3}$$

The rate of change of the moisture content in the gas phase can be obtained as:

$$\frac{dM_{gw}}{dt} = F_{ev} - F_{a\_\mathrm{out}}Y_s + F_{a\_\mathrm{in}}Y_{w\_\mathrm{in}} \tag{4}$$

The rate of change of enthalpy in the granulator is calculated as follows:

$$\frac{dh}{dt} = h_{a\_\mathrm{in}} - h_{a\_\mathrm{out}} + h_b + F_{ev}\Delta H_v - Q_{en} \tag{5}$$

Algebraic equations (explicit)

The total number of initial particles in the powder is calculated as follows:

$$N_0 = \frac{6M_0}{\pi d_0^3 \rho_s} \tag{6}$$

The coagulation half-time is given as follows (Hogg, 1992):

$$\tau = \frac{2}{KN_0} \tag{7}$$

The number of granules after time t is given by (Hogg, 1992):

$$N = \frac{N_0}{1 + t/\tau} \tag{8}$$

The water fraction in the binder solution is calculated as:

$$F_{bw} = X_{bw}F_b \tag{9}$$

The total mass of liquid in the granulator is given by:

$$M_T = M_b + M_w + M_0 \tag{10}$$

The temperature in the fluidised bed is defined as:

$$T = \frac{h}{M_T C_{PT}} \tag{11}$$

The mass flow rate of air is related to the volumetric air flow rate:

$$F_{a\_in} = f_{a\_in} \rho_g \tag{12}$$

The mass flow rate of exhaust air is related to the volumetric air flow rate:

$$F_{a\_out} = f_{a\_out} \rho_{g_e x} \tag{13}$$

The rate of heat loss to the surroundings through the granulator wall is given by:

$$Q_S = UA_S(T - T_S) \tag{14}$$

The enthalpy flow associated with the binder flow is obtained as:

$$h_b = F_b C_{Pb}(T_b - T) \tag{15}$$

The enthalpy flow associated with the outlet air is given by:

$$h_{a\_out} = F_{a\_out} C_{Pa\_out}(T_{a\_out} - T) \tag{16}$$

The enthalpy flow due to the inlet air is calculated as follows:

$$h_{a\_in} = F_{a\_in} C_{Pa\_in}(T_{a\_in} - T) \tag{17}$$

The saturation pressure is calculated by means of the Antoine equation:

$$P_{sat} = B_1 - \frac{B_2}{T + B_3} \tag{18}$$

The mass fraction of moisture in a granule is calculated as follows:

$$X_w = \frac{M_w}{M_T} \tag{19}$$

The relation between the relative humidity and the moisture content is calculated (Peglow et al., 2007):

$$X_w = X_1 \left(\frac{\Phi c}{1 - \Phi} \right) \left(\frac{1 - (N_l + 1)\Phi^{N_l} + N\Phi^{N_l+1}}{1 + (c - 1)\Phi - c\Phi^{N_l+1}} \right)$$

Assuming $N_l = 8$, $c = 10$, $X_l = 0.04$ the above equation can be reduced to a more simple form:

$$X_w = 0.04\left(\frac{10\Phi}{1-\Phi}\right)\left(\frac{1-9\Phi^8+8\Phi^9}{1+9\Phi-10\Phi^9}\right) \tag{20}$$

The equilibrium pressure is given as below:

$$P_{\text{eq}} = P_{\text{sat}}\Phi \tag{21}$$

The equilibrium moisture content can be calculated as follows (Peglow et al., 2007):

$$Y_{\text{eq}} = \frac{\tilde{M}_w}{\tilde{M}_g}\left(\frac{P_{\text{eq}}}{P-P_{\text{eq}}}\right) \tag{22}$$

The mass fraction of binder in a granule is calculated as follows:

$$X_b = \frac{M_b}{M_T} \tag{23}$$

The average density of a granule is calculated as follows:

$$\rho_p = \frac{\rho_s\rho_b\rho_w}{\rho_b\rho_w(1-X_w-X_b)+\rho_s\rho_w X_b+\rho_s\rho_b X_w} \tag{24}$$

The volume of a granule is calculated as follows

$$V_P = \frac{M_T}{N\rho_p} \tag{25}$$

The diameter of a granule is obtained as follows

$$d_p = \frac{6(V_P)^{1/3}}{\pi} \tag{26}$$

The total surface area of all granules can then be calculated:

$$A_{\text{bed}} = \pi d_p^2 N \tag{27}$$

The moisture content in the gas phase is calculated as:

$$Y_s = \frac{M_{gw}}{M_g} \tag{28}$$

The rate of evaporation from the granule surface is defined as (Peglow et al., 2007):

$$F_{ev} = \rho_g\beta_{pg}A_{\text{bed}}\left(Y_{eq}-Y_s\right)\dot{\nu} \tag{29}$$

TABLE 14 Classification of Granulation Process Model Variables

Variable types		Variables	No. of variables	Total no
To be specified	Constant	$\pi, \rho_w, \Delta H_v, \tilde{M}_w$	4	33
	Fixed by problem	$M_0, d_0, \rho_s, K, \rho_g, \beta_{pg}, X_{bw}, F_b, \rho_b, P, \dot{v}, f_{a\_in}, f_{a\_out}, \rho_{g\_ex}, T_{a\_in}, T_{a\_out}, T_b, Y_{w\_in}, \tilde{M}_g$	19	
	Fixed by system	$U, A_S, T_S, B_1, B_2, B_3, C_{PT}, C_{Pa\_in}, C_{Pa\_out}, C_{pb}$	10	
To be predicted	Explicit	$N, N_0, \tau, F_{bw}, A_{bed}, Y_{eq}, d_p, V_P, M_T, \rho_p, X_w, X_b, P_{eq}, P_{sat}, T, F_{a\_in}, F_{a\_out}, h_{a\_in}, h_{a\_out}, h_b, Q_{en}, Y_s, F_{ev}$	23	29
	Implicit	Φ	1	
	Differential	M_b, M_w, M_g, M_{gw}, h	5	
Total number of variables involved in the model				62
Total number of differential equations				5
Total number of algebraic equations				24

12.5.3. Model Analysis

The total number of equations in this case is 29. The total number of variables involved in the model is found to be 62 (see Table 14). The degrees of freedom is found to be 33. The detailed classification of the variables is given in Table 14. The incidence matrix of this model has shown a lower triangular form (see Table 15) meaning that the model equations can be solved sequentially.

12.5.4. Model Solution

The granulation process model is simulated in ICAS-MoT (Sales-Cruz, 2006). Selected simulation results are given here. Figure 16 shows the temperature profile during the granulation process. It should be noted that the binder flow rate and the inlet fluidisation air temperature are assumed to be constant. A very small variation in granulator temperature can be seen in Figure 16. The granule size increases during the granulation process that can be seen in Figure 17. It should be noted that the granule shape is assumed to be spherical.

VALUE OF KNOWN VARIABLES AND PARAMETERS

Differential variables	Symbol	Initial value
Mass of binder in granulator (kg)	M_b	0
Water contents in solid phase (kg)	M_w	2
Total mass of gas phase (kg)	M_g	5
Water contents in gas phase (kg)	M_{gw}	1E-10
Enthalpy (J)	h	83740000

Known variables	Symbol	Value
Weight of feed powder (kg)	M_0	1000
Size of powder particles (m)	d_0	1E-09
True density of the powder particles (kg/m^3)	ρ_s	2650
Air density (kg/m^3)	ρ_g	1.2
Mass fraction of water in binder	X_{bw}	0.3
Binder flow rate (kg/hr)	F_b	5
Binder density (kg/m^3)	ρ_b	1500
Pressure (atm)	P	1
Inlet air flow rate (m^3/hr)	f_{a_in}	10
Outlet air flow rate (m^3/hr)	f_{a_out}	9.5
Exhaust air density (kg/m^3)	ρ_{g_ex}	1.2
Inlet air temperature (°C)	T_{a_in}	40
Outlet air temperature (°C)	T_{a_out}	20
Binder temperature (°C)	T_b	18
Mass fraction of water in inlet air	Y_{w_in}	0
Over all heat transfer coefficient (J/ (hr m^2 K))	U	90000
Heat transfer area between granulator wall and surrounding (m)	A_S	25
Surrounding temperature (°C)	T_S	21.3

Parameters	Symbol	Value
Effective coagulation rate constant (hr^{-1})	K	1E-26
Heat capacity of bed contents (J/(kg K))	C_{PT}	5000
Heat capacity of binder (J/(kg K))	C_{pb}	4200
Heat capacity of inlet air (J/(kg K))	C_{Pa_in}	1003.5
Heat capacity of outlet air (J/(kg K))	C_{Pa_out}	1003.5
Antoine equation constant (atm)	B_1	11.67
Antoine equation constant (atm K)	B_2	3816.44
Antoine equation constant (K)	B_3	−46.13
Water density (kg/m^3)	ρ_w	1000
Mass transfer coefficient (m/s)	β_{pg}	0.1
Normalised drying rate	$\dot{\nu}$	1
Heat of vaporisation (J/kg)	ΔH_v	2500000
Molecular weight of water	$\tilde{M}_w$	18
Molecular weight of air	$\tilde{M}_g$	29

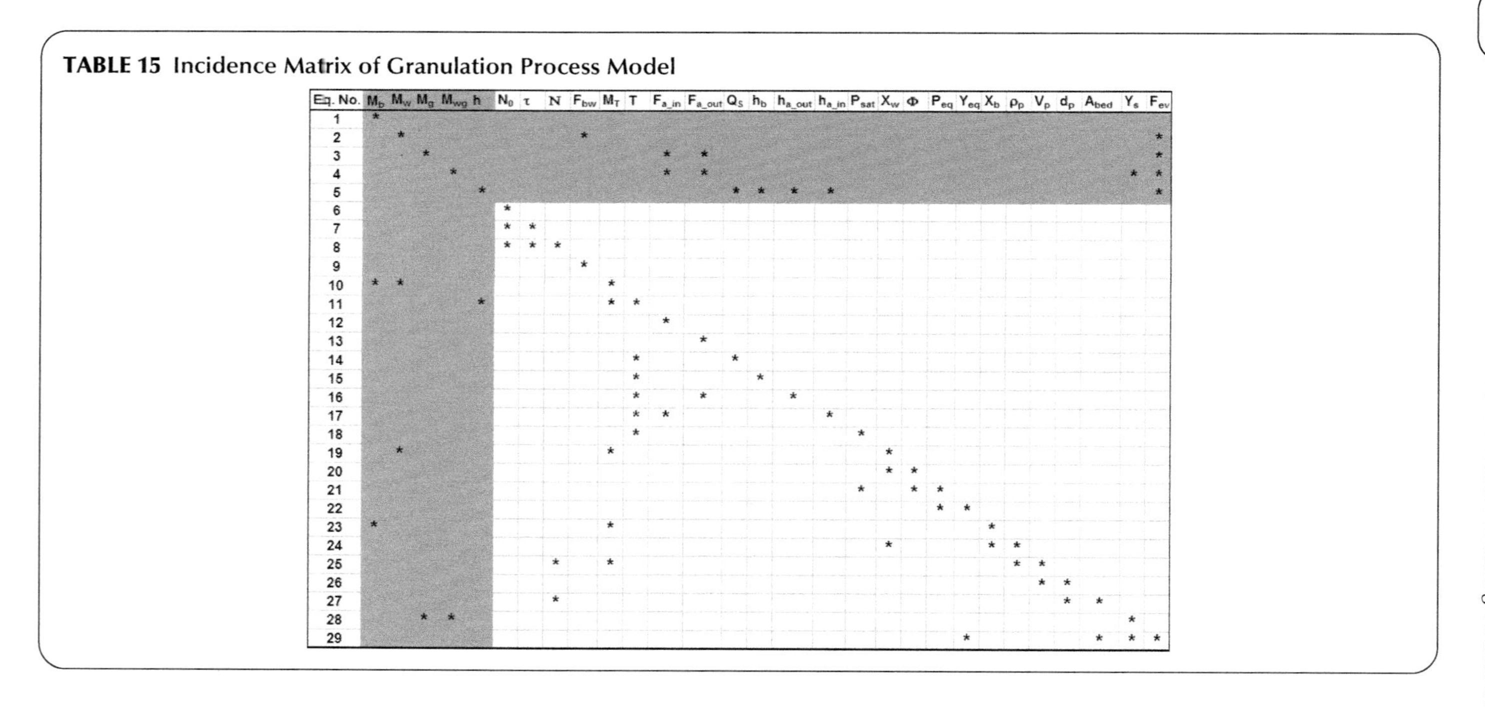

TABLE 15 Incidence Matrix of Granulation Process Model

Eq. No.	M_b	M_w	M_g	M_{wg}	h	N_0	τ	N	F_{bw}	M_T	T	F_{a_in}	F_{a_out}	Q_S	h_b	h_{a_out}	h_{a_in}	P_{sat}	X_w	Φ	P_{eq}	Y_{eq}	X_b	ρ_p	V_p	d_p	A_{bed}	Y_s	F_{ev}
1	*																												
2		*							*																				*
3			*									*	*																*
4				*								*	*															*	*
5					*									*	*	*	*												*
6						*																							
7						*	*																						
8						*	*	*																					
9									*																				
10	*	*								*																			
11					*					*	*																		
12												*																	
13													*																
14											*			*															
15											*				*														
16											*		*			*													
17											*	*					*												
18											*							*											
19		*								*									*										
20																			*	*									
21																		*		*	*								
22																					*	*							
23	*									*													*						
24																			*				*	*					
25								*		*														*	*				
26																									*	*			
27								*																		*	*		
28			*	*																								*	
29																						*					*	*	*

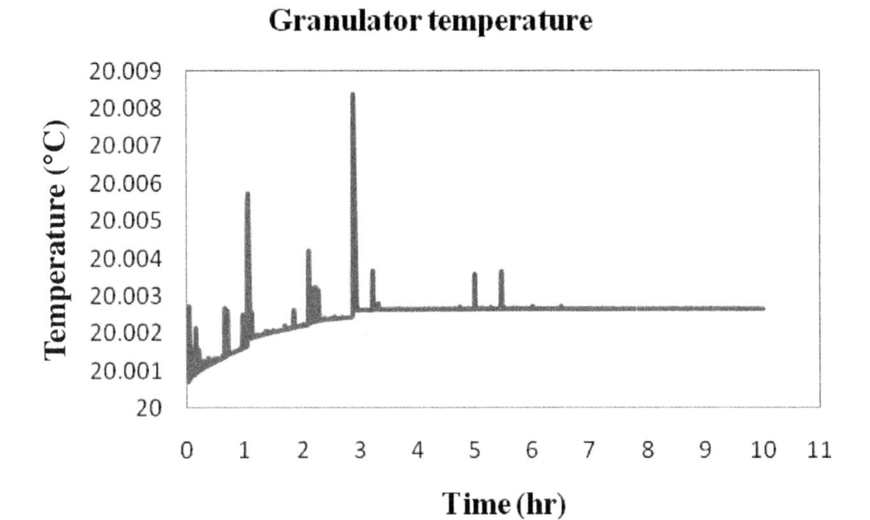

FIGURE 16 Granulator temperature as a function of time.

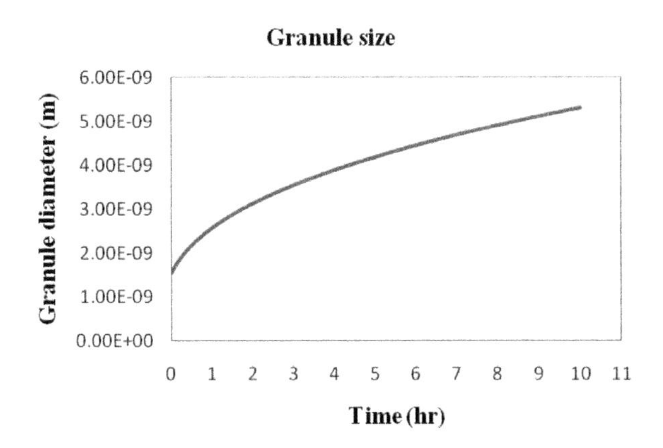

FIGURE 17 Granule size (diameter) as a function of time.

Nomenclature (for section 12.5)

A_{bed}	Particles surface area (m^2)
A_{en}	Heat transfer area between granulator and the environment (m^2)
A, B, C	Parameters in the Antoine equation
C_{p}	Specific heat capacity (J/kg K)
d	Diameter (m)
F	Mass flow rate (kg/hr)
f	Volumetric flow rate (m^3/hr)

h	Enthalpy (J)
K	Effective coagulation rate constant
M	Mass (kg)
$\tilde{M}$	Molecular weight (g mol)
N	Number of particles
P	Pressure (atm)
Q	Rate of heat exchange (J/hr)
T	Temperature (K)
t	Time (hr)

Greek letters

τ	Coagulation half-time (hr)
ρ	Density (kg/hr)
β_{pg}	Mass transfer coefficient (m/hr)
$\dot{\nu}$	Specific drying rate (per hr)

Subscript

0	Initial
p	Particle
T	Total
w	Water
bw	Water in binder
ev	Evaporation
b	Binder
eq	Equilibrium
g	Gas
sat	Saturation
a_in	Air in
a_out	Air out
en	Environment
g_ex	Gas exhaust

12.6. PHARMACEUTICAL TABLET PRESSING PROCESS MODEL

The pharmaceutical tablet manufacturing process involves many unit operations such as mixing, milling, granulation, tablet storage, tablet pressing and tablet coating. Singh et al. (2010) have proposed mathematical models for these unit operations. The model for a tablet pressing process is presented here.

12.6.1. Process (System) Description

The schematic diagram of a cross sectional view of a tablet press is shown in Figure 18 (Venkataram et al., 1996). The powder granules are filled in the die and then appropriate pressure is applied to compress the granules to obtain the tablets. The formed tablet is then sent to the tablet coater for the coating.

The modelling objective here is to study the process behaviour under different operational scenarios. This model is analysed systematically to characterise the model equations and model variables. Based on the model analysis, an incidence matrix is generated to identify an appropriate solution strategy and subsequently the model is implemented into a simulation tool.

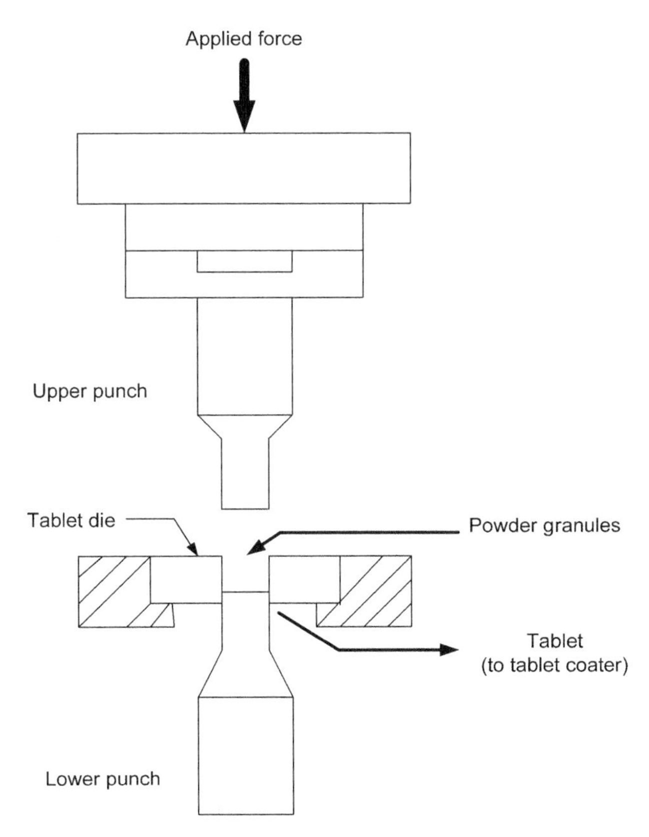

FIGURE 18 Tablet pressing process.
(adapted from Venkataram et al. (1996).

12.6.2. Model Description

Perfect control of the tablet thickness is assumed, giving thereby, a consistent tablet thickness. A disturbance in the feed volume is introduced to observe the effect of these disturbances on tablet properties. The model is simulated in open loop conditions. The tablet pressing process model is described as follows:

Differential equations

The constant tablet production rate is assumed as given below:

$$\frac{d(N_{Tab})}{dt} = r_{Tab} \tag{1}$$

Algebraic equations (explicit)

The base area of a tablet is calculated as follows, assuming that the tablet has a cylindrical shape:

$$A = \frac{3.14d^2}{4} \tag{2}$$

The volume of a tablet is calculated as follows:

$$V = AL \tag{3}$$

The pre-compression volume is given by:

$$V_{Pre} = AL_{\text{pre}} \tag{4}$$

A noise term is assumed to take into account the variation in porosity, and is implemented as given below:

$$n_{t_\varepsilon} = 0.009\left(t + t^2 - 2.5t^3 + t^4 - Sin(10t)\right) \tag{5}$$

The porosity of powder with added noise is then obtained as:

$$\varepsilon = \varepsilon_0 + n_{t_\varepsilon} \tag{6}$$

A noise term is assumed to show the variation in feed volume, as given below:

$$n_{t_F} = 3*10^{-9}\left(t - t^2 + 2t^3 - t^4 - Sin(10t)\right) \tag{7}$$

The feed volume with added noise is then:

$$V_0 = V_m + n_{t_F} \tag{8}$$

The weight of a tablet is calculated:

$$M = (1 - \varepsilon)V_0\rho \tag{9}$$

The height of the powder in the die is calculated as follows:

$$L_{\text{depth}} = \frac{V_0}{A} \tag{10}$$

The displacement of the upper punch in the compression process is calculated as follows:

$$L_{\text{punch\_displ}} = L_{\text{depth}} - L \tag{11}$$

The dwell time is calculated as follows:

$$t_{\text{dwell}} = \frac{L_{\text{punch\_displ}}}{u} \tag{12}$$

All the expressions for compression pressure (eq. 14, 18) are derived from the Kawakita compression equation (Kawakita & Lüdde, 1971). An intermediate term used for calculation of pre-compression pressure is first computed:

$$\lambda_{\text{pre}} = b\left(V_0(\varepsilon - 1) + V_{\text{pre}}\right) \tag{13}$$

The pre-compression pressure is given by:

$$C\_P_{\text{pre}} = \frac{b\left(V_0 - V_{\text{pre}}\right)}{\lambda_{pre}} \tag{14}$$

The pre-compression force is given by:

$$C\_F_{\text{pre}} = 10^6 C\_P_{\text{pre}} A \tag{15}$$

The right-hand side of the above equation is multiplied by 10^6 to adjust the unit of force.

The porosity of the powder after the pre-compression is then calculated:

$$\varepsilon_{\text{main}} = 1 - \frac{(1-\varepsilon)\mathrm{V}_0}{\mathrm{V}_{\text{pre}}} \tag{16}$$

The intermediate term used for calculation of the main compression pressure is given by:

$$\lambda_{\text{main}} = b\left(V_{\text{pre}}(\varepsilon_{\text{main}} - 1) + V\right) \tag{17}$$

The main compression pressure is given by:

$$C\_P_{\text{main}} = \frac{b\left(V_{\text{pre}} - V\right)}{\lambda_{\text{main}}} \tag{18}$$

The main compression force is obtained as:

$$C\_F_{\text{main}} = 10^6 C\_P_{\text{main}} A \tag{19}$$

The solid volume of powder is given by:

$$V_s = (1 - \varepsilon)V_0 \tag{20}$$

The relative density is defined as follows:

$$\rho_r = \frac{V_s}{V} \tag{21}$$

An intermediate term used for hardness calculation is then introduced:

$$\lambda_{\mathrm{H}} = \ln\left(\frac{1 - \rho_r}{1 - \rho_{r_c}}\right) \tag{22}$$

The hardness of a tablethardness of a tablet is calculated as (Kuentz & Leuenberger, 2000):

$$H = H_{\max}\left(1 - \exp\left(\rho_r - \rho_{r_{cr}} + \lambda_{\mathrm{H}}\right)\right) \tag{23}$$

12.6.3. Model Analysis

The total number of equations in this case is 23. The total number of variables involved in the model is found to be 35 (see Table 16). The degrees of freedom is found to be 12. The detailed classification of the variables is given in Table 16. The incidence matrix has the lower triangular form (see Table 17), so the model equations can be solved sequentially.

TABLE 16 Classification of Tablet Pressing Process Model Variables

Variable types		Variables	No. of variables	Total no
To be specified	Constant	B	1	12
	Fixed by problem	ε_0, H_{max}, L, L_{Pre}, r_{Tab}, t, u, V_m, ρ, ρ_{rcr}	10	
	Fixed by system	D	1	
To be predicted	Algebraic (explicit)	ε_0, ε_{main}, A, C_F_{main}, C_F_{Pre}, H, L_{depth}, L_{depth_displ}, M, n_t_{ε}, n_t_F, C_P_{Pre}, C_P_{main}, t_{dwell}, V, V_0, V_{Pre}, V_s, λ_H, λ_{pre}, λ_{main}, ρ_r,	22	23
	Differential	N_{Tab}	1	
Total number of variables involved in the model				35
Total number of differential equations				1
Total number of algebraic equations				22

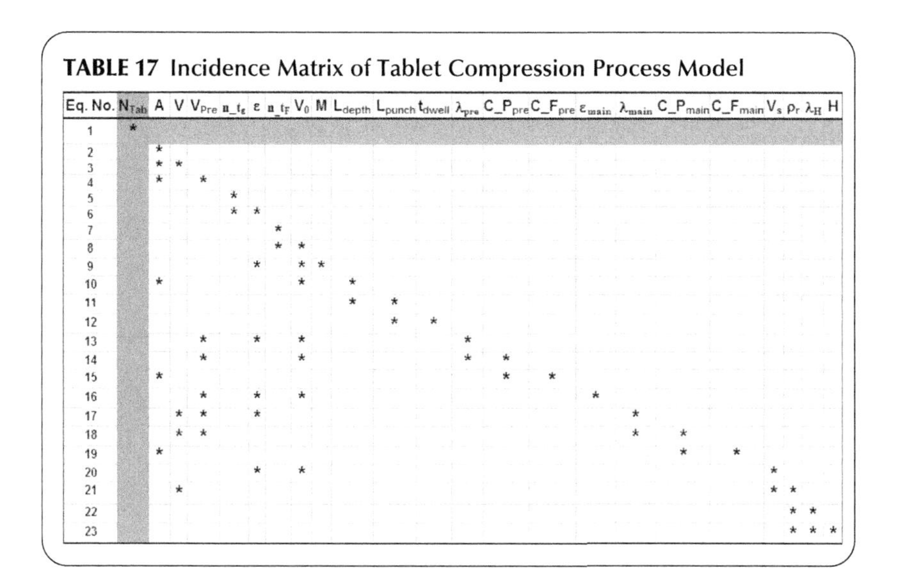

TABLE 17 Incidence Matrix of Tablet Compression Process Model

Eq. No.	N_{Tab}	A	V	V_{Pre}	n_t_g	ε	n_t_F	V_0	M	L_{depth}	L_{punch}	t_{dwell}	λ_{pre}	C_P_{pre}	C_F_{pre}	ε_{main}	λ_{main}	C_P_{main}	C_F_{main}	V_s	ρ_r	λ_H	H
1	*																						
2		*																					
3		*	*																				
4		*		*																			
5					*																		
6					*	*																	
7							*																
8							*	*															
9						*		*	*														
10		*						*		*													
11										*	*												
12											*	*											
13				*		*		*					*										
14				*				*					*	*									
15		*												*	*								
16				*		*		*								*							
17			*	*		*											*						
18			*	*													*	*					
19		*																*	*				
20						*		*												*			
21			*																	*	*		
22																					*	*	
23																					*	*	*

12.6.4. Model Solution

The tablet compression process model is simulated in ICAS-MoT (Sales-Cruz, 2006). The main disturbances in the process are due to powder porosity as well as variations in feed volume. The feed (volume of powder granules) to the press machine as a function of time is shown in Figure 19. This figure shows that there are some disturbances in the feed. It should be noted that the disturbances in the

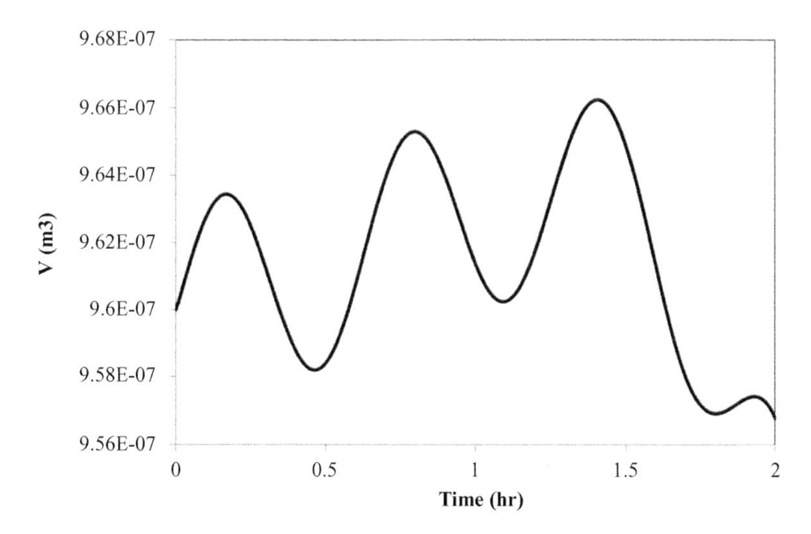

FIGURE 19 Feed to tablet press as a function of time (with disturbances).

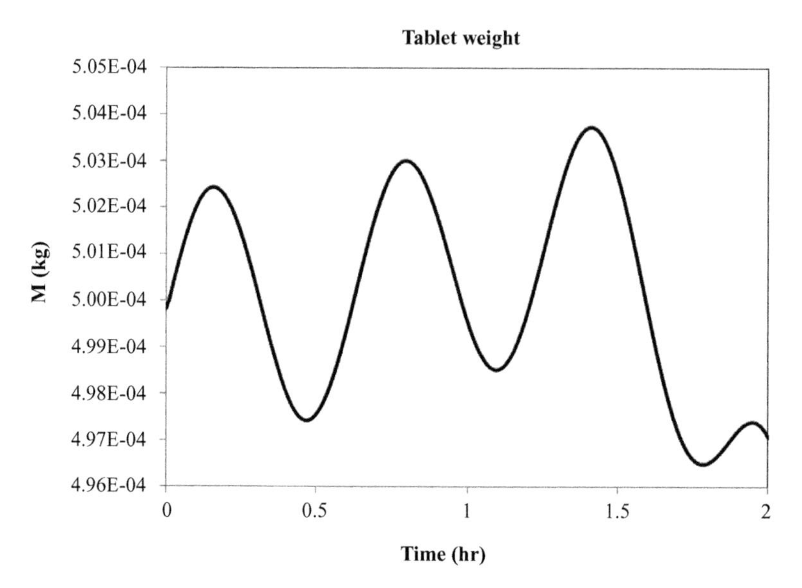

FIGURE 20 Dynamic response of the weight of a pharmaceutical tablet.

feed are very minor. The effect of process disturbances (powder porosity variations, feed volume variations) can be seen in the tablet properties. For example, dynamic responses of the tablet weight and hardness are shown in Figure 20 and 21, respectively. A minor oscillation in the tablet weight can be seen in Figure 20.

VALUE OF KNOWN VARIABLES AND PARAMETERS

Differential variables	Symbol	Initial value
Rate of production of tablet (no./hr)	N_{Tab}	0

Known variables	Symbol	Value
Powder porosity	ε_0	0.55
Tablet diameter (m)	d	0.012
Maximum tablet hardness (MPa)	H_{max}	236
Thickness of tablet (m)	L	0.004
Pre-compression thickness (m)	L_{pre}	0.007
Punch speed (m/hr)	u	2019.6
Feed volume (m^3)	V_m	9.6E-007
Powder solid density (kg/m^3)	ρ	1157
Critical relative density (kg/m^3)	$\rho_{r_{cr}}$	0.4

Parameters	Symbol	Value
Kawakita constant (MPa)	b	0.04

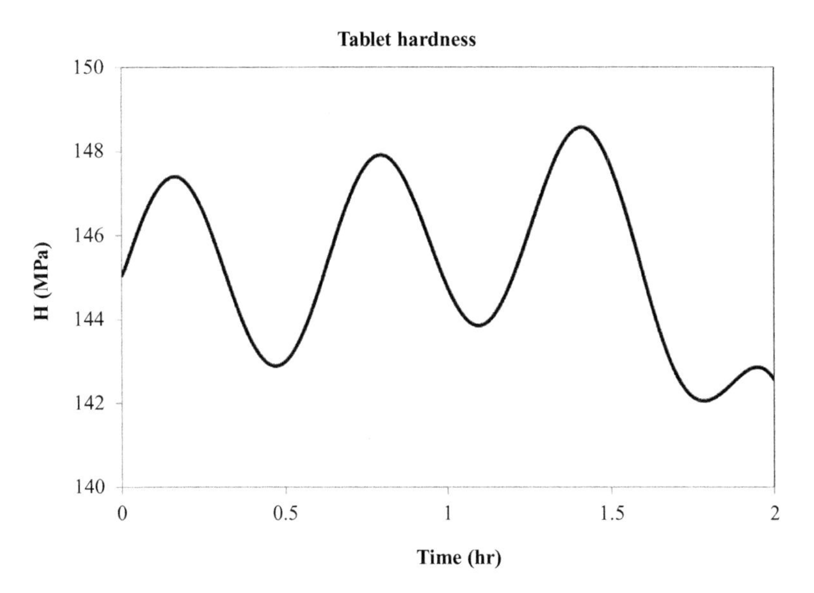

FIGURE 21 **Dynamic response of the hardness of a pharmaceutical tablet.**

Nomenclature (for section 12.6)

a	Powder porosity with noise
a_0	Powder porosity
a_{main}	Powder porosity after pre-compression
Ar	Base area (m^2)
b	Kawakita constant (MPa)
d	Diameter of tablet(m)
$error_{pre}$	Deviation of pre-compression pressure from set point
$error_H$	Deviation of tablet hardness from set point
F_{main}	Main compression force (N)
F_{Pre}	Pre-compression force (N)
F_{pre_set}	Pre-compression force set point (N)
H	Tablet hardness (MPa)
H_{max}	Maximum tablet hardness (MPa)
H_{set}	Hardness set point (MPa)
K_{C_M}	Proportional constant for tablet weight control (m3/MPa)
K_{C_H}	Proportional constant for tablet hardness control (m/MPa)
K_{I_M}	Integral constant for tablet weight control (m3/hr.MPa)
K_{I_H}	Integral constant for tablet hardness control (m/hr.MPa)
L	Thickness of tablet (m)
L_{depth}	Height of powder in die (m)
L_{depth_displ}	Displacement of punch (m)
L_0	Thickness of tablet (reference value for controller setting)(m)

L_{Pre}	Pre-compression thickness (m)
M	Tablet weight (Kg)
M_set	Tablet weight set point (kg)
n	Feed volume noise term
n_1	Porosity noise term
n_{Tab}	No. of tablet produced in time t
P_{Pre}	Pre-compression pressure (MPa)
P_{main}	Main compression pressure (MPa)
P_{pre_set}	Pre-compression pressure set point (MPa)
r_{Tab}	Rate of production of tablet (no./hr)
t	Time (hr)
t_{duell}	Dwell time (hr)
u	Punch speed (m/hr)
V	Volume of tablet
V_m	Feed volume (m^3)
V_{m_0}	Feed volume (reference value for controller setting)(m3)
V_0	Feed volume with noise (m^3)
V_{0_set}	Feed volume set point (m^3)
V_{Pre}	Pre-compression volume (m^3)
V_s	solid volume of powder (m^3)

Greek letters

ρ	Powder solid density (kg/m3)
ρ_r	Relative density (kg/m3)
ρ_{rc}	Critical relative density (kg/m3)

REFERENCES

Armenante, P.M., Leskowicz, M.A., 1990. Design of Continuous Sterilization Systems for Fermentation Media Containing Suspended Solids. Biotechnology Progress. 6, 292–306.

Angst, R., Kraume, M., 2006. Experimental investigations of stirred solid/liquid systems in three different scales: Particle distribution and power consumption. Experimental investigations of stirred solid/liquid systems in three different scales: Particle distribution and power consumption. 61, 2864–2870.

Austin, L.G., Brame, K., 1983. A Comparison of the Bond Method for Sizing Wet Tumbling Ball Mills with a Size-Mass Balance Simulation Model. Powder Technology. 34, 261–274.

Cooney, C.L., Wang, D.I.C., Mateles, R.I., 1968. Measurement of heat evolution and correlation with oxygen consumption during microbial growth. Biotechnology and Bioengineering. 11, 269–281.

De Jong, J.A.H., 1991. Tablet properties as a function of the properties of granules made in a fluidized bed process. Powder Technology. 65, 293–303.

Descoins, C., Mathlouthi, M., Le Moual, M., Hennequin, J., 2006. Carbonation monitoring of beverage in a laboratory scale unit with on-line measurement of dissolved CO2. Food Chemistry. 95, 541–553.

Doran, P.M., 2006. Bioprocess engineering principles. New York: Academic Press.

Hogg, R., 1992. Agglomeration models for process design and control. Powder Technology. 69, 69–76.

Hogg, R., 1999. Breakage mechanisms and mill performance in ultrafine grinding. Powder Technology. 105, 135–140.

Ingram, G.D., Cameron, I.T., Hangos, K.M., 2004. Classification and analysis of integrating frameworks in multiscale modeling. Chemical Engineering Science. 59, 2171–2187.

Kawakita, K., Lüdde, K.H., 1971. Some considerations on powder compression equations. Powder Technology. 4, 61–68.

Kuentz, M., Leuenberger, H., 2000. A new model for the hardness of a compacted particle system, applied to tablets of pharmaceutical polymers. Powder Technology. 111, 143–145.

Kubo, M., Harada, Y., Kawakatsu, T., Yonemoto, T., 2001. Modeling of the formation kinetics of Polyurea microcapsules with size distribution by interfacial polycondensation. Journal of Chemical Engineering of Japan. 34, 1506–1515.

Linek, V., Vacek, V., Benes, P., 1987. A critical review and experimental verification of the correct use of the dynamic method for the determination of oxygen transfer in aerated agitated vessels to water. Electrolyte solutions and viscous liquids. The Chemical Engineering Journal. 34, 11–34.

Lukaszczyk, J., Urbas, P., 1997. Influence of the parameters of encapsulation process and of the structure of diisocyanates on the release of codeine from resinate encapsulated in polyurea by interfacial water promoted polyreaction. Reactive and Functional Polymers. 33, 233–239.

McCabe, W.L., Smith, J.C., Harriott, P., 2004. Unit operations of Chemical Engineering, 7th edit. McGraw Hill.

Morales-Rodriguez, R., Gani, R., 2009. Multiscale modelling framework for chemical product-process design. Computer Aided Chemical Engineering. 26, 495–500.

Muro Suñe, N., 2005. Prediction of solubility and diffusion properties of pesticides in polymers, PhD Thesis. Lyngby, Denmark: Department of Chemical and Biochemical engineering, Technical University of Denmark.

Muro-Suñé, N., Gani, R., Bell, G., Shirley, I., 2005. Model-based computer-aided design for controlled release of pesticides. Computers & Chemical Engineering. 30, 28–41.

Musvoto, E.V., Wentzel, M.C., Loewenthal, R.E., Ekama, G.A., 2005. Integrated chemical physical processes modelling. 1. Development of a kinetic-based model for mixed weak acid/base systems. Water Research. 34, 1857–1867.

Peglow, M., Kumar, J., Heinrich, S., Warnecke, G., Tsotsas, E., Mörl, L., 2005. A generic population balance model for simultaneous agglomeration and drying in fluidized beds. Chemical Engineering Science. 62, 513–532.

Petrides, D., Sapidou, E., Calandranis, J., 2005. Computer-aided process analysis and economic evaluation for biosynthetic human insulin production—A case study. Biotechnology and Bioengineering. 48, 529–541.

Quintana-Hernàndez, P., Bolãnos-Reynoso, E., Miranda-Castro, B., Salcedo-Estrada, L., 2005. Mathematical modeling and kinetic parameter estimation in batch crystallization. AIChE Journal. 50, 1407–1417.

Sales-Cruz, M., 2005. Development of a Computer Aided Modeling System for Bio and Chemical Process and Product Design, PhD. Thesis. CAPEC, Department of Chemical Engineering, Technical University of Denmark.

Shakhashiri, B.Z., (2007)., 'Chemical of the week: Ammonia. Science is fun, in the lab of Shakhashiri.' University of Wisconsin-Madison, http://scifun.chem.wisc.edu/CHEMWEEK/Ammonia/AMMONIA.html.

Sin, G., Ödman, P., Petersen, L., Lantz, A.E., Gernaey, K.V., 2005. Matrix notation for efficient development of first-principles models within PAT applications: Integrated modeling of antibiotic production with Streptomyces coelicolor. Biotechnology and Bioengineering. 101, 153–171.

Singh, R., Gernaey, K.V., Gani, R., 2005. Model-based computer-aided framework for design of process monitoring and analysis systems. Computers & Chemical Engineering. 33, 22–42.

Singh, R., Gernaey, K.V., Gani, R., 2005. ICAS-PAT: A Software for Design. Analysis & Validation of PAT Systems. Computers & Chemical Engineering. 34, 1108–1136.

Tangsathitkulchai, C., 2005. Acceleration of particle breakage rates in wet batch ball milling. Powder Technology. 124, 67–75.

Truesdale, G.A., Downing, A.L., Lowden, G.F., 2005. The solubility of oxygen in pure water and sea water. Journal of Applied Chemistry. 5, 53–62.

Venkataram, S., Khohlokwane, M., Balakrishnan, S., Litke, A., Lepp, H., 2005. Tablet compression force measurement using strain gauges. Pharmaceutics Acta Helvetiae. 71, 329–334.

Chapter 13

Computer Aided Modelling – Opportunities and Challenges

Modelling is needed in various forms in almost all disciplines. The mathematical form of modelling requires the representation of the system being modelled by a set of equations, the solution of which helps to understand and evaluate the system. This is the form predominantly used in development of products and/or processes of different types involving different industrial sectors. Also, in research involving the development of products-processes, there is an increasing trend to use mathematical models for a wide variety of applications. However, are the models being developed and used in the most efficient manner? The answer is probably no because except for isolated examples, there is a general lack of systematic methods and tools for modelling, and, among those that are available, only a few are being used. Among the modelling tools that are being used on a regular basis are process simulators and numerical solvers. A common feature of these tools is that users select models/solvers from a library for their specific problems. While this is all right when the needed models or solvers are available, the problem comes when the library does not contain what is needed. Also, because of their nature, these tools are efficient for the problems they are designed to solve but often not very flexible to adapt to new problems.

If one would like to develop new models and/or find new uses of available models, what is the best way to do this? In this case, it is necessary to understand the modelling steps (see chapters 1-3 of this book) and identify the best way to perform them. For example, describing the modelling objectives for a specific model-based project and linking them to the system description is the task of the model developer. However, the task of representing a system by its identified mathematical form in terms of the model equations, their translation, analysis, and solution could be done efficiently via computer. That is, through computer aided modelling tools, the model developer and/or user is able to perform specific modelling tasks significantly faster and more reliably than without these tools. A key question here is which modelling tasks should be done by the human model developer or user and which modelling tasks should be done by the computer-aided model developer or user? The framework for such a distribution is given in Figure 1 of Chapter 3, where the distribution of the tasks has been made on the basis of who can do these tasks more efficiently and reliably. That is, the objective of a computer aided modelling framework is to aid and guide the human model developer or user in performing specific modelling tasks.

The current and future trend of addressing issues of sustainability, safety-hazards, reliability, flexibility and economics in the design of products and the corresponding processes to manufacture them, has made these problems multi-dimensional, multi scale, and multi-disciplinary. Therefore, opportunities exist for developing model-based systems to solve the product-process design problems of current and future interest. The challenge is not only to develop reliable, efficient and flexible models but to also develop modelling tools that can help to develop the necessary models in a systematic and efficient manner. The idea of a computer aided modelling framework becomes interesting as it can provide the human model developer/user with established work-flows, their corresponding data-flows, and the set of available modelling tools that could be used. Examples of work-flow and data-flow for single-scale modelling, multi scale modelling and model-identification are given in Figures 1-3 in Chapter 4. The challenge is to make them as generic as possible to cover a wide range of modelling challenges. That is, how to generate the required mathematical models with the least effort without sacrificing the accuracy, reliability and efficiency of the models? Also, how do we use the model application area, development time constraints and available resources to drive model creation and use? These are important industrial considerations as noted in international surveys given in Chapter 1 (Cameron & Ingram 2008, Foss et al. 1998).

A specific example of a generic model is given in Chapter 10 involving the modelling of batch crystallization processes. What this example highlights are issues such as model structure seen in the decomposition of the model, model re-use through use of reference models, model identification via data-based modelling, and model solution using appropriate simulation strategies. Each of the above issues has also been discussed in Chapter 3. The wide range of problems where potential use of a model based solution strategy could be envisaged provides the opportunity for developing computer aided modelling frameworks that incorporate the above features. The challenge is to make the framework and associated tools generic, having wide application range and providing reliable models.

The architecture of a computer aided modelling framework is very important and here knowledge representation and management play an important role. For example, how to efficiently store models in a model library so that they can be retrieved as and when needed? Clearly, success of a model re-use strategy depends very much on how the specific model information is stored and used by search engines. An example ontological representation of a specific modelling system is given in Figure 6 of Chapter 3. Much work is needed, however, in the development and use of appropriate knowledge representation and management systems in model-based solution techniques for current and future problems of product-process design. Here models of different types, forms, and dimension may be needed together with different sources and types of data requiring a suite of solution and data-analysis techniques. Specific model information normally not recorded during model development is very important here. For example, information related to model objectives and modelling assumptions. Again this is a recognised challenge in industrial model development (Cameron & Ingram 2008).

A central modelling tool within any computer aided modelling framework is a tool for model generation or creation. That is, how to convert a system description in terms of mechanisms and balance volumes to a representative set of mathematical equations? Again, opportunities exist for the further development of such systems (see Chapter 3 for more information on currently available tools), while the challenge is to make their application range cover modelling problems where effortless generation of reliable models would make a real difference in modelling practice. Such an example could be the generation of a unit operation model that is not available in any model library. A feature of any model generation and creation system should be a "plug and play" feature. That is, can partial models developed by one modelling tool be used by another tool? Specific protocols have been developed and continue to be developed in a generic way for model interoperability – for example, the CAPE-OPEN interface standards. Model generation tools, however, can only generate or create mechanistic models where the mechanisms or phenomena being applied are already developed. That is, the constitutive relations that are needed already exist. If, however, the parameters for the constitutive relations need to be regressed and/or new constitutive relations need to be developed, a model identification tool is also necessary. Chapter 6 has highlighted the use of models from existing model libraries while Chapters 7, 8, 9, 10 and 12 have discussed generation of models for specific modelling objectives.

Model reliability and application range is intimately related to the reliability and application range of constitutive models. The opportunity exists for modelling tools that can discriminate between constitutive models to find the optimal model-data match. Here, uncertainties in the measured data need to be incorporated in the model parameter regression step. The challenge here is to also make these models as predictive as possible. That is, provide a safe extrapolation of the model parameters and an uncertainty estimate of the model based predictions. Also, how do we predict the missing constitutive model parameters without the need for new experimentally measured data? Case studies in Chapters 5 and 11 have highlighted some of the issues. The wide range of potential problems requiring model-based solutions that come from the traditional chemical and petrochemical sectors as well as from the pharma, biochemical, food and biomedical sectors, to name a few, make this issue a very attractive opportunity as well as challenge for developers of computer aided modelling tools.

Another very important computer aided tool within any modelling framework is the tool for model solution. This is perhaps the area where most developments have taken place, for example, process simulators, chemical and molecular simulators, equation solvers and many more. Many of these exist as stand-alone software with little integration with external software. However, as the nature of current and future model based solution strategies imply the need for multiple models with their different sources of data, size, and the like, the tools for model solution also need to be integrated (Hamid et al. 2010, Conte et al. 2011). A typical example of a working example is the integration of external constitutive (property) models in process simulators, which has built-in equation solvers. Another example is the introduction of model equations to equation solving tools. However, the integration needs to be wider and more generic, requiring less customized effort from the model user. The main model solution tool used throughout this book is the ICAS-MoT modelling tool. It has the option to accept imported models (transfer of model equations); translate and analyze the model; characterize the model equations and variables into their different types; and finally, solve the model equations according to an appropriate solution strategy. Therefore, it also has a library of numerical solvers. The ICAS-MoT can be used for model-based simulation; for model-based product-process optimization; for model identification and for model discrimination. It has developed templates for single scale as well as multi scale models and established work-flows for simulation, optimization, and model identification. Sufficient information is given, however, for the reader to try any model solution tool of their preference. The models considered in this book as well as other models to be added in future to the model library website for this book could serve as tests for any new model solution tool. Chapter 4 gives the website address. Despite theoretical and practical advances in computer aided modelling tools there still exists significant opportunities for the development and solution of multi scale, multiform models for industrial applications.

Finally, several questions arise. With the best architecture of a modelling framework and the best suite of integrated modelling tools, what are the advantages of using a model-based solution approach as opposed to not doing so? Also, when should a model-based approach be used and what should be the role of the models and the modelling framework? The most difficult issue is providing a quantitative indicator as an argument for using a computer aided modelling framework for the model development task. In economic terms, such data are not routinely measured or collected to provide any reliable estimate. However, from the point of view of time and resources employed, the advantages become obvious. For example, reduction of time and therefore human resources could be orders of magnitude; that is, from double digit months or more to single digit days or less. Use of the framework for simulation, optimization and identification, can provide similar benefits, provided the appropriate models are used. The advantage is not only that the model development and solution is faster through the appropriate use of computer aided tools but also duplication of work is avoided via efficient re-use of models and better data-flow with redundant or unnecessary steps being avoided through the use of excellent work-flows and generic models. Even though many products-processes are designed and developed through experiment based trial-and-error approaches, model-based solution approaches have an important role. It should not be to simply replace the experiment with model-based simulation. In this case, the only advantage, provided the models represent the experiments with sufficient accuracy, is the reduction of time but a better optimal solution is not necessarily obtained because the model is not predictive and a trial-and error solution approach is still being used. For a significant benefit, the experiments need to be replaced with truly predictive models and solution approaches that can find a design target, if such a solution exists within the solution space. In this case, the model-based approach will quickly reduce the search space and identify a few potential solutions or designs so that they can be verified through experiments. In this way, by combining a computer aided modelling framework, its resident tools and libraries, a model-based solution could be developed that can provide significant technical and economic advantages. The key issue is that the model-based solution approach is employed first to identify potential solutions and experiments are focused only on the potential candidates. The success of this approach depends on development and/or availability of reliable models. The advantage would be better and more sustainable and safer products-processes coming to the market faster and at lower cost. Hopefully the modelling methods and tools, and the case studies discussed in this book can motivate interested readers to take the opportunities that are available to meet the current and future challenges in what is an important and dynamic area.

REFERENCES

Cameron, I.T., Ingram, G.D., 2008. A survey of industrial process modelling across the product and process lifecycle. Computers & Chemical Engineering. 32 (3), 420–438.

Conte, E., Gani, R., Ng, K.M., 2011. Design of formulated products: a systematic methodology. AIChE Journal., DOI: 10.1002/aic.12458.

Foss, B.A., Lohmann, B., Marquardt, W., 1998. A field study of the industrial modelling process. Journal of Process Control. 8, 325.

Hamid, M.K.A., Sin, G., Gani, R., 2010. Integration of process design and controller design for chemical processes using model-based methodology. Computers and Chemical Engineering. 34, 1567–1579.

Appendix-1

In this appendix, the equations for all models presented in Chapters 5–12 and solved with ICAS-MoT are given in the form used in this software. To use ICAS-MoT, simply copy and save the model equations as a txt-file and then insert the equations into ICAS-MoT through the "import" option. Alternatively, the interested reader can use the listed model equations to create their own model object. The equations, in most cases, are already arranged in the lower tridiagonal form. Copies of the ICAS-MoT files can also be downloaded from the book website (http://www.elsevierdirect.com/companion.jsp?ISBN=9780444531612)

A1-5: MODELS FROM CHAPTER 5

A1.5.1 Constitutive (Property) Models

5.1.1 SRK Equation of State (pure component)

The model equations are given in Chapters 5–12

5.1.2 Wilson activity coefficient model (binary mixture): ch-5-1-2-wilson.mot

```
# Wilson model for a binary mixture
# Specified T, X1, X2
# Parameters V1_in, V2_in, A1_2in, A2_1in
# Properties LnGamma_1, LnGamma_2

#Calculate Mol fraction 2
X2 = 1 - X1

# Initialize parameters
V1 = V1_in
V2 = V2_in
A1_1 = 0
A2_2 = 0
A1_2 = A12_in
A2_1 = A21_in

# Calculate interaction term

Del1_1= (V1/V1)*exp(-A1_1/T)
Del1_2= (V2/V1)*exp(-A1_2/T)
Del2_1= (V1/V2)*exp(-A2_1/T)
Del2_2= (V2/V2)*exp(-A2_2/T)
```

```
# Calculate E(i)

E1= X1*Del1_1 + X2*Del1_2
E2= X1*Del2_1 + X2*Del2_2

#Calculate Ln(Gamma(i))
LnGamma_1= 1 - ln(E1) - (X1*Del1_1/E1 + X2*Del2_1/E2)
LnGamma_2= 1 - ln(E2) - (X1*Del1_2/E1 + X2*Del2_2/E2)

Gamma_1 = exp(LnGamma_1)
Gamma_2 = exp(LnGamma_2)
```

A1.5.2 Parameter Estimation

5.2.1 Antoine correlation: ch-5-2-1-antoine-par-fit.mot

```
# Parameter estimation in Antione equation for vapour
pressure calculation
#Antione equation

P[k]=10^(A-B/(C+T[k]))

#Calculated value of vapour pressure

pcalc[k]=log(P[k])

#Experimental value of vapour pressure

pexp[k]= log(pe[k])

#Objective function

ssum=sum_k((pexp[k]-pcalc[k])^2)

Fobj=ssum/N
```

5.2.2 Wilson-Gamma

Here the same model as given in chapter 5.1.2 is used to generate the data points and an objective function minimizing the sum of errors is used to regress the two Wilson model parameters (A1_2 and A2_1)

5.2.3 Elec-NRTL: ch-5-2-3-elec-nrtl-fit.mot

```
# eNRTL model for binary mixture

# Properties
#* A_phi    Debye-Hückel parameter
```

```
#* I        ionic strength
#* m[i]     Molality (moles ions/kg solvent)
#* Za       Absolute value of charge of  anion
#* Zc       Absolute value of charge of cation
#* LnGammaDH_j[i]     Debye-Hückel term for ions
#* LnGamma_j[i]       NRTL term for ions
#* LnGamma_tot_j[i]   Total Activity for ions
#* LnGamma_mac[i]     Mean activity coefficient of ions
#* Ms       Solvent molecular weight
#* v        Salt stoichiometric coefficient

T=298.15
R=8.3145

#Calculate Mol fractions


v = v_c+v_a
Xa[j]=v_a*m[j]/((1000/Ms)+(v*m[j]))
Xc[j]=v_c*m[j]/((1000/Ms)+(v*m[j]))
Xm[j]=1-Xa[j]-Xc[j]


# Calculate Tau and G
C_m_ca=0.2
C_ca_m=0.2

Tau_ca_m = A_ca_m/(R*T)
Tau_m_ca = A_m_ca/(R*T)

G_m_ca = exp(-C_m_ca*Tau_m_ca)
G_ca_m = exp(-C_ca_m*Tau_ca_m)

#Debye-Hückel parameter

Na=6.0221E+23
d=1000
e=1.6022E-19
eps0=8.8542E-12
eps=80
kappa=1.3807E-23

A_phi=1/3*(2*3.1415*Na*d/1000)^(1/2)*((e^2)/
(eps0*eps*kappa*T))^(3/2)

# Ionic strength
I[j] = 0.5*(Xa[j]*Za^2+Xc[j]*Zc^2)
```

```
#Calculate Ln(Gamma(i))

lnGamma_c[j]=(((Xm[j])^2)*Tau_ca_m*G_ca_m)/(Xa[j]
*G_ca_m+Xc[j]*G_ca_m+Xm[j])^2+(Tau_m_ca*Zc*Xm[j]
*G_m_ca)/(Xa[j]+Xm[j]*G_m_ca)-
(Tau_m_ca*Za*Xa[j]*Xm[j]*G_m_ca)/((Xm[j]*G_m_ca+Xc
[j])^2)-Tau_ca_m*G_ca_m-Tau_m_ca*Zc
lnGamma_a[j]=(((Xm[j])^2)*Tau_ca_m*G_ca_m)/(Xa[j]
*G_ca_m+Xc[j]*G_ca_m+Xm[j])^2+(Tau_m_ca*Za*Xm[j]
*G_m_ca)/(Xc[j]+Xm[j]*G_m_ca)-
(Tau_m_ca*Zc*Xc[j]*Xm[j]*G_m_ca)/((Xm[j]*G_m_ca+Xa
[j])^2)-Tau_ca_m*G_ca_m-Tau_m_ca*Za



lnGammaDH_a[j]=-
((1000/Ms)^(0.5))*A_phi*((2/ro)*(Za)^2)*(ln(1+ro*
(I[j]^0.5)))+((Za^2)*(I[j]^0.5)-2*(I[j]^1.5))/
(1+ro*(I[j]^0.5))
lnGammaDH_c[j]=-
((1000/Ms)^(0.5))*A_phi*((2/ro)*(Zc)^2)*(ln(1+ro*
(I[j]^0.5)))+((Zc^2)*(I[j]^0.5)-2*(I[j]^1.5))/
(1+ro*(I[j]^0.5))

lnGammatot_a[j]=lnGamma_a[j]+lnGammaDH_a[j]
lnGammatot_c[j]=lnGamma_c[j]+lnGammaDH_c[j]


# Mean molal activity coefficient
lnGamma_mac[j]=(1/v)*(v_c*lnGammatot_c[j]
+v_a*lnGammatot_a[j])-ln(1+0.001*Ms*m[j]*v)

Gamma_mac[j]=exp(lnGamma_mac[j])

#Objective function
ssum= sum_j((Gamma_mac_ex[j]-Gamma_mac[j])^2)
```

A1.5.3 Use of constitutive models

5.3.1 Parameter estimation – VLE data: ch-5-3-1-wilson-vle-fit.mot

```
# Wilson model for a binary mixture
# Specified T[k], X1[k], X2[k]
# Parameters V1_in, V2_in, A1_2in, A2_1in
# Properties LnGamma_1, LnGamma_2

#Calculate Mol fraction 2
X2[k] = 1 - X1[k]
```

```
# Initialize parameters
V1 = V1_in
V2 = V2_in
A1_1 = 0
A2_2 = 0
A1_2 = A12_in
A2_1 = A21_in

# Calculate interaction term

Del1_1[k]= (V1/V1)*exp(-A1_1/T[k])
Del1_2[k]= (V2/V1)*exp(-A1_2/T[k])
Del2_1[k]= (V1/V2)*exp(-A2_1/T[k])
Del2_2[k]= (V2/V2)*exp(-A2_2/T[k])

# Calculate E(i)

E1[k]=X1[k]*Del1_1[k] + X2[k]*Del1_2[k]
E2[k]=X1[k]*Del2_1[k] + X2[k]*Del2_2[k]

#Calculate Ln(Gamma(i))
LnGamma_1[k]= 1 - ln(E1[k]) - (X1[k]*Del1_1[k]/E1[k] + X2
[k]*Del2_1[k]/E2[k])
LnGamma_2[k]= 1 - ln(E2[k]) - (X1[k]*Del1_2[k]/E1[k] + X2
[k]*Del2_2[k]/E2[k])

Gamma_1[k] = exp(LnGamma_1[k])
Gamma_2[k] = exp(LnGamma_2[k])

# Saturation pressure calculation
#Antoine equation

P_sat[k]=10^((A-B/(C+T[k])))

# Vapor phase composition

Y1[k]=(P_sat[k]*X1[k])*Gamma_1[k]/P

#Objective function

ssum= sum_k((Y1_ex[k]-Y1[k])^2)

Fobj= ssum/N
```

5.3.2 NRTL-SLE calculation: ch-5-3-2-nrtl-sle.mot

```
#
# NRTL (ternary) model as given in DECHEMA DAta Series
# R is not used as bij have units of K
# RR = 8.314; hj = heat of fusion; TM = melting point (for
fixed af=0.3)
#


x2 = 1-x1

#Note the equation below is removed in order to use the model
for a binary
# mixture. For a ternary mixture, replace this equation
with the one used below

#x3 = 1 - x1 - x2
x3 = 0

# Computation of gamma1 (given T, NRTL-model parameters,
RR, TM1, hf1
t12=b12/(R*T)
t21=b21/(R*T)
t13=b13/(R*T)
t31=b31/(R*T)
t23=b23/(R*T)
t32=b32/(R*T)

G12=exp(-af*t12)
G21=exp(-af*t21)
G13=exp(-af*t13)
G31=exp(-af*t31)
G23=exp(-af*t23)
G32=exp(-af*t32)

denom1 =x1 +G21*x2 +G31*x3
denom12 = G12*x1 + x2 + G32*x3
denom2 =G12*x1 +x2 +G32*x3
denom3 =G13*x1 +G23*x2 +x3
term1=(t21*G21*x2 +t31*G31*x3)/(denom1)
term12= (t12*G12*x1 +t32*G32*x3)/(denom12)
term2=(x1/(denom1)*(-x2*t21*G21 - x3*t31*G31)/(denom1)
term22=(x1*G21/(denom1)*(t21-(x2*t21*G21
+x3*t31*G31)/(denom1)
term3 =((x2*G12)/(denom2))*(t12-(x1*t12*G12
+x3*t32*G32)/(denom2))
```

```
term32=((x2)/(denom2))*(-(x1*t12*G12+x3*t32*G32)/
(denom2))
term4 =((x3*G13)/(denom3))*(t13-(x1*t13*G13
+x2*t23*G23)/(denom3))
term42=((x3*G23)/(denom3))*(t23-(x1*t13*G13
+x2*t23*G23)/(denom3))
lngam1 = term1 + term2 + term3 + term4

# Note: only value for gamma1 is needed since only one
compound precipitates
#lngam2 = term12 + term22 + term32 + term42

gamma1 = exp(lngam1)
#gamma2 = exp(lngam2)

# Compute x1
D = exp(-hf1/(RR*TM1)*(1 - TM1/T))
0 = 1 - gamma1*x1*exp(-hf1/(RR*TM1)*(1 - TM1/T))
```

5.3.3 Evaporation from a droplet: ch-5-3-3-drop-evap.mot

```
# Chapter 5-3-3
# Evaporation of droplet containing binary mixture of
methanol and water
# Wilson activity coefficient model

# part 1 of AEs calculation of T and y
#Calculate Mol fraction
X1=n1/(n1+n2)
X2=n2/(n1+n2)
# Calculate interaction term
Del1_1=(V1/V1)*exp(-A1_1/T)
Del1_2=(V2/V1)*exp(-A1_2/T)
Del2_1=(V1/V2)*exp(-A2_1/T)
Del2_2=(V2/V2)*exp(-A2_2/T)
# Calculate E(i)
E1= X1*Del1_1 + X2*Del1_2
E2= X1*Del2_1 + X2*Del2_2
#Calculate Ln(Gamma(i))
LnGamma_1= 1 - ln(E1) - (X1*Del1_1/E1 + X2*Del2_1/E2)
LnGamma_2= 1 - ln(E2) - (X1*Del1_2/E1 + X2*Del2_2/E2)
Gamma_1 = exp(LnGamma_1)
Gamma_2 = exp(LnGamma_2)
# Saturation pressure calculation
#Antoine equation
P_sat=10^((A-B/(C+T)))
P_sat2=10^((A2-B2/(C2+T)))
# Vapor phase composition
```

```
Y1=(P_sat*X1)*Gamma_1/P
Y2=(P_sat2*X2)*Gamma_2/P

0=1-Y1-Y2

#rest of constitutiv equations
# cp values assume that constant with temperature in J/molK
cp1_liq=80.2
cp2_liq=75.5
cp1_vap=43.7
cp2_vap=33.6
cpmix_liq=X1*cp1_liq+X2*cp2_liq
cpmix_vap=Y1*cp1_vap+Y2*cp2_vap
delta_HV=Y1*delta_HV1+Y2*delta_HV2
#density for gas
#Radius
r=((n1+n2)/(1/32*1000)*3/2/3.14))^(1/3)
Area=2*3.14*(((n1+n2)/(1/32*1000)*3/2/3.14))^(1/3))^2
V = if((P < 1) then (0) else ((n1+n2)*k*(P-1)/T))
#ODEs
dn1=-V*Y1
dn2=-V*Y2
dT=(-V*(cpmix_vap*(T)+delta_HV)+QR+T*V*cpmix_liq)/
(cpmix_liq*(n1+n2))
```

5.3.4 Kinetic models for computation of attainable region: ch-5-3-4-km-ar.mot

```
#        -- Reversible van de Vusse Reactions --

#                        k1f    k2f         k3f
# Reaction system A <---> B ---> C; 2A ---> D
#                     k1b
# k1f = 0.01, k1b = 5.0
# k2f = 10.0, k3f = 100
#
# -- Design Equation for a PFR Reactor (Eq. 4)--
# Initial Conditions:
# Ca = 1 & Cb = 0


dCb = (k1f*Ca - (k1b+k2f)*Cb)/(k1b*Cb-k1f*Ca*(1+(k3f/
k1f)*Ca))
```

A1-6: MODELS FROM CHAPTER 6

A1.6.1 Equation-oriented process simulation – Williams-Otto plant: ch-6-1-williams-otto-biegler.mot

```
# Reactor

0=FA+FrecycleA-FeffA-k1*XA*XB*V*rho
0=FB+FrecycleB-FeffB-(k1*XA+k2*XC)*XB*V*rho
0=FrecycleC-FeffC+(2*k1*XA*XB-2*k2*XB*XC-k3*XP*XC)
*V*rho
0=FrecycleE-FeffE+(2*k2*XB*XC)*V*rho
0=FrecycleP-FeffP+(k2*XB*XC-0.5*k3*XP*XC)*V*rho
0=FrecycleG-FeffG+(1.5*k3*XP*XC)*V*rho

XA=FeffA/(FeffA+FeffB+FeffC+FeffE+FeffP+FeffG)
XB=FeffB/(FeffA+FeffB+FeffC+FeffE+FeffP+FeffG)
XC=FeffC/(FeffA+FeffB+FeffC+FeffE+FeffP+FeffG)
XE=FeffE/(FeffA+FeffB+FeffC+FeffE+FeffP+FeffG)
XG=FeffG/(FeffA+FeffB+FeffC+FeffE+FeffP+FeffG)
XP=FeffP/(FeffA+FeffB+FeffC+FeffE+FeffP+FeffG)

k1=5.9755e9*exp(-12000/T)
k2=2.5962e12*exp(-15000/T)
k3=9.6283e15*exp(-20000/T)

#Decanter
0=FDecantA-FeffA
0=FDecantB-FeffB
0=FDecantC-FeffC
0=FDecantE-FeffE
0=FDecantP-FeffP
0=FDecantG
0=FWasteG-FeffG
0=FWasteA
0=FWasteB
0=FWasteC
0=FWasteE
0=FWasteP

#Distillation Column
0=FBottomA-FDecantA
0=FBottomB-FDecantB
0=FBottomC-FDecantC
0=FBottomE-FDecantE
0=FBottomP-0.1*FDecantE
0=FDecantP-FProdP-0.1*FDecantE
```

```
#Flow splitter
0=etha*FBottomA-FPurgeA
0=etha*FBottomB-FPurgeB
0=etha*FBottomC-FPurgeC
0=etha*FBottomE-FPurgeE
0=etha*FBottomP-FPurgeP

0=(1-etha)*FBottomA-FrecycleA
0=(1-etha)*FBottomB-FrecycleB
0=(1-etha)*FBottomC-FrecycleC
0=(1-etha)*FBottomE-FrecycleE
0=(1-etha)*FBottomP-FrecycleP
0=FrecycleG-FDecantG
```

A1.6.2 Williams-Otto plant optimisation

```
# Chapter 6-2
# Williams-Otto plant optimization
#


# Reactor

0=FA+FrecycleA-FeffA-k1*XA*XB*V*rho
0=FB+FrecycleB-FeffB-(k1*XA+k2*XC)*XB*V*rho
0=FrecycleC-FeffC+(2*k1*XA*XB-2*k2*XB*XC-k3*XP*XC)
*V*rho
0=FrecycleE-FeffE+(2*k2*XB*XC)*V*rho
0=FrecycleP-FeffP+(k2*XB*XC-0.5*k3*XP*XC)*V*rho
0=FrecycleG-FeffG+(1.5*k3*XP*XC)*V*rho

XA=FeffA/(FeffA+FeffB+FeffC+FeffE+FeffP+FeffG)
XB=FeffB/(FeffA+FeffB+FeffC+FeffE+FeffP+FeffG)
XC=FeffC/(FeffA+FeffB+FeffC+FeffE+FeffP+FeffG)
XE=FeffE/(FeffA+FeffB+FeffC+FeffE+FeffP+FeffG)
XG=FeffG/(FeffA+FeffB+FeffC+FeffE+FeffP+FeffG)
XP=FeffP/(FeffA+FeffB+FeffC+FeffE+FeffP+FeffG)

k1=5.9755e9*exp(-12000/T)
k2=2.5962e12*exp(-15000/T)
k3=9.6283e15*exp(-20000/T)

#Decanter
0=FDecantA-FeffA
0=FDecantB-FeffB
0=FDecantC-FeffC
0=FDecantE-FeffE
```

```
0=FDecantP-FeffP
0=FDecantG
0=FWasteG-FeffG
0=FWasteA
0=FWasteB
0=FWasteC
0=FWasteE
0=FWasteP

#Distillation Column
0=FBottomA-FDecantA
0=FBottomB-FDecantB
0=FBottomC-FDecantC
0=FBottomE-FDecantE
0=FBottomP-0.1*FDecantE
0=FDecantP-FProdP-0.1*FDecantE

#Flow splitter
0=etha*FBottomA-FPurgeA
0=etha*FBottomB-FPurgeB
0=etha*FBottomC-FPurgeC
0=etha*FBottomE-FPurgeE
0=etha*FBottomP-FPurgeP

0=(1-etha)*FBottomA-FrecycleA
0=(1-etha)*FBottomB-FrecycleB
0=(1-etha)*FBottomC-FrecycleC
0=(1-etha)*FBottomE-FrecycleE
0=(1-etha)*FBottomP-FrecycleP
0=FrecycleG-FDecantG


FRecycle = FrecycleA+FrecycleB+FrecycleC+FrecycleE
+FrecycleP
FWaste = FWasteG
FPurge=FPurgeA+FPurgeB+FPurgeC+FPurgeE+FPurgeP

#Objective Function

ROI = -(2207*FProdP+50*FPurge-168*FA-252*FB-
2.22*FRecycle-84*FWaste+600*V*rho-1041.6)/(6*rho*V)

Product = FProdP
```

A1-7: MODELS FROM CHAPTER 7

A1.7.1 Dynamic blending tank

This model was not solved with ICAS-MoT

A1.7.2 Fuel-Cell (DMFC) model: ch-7-2-fuel-cell-dmfc.mot

```
#Fuel Cell dynamic mode

#Components Index
#0->CH3OH, 1->CO2

#**************************************************
#*            Constants           *
#**************************************************

# Faraday Constant [C/mol]
F = 96485

# Universal gas constant [J/ mol K]
R = 8.314

# Standard pressure
Pstd = 101325

#******************** Fuel Cell Model ****************

kls = (2.2e-13*CFCH3OH^4-8.56765e-10*CFCH3OH^3+7.755e-
7*CFCH3OH^2+0.00019542*CFCH3OH+0.281648)*1.0e-5

r5 = (k5*exp(alpha*F*ethaC/R/T)*(1-exp(-F*ethaC/R/T)*
(PO2/Pstd)^1.5))

r1 = (k1*thetaPt^3*( CCLCH3OH*exp(alpha*F*ethaA/R/T)-
CCLCO2/Keq1/Keq2^3/Keq3/Keq4/(exp(alpha*F*ethaA/R/
T))^7))

dethaA = (icell - 6*F*r1)/Ca

dpdz=(Pcathode-Panode)/dm
temp1=kp/mu*dpdz
temp2=(Kphi/mu)*(icell/F+cmproton*kp/mu*dpdz)/
(dmproton/R/T+cmproton*(Kphi/mu))
vcrossover=temp2-temp1
```

```
Pe=vcrossover*dm/dmch3oh

nmCH3OH=dmch3oh/dm*(Pe*exp(Pe)/(exp(Pe)-1))*CCLCH3OH

PO2=Pcathode*0.2095

dCCH3OH=(CFCH3OH - CCH3OH)/tau - kls*As/Va*(CCH3OH-
CCLCH3OH)

dCCLCH3OH=(kls*(CCH3OH-CCLCH3OH)-nmCH3OH - r1)* (As/
Vcla)

dCCO2=(CFCO2-CCO2)/tau - kls*As/Va*(CCO2-CCLCO2)

dCCLCO2=(As/Vcla)*(kls*(CCO2-CCLCO2)+r1)

dethaC=(-icell - 6*F*(r5+nmCH3OH))/Cc

# Calculate Ucell
Ucell=Ustdcell - ethaA + ethaC- icell*dm/kappam
```

A1.7.3 Fluidised-bed multiscale reactor model

7.3.1 Dynamic model: ch-7-3-fluidized-bed-multiscale-dyn.mot

```
# fluidized batch reactor
# simple multiscale scenario

# meso-micro

# dimension-less version


# mass balance
dP=Pe-P+Hg*(Pp-P)
dT=Te-T+Hw*(Tw-T)+Ht*(Tp-T)

# micro balance
dPp=(Hg*(P-Pp)-Hg*Kk*Pp)/A
dTp=(Ht*(T-Tp)+Ht*F*Kk*Pp)/C

Kk=0.0006*exp(20.7-15000/Tp)
```

7.3.2 Steady-state model: ch-7-3-fluidised-bed-multiscale-ss.mot

```
# fluidized batch reactor (steady state)
# simple multiscale scenario

# meso micro
```

```
# dimension-less version

# Kk=0.0006*exp(20.7-15000/Tp)


# meso

# mass balance
0=Pe-P+Hg*(Pp-P)
0=Te-T+Ht*(Tp-T)+Hw*(Tw-T)

# micro
0=(-Hg*0.0006*exp(20.7-15000/Tp)*Pp+Hg*(P-Pp))/A
0=(Ht*F*0.0006*exp(20.7-15000/Tp)*Pp+Ht*(T-Tp))/C
```

7.3.3 Reduced model: ch-7-3-fluidized-bed-reduced.mot

```
# mass and energy balance for overall reactor

dP=(Pe-P-Hg*Kk*P)/(1+A)
dT=(Te-T+Hw*(Tw-T)+Ht*F*Kk*P)/(1+C)

Kk=0.0006*exp(20.7-15000/T)
```

A1.7.4: Chemical reactor model: ch-7-4-chem-reactor.mot

```
#**********************

#The series reactions:

#        k1A      K2B
#    2A -----> B -----> 3C
#        (1)       (2)

#are catalyzed by H2SO4. All reactions are first order
#in the reactant concentration. The reaction is to be
#carried out in a semi-batch reactor that has a heat
#exchanger inside with UA = 35,000.0 cal/h K and ambient
#temperature of 298 K. Pure A enters at a concentration
#of 4 mol/dm3, a volumetric flow rate of 240 dm3/hr, and
#a temperature of 305 K. Initially there is a total of
#100 dm3 in the reactor, which contains 1.0 mol/dm3 of A
#and 1.0 mol/dm3 of the catalyst H2SO4. The reaction rate
#is independent of the catalyst concentration. The initial
#Temperature of the reactor is 290 K.
```

```
#Given:
;CA0     = 4.0
;CH2SO40 = 1.0
;vo      = 240.0
;V0      = 100.0
;UA      = 35000.0
;Ta      = 298.0
;T0      = 305.0

#Aditional Information:
;R =1.987

;T1A0 = 320
;T2B0 = 300
;k1A0 = 1.25
;k2B0 = 0.08
;E1A  = 9500.0
;E2B  = 7000.0

;CpA   = 30.0
;CpB   = 60.0
;CpC   = 20.0
;CpH2SO4 = 35.0

;DHRx1A = -6500.00
;DHRx2B = +8000.00
#***********************

#*************
#Model equations
#*************

#Kinetic
k1A = k1A0*exp((E1A/R)*(1/T1A0 - 1/T))
k2B = k2B0*exp((E2B/R)*(1/T2B0 - 1/T))

#Rate Laws
r1A = -k1A*CA
r2B = -k2B*CB

rA = r1A
rB = k1A*CA/2-k2B*CB
rC = 3*k2B*CB

#Reactor volume
V = V0 + vo*t
```

```
FA0 = CA0*vo
Cpmix = CA*CpA + CB*CpB + CC*CpC

#Mol balances
dCA = rA + (CA0 - CA)*vo/V
dCB = rB - CB*vo/V
dCC = rC - CC*vo/V

#Energy Balance
dT = (UA*(Ta-T) - FA0*CpA*(T-T0) + (DHRx1A*r1A +
DHRx2B*r2B)*V)/(Cpmix*V + CH2SO40*V0*CpH2SO4)
```

A1.7.5 Polymerisation reactor: ch-7-5-poly-reactor.mot

```
# Chapter 7.5
#Polymerization Reactor

kp  = Ap*exp(-Ep/R/T)
kfm = Afm*exp(-Efm/R/T)
kI  = AI*exp(-EI/R/T)
ktd = Atd*exp(-Etd/R/T)
ktc = Atc*exp(-Etc/R/T)

P0 = sqrt(2.0*fe*CI*kI/(ktd+ktc))

#Dynamical Model
dCm = -(kp+kfm)*Cm*P0+F*(Cmin-Cm)/V
dCI = -kI*CI+ (FI*CIin-F*CI)/V
dT  = mDH*kp*Cm*P0/(ro*Cp) - U*A*(T-Tj)/(ro*Cp*V)+F*
(Tin-T)/V
dD0 = -(0.5*ktc-ktd)*P0*P0 + kfm*Cm*P0 - F*D0/V
dD1 = Mm*(kp+kfm)*Cm*P0-F*D1/V
DTj = Fcw*(Tw0-Tj)/V0 + U*A*(T-Tj)/(row*Cpw*V0)

# % monomer conversion
X=(Cmin-Cm)/Cmin *100

#Average molecular weight
PM =D1/D0
```

A1-8: MODELS FROM CHAPTER 8

A1.8-1 Oil-shale pre-heating cooling unit

This model was not solved with ICAS-MoT

A1.8.2 Dynamic model for complex granulator

This model was not solved with ICAS-MoT

A1.8.3 Short-path evaporation: ch-8-3-short-path-evap.mot

```
#*******************************
#Short Path Evaporator     *
*
#***********************************************

r1 = diameter/2.0
Ah = PI*diameter*L
Ak = PI*(diameter - 2.0*d)*L
P=Patm*101325.0
Ts = T_10
Tf1 = Tf
Pout=P/101325
#Thermodynimic properties
Pvap[i]   = exp(PVAPA[i] + PVAPB[i]/Ts + PVAPC[i]*ln(Ts) +
PVAPD[i]*Ts^2)
DHvap[i]  = (HVAPA[i]*(1.0 - (Ts/Tc[i]))^(HVAPB[i] +
HVAPC[i]*(Ts/Tc[i]) + HVAPD[i]*(Ts/Tc[i])^2)) /1.0E+6
Cp[i]     = (CPA[i] + CPB[i]*Ts + CPC[i]*Ts^2 + CPD[i]
*Ts^3 + CPE[i]*Ts^4) / 1.0E+3
rho[i]    = (RHOA[i]/RHOB[i]^(1.0 + ( 1.0 - Ts/RHOC[i])
^RHOD[i]))*MW[i]*1000.0
Eta[i]    = exp(VISCOA[i]+ VISCOB[i]/Ts + VISCOC[i]*ln
(Ts) + VISCOD[i]*Ts^VISCOE[i])
Lambda[i] = THERMALA[i] + THERMALB[i]*Ts + THERMALC[i]
*Ts^2+ THERMALD[i]*Ts^3
nu[i]   = Eta[i]/rho[i]

;Mixture properties
Q[i]      = Flow[i]*MW[i]/rho[i]
Qtot      = sum_i(Q[i])
C[i]      = Flow[i]/Qtot
Ctot      = sum_i(C[i])
FlowTot   = sum_i(Flow[i])
FlowVT    = sum_i(Flow_V[i])


x[i]        = Flow[i]/FlowTot
y[i]        = Flow_V[i]/FlowVT
rhoMix      = sum_i(x[i]*rho[i])
LambdaMix = sum_i(x[i]*Lambda[i])
CpMix       = sum_i(x[i]*Cp[i])
MWMix       = sum_i(x[i]*MW[i])
```

```
EtaMix    =sum_i(x[i]*Eta[i])
DHvapMix  =sum_i(x[i]*DHvap[i])
nuMix     =EtaMix/rhoMix

#Initial condition
Flux0 =2.0E0*PI*r1*Gamma*rhoMix/(3.6E05*MWMix)

#Evaporation rate efficiency
beta=d/4.0
F   =Ak/(Ak+Ah)
kappa=10.0^(0.2*F+1.38*(F+0.1)^4.0)
EREF =1.0-(1.0-F)*(1.0-exp(-d/kappa/beta))^nmc
EREF =EREF*(P/Pref)

#*********************************************
h1    =(3.0*nuMix*FlowTot/(2.0*PI*r1*g*Ctot))^(1.0/3.0)
by    =h1
Dy    =(by-ay)/N

Y_0   =0.0*Dy
Y_1   =1.0*Dy
Y_2   =2.0*Dy
Y_3   =3.0*Dy
Y_4   =4.0*Dy
Y_5   =5.0*Dy
Y_6   =6.0*Dy
Y_7   =7.0*Dy
Y_8   =8.0*Dy
Y_9   =9.0*Dy
Y_10  =10.0*Dy

Vz_0  =g*h1^2/nuMix*(Y_0/h1-0.5*Y_0*Y_0/h1^2)
Vz_1  =g*h1^2/nuMix*(Y_1/h1-0.5*Y_1*Y_1/h1^2)
Vz_2  =g*h1^2/nuMix*(Y_2/h1-0.5*Y_2*Y_2/h1^2)
Vz_3  =g*h1^2/nuMix*(Y_3/h1-0.5*Y_3*Y_3/h1^2)
Vz_4  =g*h1^2/nuMix*(Y_4/h1-0.5*Y_4*Y_4/h1^2)
Vz_5  =g*h1^2/nuMix*(Y_5/h1-0.5*Y_5*Y_5/h1^2)
Vz_6  =g*h1^2/nuMix*(Y_6/h1-0.5*Y_6*Y_6/h1^2)
Vz_7  =g*h1^2/nuMix*(Y_7/h1-0.5*Y_7*Y_7/h1^2)
Vz_8  =g*h1^2/nuMix*(Y_8/h1-0.5*Y_8*Y_8/h1^2)
Vz_9  =g*h1^2/nuMix*(Y_9/h1-0.5*Y_9*Y_9/h1^2)
Vz_10 =g*h1^2/nuMix*(Y_10/h1-0.5*Y_10*Y_10/h1^2)

Theta=MWMix*LambdaMix/(rhoMix*CpMix)
k[i] =Act[i]*Pvap[i]/sqrt(2.0*PI*Rg*MW[i]*Ts)
ktot =sum_i(x[i]*k[i])
```

```
#Discretised Temperature equation (PDE) to ODEs
;boundary condition at y = 0
dT_0 = 0.0

; internal discretisation points
dT_1 = (Theta/Vz_1)*(T_2 - 2.0*T_1 + T_0)/Dy/Dy
dT_2 = (Theta/Vz_2)*(T_3 - 2.0*T_2 + T_1)/Dy/Dy
dT_3 = (Theta/Vz_3)*(T_4 - 2.0*T_3 + T_2)/Dy/Dy
dT_4 = (Theta/Vz_4)*(T_5 - 2.0*T_4 + T_3)/Dy/Dy
dT_5 = (Theta/Vz_5)*(T_6 - 2.0*T_5 + T_4)/Dy/Dy
dT_6 = (Theta/Vz_6)*(T_7 - 2.0*T_6 + T_5)/Dy/Dy
dT_7 = (Theta/Vz_7)*(T_8 - 2.0*T_7 + T_6)/Dy/Dy
dT_8 = (Theta/Vz_8)*(T_9 - 2.0*T_8 + T_7)/Dy/Dy
dT_9 = (Theta/Vz_9)*(T_10 - 2.0*T_9 + T_8)/Dy/Dy

;boundary condition at y = 1
0 = T_10 - T_9 + Dy*DHvapMix*ktot/LambdaMix


#Flowrate in the liquid and vapour phase (mol/s)
dFlow[i]   = - 2.0*PI*r1*k[i]*x[i]* EREF
Flow_L[i]  = Flow[i]

dFlow_V[i] =  2.0*PI*r1*k[i]*x[i]* EREF

T_out=(T_1+T_2+T_3+T_4+T_5+T_6+T_7+T_8+T_9+T_10)/10

Q = 0.0
```

A1-9: MODELS FROM CHAPTER 9

A1.9.1 Tennessee Eastman Challenge Process - Simple model: ch-9-1-te-dynamic-ricker.mot

```
# Simplified Tennessee Eastman Process
#
# A. Mauricio Sales-Cruz
# CAPEC, Chemical ingennering Departmen
# Technical University of Denamrk
# Bulding 299, 2800 kgs, Lyngby Denamrk
#
# Reference:
; [1] N.L Ricker and J. H. Lee
;   Nonlinear Modeling and State Estimation for the
Tennessee Eastman Challenge Process
;   Computer Chem. Engng. (1995), Vol. 19, No. 9, pp. 983-1005.
; [2] J.J. Downs and E. F. Vogel
;   A Plant-Wide Industrial Process Control Problem
```

```
;   Computer Chem Engng. (1993), Vol. 17, No. 3, pp. 245-255

#Reaction Stoichiometric is:

# A(g)     + C(g) +     D(g) --> G(l)
# A(g)     + C(g) +     E(g) --> H(l)
# 1/3A(g) + D(g) + 1/3E(g) --> F(l)

# A => 1
# B => 2
# C => 3
# D => 4
# E => 5
# F => 6
# G => 7
# H => 8
#####################################################

#****************
# STATE VARIABLES*
#****************

#Table 4 form Ricker and Lee, 1995 [1]
# Base case at four stedy-state conditions
#(all units are [kmol] (molar holdups in a certain
location)

 ; x_1  A in reactor (NAr)
 ; x_2  B in reactor (NBr)
 ; x_3  C in reactor (NCr)
 ; x_4  D in reactor (NDr)
 ; x_5  E in reactor (NEr)
 ; x_6  F in reactor (NFr)
 ; x_7  G in reactor (NGr)
 ; x_8  H in reactor (NHr)

 ; x_9  A in separator (NAs)
 ; x_10 B in separator (NBs)
 ; x_11 C in separator (NCs)
 ; x_12 D in separator (NDs)
 ; x_13 E in separator (NEs)
 ; x_14 F in separator (NFs)
 ; x_15 G in separator (NGs)
 ; x_16 H in separator (NHs)

 ; x_17 A in feed-mixing zone (NAm)
 ; x_18 B in feed-mixing zone (NBm)
```

```
; x_19 C in feed-mixing zone (NCm)
; x_20 D in feed-mixing zone (NDm)
; x_21 E in feed-mixing zone (NEm)
; x_22 F in feed-mixing zone (NFm)
; x_23 G in feed-mixing zone (NGm)
; x_24 H in feed-mixing zone (NHm)

; x_25 G in product reservoir (stripper bottoms) (NGp)
; x_26 H in product reservoir (NHp)


#**********************************************
#INPUTS are flows [kmol/h] unless noted otherwise*
#**********************************************
#Table 3. Manipulated variables at four steady-state
conditions
#base case
;Feed 1 ([F1] - pure A )
 u_1 = 11.2

;Feed 2 ([F2] - pure D )
 u_2 = 114.5

;Feed 3 ([F3] - pure E )
 u_3 = 98.0

;Feed 4 ([F4] - A and C)
 u_4 = 417.5

;Recycle ([F8] - stream 8)
 u_5 = 1201.5

;Purge ([F9] - stream 9)
 u_6 = 15.1

;Separator underflow ([F10] - stream 10)
 u_7 = 259.5

;Product rate     ([F11] - stream 11)
 u_8 = 211.3

;Reactor temperature (TCR - [deg C])
 u_9 = 120.4

;Separator temperature (TCS - [deg C]
 u_10 = 80.109
```

```
#**********************************************
#Table 2. Variables used to eliminate bias for four
#stedy-state conditions - base case


;A in stream 4 [mole %] (formula 100zA4)
 u_11 = 48.17993608770129

;B in stream 4 [mole %]
 u_12 = 0.4999456287425161

;Reaction 1 activity factor [%].
 u_13 = 103.0019053957069

;Reaction 2 activity factor [%].
 u_14 = 100.8789972898921

;Bias subtracted from u_7 to get stream 10 flow [kmol/h].
 u_15 = 0.9349498036041268

;Reactor/Separator flow parameter [%].
 u_16 = 99.28395653130411

;Feed/Reactor flow parameter [%].
 u_17 = 99.72033077752585

;Product G + H purity parameter [%].
 u_18 = 97.552

;Adjustment to VLE of D to G in separator [%]
 u_19 = 100.1125401829961

;Adjustment to H VLE in separator [%]
 u_20 = 98.59314504463755

;C bias flow to feed zone [kmol/h]
 u_21 = -3.434967633847123

;D bias flow to feed zone [kmol/h]
 u_22 = 0.4466146860446880

;E bias flow to feed zone [kmol/h]
 u_23 = -2.070191433847128

;F bias flow to feed zone [kmol/h]
 u_24 = -0.1649292860446963

;VLE correction to D-H in reactor [%]
 u_25 = 99.60083735827341
```

```
#*****************************************
#Calculates outputs and rates of change of states.*
#*****************************************

 ;Parameters
 nc    =8
 cf2cm=0.028317
 TauTr=0.08333
 TauTs=0.08333
 TauF  =0.005
 TauC  =0.08333
 zero  =0.0

 ;A in Antoine eqn.
 AVP_1=0.0
 AVP_2=0.0
 AVP_3=0.0
 AVP_4=20.81
 AVP_5=21.24
 AVP_6=21.24
 AVP_7=21.32
 AVP_8=22.10

 ;B in Antoine eqn.
 BVP_1=0.0
 BVP_2=0.0
 BVP_3=0.0
 BVP_4=-1444.0
 BVP_5=-2114.0
 BVP_6=-2114.0
 BVP_7=-2748.0
 BVP_8=-3318.0

 ;C in Antoine eqn.
 CVP_1=0.0
 CVP_2=0.0
 CVP_3=0.0
 CVP_4=259.0
 CVP_5=266.0
 CVP_6=266.0
 CVP_7=233.0
 CVP_8=250.0

 ;molec. wts.
 mwts_1=2.0
 mwts_2=25.4
```

```
mwts_3 = 28.0
mwts_4 = 32.0
mwts_5 = 46.0
mwts_6 = 48.0
mwts_7 = 62.0
mwts_8 = 76.0

;Liquid densisty [kg/m3]
rhol_1 = 0.0
rhol_2 = 0.0
rhol_3 = 0.0
rhol_4 = 299.0
rhol_5 = 365.0
rhol_6 = 328.0
rhol_7 = 612.0
rhol_8 = 617.0


;Molar volume [m3/kmol]
molvol_1 = 0.0
molvol_2 = 0.0
molvol_3 = 0.0
molvol_4 = mwts_4/rhol_4
molvol_5 = mwts_5/rhol_5
molvol_6 = mwts_6/rhol_6
molvol_7 = mwts_7/rhol_7
molvol_8 = mwts_8/rhol_8

;stripping factors
sfr_1 = 1.0
sfr_2 = 1.0
sfr_3 = 1.0
sfr_4 = 1.0
sfr_5 = 1.0
sfr_6 = 1.0
sfr_7 = 0.07
sfr_8 = 0.04

;sep. activ. coeffs.
gami_1 = 1.0
gami_2 = 1.0
gami_3 = 1.0
gami_4 = 1.0
gami_5 = 1.0
gami_6 = 1.0
gami_7 = 1.0
gami_8 = 1.0
```

```
;react. activ. coeffs.
gamr_1 = 1.0
gamr_2 = 1.0
gamr_3 = 1.0
gamr_4 = 1.0
gamr_5 = 1.0
gamr_6 = 1.0
gamr_7 = 1.0
gamr_8 = 1.0

#Volumes in reactor, separator, & feed zone

;total reactor volume [m3]
 VTR=36.8
;total separator volume [m3]
 VTS=99.1

;total feed vapor holdup volume [m3]
 VTV=150.0

#Constants
 ;gas constant [kPa*m3/kmol-K]=[kJ/kmol-K] and [cal/
gmol-K]
   Rg = 8.314
   Rg1= 1.987

#*****************************************
#*****************************************

#**************************************************
#  Get disturbances from input vector. Flow disturbances *
#  are scaled so they're comparable to manipulated       *
#   variables (0-100% basis)                             *
#**************************************************

 ;mol frac B in stream 4 (A + C feed)
   xA4 = u_11/100.

 ;mol frac C in stream 4 (A + C feed)
   xB4 = u_12/100.

 ;get A in stream 4 by difference
   xC4 = 1.0 - xA4 - xB4

 ;Reactivity factor, reaction 1
   R1F = u_13/100.
```

```
  ;Reactivity factor, reaction 2
    R2F = u_14/100.

  ;Bias adjustment for stream 10 flow [kmol/h]
    F10b = u_15

  ;Adjusts flow between reactor and separator
    RSflow = u_16/100.

  ;Adjusts flow between feed zone and reactor
    FRflow = u_17/100.

  ;Adjusts for impurity in product (components other than G
and H)
    fracGH = u_18/100.

  ;Adjusts VLE of D to G in separator

   gami_4 = u_19/100.
   gami_5 = u_19/100.
   gami_6 = u_19/100.
   gami_7 = u_19/100.

  ;Adjusts VLE of H in separator
     gami_8 = u_20/100.

  ;C, D, E, abd F flow bias at feed point
     Cbias = u_21
     Dbias = u_22
     Ebias = u_23
     Fbias = u_24

  ;Adjusts VLE in reactor
     gamr_4 = u_25/100.0
     gamr_5 = u_25/100.0
     gamr_6 = u_25/100.0
     gamr_7 = u_25/100.0
     gamr_8 = u_25/100.0

#*********************************
#Temperatures in reactor & separator*
#*********************************

  ;reactor and separator temperatures in [C]
    TCR = u_9
    TCS = u_10
```

```
;reactor and separator temperatures in [K]
  TKR = TCR + 273.2
  TKS = TCS + 273.2

#Product reservoir states (stripper bottoms)

;G and H holdup [kmol]
  plG = max(x_25,zero)
  plH = max(x_26,zero)

;Volume of liquid [m3]
  VLP = plG*molvol_7 + plH*molvol_8

;mol frac G in product.
  xGp = fracGH*(plG/(plG + plH) )

;adjusted for assumed impurity level (fracGH)
;mol frac H in product
  xHp = fracGH*(plH/(plG + plH) )

#Get molar holdups in reactor, separator, and feed vapor.
#Assumes A, B, C insoluble in liquid.

rlsum = 0.0
slsum = 0.0
rvsum = 0.0
svsum = 0.0
VLR  = 0.0
VLS  = 0.0


rlmol_1 = 0.0
rlmol_2 = 0.0
rlmol_3 = 0.0

slmol_1 = 0.0
slmol_2 = 0.0
slmol_3 = 0.0

;gets liq. mol from state vector
 rlmol_4 = max(x_4,zero)
 rlmol_5 = max(x_5,zero)
 rlmol_6 = max(x_6,zero)
 rlmol_7 = max(x_7,zero)
 rlmol_8 = max(x_8,zero)
```

```
  ;accumulate total moles
   slmol_4 = max(x_12,zero)
   slmol_5 = max(x_13,zero)
   slmol_6 = max(x_14,zero)
   slmol_7 = max(x_15,zero)
   slmol_8 = max(x_16,zero)


   rlsum = rlsum + rlmol_4 + rlmol_5 + rlmol_6 + rlmol_7 +
rlmol_8
   slsum = slsum + slmol_4 + slmol_5 + slmol_6 + slmol_7 +
slmol_8

     ;accumulate liquid volume [m3]
   VLR = VLR + rlmol_4*molvol_4 + rlmol_5*molvol_5 +
rlmol_6*molvol_6 + rlmol_7*molvol_7 + rlmol_8*molvol_8
   VLS = VLS + slmol_4*molvol_4 + slmol_5*molvol_5 +
slmol_6*molvol_6 + slmol_7*molvol_7 + slmol_8*molvol_8

    ;molar volume of product [m3/kmol]
     Pvol = VLS/slsum

     fvsum = 0.0

  ;gets feed vapor from state vector
   fvmol_1 = max (x_17, zero)
   fvmol_2 = max (x_18, zero)
   fvmol_3 = max (x_19, zero)
   fvmol_4 = max (x_20, zero)
   fvmol_5 = max (x_21, zero)
   fvmol_6 = max (x_22, zero)
   fvmol_7 = max (x_23, zero)
   fvmol_8 = max (x_24, zero)

  ;total feed vapor moles
   fvsum = fvsum + fvmol_1 + fvmol_2 + fvmol_3 + fvmol_4 +
fvmol_5 + fvmol_6 + fvmol_7 + fvmol_8


  #Vapor volumes in reactor, separator, & feed zone
   ;vapor in reactor [m3]
     VVR = VTR - VLR

  ;vapor in separator [m3]
     VVS = VTS - VLS
```

```
;Mole fractions, partial pressures, etc.
   PTR = 0.0
   PTS = 0.0

 ;Total pressure in feed zone [kPa] (assumes constant
temperature of 86.1 C)
   PTV = fvsum*Rg*(273.2 + 86.1)/VTV

 #For A, B, C


 ;Auxiliar variables
   Axppr_1 = max(x_1,zero)
   Axppr_2 = max(x_2,zero)
   Axppr_3 = max(x_3,zero)

   Axpps_1 = max(x_9,zero)
   Axpps_2 = max(x_10,zero)
   Axpps_3 = max(x_11,zero)

 ;pp in reactor [kPa]
   ppr_1 = Axppr_1*Rg*TKR/VVR
   ppr_2 = Axppr_2*Rg*TKR/VVR
   ppr_3 = Axppr_3*Rg*TKR/VVR

 ;pp in separator [kPa]
   pps_1 = Axpps_1*Rg*TKS/VVS
   pps_2 = Axpps_2*Rg*TKS/VVS
   pps_3 = Axpps_3*Rg*TKS/VVS


 ;total P in reactor [kPa]
   PTR = PTR + ppr_1 + ppr_2 + ppr_3

 ;total P in separator [kPa]
   PTS = PTS + pps_1 + pps_2 + pps_3

  ;reactor liquid mol frac
  xlr_1 = 0.0
  xlr_2 = 0.0
  xlr_3 = 0.0

  xls_1 = 0.0
  xls_2 = 0.0
  xls_3 = 0.0
```

```
#For D,E,F,G,H

;reactor liquid mol frac
 xlr_4 = rlmol_4/rlsum
 xlr_5 = rlmol_5/rlsum
 xlr_6 = rlmol_6/rlsum
 xlr_7 = rlmol_7/rlsum
 xlr_8 = rlmol_8/rlsum

;separator liquid mol frac
 xls_4 = slmol_4/slsum
 xls_5 = slmol_5/slsum
 xls_6 = slmol_6/slsum
 xls_7 = slmol_7/slsum
 xls_8 = slmol_8/slsum


 ;reactor vap. pres. [kPa]
 pvapr_4 = 1.e-3*exp( AVP_4 + BVP_4/(CVP_4 + TCR))
 pvapr_5 = 1.e-3*exp( AVP_5 + BVP_5/(CVP_5 + TCR))
 pvapr_6 = 1.e-3*exp( AVP_6 + BVP_6/(CVP_6 + TCR))
 pvapr_7 = 1.e-3*exp( AVP_7 + BVP_7/(CVP_7 + TCR))
 pvapr_8 = 1.e-3*exp( AVP_8 + BVP_8/(CVP_8 + TCR))

;separator vap. pres. [kPa]
pvaps_4 = 1.e-3*exp( AVP_4 + BVP_4/(CVP_4 + TCS))
 pvaps_5 = 1.e-3*exp( AVP_5 + BVP_5/(CVP_5 + TCS))
 pvaps_6 = 1.e-3*exp( AVP_6 + BVP_6/(CVP_6 + TCS))
 pvaps_7 = 1.e-3*exp( AVP_7 + BVP_7/(CVP_7 + TCS))
 pvaps_8 = 1.e-3*exp( AVP_8 + BVP_8/(CVP_8 + TCS))

;VLE
 ppr_4 = pvapr_4*xlr_4*gamr_4
 ppr_5 = pvapr_5*xlr_5*gamr_5
 ppr_6 = pvapr_6*xlr_6*gamr_6
 ppr_7 = pvapr_7*xlr_7*gamr_7
 ppr_8 = pvapr_8*xlr_8*gamr_8

;Note use of VLE adjustment
 pps_4 = pvaps_4*xls_4*gami_4
 pps_5 = pvaps_5*xls_5*gami_5
 pps_6 = pvaps_6*xls_6*gami_6
 pps_7 = pvaps_7*xls_7*gami_7
 pps_8 = pvaps_8*xls_8*gami_8

 ;Reactor total press [kPa]
 PTR = PTR + ppr_4 + ppr_5 + ppr_6 + ppr_7 + ppr_8
```

```
  ;Separator total press [kPa]
 PTS = PTS + pps_4 + pps_5 + pps_6 + pps_7 + pps_8


  wtmolr=0.0
  wtmolf=0.0

;reactor vapor mol frac
  xvr_1 = ppr_1/PTR
  xvr_2 = ppr_2/PTR
  xvr_3 = ppr_3/PTR
  xvr_4 = ppr_4/PTR
  xvr_5 = ppr_5/PTR
  xvr_6 = ppr_6/PTR
  xvr_7 = ppr_7/PTR
  xvr_8 = ppr_8/PTR

;separator vapor mol frac
  xvs_1 = pps_1/PTS
  xvs_2 = pps_2/PTS
  xvs_3 = pps_3/PTS
  xvs_4 = pps_4/PTS
  xvs_5 = pps_5/PTS
  xvs_6 = pps_6/PTS
  xvs_7 = pps_7/PTS
  xvs_8 = pps_8/PTS

;feed vapor mol frac
  xvf_1 = fvmol_1/fvsum
  xvf_2 = fvmol_2/fvsum
  xvf_3 = fvmol_3/fvsum
  xvf_4 = fvmol_4/fvsum
  xvf_5 = fvmol_5/fvsum
  xvf_6 = fvmol_6/fvsum
  xvf_7 = fvmol_7/fvsum
  xvf_8 = fvmol_8/fvsum

;average mol wt of reactor vapor
  wtmolr = wtmolr + xvr_1*mwts_1 + xvr_2*mwts_2 +
xvr_3*mwts_3 + xvr_4*mwts_4 + xvr_5*mwts_5 + xvr_6*mwts_6
+ xvr_7*mwts_7 + xvr_8*mwts_8

;average mol wt of feed vapor
  wtmolf = wtmolf + xvf_1*mwts_1 + xvf_2*mwts_2 +
xvf_3*mwts_3 + xvf_4*mwts_4 + xvf_5*mwts_5 + xvf_6*mwts_6
+ xvf_7*mwts_7 + xvf_8*mwts_8
```

```
#******************************
#Rate Laws. All rates in [kmol/h].*
#******************************

RR_1 = VVR*exp( 44.06 -
42600.0/Rg1/TKR)*R1F*(ppr_1^1.08)*(ppr_3^0.311)*(ppr_4^0.874)
RR_2 = VVR*exp( 10.27 -
19500.0/Rg1/TKR)*R2F*(ppr_1^1.15)*(ppr_3^0.370)*(ppr_5^1.00)
RR_3 = VVR*exp( 59.50 - 59500.0/Rg1/TKR)*ppr_1*
(0.77*ppr_4 + ppr_5)


#Define total flows [kmol/h]
 F1 = u_1
 F2 = u_2
 F3 = u_3
 F4 = u_4
 F8 = u_5
 F9 = u_6
 F10= u_7 - F10b
 F11= u_8

#   Stripper balance to get combined feeds (streams 1+2+3+5)
#   Also account for flow biases.

 Fcmol_1 = F1 + xA4*F4
 Fcmol_2 = xB4*F4
 Fcmol_3 = xC4*F4 + Cbias
 Fcmol_4 = F2 + sfr_4*F10*xls_4 + Dbias
 Fcmol_5 = F3 + sfr_5*F10*xls_5 + Ebias
 Fcmol_6 = sfr_6*F10*xls_6 + Fbias
 Fcmol_7 = sfr_7*F10*xls_7
 Fcmol_8 = sfr_8*F10*xls_8
 Fcomb=0.0


  Fcomb = Fcomb + Fcmol_1 + Fcmol_2 + Fcmol_3 + Fcmol_4 +
Fcmol 5 + Fcmol_6 + Fcmol_7 + Fcmol_0

# Pressure drop vs. flow to get flows of streams 6 & 7.

 F6= (2413.7/wtmolf)*sqrt(abs(PTV - PTR))*FRflow

 ;if (ptr .gt. ptv) then
 ;F6=-F6
 ;end if
```

```
F7 = (5722.0/wtmolr)*sqrt(abs(PTR - PTS))*RSflow

;if (PTS .gt. PTR) then
;F7=-F7
;end if

;F6 = 1889.91460017435
;F7 = 1475.16538329608


# Balances to get rates of change of states.  First do
convection terms (in - out).

;Reactor
dx_1 = F6*xvf_1 - F7*xvr_1  - RR_1 - RR_2 - 0.333*RR_3
dx_2 = F6*xvf_2 - F7*xvr_2
dx_3 = F6*xvf_3 - F7*xvr_3 - RR_1 - RR_2
dx_4 = F6*xvf_4 - F7*xvr_4 - RR_1 - RR_3
dx_5 = F6*xvf_5 - F7*xvr_5 - RR_2 - 0.333*RR_3
dx_6 = F6*xvf_6 - F7*xvr_6 + RR_3
dx_7 = F6*xvf_7 - F7*xvr_7 + RR_1
dx_8 = F6*xvf_8 - F7*xvr_8 + RR_2

;Separator
dx_9  = F7*xvr_1 - (F8 + F9)*xvs_1 - F10*xls_1
dx_10 = F7*xvr_2 - (F8 + F9)*xvs_2 - F10*xls_2
dx_11 = F7*xvr_3 - (F8 + F9)*xvs_3 - F10*xls_3
dx_12 = F7*xvr_4 - (F8 + F9)*xvs_4 - F10*xls_4
dx_13 = F7*xvr_5 - (F8 + F9)*xvs_5 - F10*xls_5
dx_14 = F7*xvr_6 - (F8 + F9)*xvs_6 - F10*xls_6
dx_15 = F7*xvr_7 - (F8 + F9)*xvs_7 - F10*xls_7
dx_16 = F7*xvr_8 - (F8 + F9)*xvs_8 - F10*xls_8

;Feed(Mixing) zone
dx_17 = Fcmol_1 + F8*xvs_1 - F6*xvf_1
dx_18 = Fcmol_2 + F8*xvs_2 - F6*xvf_2
dx_19 = Fcmol_3 + F8*xvs_3 - F6*xvf_3
dx_20 = Fcmol_4 + F8*xvs_4 - F6*xvf_4
dx_21 = Fcmol_5 + F8*xvs_5 - F6*xvf_5
dx_22 = Fcmol_6 + F8*xvs_6 - F6*xvf_6
dx_23 = Fcmol_7 + F8*xvs_7 - F6*xvf_7
dx_24 = Fcmol_8 + F8*xvs_8 - F6*xvf_8

;G in product
dx_25 =(1.0 - sfr_7)*F10*xls_7 - F11*xGp
```

```
;H in product
 dx_26 = (1.0 - sfr_8)*F10*xls_8 - F11*xHp


#Calculate the outputs.
#Table 1. Elements of the output vector y

;Reactor pressure [kPa gage]
 y_1 = PTR - 101

;Reactor liq. [%]
 y_2 = 5.263*VLR - 12.105

;Separator pressure [kPa gage]
 y_3 = PTS - 101
;Separator liquid [%]
 y_4 = 12.28*VLS - 10.53

;Product liq. holdup [%]
 y_5 = 22.58*VLP - 49.03

;Pressure in feed zone [kPa guage]
 y_6 = PTV - 101

;Stream 6 flowrate [kscmh]
 y_7 = F6/44.79

;Reactor feed [mol %]
 y_8  = xvf_1*100
 y_9  = xvf_2*100
 y_10 = xvf_3*100
 y_11 = xvf_4*100
 y_12 = xvf_5*100
 y_13 = xvf_6*100

;Purge [mol %]
 y_14 = xvs_1*100
 y_15 = xvs_2*100
 y_16 = xvs_3*100
 y_17 = xvs_4*100
 y_18 = xvs_5*100
 y_19 = xvs_6*100

;G in purge [mol %]
 y_20 = xvs_7*100
```

```
;H in purge [mol %]
 y_21 = xvs_8*100

;G in product [mol %]
 y_22 = xGp*100

;H in product [mol %]
 y_23 = xHp*100

;Molar production [kmol/h]
 y_24 = F11

;[cents/kmol]
 y_25 = (F9/(F11+0.01))*(221*xvs_1 + 618*xvs_3 +
2210*xvs_4 + 1460*xvs_5 + 1790*xvs_6 + 3040*xvs_7 +
2290*xvs_8 )

;G production [kmol/h]
 y_26 = RR_1

;H production [kmol/h]
 y_27 = RR_2

;F production [kmol/h]
 y_28 = RR_3

;A PP in reactor [kPa]
 y_29 = ppr_1

;C PP in reactor [kPa]
 y_30 = ppr_3

;D PP in reactor [kPa]
 y_31 =ppr_4

;F PP in reactor [kPa]
 y_32 = ppr_5


XMEAS_7  = y_1
XMEAS_8  = y_2
XMEAS_13 = y_3
XMEAS_12 = y_4
XMEAS_15 = y_5
XMEAS_16 = y_6
XMEAS_6  = y_7
XMEAS_23 = y_8
```

```
XMEAS_24 = y_9
XMEAS_25 = y_10
XMEAS_26 = y_11
XMEAS_27 = y_12
XMEAS_28 = y_13
XMEAS_29 = y_14
XMEAS_30 = y_15
XMEAS_31 = y_16
XMEAS_32 = y_17
XMEAS_33 = y_18
XMEAS_34 = y_19
XMEAS_35 = y_20
XMEAS_36 = y_21
XMEAS_40 = y_22
XMEAS_41 = y_23
```

A1.9.2 Tennessee Eastman Challenge Process - Complete model: ch-9-2-te-dynamic-complete.mot

```
# Tennessee Eastman Process
#
# A. Mauricio Sales-Cruz
# CAPEC, Chemical ingennering Departmen
# Technical University of Denamrk
# Bulding 299, 2800 kgs, Lyngby Denamrk
#
# Reference:
; [1] Tobias Jockenhövel, Lorenz T. Bigler, Andreas
Wächter
;   Dynamic Optimization of the TennesseeEastman Process
Using OptControlCentre.
;   Computers and Chemical Engineering 27(2003), pp.
1513-1531.
; [2] N.L Ricker and J. H. Lee
;   Nonlinear Modeling and State Estimation for the
Tennessee Eastman Challenge Process
;   Computer Chem. Engng (1995), Vol. 19, No. 9, pp. 983-
1005.
; [3] J.J. Downs and E. F. Vogel
;   A Plant-Wide Industrial Process Control Problem
;   Computer ChemEngng. (1993), Vol. 17, No. 3, pp. 245-255

#Reaction Stoichiometric is:
; A(g) + C(g) + D(g) --> G(L),  Product 1
; A(g) + C(g) + E(g) --> H(L),  Product 2
;        A(g) + E(g) --> F(L),  Byproduct
;             3D(g) --> 2F(L),  Byroduct 1
```

```
#####################################################

#*************************************************
# Stream property data - Taken from Table 2, Downs & Vogel*
#*************************************************

 ;Molecular weight
 M_A = 2.0
 M_B = 25.4
 M_C = 28.0
 M_D = 32.0
 M_E = 46.0
 M_F = 48.0
 M_G = 62.0
 M_H = 76.0

 ;Vapor heat capacity [KJ/(Kg C)]
 cp_vap_A = 14.6  * M_A
 cp_vap_B = 2.04  * M_B
 cp_vap_C = 1.05  * M_C
 cp_vap_D = 1.85  * M_D
 cp_vap_E = 1.87  * M_E
 cp_vap_F = 2.02  * M_F
 cp_vap_G = 0.712* M_G
 cp_vap_H = 0.628* M_H

 ;Liquid heat capacity (A-C not condensable) [KJ/(Kg C)]
 cp_liq_A = 0     * M_A
 cp_liq_B = 0     * M_B
 cp_liq_C = 0     * M_C
 cp_liq_D = 7.66  * M_D
 cp_liq_E = 4.17  * M_E
 cp_liq_F = 4.45  * M_F
 cp_liq_G = 2.55  * M_G
 cp_liq_H = 2.45  * M_H

 ;Heat of vaporization [KJ/Kg]
 H_vap_A = 0     * M_A
 H_vap_B = 0     * M_B
 H_vap_C = 0     * M_C
 H_vap_D = 202   * M_D
 H_vap_E = 372   * M_E
 H_vap_F = 372   * M_F
 H_vap_G = 523   * M_G
 H_vap_H = 486   * M_H
```

```
#***********
# Constants *
#***********

;Reference Temperature
 Tref = 273.15

;Universal gas constant
 ;RkJ [=] [kPa m3/(kmol K)] or [kJ/(kmol K)] and Rkcal [=]
kcal/(kmol K)
 RkJ  = 8.31451
 Rkcal = 1.987

#****************************
#Cooling water heat capacity*
#****************************
cp_cw = 4.18

#Reaction stochiometric coeficients
 mue_A_1 = -1
 mue_A_2 = -1
 mue_A_3 = -1/3

 mue_B_1 =  0
 mue_B_2 =  0
 mue_B_3 =  0

 mue_C_1 = -1
 mue_C_2 = -1
 mue_C_3 =  0

 mue_D_1 = -1
 mue_D_2 =  0
 mue_D_3 = -1

 mue_E_1 =  0
 mue_E_2 = -1
 mue_E_3 = -1/3

 mue_F_1 =  0
 mue_F_2 =  0
 mue_F_3 =  1

 mue_G_1 =  1
 mue_G_2 =  0
 mue_G_3 =  0
```

```
 mue_H_1 = 0
 mue_H_2 = 1
 mue_H_3 = 0

# Antoine Constants ref [3] Table 2.
#P [=] Pa and T [=] C
 A_D = 20.81
 B_D = -1444
 C_D = 259.0

 A_E = 21.24
 B_E = -2114
 C_E = 265.5

 A_F = 21.24
 B_F = -2144
 C_F = 265.5

 A_G = 21.32
 B_G = -2748
 C_G = 232.9

 A_H = 22.10
 B_H = -3318
 C_H = 249.6

#************************************
# Feed streams; Table 1 - Downs and Vogel*
#************************************

;Molar feed flow [Kmol/s]
 stream_1_flow  = 11.2   / 3600
 stream_2_flow  = 114.5  / 3600
 stream_3_flow  = 98.0   / 3600
 stream_4_flow  = 417.5  / 3600
 stream_8_flow  = 1201.5/ 3600
 stream_9_flow  = 15.1   / 3600
 stream_10_flow = 259.5  / 3600
 stream_11_flow = 211.3  / 3600

 ;Temperatures ; 318.15 K

 stream_1_T = 45 + Tref
 stream_2_T = 45 + Tref
 stream_3_T = 45 + Tref
 stream_4_T = 45 + Tref
```

```
;Composition of Feed Stream A
stream_1_conc_A = 0.99990
stream_1_conc_B = 0.00010
stream_1_conc_C = 0.00000
stream_1_conc_D = 0.00000
stream_1_conc_E = 0.00000
stream_1_conc_F = 0.00000
stream_1_conc_G = 0.00000
stream_1_conc_H = 0.00000

;Composition of Feed Stream D
stream_2_conc_A = 0.00000
stream_2_conc_B = 0.00010
stream_2_conc_C = 0.00000
stream_2_conc_D = 0.99990
stream_2_conc_E = 0.00000
stream_2_conc_F = 0.00000
stream_2_conc_G = 0.00000
stream_2_conc_H = 0.00000

;Composition of Feed Stream E
stream_3_conc_A = 0.00000
stream_3_conc_B = 0.00000
stream_3_conc_C = 0.00000
stream_3_conc_D = 0.00000
stream_3_conc_E = 0.99990
stream_3_conc_F = 0.00010
stream_3_conc_G = 0.00000
stream_3_conc_H = 0.00000

;Composition of Feed Stream C
stream_4_conc_A = 0.48500
stream_4_conc_B = 0.00500
stream_4_conc_C = 0.51000
stream_4_conc_D = 0.00000
stream_4_conc_E = 0.00000
stream_4_conc_F = 0.00000
stream_4_conc_G = 0.00000
stream_4_conc_H = 0.00000

#*************************
#Feed stream heat capacities*
#*************************

stream_1_cp = stream_1_conc_A*cp_vap_A +
stream_1_conc_B*cp_vap_B + stream_1_conc_C*cp_vap_C +
stream_1_conc_D*cp_vap_D + stream_1_conc_E*cp_vap_E +
```

```
stream_1_conc_F*cp_vap_F + stream_1_conc_G*cp_vap_G +
stream_1_conc_H*cp_vap_H
stream_2_cp = stream_2_conc_A*cp_vap_A +
stream_2_conc_B*cp_vap_B + stream_2_conc_C*cp_vap_C +
stream_2_conc_D*cp_vap_D + stream_2_conc_E*cp_vap_E +
stream_2_conc_F*cp_vap_F + stream_2_conc_G*cp_vap_G +
stream_2_conc_H*cp_vap_H
stream_3_cp = stream_3_conc_A*cp_vap_A +
stream_3_conc_B*cp_vap_B + stream_3_conc_C*cp_vap_C +
stream_3_conc_D*cp_vap_D + stream_3_conc_E*cp_vap_E +
stream_3_conc_F*cp_vap_F + stream_3_conc_G*cp_vap_G +
stream_3_conc_H*cp_vap_H
stream_4_cp = stream_4_conc_A*cp_vap_A +
stream_4_conc_B*cp_vap_B + stream_4_conc_C*cp_vap_C +
stream_4_conc_D*cp_vap_D + stream_4_conc_E*cp_vap_E +
stream_4_conc_F*cp_vap_F + stream_4_conc_G*cp_vap_G +
stream_4_conc_H*cp_vap_H

#**********
#Utilities*
#**********
;cooling water flow mol/s -Ref [3] Table 1
 reactor_m_CWS  = 93.37/3600*1000/18
 separator_m_CWS = 49.37/3600*1000/18

;Stripper stream flow [Kg/s] equivalent to 47.446% of
stripper valve
 stripper_m_CWS  = 230.31/3600

#***************
# Volumes  [=] m3*
#***************
 ;Mixing zone volume
 Vm = 141.53

 ;Total Reactor volume
 Vr = 36.8117791

 ;Total separator volume
 Vs = 99.1

#**********
#Densities*
#**********
 ;Molar liquid density [=] kmol/m3
 rho_liq_reactor  = 9.337145754
 rho_liq_separator = 10.29397546
```

```
#***********************
#Activity coefficients VLE*
#***********************
 ;Reactor
 gamma_D_r = 0.996011
 gamma_E_r = 1
 gamma_F_r = 1.078
 gamma_G_r = 0.999
 gamma_H_r = 0.999

 ;Separator
 gamma_D_s = 1.001383
 gamma_E_s = 1.001383
 gamma_F_s = 1.091383
 gamma_G_s = 1.001383
 gamma_H_s = 0.992188



 alpha_1 = 1.0399157
 alpha_2 = 1.011373129
 alpha_3 = 1

#Heat transfer exchange
 UA    = 127.6
 UA_sep = 152.7

#Compressor

s_kap = .7166374645

#Stripper

;Density
 rho_liq_stripper = 8.6496e0


# Eq. 5.22 - Split factors Stripper
 phi_A = 1
 phi_B = 1
 phi_C = 1

PC_1  = 3.51e-7
PC_6  = -0.00011
PC_11 = 0.011351
PC_16 = 0.548012
```

```
PC_2  = 3.69e-7
PC_7  = -0.0001
PC_12 = 0.010197
PC_17 = 0.620794

PC_3  = 3.69e-7
PC_8  = -0.0001
PC_13 = 0.010049
PC_18 = 0.628854

PC_4  = 1.84e-9
PC_9  = 1.37e-5
PC_14 = 0.000217
PC_19 = 0.001393

PC_5  = 8.32e-8
PC_10 = -7.64e-6
PC_15 = 0.000976
PC_20 = -0.01568


###################################################
#############
#From Mixer zone
#############
# Eq. 5.46 - Pressure in the mixing zone

  mixing_zone_N = mixing_zone_NA + mixing_zone_NB +
mixing_zone_NC + mixing_zone_ND + mixing_zone_NE +
mixing_zone_NF + mixing_zone_NG + mixing_zone_NH
  pm_MPa = mixing_zone_N * (RkJ * Tm / Vm) / 1000

# Eq. 5.47
  stream_6_conc_A = mixing_zone_NA / mixing_zone_N
  stream_6_conc_B = mixing_zone_NB / mixing_zone_N
  stream_6_conc_C = mixing_zone_NC / mixing_zone_N
  stream_6_conc_D = mixing_zone_ND / mixing_zone_N
  stream_6_conc_E = mixing_zone_NE / mixing_zone_N
  stream_6_conc_F = mixing_zone_NF / mixing_zone_N
  stream_6_conc_G = mixing_zone_NG / mixing_zone_N
  stream_6_conc_H = mixing_zone_NH / mixing_zone_N

###################################################
#############
```

```
######################
#From reactor zone
######################
# Eq. 5.17  (A, B, C not condensable)
 reactor_x_A = 0.0
 reactor_x_B = 0.0
 reactor_x_C = 0.0

# Eq. 5.16
 reactor_x_D = reactor_ND / (reactor_ND + reactor_NE
+ reactor_NF + reactor_NG + reactor_NH)
 reactor_x_E = reactor_NE / (reactor_ND + reactor_NE
+ reactor_NF + reactor_NG + reactor_NH)
 reactor_x_F = reactor_NF / (reactor_ND + reactor_NE
+ reactor_NF + reactor_NG + reactor_NH)
 reactor_x_G = reactor_NG / (reactor_ND + reactor_NE
+ reactor_NF + reactor_NG + reactor_NH)
 reactor_x_H = reactor_NH / (reactor_ND + reactor_NE
+ reactor_NF + reactor_NG + reactor_NH)

# Eq. 5.18
 Vr_liq = (reactor_ND + reactor_NE + reactor_NF +
reactor_NG + reactor_NH) / rho_liq_reactor

# Equation 5.19
  Vr_vap = Vr - Vr_liq

# Eq. 5.11
 p_A_r = reactor_NA * RkJ * Tr /  Vr_vap / 1000
 p_B_r = reactor_NB * RkJ * Tr /  Vr_vap / 1000
 p_C_r = reactor_NC * RkJ * Tr /  Vr_vap / 1000

# Eq. 5.12
   pr_sat_D = 1e-3 * exp(A_D+ (B_D / (C_D + Tr - Tref) ) ) / 1000
   pr_sat_E = 1e-3 * exp(A_E+ (B_E / (C_E + Tr - Tref) ) ) / 1000
   pr_sat_F = 1e-3 * exp(A_F+ (B_F / (C_F + Tr - Tref) ) ) / 1000
   pr_sat_G = 1e-3 * exp(A_G+ (B_G / (C_G + Tr - Tref) ) ) / 1000
   pr_sat_H = 1e-3 * exp(A_H+ (B_H / (C_H + Tr - Tref) ) ) / 1000

# Eq. 5.13
 p_D_r = gamma_D_r * reactor_x_D * pr_sat_D
 p_E_r = gamma_E_r * reactor_x_E * pr_sat_E
 p_F_r = gamma_F_r * reactor_x_F * pr_sat_F
 p_G_r = gamma_G_r * reactor_x_G * pr_sat_G
 p_H_r = gamma_H_r * reactor_x_H * pr_sat_H
```

```
#Eq. 5.14
 pr_MPa=p_A_r+p_B_r+p_C_r+p_D_r+p_E_r+p_F_r+p_G_r
 +p_H_r

#Eq. 5.24 - Flow rates
 press_m_r_diff=sqrt(pm_MPa -pr_MPa)

 stream_6_flow=0.8333711713*press_m_r_diff

#Eq. 5.25 - assumption base case
 press_r_s_diff=sqrt( pr_MPa-ps_MPa)
 stream_7_flow=1.53546206685993*press_r_s_diff

#Eq. 5.15
 stream_7_conc_A =p_A_r/pr_MPa
 stream_7_conc_B =p_B_r/pr_MPa
 stream_7_conc_C =p_C_r/pr_MPa
 stream_7_conc_D =p_D_r/pr_MPa
 stream_7_conc_E =p_E_r/pr_MPa
 stream_7_conc_F =p_F_r/pr_MPa
 stream_7_conc_G =p_G_r/pr_MPa
 stream_7_conc_H =p_H_r/pr_MPa

#######################################################
#####

#####################
#From Separator zone
#####################
 ;Eq. 5.32  (A, B, C not condensable)
 separator_x_A=0.0
 separator_x_B=0.0
 separator_x_C=0.0

 separator_x_D=separator_ND/(separator_ND+separator_NE
+separator_NF+separator_NG+separator_NH)
 separator_x_E=separator_NE/(separator_ND+separator_NE
+separator_NF+separator_NG+separator_NH)
 separator_x_F=separator_NF/(separator_ND+separator_NE
+separator_NF+separator_NG+separator_NH)
 separator_x_G=separator_NG/(separator_ND+separator_NE
+separator_NF+separator_NG+separator_NH)
 separator_x_H=separator_NH/(separator_ND+separator_NE
+separator_NF+separator_NG+separator_NH)
```

```
#Eq. 5.34

Vs_liq=(separator_ND+separator_NE+separator_NF+
separator_NG+separator_NH)/rho_liq_separator

#Equation 5.19
 Vs_vap=Vs-Vs_liq

#Eq. 5.28
p_A_s=separator_NA*RkJ*Ts/Vs_vap /1000
p_B_s=separator_NB*RkJ*Ts/Vs_vap /1000
p_C_s=separator_NC*RkJ*Ts/Vs_vap /1000

#Eq. 5.29
ps_sat_D=1e-3*exp(A_D+(B_D/(C_D+Ts-Tref)))/1000
ps_sat_E=1e-3*exp(A_E+(B_E/(C_E+Ts-Tref)))/1000
ps_sat_F=1e-3*exp(A_F+(B_F/(C_F+Ts-Tref)))/1000
ps_sat_G=1e-3*exp(A_G+(B_G/(C_G+Ts-Tref)))/1000
ps_sat_H=1e-3*exp(A_H+(B_H/(C_H+Ts-Tref)))/1000

 p_D_s=gamma_D_s*separator_x_D*ps_sat_D
 p_E_s=gamma_E_s*separator_x_E*ps_sat_E
 p_F_s=gamma_F_s*separator_x_F*ps_sat_F
 p_G_s=gamma_G_s*separator_x_G*ps_sat_G
 p_H_s=gamma_H_s*separator_x_H*ps_sat_H

#Eq. 5.30
ps_MPa=p_A_s+p_B_s+p_C_s+p_D_s+p_E_s+p_F_s+p_G_s+
p_H_s

#Eq. 5.31
stream_8_conc_A=p_A_s/ps_MPa
stream_8_conc_B=p_B_s/ps_MPa
stream_8_conc_C=p_C_s/ps_MPa
stream_8_conc_D=p_D_s/ps_MPa
stream_8_conc_E=p_E_s/ps_MPa
stream_8_conc_F=p_F_s/ps_MPa
stream_8_conc_G=p_G_s/ps_MPa
stream_8_conc_H=p_H_s/ps_MPa
```

```
#########################################################
#######
###########
#  Purge   #
###########

# Introduced as independent variable instead of valve
position
;define_control(stream_9_flow)

stream_9_conc_A = stream_8_conc_A
stream_9_conc_B = stream_8_conc_B
stream_9_conc_C = stream_8_conc_C
stream_9_conc_D = stream_8_conc_D
stream_9_conc_E = stream_8_conc_E
stream_9_conc_F = stream_8_conc_F
stream_9_conc_G = stream_8_conc_G
stream_9_conc_H = stream_8_conc_H

###############
#    Compressor#
###############

# Instead of compressor recycle valve

  stream_8_T = Ts * (pm_MPa / ps_MPa)^((1 - s_kap) / s_kap)


####################
#From Stripper zone
####################

Vstr_liq= (stripper_NG + stripper_NH) / rho_liq_stripper /
(stream_11_conc_G + stream_11_conc_H)

phi_D = PC_1 * (Tstr - Tref)^3 + PC_6 * (Tstr - Tref)^2
+ PC_11 * (Tstr - Tref) + PC_16
phi_E = PC_2 * (Tstr - Tref)^3 + PC_7 * (Tstr - Tref)^2
+ PC_12 * (Tstr - Tref) + PC_17
phi_F = PC_3 * (Tstr - Tref)^3 + PC_8 * (Tstr - Tref)^2
+ PC_13 * (Tstr - Tref) + PC_18
phi_G = PC_4 * (Tstr - Tref)^3 + PC_9 * (Tstr - Tref)^2
+ PC_14 * (Tstr - Tref) + PC_19
phi_H = PC_5 * (Tstr - Tref)^3 + PC_10 * (Tstr - Tref)^2
+ PC_15 * (Tstr - Tref) + PC_20
```

```
dNstr_G = (1 - phi_G) * (separator_x_G * stream_10_flow +
stream_4_conc_G * stream_4_flow) - stream_11_conc_G *
stream_11_flow
dNstr_H = (1 - phi_H) * (separator_x_H * stream_10_flow +
stream_4_conc_H * stream_4_flow) - stream_11_conc_H *
stream_11_flow

# Eq. 5.54
stream_5_flow = stream_10_flow + stream_4_flow -
stream_11_flow - (dNstr_G + dNstr_H)

# Eq. 5.55
stream_5_conc_A = phi_A * (stream_4_conc_A *
stream_4_flow + separator_x_A * stream_10_flow) /
stream_5_flow
stream_5_conc_B = phi_B * (stream_4_conc_B *
stream_4_flow + separator_x_B * stream_10_flow) /
stream_5_flow
stream_5_conc_C = phi_C * (stream_4_conc_C *
stream_4_flow + separator_x_C * stream_10_flow) /
stream_5_flow
stream_5_conc_D = phi_D * (stream_4_conc_D *
stream_4_flow + separator_x_D * stream_10_flow) /
stream_5_flow
stream_5_conc_E = phi_E * (stream_4_conc_E *
stream_4_flow + separator_x_E * stream_10_flow) /
stream_5_flow
stream_5_conc_F = phi_F * (stream_4_conc_F *
stream_4_flow + separator_x_F * stream_10_flow) /
stream_5_flow
stream_5_conc_G = phi_G * (stream_4_conc_G *
stream_4_flow + separator_x_G * stream_10_flow) /
stream_5_flow
stream_5_conc_H = phi_H * (stream_4_conc_H *
stream_4_flow + separator_x_H * stream_10_flow) /
stream_5_flow

# Eq. 5.56
stream_11_conc_A = 0
stream_11_conc_B = 0
stream_11_conc_C = 0

# Conc D - F
stream_11_conc_D = (stream_4_conc_D * stream_4_flow +
separator_x_D * stream_10_flow - stream_5_conc_D *
stream_5_flow ) / stream_11_flow
```

```
stream_11_conc_E = (stream_4_conc_E * stream_4_flow +
separator_x_E * stream_10_flow - stream_5_conc_E *
stream_5_flow ) / stream_11_flow
stream_11_conc_F = (stream_4_conc_F * stream_4_flow +
separator_x_F * stream_10_flow - stream_5_conc_F *
stream_5_flow ) / stream_11_flow

# Eq. 5.57

# Conc G, H
stream_11_conc_G = ((1 - stream_11_conc_D -
stream_11_conc_E - stream_11_conc_F) * stripper_NG) /
(stripper_NG + stripper_NH)
stream_11_conc_H = ((1 - stream_11_conc_D -
stream_11_conc_E - stream_11_conc_F) * stripper_NH) /
(stripper_NG + stripper_NH)

#####################################################
######################################################


##########################
#                        #
# Mixing Unit            #
#                        #
##########################

# Base case initial values - 151.7272279 (Calculated by
Temp, press, Vm and eq. 5.46
 #initial conditions - From Table 1
 ;mixing_zone_N  = 48.83796012
 ;mixing_zone_NB = 13.49310238
 ;mixing_zone_NC = 40.03019454
 ;mixing_zone_ND = 10.44186782
 ;mixing_zone_NE = 28.48830431
 ;mixing_zone_NF = 2.514120166
 ;mixing_zone_NG = 5.403006586
 ;mixing_zone_NH = 2.517154711

# Molar Balance for components A-H (5.44)
dmixing_zone_NA = stream_1_flow * stream_1_conc_A +
stream_2_flow * stream_2_conc_A + stream_3_flow *
stream_3_conc_A + stream_5_flow * stream_5_conc_A +
stream_8_flow * stream_8_conc_A - stream_6_flow *
stream_6_conc_A
dmixing_zone_NB = stream_1_flow * stream_1_conc_B +
stream_2_flow * stream_2_conc_B + stream_3_flow *
```

```
stream_3_conc_B + stream_5_flow * stream_5_conc_B +
stream_8_flow * stream_8_conc_B - stream_6_flow *
stream_6_conc_B
dmixing_zone_NC = stream_1_flow * stream_1_conc_C +
stream_2_flow * stream_2_conc_C + stream_3_flow *
stream_3_conc_C + stream_5_flow * stream_5_conc_C +
stream_8_flow * stream_8_conc_C - stream_6_flow *
stream_6_conc_C
dmixing_zone_ND = stream_1_flow * stream_1_conc_D +
stream_2_flow * stream_2_conc_D + stream_3_flow *
stream_3_conc_D + stream_5_flow * stream_5_conc_D +
stream_8_flow * stream_8_conc_D - stream_6_flow *
stream_6_conc_D
dmixing_zone_NE = stream_1_flow * stream_1_conc_E +
stream_2_flow * stream_2_conc_E + stream_3_flow *
stream_3_conc_E + stream_5_flow * stream_5_conc_E +
stream_8_flow * stream_8_conc_E - stream_6_flow *
stream_6_conc_E
dmixing_zone_NF = stream_1_flow * stream_1_conc_F +
stream_2_flow * stream_2_conc_F + stream_3_flow *
stream_3_conc_F + stream_5_flow * stream_5_conc_F +
stream_8_flow * stream_8_conc_F - stream_6_flow *
stream_6_conc_F
dmixing_zone_NG = stream_1_flow * stream_1_conc_G +
stream_2_flow * stream_2_conc_G + stream_3_flow *
stream_3_conc_G + stream_5_flow * stream_5_conc_G +
stream_8_flow * stream_8_conc_G - stream_6_flow *
stream_6_conc_G
dmixing_zone_NH = stream_1_flow * stream_1_conc_H +
stream_2_flow * stream_2_conc_H + stream_3_flow *
stream_3_conc_H + stream_5_flow * stream_5_conc_G +
stream_8_flow * stream_8_conc_H - stream_6_flow *
stream_6_conc_H


#Energy balance for the mixing zone (5.45)
 mixing_zone_Ncp =  mixing_zone_NA * cp_vap_A +
mixing_zone_NB * cp_vap_B + mixing_zone_NC * cp_vap_C +
mixing_zone_ND * cp_vap_D + mixing_zone_NE * cp_vap_E +
mixing_zone_NF * cp_vap_F + mixing_zone_NG * cp_vap_G +
mixing_zone_NH * cp_vap_H

#Streem Cp
 stream_5_cp = stream_5_conc_A * cp_vap_A +
stream_5_conc_B * cp_vap_B + stream_5_conc_C * cp_vap_C +
stream_5_conc_D * cp_vap_D + stream_5_conc_E * cp_vap_E +
stream_5_conc_F * cp_vap_F + stream_5_conc_G * cp_vap_G +
stream_5_conc_H * cp_vap_H
```

```
 stream_8_cp = stream_8_conc_A * cp_vap_A +
stream_8_conc_B * cp_vap_B + stream_8_conc_C * cp_vap_C +
stream_8_conc_D * cp_vap_D + stream_8_conc_E * cp_vap_E +
stream_8_conc_F * cp_vap_F + stream_8_conc_G * cp_vap_G +
stream_8_conc_H * cp_vap_H

# parts of right hand side of equation 5.45
 mix_enth_1=stream_1_flow*stream_1_cp * (stream_1_T-Tm)
 mix_enth_2=stream_2_flow*stream_2_cp * (stream_2_T-Tm)
 mix_enth_3=stream_3_flow*stream_3_cp * (stream_3_T-Tm)
 mix_enth_5=stream_5_flow*stream_5_cp * (Tstr       -Tm)
 mix_enth_8=stream_8_flow*stream_8_cp * (stream_8_T-Tm)

# Right hand side of equation 5.45
 mix_rhs = (mix_enth_1 + mix_enth_2 + mix_enth_3 + mix_enth_5
+ mix_enth_8) / mixing_zone_Ncp

# Temperature in the mixing zone
  dTm = mix_rhs
     #initial conditions - From Table 1
     ;Tm = 359.25

###################################################
##################################
#                                #
#   Reactor                      #
#                                #
##################################

# Eq. 5.6
reactor_R1 = alpha_1 * Vr_vap * exp(44.06 - (42600/(Rkcal *
Tr)) ) * ((p_A_r*1000)^1.080) * ((p_C_r*1000)^0.311) *
((p_D_r*1000)^0.874) / 3600

# Eq. 5.7
reactor_R2 = alpha_2 * Vr_vap * exp(10.27 - (19500/(Rkcal *
Tr)) ) * ((p_A_r*1000)^1.150) * ((p_C_r*1000)^0.370) *
((p_E_r*1000)^1.000) / 3600

# Eq. 5.8
reactor_R3 = alpha_3 * Vr_vap * exp(59.50 - (59500/(Rkcal *
Tr)) ) * (p_A_r*1000) * ( 0.77 * (p_D_r*1000) +
(p_E_r*1000)) / 3600
```

```
reactor_conv_rate_A = mue_A_1 * reactor_R1 + mue_A_2 *
reactor_R2 + mue_A_3 * reactor_R3
reactor_conv_rate_B = mue_B_1 * reactor_R1 + mue_B_2 *
reactor_R2 + mue_B_3 * reactor_R3
reactor_conv_rate_C = mue_C_1 * reactor_R1 + mue_C_2 *
reactor_R2 + mue_C_3 * reactor_R3
reactor_conv_rate_D = mue_D_1 * reactor_R1 + mue_D_2 *
reactor_R2 + mue_D_3 * reactor_R3
reactor_conv_rate_E = mue_E_1 * reactor_R1 + mue_E_2 *
reactor_R2 + mue_E_3 * reactor_R3
reactor_conv_rate_F = mue_F_1 * reactor_R1 + mue_F_2 *
reactor_R2 + mue_F_3 * reactor_R3
reactor_conv_rate_G = mue_G_1 * reactor_R1 + mue_G_2 *
reactor_R2 + mue_G_3 * reactor_R3
reactor_conv_rate_H = mue_H_1 * reactor_R1 + mue_H_2 *
reactor_R2 + mue_H_3 * reactor_R3


#Eq5.1 - Base case initial values - Taken from TE_20_10.cmp
(Andreas Waechter)
 ;reactor_NA = 0.520462483681871e+01
 ;reactor_NB = 0.228951164244119e+01
 ;reactor_NC = 0.465082827002657e+01
 ;reactor_ND = 0.115341662452407e+00
 ;reactor_NE = 0.744601110373549e+01
 ;reactor_NF = 0.114649137635668e+01
 ;reactor_NG = 0.560529035378437e+02
 ;reactor_NH = 0.598256416887065e+02

dreactor_NA = stream_6_flow * stream_6_conc_A -
stream_7_flow * stream_7_conc_A + reactor_conv_rate_A
dreactor_NB = stream_6_flow * stream_6_conc_B -
stream_7_flow * stream_7_conc_B + reactor_conv_rate_B
dreactor_NC = stream_6_flow * stream_6_conc_C -
stream_7_flow * stream_7_conc_C + reactor_conv_rate_C
dreactor_ND = stream_6_flow * stream_6_conc_D -
stream_7_flow * stream_7_conc_D + reactor_conv_rate_D
dreactor_NE = stream_6_flow * stream_6_conc_E -
stream_7_flow * stream_7_conc_E + reactor_conv_rate_E
dreactor_NF = stream_6_flow * stream_6_conc_F -
stream_7_flow * stream_7_conc_F + reactor_conv_rate_F
dreactor_NG = stream_6_flow * stream_6_conc_G -
stream_7_flow * stream_7_conc_G + reactor_conv_rate_G
dreactor_NH = stream_6_flow * stream_6_conc_H -
stream_7_flow * stream_7_conc_H + reactor_conv_rate_H
```

```
# Eq. 5.2
 reactor_Ncp =  reactor_NA * cp_vap_A + reactor_NB *
cp_vap_B + reactor_NC * cp_vap_C + reactor_ND * cp_liq_D +
reactor_NE * cp_liq_E + reactor_NF * cp_liq_F + reactor_NG
* cp_liq_G + reactor_NH * cp_liq_H
 stream_6_cp = stream_6_conc_A * cp_vap_A +
stream_6_conc_B * cp_vap_B + stream_6_conc_C * cp_vap_C +
stream_6_conc_D * cp_vap_D + stream_6_conc_E * cp_vap_E +
stream_6_conc_F * cp_vap_F + stream_6_conc_G * cp_vap_G +
stream_6_conc_H * cp_vap_H


# Eq. 5.10
 H_A = cp_vap_A * (Tr - Tref)
 H_B = cp_vap_B * (Tr - Tref)
 H_C = cp_vap_C * (Tr - Tref)
 H_D = cp_vap_D * (Tr - Tref)
 H_E = cp_vap_E * (Tr - Tref)
 H_F = cp_vap_F * (Tr - Tref)
 H_G = cp_vap_G * (Tr - Tref)
 H_H = cp_vap_H * (Tr - Tref)

delt_Hr_1 = (mue_A_1 * H_A + mue_C_1 * H_C + mue_D_1 * H_D +
mue_G_1 * H_G - 136033.04e0) / 1000
delt_Hr_2 = (mue_A_2 * H_A + mue_C_2 * H_C + mue_E_2 * H_E +
mue_H_2 * H_H - 93337.9616e0) / 1000
delt_Hr_3 = (mue_A_3 * H_A + mue_D_3 * H_D + mue_E_3 * H_E +
mue_F_3 * H_F + 0) / 1000


# Eq. 5.9
reactor_exoth_heat = reactor_R1 * delt_Hr_1 + reactor_R2 *
delt_Hr_2 + reactor_R3 * delt_Hr_3

# --- Heat Exchanger ------------------
# Heat exchange with cooling water
 T_CWSr_in = 0.308000000000000e+03

# Assumption - from measurement data
  T_CWSr_out = 0.367599000000000e+03

# Preparation for delt_T_log
 delt_T1_reactor = Tr - T_CWSr_in
 delt_T2_reactor = Tr - T_CWSr_out

#  delt_T1_reactor/(delt_T2_reactor+0.000000001))
^delt_T_log = exp(delt_T1_reactor - delt_T2_reactor)
```

```
 delt_T_log=(delt_T1_reactor-delt_T2_reactor)/(ln
(delt_T1_reactor/(delt_T2_reactor)))

#Eq.5.22
 Qr=reactor_m_CWS*cp_cw*(T_CWSr_out-T_CWSr_in)/1000

#Delt T log try
  Qr=UA*delt_T_log/1000

reactor_rhs=((1/1000)*stream_6_flow*stream_6_cp*
(Tm-Tr)-Qr-reactor_exoth_heat)/reactor_Ncp

#Temperature in the reaction zone
 dTr=reactor_rhs
 ;Initial condition Tr= 393.55 (From Table 1)


#####################################################
###
#################################
#                               #
#  Separator                    #
#                               #
#################################

#Eq 5.26 - Base case initial values - Taken from TE_20_10.
cmp (Andreas Waechter)

#Initial Condition
  ;separator_NA=0.274979760720663e+02
  ;separator_NB=0.120963446339331e+02
  ;separator_NC=0.245720622755470e+02
  ;separator_ND=0.836049078688156e-01
  ;separator_NE=0.586405721828844e+01
  ;separator_NF=0.901124942834656e+00
  ;separator_NG=0.241376638377975e+02
  ;separator_NH=0.197525541355506e+02

dseparator_NA=stream_7_flow*stream_7_conc_A-
(stream_8_flow+stream_9_flow)*stream_8_conc_A-
stream_10_flow*separator_x_A
dseparator_NB=stream_7_flow*stream_7_conc_B-
(stream_8_flow+stream_9_flow)*stream_8_conc_B-
stream_10_flow*separator_x_B
dseparator_NC=stream_7_flow*stream_7_conc_C-
(stream_8_flow+stream_9_flow)*stream_8_conc_C-
stream_10_flow*separator_x_C
```

```
dseparator_ND = stream_7_flow * stream_7_conc_D -
(stream_8_flow + stream_9_flow) * stream_8_conc_D -
stream_10_flow * separator_x_D
dseparator_NE = stream_7_flow * stream_7_conc_E -
(stream_8_flow + stream_9_flow) * stream_8_conc_E -
stream_10_flow * separator_x_E
dseparator_NF = stream_7_flow * stream_7_conc_F -
(stream_8_flow + stream_9_flow) * stream_8_conc_F -
stream_10_flow * separator_x_F
dseparator_NG = stream_7_flow * stream_7_conc_G -
(stream_8_flow + stream_9_flow) * stream_8_conc_G -
stream_10_flow * separator_x_G
dseparator_NH = stream_7_flow * stream_7_conc_H -
(stream_8_flow + stream_9_flow) * stream_8_conc_H -
stream_10_flow * separator_x_H

# Eq. 5.27
separator_Ncp = separator_NA * cp_vap_A + separator_NB *
cp_vap_B + separator_NC * cp_vap_C + separator_ND *
cp_liq_D + separator_NE * cp_liq_E + separator_NF *
cp_liq_F + separator_NG * cp_liq_G + separator_NH *
cp_liq_H

stream_7_cp  = stream_7_conc_A * cp_vap_A +
stream_7_conc_B * cp_vap_B + stream_7_conc_C * cp_vap_C +
stream_7_conc_D * cp_vap_D + stream_7_conc_E * cp_vap_E +
stream_7_conc_F * cp_vap_F + stream_7_conc_G * cp_vap_G +
stream_7_conc_H * cp_vap_H

# Eq. 5.37
 HoVs = (separator_x_D * stream_10_flow * M_D * H_vap_D +
separator_x_E * stream_10_flow * M_E * H_vap_E +
separator_x_F * stream_10_flow * M_F * H_vap_F +
separator_x_G * stream_10_flow * M_G * H_vap_G +
separator_x_H * stream_10_flow * M_H * H_vap_H) / 1000

# Heat exchange with cooling water
T_CWSs_in = 0.313000000000000e+03

# Assumption - from measurement data
  T_CWSs_out = 350.447

# Eq. 5.38

Qs = separator_m_CWS * cp_cw * (T_CWSs_out - T_CWSs_in) / 1000
```

```
# Preparation for delt_T_log
  delt_T1_separator = Ts - T_CWSs_in
  delt_T2_separator = Ts - T_CWSs_out

 delt_T_log_sep = (delt_T1_separator - delt_T2_separator)/
(ln(delt_T1_separator/(delt_T2_separator)))

Qs = UA_sep * (delt_T_log_sep) / 1000

separator_rhs = ( (1/1000) * stream_7_flow * stream_7_cp *
(Tr - Ts) + HoVs - Qs)/ separator_Ncp

# Temperature in the separation zone
   dTs = separator_rhs

# Initial condition Ts = 353.25 (Table 1)

####################################################
########

#################################
#                               #
#   Stripper                    #
#                               #
#################################

# Eq 5.48
# Base case initial values - accumulation of A-F is neglected
; Initial conditions
;  stripper_NG = 0.203813521021438e+02
;  stripper_NH = 0.173130910905421e+02

dstripper_NG = dNstr_G
dstripper_NH = dNstr_H


# Eq. 5.27
stripper_Ncp = stripper_NG * cp_liq_G + stripper_NH *
cp_liq_H
stream_10_cp = separator_x_A * cp_liq_A + separator_x_B *
cp_liq_B + separator_x_C * cp_liq_C + separator_x_D *
cp_liq_D + separator_x_E * cp_liq_E + separator_x_F *
cp_liq_F + separator_x_G * cp_liq_G + separator_x_H *
cp_liq_H
```

```
HoVstr = ((stream_5_conc_D * stream_5_flow -
stream_4_conc_D * stream_4_flow) * M_D * H_vap_D +
(stream_5_conc_E * stream_5_flow - stream_4_conc_E *
stream_4_flow) * M_E * H_vap_E + (stream_5_conc_F *
stream_5_flow - stream_4_conc_F * stream_4_flow) * M_F *
H_vap_F + (stream_5_conc_G * stream_5_flow -
stream_4_conc_G * stream_4_flow) * M_G * H_vap_G +
(stream_5_conc_H * stream_5_flow - stream_4_conc_H *
stream_4_flow) * M_H * H_vap_H) / 1000

# Eq. 5.58
 Qstr = 2.258717 * stripper_m_CWS

stripper_rhs =( (1/1000) * stream_10_flow * stream_10_cp *
(Ts - Tstr) + (1/1000) * stream_4_flow * stream_4_cp *
(stream_4_T - Tstr) - HoVstr + Qstr)/ stripper_Ncp

# Temperature in the stripper zone
  dTstr = stripper_rhs
 ;Initial Condition Tstr = 338.85

XMEAS_6  = stream_6_flow / 44.79*3600
XMEAS_7  = pr_MPa*1000 - 101
XMEAS_8  = 5.263*Vr_liq - 12.105
XMEAS_9  = Tr - Tref
```

A1-10: MODELS FROM CHAPTER 10

A1.10.1 One-dimensional batch crystallisation model (method of moments)

```
# Chapter 10-1
# Sucrose Crystallization (Quintana-Hernandez et al.)
# One-dimensional batch crystallization model (mthod of
moments)

#Solubility [73]

Csat= (0.7533+0.00225*(T-65))/(1-(0.7533+0.00225*(T-
65)))

#Supersaturation

S=if((C>Csat) then ((C-Csat)/Csat) else (0.0))

#Stirrer speed

N1=if((t<30) then (600) else (N1-0.08707))
```

```
# Growth rate

G=Kg*S^g*N1^q

#Nucleation rate

B=Kb*S^b*M_c^j*N1^p

# Production reduction rate

alpha=Ka*S^a*M_c^k*N1^r

# Mass of the crystal

M_c=rho_c*Kv*mue_3*2230

dmue_0=B + alpha
dmue_1=G*mue_0
dmue_2=2*G*mue_1
dmue_3=3*G*mue_2
dmue_4=4*G*mue_3

diameter=(mue_4/mue_3)*10000


#mass balance

dC=(-rho_c*Kv*V*(3*G*mue_2+B*Lo^3)/Mw)

H_c=if((t<20) then (-12.2115-0.7937*T) else (0))

dT=-((H_c)*rho_c*Kv*V*(3*G*mue_2+B*Lo^3)+U1*A1*(T-Tw))/
(rho*V*Cp)

Fwin = if((t<5) then (8100) else (0))

dTw=(rho_w*Fwin*Cp_w*(Tw_in-Tw)+U1*A1*(T-Tw)+U2*A2*
(Tex-Tw))/(rho_w*Cp_w*Vw)
```

A1.10.2 Batch distillation

This model was not solved with ICAS-MoT

A1.11 PARAMETER ESTIMATION

A1.11.1 Modelling simple reaction kinetics

See the model given in A1.11-2

A1.11.2 Kinetic modelling parameter estimation: ch-11-2-kin-fit-ode-1.mot

```
#Example DOPT-2
#Optimization Problem
#First-Order Reversible chain reaction.
#************************************
# If the reaction in problem 1 is extended to reversible
# reaction
#
#          k1    k3
#       A  <--> B <-- > C
#          k2    k4
#
# In this case all of the components are measured, and
# therefore their concentrations are include in the model
# used for estimation. The differential equation model
# takes the form:

#Calcualted model
  dz1 = - k1*z1 + k2*z2
  dz2 =  k1*z1 - (k2 + k3)*z2 + k4*z3
  dz3 =  k3*z2 - k4*z3

   ;Where z1 = [A], z2= [B] and z3 = [C]


#Objective function.
 ;Ordinary least square(implicit)
```

A1.11.3 Kinetic modelling – parameter estimation (2): ch-11-3-kin-fit-ode-2.mot

```
#Dynamic Optimization Problem
#Catalytic Cracking of Gas Oil.
#*******************************
# This problem describes an overall reaction of catalytic
# cracking of gas oil(A)
# to gasoline (Q) and other side products (S).
#       k1
#    A -------> Q
#              /
#    k2      / k3
#          /
#         S
#
#Only the concentration of A an Q were measured, therefore,
# the concentration of S does not appear in the model for
# estimation.
```

```
#The model
  dz1 = - (k1 + k3)*z1^2
  dz2 =   k1*z1^2 - k2*z2

 ;Where z1 = [A] and z2= [Q]

#Objective function.
 ;Ordinary least square(implicit)
```

A1.11.4 Kinetic modelling – parameter estimation (3)

11.4.1 Linear least squares (model-1): ch-11-4-kin-fit-pl-1.mot

```
# Define variables for linear model
y[j]  = ln(r_exp[j])
x1[j] = 1/T[j]
x2[j] = ln(Pa[j])
x3[j] = ln(Pb[j])
x4[j] = ln(Pc[j])

SUMY    = sum_j(y[j])
SUMX1   = sum_j(x1[j])
SUMX2   = sum_j(x2[j])
SUMX3   = sum_j(x3[j])
SUMX4   = sum_j(x4[j])

SUMYX1  = sum_j(y[j]*x1[j])
SUMYX2  = sum_j(y[j]*x2[j])
SUMYX3  = sum_j(y[j]*x3[j])
SUMYX4  = sum_j(y[j]*x4[j])

SUMX1X1 = sum_j(x1[j]*x1[j])
SUMX2X1 = sum_j(x2[j]*x1[j])
SUMX3X1 = sum_j(x3[j]*x1[j])
SUMX4X1 = sum_j(x4[j]*x1[j])

SUMX1X2 = sum_j(x1[j]*x2[j])
SUMX2X2 = sum_j(x2[j]*x2[j])
SUMX3X2 = sum_j(x3[j]*x2[j])
SUMX4X2 = sum_j(x4[j]*x2[j])

SUMX1X3 = sum_j(x1[j]*x3[j])
SUMX2X3 = sum_j(x2[j]*x3[j])
SUMX3X3 = sum_j(x3[j]*x3[j])
SUMX4X3 = sum_j(x4[j]*x3[j])

SUMX1X4 = sum_j(x1[j]*x4[j])
SUMX2X4 = sum_j(x2[j]*x4[j])
```

```
SUMX3X4 = sum_j(x3[j]*x4[j])
SUMX4X4 = sum_j(x4[j]*x4[j])

# Solve linear system of equations
0=a0*N      +a1*SUMX1   +a2*SUMX2   +a3*SUMX3   +a4*SUMX4   -SUMY
0=a0*SUMX1  +a1*SUMX1X1+a2*SUMX1X2+a3*SUMX1X3+a4*SUMX1X4-SUMYX1
0=a0*SUMX2  +a1*SUMX2X1+a2*SUMX2X2+a3*SUMX2X3+a4*SUMX2X4-SUMYX2
0=a0*SUMX3  +a1*SUMX3X1+a2*SUMX3X2+a3*SUMX3X3+a4*SUMX3X4-SUMYX3
0=a0*SUMX4  +a1*SUMX4X1+a2*SUMX4X2+a3*SUMX4X3+a4*SUMX4X4-SUMYX4

# Insert parameters into model
R = 82.05

k40   = exp(a0)
E4    = - a1*R
alpha = a2
beta  = a3
gama  = a4

k4[j]  = k40*exp(-E4/R/T[j])

r_calc[j] = k4[j]*(Pa[j]^alpha)*(Pb[j]^beta)*Pc[j]
^gama


# Compute errors
error[j] = (r_exp[j] - r_calc[j])^2
SUMerror  = sum_j( w[j]*error[j] )
```

11.4.2 Non-linear least squares (model-1): ch-11-4-kin-fit-pl-2.mot

```
# Model
R = 82.05

k4[j]   = k40*exp(-E4/R/T[j])
f_a[j] = Pa[j]^alpha
f_b[j] = Pb[j]^beta
f_c[j] = Pc[j]^gama

r_calc[j] = k4[j]*f_a[j]*f_b[j]*f_c[j]

# Calculate objective function
error[j] = (r_exp[j] - r_calc[j])^2
OBJ      = sum_j( w[j]*error[j] )
```

11.4.3 Non-linear least squares (model-2): ch-11-4-kin-fit-lh-1.mot

```
# Model
R = 82.05

K1[j]  = K10*exp(-H1/R/T[j])
K3[j]  = K30*exp(-H3/R/T[j])
k4[j]  = k40*exp(-E4/R/T[j])

f_a[j] = K1[j]*Pa[j]
f_c[j] = K3[j]*Pc[j]

r_calc[j] = k4[j]*K1[j]*Pa[j]*Pb[j] / (1+f_a[j] +f_c[j])

# Calculate objective function
error[j] = (r_exp[j] - r_calc[j])^2
OBJ      = sum_j( w[j]*error[j] )
```

A1.11.5 Kinetic modelling – parameter estimation with maximum likelihood method: ch-11-5-kin-fit-lh-1.mot

```
#Constant variables

PI   = acos(-1)
N    = 32
SEED = 2
R    = 82.05
rmax = 3.676662

#Reaction rate model 1
;kinetic parameter - Arrhenius dependence

K1[j]  = K10*exp(-H1/R/T[j])
K3[j]  = K30*exp(-H3/R/T[j])
k4[j]  = k40*exp(-E4/R/T[j])

f_a[j] = K1[j]*Pa[j]
f_c[j] = K3[j]*Pc[j]

r_model[j] = k4[j]*K1[j]*Pa[j]*Pb[j] / (1+f_a[j] +f_c[j])


#Probability Distribution Function
EPSILON = random(SEED)*OMMEGA*rmax
y_model[j] = r_model[j]*(1 - r_model[j]^(GAMMA-1)
*EPSILON)
```

```
#Maximum likelihood funtion for a normal distribution
  error[j] = (y_exp[j] - y_model[j])^2
  SSUM   = sum_j(error[j])

 Obj_lnf = -(-0.5*N*ln(2*PI) - N*ln(SIGMA) - SSUM/
(2*SIGMA^2))
```

A1. 11.6 Application of orthogonal collocation for dynamic optimisation (parameter estimation)

11.6.1 Finite difference: ch-11-6-finite-element.mot

Model equations (MoT) for solving the optimisation problem by finite difference approximation between the collocation points (4 elements whith 20 collocation points each)

```
# use collocation points from orthogonal collocation and
do finite difference
# calculate delta_tj of discretization
delta_t1=t[1]-t0
delta_t2=t[2]-t[1]
delta_t3=t[3]-t[2]
delta_t4=t[4]-t[3]
delta_t5=t[5]-t[4]
delta_t6=t[6]-t[5]
delta_t7=t[7]-t[6]
delta_t8=t[8]-t[7]
delta_t9=t[9]-t[8]
delta_t10=t[10]-t[9]
delta_t11=t[11]-t[10]
delta_t12=t[12]-t[11]
delta_t13=t[13]-t[12]
delta_t14=t[14]-t[13]
delta_t15=t[15]-t[14]
delta_t16=t[16]-t[15]
delta_t17=t[17]-t[16]
delta_t18=t[18]-t[17]
delta_t19=t[19]-t[18]
delta_t20=t[20]-t[19]
delta_t21=t[21]-t[20]
delta_t22=t[22]-t[21]
delta_t23=t[23]-t[22]
delta_t24=t[24]-t[23]
delta_t25=t[25]-t[24]
delta_t26=t[26]-t[25]
delta_t27=t[27]-t[26]
delta_t28=t[28]-t[27]
delta_t29=t[29]-t[28]
```

```
delta_t30=t[30]-t[29]
delta_t31=t[31]-t[30]
delta_t32=t[32]-t[31]
delta_t33=t[33]-t[32]
delta_t34=t[34]-t[33]
delta_t35=t[35]-t[34]
delta_t36=t[36]-t[35]
delta_t37=t[37]-t[36]
delta_t38=t[38]-t[37]
delta_t39=t[39]-t[38]
delta_t40=t[40]-t[39]
delta_t41=t[41]-t[40]
delta_t42=t[42]-t[41]
delta_t43=t[43]-t[42]
delta_t44=t[44]-t[43]
delta_t45=t[45]-t[44]
delta_t46=t[46]-t[45]
delta_t47=t[47]-t[46]
delta_t48=t[48]-t[47]
delta_t49=t[49]-t[48]
delta_t50=t[50]-t[49]
delta_t51=t[51]-t[50]
delta_t52=t[52]-t[51]
delta_t53=t[53]-t[52]
delta_t54=t[54]-t[53]
delta_t55=t[55]-t[54]
delta_t56=t[56]-t[55]
delta_t57=t[57]-t[56]
delta_t58=t[58]-t[57]
delta_t59=t[59]-t[58]
delta_t60=t[60]-t[59]
delta_t61=t[61]-t[60]
delta_t62=t[62]-t[61]
delta_t63=t[63]-t[62]
delta_t64=t[64]-t[63]
delta_t65=t[65]-t[64]
delta_t66=t[66]-t[65]
delta_t67=t[67]-t[66]
delta_t68=t[68]-t[67]
delta_t69=t[69]-t[68]
delta_t70=t[70]-t[69]
delta_t71=t[71]-t[70]
delta_t72=t[72]-t[71]
delta_t73=t[73]-t[72]
delta_t74=t[74]-t[73]
delta_t75=t[75]-t[74]
delta_t76=t[76]-t[75]
```

```
delta_t77=t[77]-t[76]
delta_t78=t[78]-t[77]
delta_t79=t[79]-t[78]
delta_t80=t[80]-t[79]

# discretized equations
ya0=1
yb0=0
yc0=0

ya1=ya0+(-k1*ya0+k2*yb0)*delta_t1
yb1=yb0+(k1*ya0-(k2+k3)*yb0+k4*yc0)*delta_t1
yc1=yc0+(k3*yb0-k4*yc0)*delta_t1

ya2=ya1+(-k1*ya1+k2*yb1)*delta_t2
yb2=yb1+(k1*ya1-(k2+k3)*yb1+k4*yc1)*delta_t2
yc2=yc1+(k3*yb1-k4*yc1)*delta_t2

ya3=ya2+(-k1*ya2+k2*yb2)*delta_t3
yb3=yb2+(k1*ya2-(k2+k3)*yb2+k4*yc2)*delta_t3
yc3=yc2+(k3*yb2-k4*yc2)*delta_t3

ya4=ya3+(-k1*ya3+k2*yb3)*delta_t4
yb4=yb3+(k1*ya3-(k2+k3)*yb3+k4*yc3)*delta_t4
yc4=yc3+(k3*yb3-k4*yc3)*delta_t4

ya5=ya4+(-k1*ya4+k2*yb4)*delta_t5
yb5=yb4+(k1*ya4-(k2+k3)*yb4+k4*yc4)*delta_t5
yc5=yc4+(k3*yb4-k4*yc4)*delta_t5

ya6=ya5+(-k1*ya5+k2*yb5)*delta_t6
yb6=yb5+(k1*ya5-(k2+k3)*yb5+k4*yc5)*delta_t6
yc6=yc5+(k3*yb5-k4*yc5)*delta_t6

ya7=ya6+(-k1*ya6+k2*yb6)*delta_t7
yb7=yb6+(k1*ya6-(k2+k3)*yb6+k4*yc6)*delta_t7
yc7=yc6+(k3*yb6-k4*yc6)*delta_t7

ya8=ya7+(-k1*ya7+k2*yb7)*delta_t8
yb8=yb7+(k1*ya7-(k2+k3)*yb7+k4*yc7)*delta_t8
yc8=yc7+(k3*yb7-k4*yc7)*delta_t8

ya9=ya8+(-k1*ya8+k2*yb8)*delta_t9
yb9=yb8+(k1*ya8-(k2+k3)*yb8+k4*yc8)*delta_t9
yc9=yc8+(k3*yb8-k4*yc8)*delta_t9
```

```
ya10=ya9+(-k1*ya9+k2*yb9)*delta_t10
yb10=yb9+(k1*ya9-(k2+k3)*yb9+k4*yc9)*delta_t10
yc10=yc9+(k3*yb9-k4*yc9)*delta_t10

ya11=ya10+(-k1*ya10+k2*yb10)*delta_t11
yb11=yb10+(k1*ya10-(k2+k3)*yb10+k4*yc10)*delta_t11
yc11=yc10+(k3*yb10-k4*yc10)*delta_t11

ya12=ya11+(-k1*ya11+k2*yb11)*delta_t12
yb12=yb11+(k1*ya11-(k2+k3)*yb11+k4*yc11)*delta_t12
yc12=yc11+(k3*yb11-k4*yc11)*delta_t12

ya13=ya12+(-k1*ya12+k2*yb12)*delta_t13
yb13=yb12+(k1*ya12-(k2+k3)*yb12+k4*yc12)*delta_t13
yc13=yc12+(k3*yb12-k4*yc12)*delta_t13

ya14=ya13+(-k1*ya13+k2*yb13)*delta_t14
yb14=yb13+(k1*ya13-(k2+k3)*yb13+k4*yc13)*delta_t14
yc14=yc13+(k3*yb13-k4*yc13)*delta_t14

ya15=ya14+(-k1*ya14+k2*yb14)*delta_t15
yb15=yb14+(k1*ya14-(k2+k3)*yb14+k4*yc14)*delta_t15
yc15=yc14+(k3*yb14-k4*yc14)*delta_t15

ya16=ya15+(-k1*ya15+k2*yb15)*delta_t16
yb16=yb15+(k1*ya15-(k2+k3)*yb15+k4*yc15)*delta_t16
yc16=yc15+(k3*yb15-k4*yc15)*delta_t16

ya17=ya16+(-k1*ya16+k2*yb16)*delta_t17
yb17=yb16+(k1*ya16-(k2+k3)*yb16+k4*yc16)*delta_t17
yc17=yc16+(k3*yb16-k4*yc16)*delta_t17

ya18=ya17+(-k1*ya17+k2*yb17)*delta_t18
yb18=yb17+(k1*ya17-(k2+k3)*yb17+k4*yc17)*delta_t18
yc18=yc17+(k3*yb17-k4*yc17)*delta_t18

ya19=ya18+(-k1*ya18+k2*yb18)*delta_t19
yb19=yb18+(k1*ya18-(k2+k3)*yb18+k4*yc18)*delta_t19
yc19=yc18+(k3*yb18-k4*yc18)*delta_t19

ya20=ya19+(-k1*ya19+k2*yb19)*delta_t20
yb20=yb19+(k1*ya19-(k2+k3)*yb19+k4*yc19)*delta_t20
yc20=yc19+(k3*yb19-k4*yc19)*delta_t20

ya21=ya20+(-k1*ya20+k2*yb20)*delta_t21
yb21=yb20+(k1*ya20-(k2+k3)*yb20+k4*yc20)*delta_t21
yc21=yc20+(k3*yb20-k4*yc20)*delta_t21
```

```
ya22=ya21+(-k1*ya21+k2*yb21)*delta_t22
yb22=yb21+(k1*ya21-(k2+k3)*yb21+k4*yc21)*delta_t22
yc22=yc21+(k3*yb21-k4*yc21)*delta_t22

ya23=ya22+(-k1*ya22+k2*yb22)*delta_t23
yb23=yb22+(k1*ya22-(k2+k3)*yb22+k4*yc22)*delta_t23
yc23=yc22+(k3*yb22-k4*yc22)*delta_t23

ya24=ya23+(-k1*ya23+k2*yb23)*delta_t24
yb24=yb23+(k1*ya23-(k2+k3)*yb23+k4*yc23)*delta_t24
yc24=yc23+(k3*yb23-k4*yc23)*delta_t24

ya25=ya24+(-k1*ya24+k2*yb24)*delta_t25
yb25=yb24+(k1*ya24-(k2+k3)*yb24+k4*yc24)*delta_t25
yc25=yc24+(k3*yb24-k4*yc24)*delta_t25

ya26=ya25+(-k1*ya25+k2*yb25)*delta_t26
yb26=yb25+(k1*ya25-(k2+k3)*yb25+k4*yc25)*delta_t26
yc26=yc25+(k3*yb25-k4*yc25)*delta_t26

ya27=ya26+(-k1*ya26+k2*yb26)*delta_t27
yb27=yb26+(k1*ya26-(k2+k3)*yb26+k4*yc26)*delta_t27
yc27=yc26+(k3*yb26-k4*yc26)*delta_t27

ya28=ya27+(-k1*ya27+k2*yb27)*delta_t28
yb28=yb27+(k1*ya27-(k2+k3)*yb27+k4*yc27)*delta_t28
yc28=yc27+(k3*yb27-k4*yc27)*delta_t28

ya29=ya28+(-k1*ya28+k2*yb28)*delta_t29
yb29=yb28+(k1*ya28-(k2+k3)*yb28+k4*yc28)*delta_t29
yc29=yc28+(k3*yb28-k4*yc28)*delta_t29

ya30=ya29+(-k1*ya29+k2*yb29)*delta_t30
yb30=yb29+(k1*ya29-(k2+k3)*yb29+k4*yc29)*delta_t30
yc30=yc29+(k3*yb29-k4*yc29)*delta_t30

ya31=ya30+(-k1*ya30+k2*yb30)*delta_t31
yb31=yb30+(k1*ya30-(k2+k3)*yb30+k4*yc30)*delta_t31
yc31=yc30+(k3*yb30-k4*yc30)*delta_t31

ya32=ya31+(-k1*ya31+k2*yb31)*delta_t32
yb32=yb31+(k1*ya31-(k2+k3)*yb31+k4*yc31)*delta_t32
yc32=yc31+(k3*yb31-k4*yc31)*delta_t32

ya33=ya32+(-k1*ya32+k2*yb32)*delta_t33
yb33=yb32+(k1*ya32-(k2+k3)*yb32+k4*yc32)*delta_t33
yc33=yc32+(k3*yb32-k4*yc32)*delta_t33
```

```
ya34=ya33+(-k1*ya33+k2*yb33)*delta_t34
yb34=yb33+(k1*ya33-(k2+k3)*yb33+k4*yc33)*delta_t34
yc34=yc33+(k3*yb33-k4*yc33)*delta_t34

ya35=ya34+(-k1*ya34+k2*yb34)*delta_t35
yb35=yb34+(k1*ya34-(k2+k3)*yb34+k4*yc34)*delta_t35
yc35=yc34+(k3*yb34-k4*yc34)*delta_t35

ya36=ya35+(-k1*ya35+k2*yb35)*delta_t36
yb36=yb35+(k1*ya35-(k2+k3)*yb35+k4*yc35)*delta_t36
yc36=yc35+(k3*yb35-k4*yc35)*delta_t36

ya37=ya36+(-k1*ya36+k2*yb36)*delta_t37
yb37=yb36+(k1*ya36-(k2+k3)*yb36+k4*yc36)*delta_t37
yc37=yc36+(k3*yb36-k4*yc36)*delta_t37

ya38=ya37+(-k1*ya37+k2*yb37)*delta_t38
yb38=yb37+(k1*ya37-(k2+k3)*yb37+k4*yc37)*delta_t38
yc38=yc37+(k3*yb37-k4*yc37)*delta_t38

ya39=ya38+(-k1*ya38+k2*yb38)*delta_t39
yb39=yb38+(k1*ya38-(k2+k3)*yb38+k4*yc38)*delta_t39
yc39=yc38+(k3*yb38-k4*yc38)*delta_t39

ya40=ya39+(-k1*ya39+k2*yb39)*delta_t40
yb40=yb39+(k1*ya39-(k2+k3)*yb39+k4*yc39)*delta_t40
yc40=yc39+(k3*yb39-k4*yc39)*delta_t40

ya41=ya40+(-k1*ya40+k2*yb40)*delta_t41
yb41=yb40+(k1*ya40-(k2+k3)*yb40+k4*yc40)*delta_t41
yc41=yc40+(k3*yb40-k4*yc40)*delta_t41

ya42=ya41+(-k1*ya41+k2*yb41)*delta_t42
yb42=yb41+(k1*ya41-(k2+k3)*yb41+k4*yc41)*delta_t42
yc42=yc41+(k3*yb41-k4*yc41)*delta_t42

ya43=ya42+(-k1*ya42+k2*yb42)*delta_t43
yb43=yb42+(k1*ya42-(k2+k3)*yb42+k4*yc42)*delta_t43
yc43=yc42+(k3*yb42-k4*yc42)*delta_t43

ya44=ya43+(-k1*ya43+k2*yb43)*delta_t44
yb44=yb43+(k1*ya43-(k2+k3)*yb43+k4*yc43)*delta_t44
yc44=yc43+(k3*yb43-k4*yc43)*delta_t44

ya45=ya44+(-k1*ya44+k2*yb44)*delta_t45
yb45=yb44+(k1*ya44-(k2+k3)*yb44+k4*yc44)*delta_t45
yc45=yc44+(k3*yb44-k4*yc44)*delta_t45
```

```
ya46=ya45+(-k1*ya45+k2*yb45)*delta_t46
yb46=yb45+(k1*ya45-(k2+k3)*yb45+k4*yc45)*delta_t46
yc46=yc45+(k3*yb45-k4*yc45)*delta_t46

ya47=ya46+(-k1*ya46+k2*yb46)*delta_t47
yb47=yb46+(k1*ya46-(k2+k3)*yb46+k4*yc46)*delta_t47
yc47=yc46+(k3*yb46-k4*yc46)*delta_t47

ya48=ya47+(-k1*ya47+k2*yb47)*delta_t48
yb48=yb47+(k1*ya47-(k2+k3)*yb47+k4*yc47)*delta_t48
yc48=yc47+(k3*yb47-k4*yc47)*delta_t48

ya49=ya48+(-k1*ya48+k2*yb48)*delta_t49
yb49=yb48+(k1*ya48-(k2+k3)*yb48+k4*yc48)*delta_t49
yc49=yc48+(k3*yb48-k4*yc48)*delta_t49

ya50=ya49+(-k1*ya49+k2*yb49)*delta_t50
yb50=yb49+(k1*ya49-(k2+k3)*yb49+k4*yc49)*delta_t50
yc50=yc49+(k3*yb49-k4*yc49)*delta_t50

ya51=ya50+(-k1*ya50+k2*yb50)*delta_t51
yb51=yb50+(k1*ya50-(k2+k3)*yb50+k4*yc50)*delta_t51
yc51=yc50+(k3*yb50-k4*yc50)*delta_t51

ya52=ya51+(-k1*ya51+k2*yb51)*delta_t52
yb52=yb51+(k1*ya51-(k2+k3)*yb51+k4*yc51)*delta_t52
yc52=yc51+(k3*yb51-k4*yc51)*delta_t52

ya53=ya52+(-k1*ya52+k2*yb52)*delta_t53
yb53=yb52+(k1*ya52-(k2+k3)*yb52+k4*yc52)*delta_t53
yc53=yc52+(k3*yb52-k4*yc52)*delta_t53

ya54=ya53+(-k1*ya53+k2*yb53)*delta_t54
yb54=yb53+(k1*ya53-(k2+k3)*yb53+k4*yc53)*delta_t54
yc54=yc53+(k3*yb53-k4*yc53)*delta_t54

ya55=ya54+(-k1*ya54+k2*yb54)*delta_t55
yb55=yb54+(k1*ya54-(k2+k3)*yb54+k4*yc54)*delta_t55
yc55=yc54+(k3*yb54-k4*yc54)*delta_t55

ya56=ya55+(-k1*ya55+k2*yb55)*delta_t56
yb56=yb55+(k1*ya55-(k2+k3)*yb55+k4*yc55)*delta_t56
yc56=yc55+(k3*yb55-k4*yc55)*delta_t56

ya57=ya56+(-k1*ya56+k2*yb56)*delta_t57
yb57=yb56+(k1*ya56-(k2+k3)*yb56+k4*yc56)*delta_t57
yc57=yc56+(k3*yb56-k4*yc56)*delta_t57
```

```
ya58=ya57+(-k1*ya57+k2*yb57)*delta_t58
yb58=yb57+(k1*ya57-(k2+k3)*yb57+k4*yc57)*delta_t58
yc58=yc57+(k3*yb57-k4*yc57)*delta_t58

ya59=ya58+(-k1*ya58+k2*yb58)*delta_t59
yb59=yb58+(k1*ya58-(k2+k3)*yb58+k4*yc58)*delta_t59
yc59=yc58+(k3*yb58-k4*yc58)*delta_t59

ya60=ya59+(-k1*ya59+k2*yb59)*delta_t60
yb60=yb59+(k1*ya59-(k2+k3)*yb59+k4*yc59)*delta_t60
yc60=yc59+(k3*yb59-k4*yc59)*delta_t60

ya61=ya60+(-k1*ya60+k2*yb60)*delta_t61
yb61=yb60+(k1*ya60-(k2+k3)*yb60+k4*yc60)*delta_t61
yc61=yc60+(k3*yb60-k4*yc60)*delta_t61

ya62=ya61+(-k1*ya61+k2*yb61)*delta_t62
yb62=yb61+(k1*ya61-(k2+k3)*yb61+k4*yc61)*delta_t62
yc62=yc61+(k3*yb61-k4*yc61)*delta_t62

ya63=ya62+(-k1*ya62+k2*yb62)*delta_t63
yb63=yb62+(k1*ya62-(k2+k3)*yb62+k4*yc62)*delta_t63
yc63=yc62+(k3*yb62-k4*yc62)*delta_t63

ya64=ya63+(-k1*ya63+k2*yb63)*delta_t64
yb64=yb63+(k1*ya63-(k2+k3)*yb63+k4*yc63)*delta_t64
yc64=yc63+(k3*yb63-k4*yc63)*delta_t64

ya65=ya64+(-k1*ya64+k2*yb64)*delta_t65
yb65=yb64+(k1*ya64-(k2+k3)*yb64+k4*yc64)*delta_t65
yc65=yc64+(k3*yb64-k4*yc64)*delta_t65

ya66=ya65+(-k1*ya65+k2*yb65)*delta_t66
yb66=yb65+(k1*ya65-(k2+k3)*yb65+k4*yc65)*delta_t66
yc66=yc65+(k3*yb65-k4*yc65)*delta_t66

ya67=ya66+(-k1*ya66+k2*yb66)*delta_t67
yb67=yb66+(k1*ya66-(k2+k3)*yb66+k4*yc66)*delta_t67
yc67=yc66+(k3*yb66-k4*yc66)*delta_t67

ya68=ya67+(-k1*ya67+k2*yb67)*delta_t68
yb68=yb67+(k1*ya67-(k2+k3)*yb67+k4*yc67)*delta_t68
yc68=yc67+(k3*yb67-k4*yc67)*delta_t68

ya69=ya68+(-k1*ya68+k2*yb68)*delta_t69
yb69=yb68+(k1*ya68-(k2+k3)*yb68+k4*yc68)*delta_t69
yc69=yc68+(k3*yb68-k4*yc68)*delta_t69
```

```
ya70=ya69+(-k1*ya69+k2*yb69)*delta_t70
yb70=yb69+(k1*ya69-(k2+k3)*yb69+k4*yc69)*delta_t70
yc70=yc69+(k3*yb69-k4*yc69)*delta_t70

ya71=ya70+(-k1*ya70+k2*yb70)*delta_t71
yb71=yb70+(k1*ya70-(k2+k3)*yb70+k4*yc70)*delta_t71
yc71=yc70+(k3*yb70-k4*yc70)*delta_t71

ya72=ya71+(-k1*ya71+k2*yb71)*delta_t72
yb72=yb71+(k1*ya71-(k2+k3)*yb71+k4*yc71)*delta_t72
yc72=yc71+(k3*yb71-k4*yc71)*delta_t72

ya73=ya72+(-k1*ya72+k2*yb72)*delta_t73
yb73=yb72+(k1*ya72-(k2+k3)*yb72+k4*yc72)*delta_t73
yc73=yc72+(k3*yb72-k4*yc72)*delta_t73

ya74=ya73+(-k1*ya73+k2*yb73)*delta_t74
yb74=yb73+(k1*ya73-(k2+k3)*yb73+k4*yc73)*delta_t74
yc74=yc73+(k3*yb73-k4*yc73)*delta_t74

ya75=ya74+(-k1*ya74+k2*yb74)*delta_t75
yb75=yb74+(k1*ya74-(k2+k3)*yb74+k4*yc74)*delta_t75
yc75=yc74+(k3*yb74-k4*yc74)*delta_t75

ya76=ya75+(-k1*ya75+k2*yb75)*delta_t76
yb76=yb75+(k1*ya75-(k2+k3)*yb75+k4*yc75)*delta_t76
yc76=yc75+(k3*yb75-k4*yc75)*delta_t76

ya77=ya76+(-k1*ya76+k2*yb76)*delta_t77
yb77=yb76+(k1*ya76-(k2+k3)*yb76+k4*yc76)*delta_t77
yc77=yc76+(k3*yb76-k4*yc76)*delta_t77

ya78=ya77+(-k1*ya77+k2*yb77)*delta_t78
yb78=yb77+(k1*ya77-(k2+k3)*yb77+k4*yc77)*delta_t78
yc78=yc77+(k3*yb77-k4*yc77)*delta_t78

ya79=ya78+(-k1*ya78+k2*yb78)*delta_t79
yb79=yb78+(k1*ya78-(k2+k3)*yb78+k4*yc78)*delta_t79
yc79=yc78+(k3*yb78-k4*yc78)*delta_t79

ya80=ya79+(-k1*ya79+k2*yb79)*delta_t80
yb80=yb79+(k1*ya79-(k2+k3)*yb79+k4*yc79)*delta_t80
yc80=yc79+(k3*yb79-k4*yc79)*delta_t80

# calculate y_calc for obj fct
ya_calc[1]=ya0+(texp[1]-t0)*(ya1-ya0)/(t[1]-t0)
ya_calc[2]=ya5+(texp[2]-t[5])*(ya6-ya5)/(t[6]-t[5])
ya_calc[3]=ya9+(texp[3]-t[9])*(ya10-ya9)/(t[10]-t[9])
```

```
ya_calc[4]=ya11+(texp[4]-t[11])*(ya12-ya11)/(t[12]-t[11])
ya_calc[5]=ya14+(texp[5]-t[14])*(ya15-ya14)/(t[15]-t[14])
ya_calc[6]=ya18+(texp[6]-t[18])*(ya19-ya18)/(t[19]-t[18])
ya_calc[7]=ya26+(texp[7]-t[26])*(ya27-ya26)/(t[27]-t[26])
ya_calc[8]=ya28+(texp[8]-t[28])*(ya29-ya28)/(t[29]-t[28])
ya_calc[9]=ya31+(texp[9]-t[31])*(ya32-ya31)/(t[32]-t[31])
ya_calc[10]=ya34+(texp[10]-t[34])*(ya35-ya34)/(t[35]-t[34])
ya_calc[11]=ya41+(texp[11]-t[41])*(ya42-ya41)/(t[42]-t[41])
ya_calc[12]=ya46+(texp[12]-t[46])*(ya47-ya46)/(t[47]-t[46])
ya_calc[13]=ya49+(texp[13]-t[49])*(ya50-ya49)/(t[50]-t[49])
ya_calc[14]=ya52+(texp[14]-t[52])*(ya53-ya52)/(t[53]-t[52])
ya_calc[15]=ya55+(texp[15]-t[55])*(ya56-ya55)/(t[56]-t[55])
ya_calc[16]=ya61+(texp[16]-t[61])*(ya62-ya61)/(t[62]-t[61])
ya_calc[17]=ya75+(1-t[75])*(ya76-ya75)/(t[76]-t[75])

yb_calc[1]=yb0+(texp[1]-t0)*(yb1-yb0)/(t[1]-t0)
yb_calc[2]=yb5+(texp[2]-t[5])*(yb6-yb5)/(t[6]-t[5])
yb_calc[3]=yb9+(texp[3]-t[9])*(yb10-yb9)/(t[10]-t[9])
yb_calc[4]=yb11+(texp[4]-t[11])*(yb12-yb11)/(t[12]-t[11])
yb_calc[5]=yb14+(texp[5]-t[14])*(yb15-yb14)/(t[15]-t[14])
yb_calc[6]=yb18+(texp[6]-t[18])*(yb19-yb18)/(t[19]-t[18])
yb_calc[7]=yb26+(texp[7]-t[26])*(yb27-yb26)/(t[27]-t[26])
yb_calc[8]=yb28+(texp[8]-t[28])*(yb29-yb28)/(t[29]-t[28])
yb_calc[9]=yb31+(texp[9]-t[31])*(yb32-yb31)/(t[32]-t[31])
yb_calc[10]=yb34+(texp[10]-t[34])*(yb35-yb34)/(t[35]-t[34])
yb_calc[11]=yb41+(texp[11]-t[41])*(yb42-yb41)/(t[42]-t[41])
yb_calc[12]=yb46+(texp[12]-t[46])*(yb47-yb46)/(t[47]-t[46])
yb_calc[13]=yb49+(texp[13]-t[49])*(yb50-yb49)/(t[50]-t[49])
yb_calc[14]=yb52+(texp[14]-t[52])*(yb53-yb52)/(t[53]-t[52])
yb_calc[15]=yb55+(texp[15]-t[55])*(yb56-yb55)/(t[56]-t[55])
yb_calc[16]=yb61+(texp[16]-t[61])*(yb62-yb61)/(t[62]-t[61])
yb_calc[17]=yb75+(1-t[75])*(yb76-yb75)/(t[76]-t[75])

yc_calc[1]=yc0+(texp[1]-t0)*(yc1-yc0)/(t[1]-t0)
yc_calc[2]=yc5+(texp[2]-t[5])*(yc6-yc5)/(t[6]-t[5])
yc_calc[3]=yc9+(texp[3]-t[9])*(yc10-yc9)/(t[10]-t[9])
yc_calc[4]=yc11+(texp[4]-t[11])*(yc12-yc11)/(t[12]-t[11])
yc_calc[5]=yc14+(texp[5]-t[14])*(yc15-yc14)/(t[15]-t[14])
yc_calc[6]=yc18+(texp[6]-t[18])*(yc19-yc18)/(t[19]-t[18])
yc_calc[7]=yc26+(texp[7]-t[26])*(yc27-yc26)/(t[27]-t[26])
yc_calc[8]=yc28+(texp[8]-t[28])*(yc29-yc28)/(t[29]-t[28])
yc_calc[9]=yc31+(texp[9]-t[31])*(yc32-yc31)/(t[32]-t[31])
yc_calc[10]=yc34+(texp[10]-t[34])*(yc35-yc34)/(t[35]-t[34])
yc_calc[11]=yc41+(texp[11]-t[41])*(yc42-yc41)/(t[42]-t[41])
yc_calc[12]=yc46+(texp[12]-t[46])*(yc47-yc46)/(t[47]-t[46])
yc_calc[13]=yc49+(texp[13]-t[49])*(yc50-yc49)/(t[50]-t[49])
yc_calc[14]=yc52+(texp[14]-t[52])*(yc53-yc52)/(t[53]-t[52])
yc_calc[15]=yc55+(texp[15]-t[55])*(yc56-yc55)/(t[56]-t[55])
```

```
yc_calc[16]=yc61+(texp[16]-t[61])*(yc62-yc61)/(t[62]-t[61])
yc_calc[17]=yc75+(1-t[75])*(yc76-yc75)/(t[76]-t[75])

#calculate objective function
help1=sum_j((ya_calc[d]-ya_exp[d])^2)
help2=sum_j((yb_calc[d]-yb_exp[d])^2)
help3=sum_j((yc_calc[d]-yc_exp[d])^2)

Obj=(help1+help2+help3)/(3*NDAT)
```

11.6.2 Orthogonal collocation: ch-11-6-ortho-colloc.mot

Solution of optimization problem with orthogonal collocation (first 3 elements have 5 collocation points, last element has 2 collocation points)

```
#orthogonal collocation
# residuals at collocation points

# phi values for different tauj
# do not differ for different equations, elements i
phi0[j]=((tau[j]-tau[1])/(tau0-tau[1]))*((tau[j]-tau
[2])/(tau0-tau[2]))*((tau[j]-tau[3])/(tau0-tau[3]))*
((tau[j]-tau[4])/(tau0-tau[4]))*((tau[j]-tau[5])/
(tau0-tau[5]))
phi1[j]=((tau[j]-tau0)/(tau[1]-tau0))*((tau[j]-tau
[2])/(tau[1]-tau[2]))*((tau[j]-tau[3])/(tau[1]-tau
[3]))*((tau[j]-tau[4])/(tau[1]-tau[4]))*((tau[j]-tau
[5])/(tau[1]-tau[5]))
phi2[j]=((tau[j]-tau0)/(tau[2]-tau0))*((tau[j]-tau
[1])/(tau[2]-tau[1]))*((tau[j]-tau[3])/(tau[2]-tau
[3]))*((tau[j]-tau[4])/(tau[2]-tau[4]))*((tau[j]-tau
[5])/(tau[2]-tau[5]))
phi3[j]=((tau[j]-tau0)/(tau[3]-tau0))*((tau[j]-tau
[1])/(tau[3]-tau[1]))*((tau[j]-tau[2])/(tau[3]-tau
[2]))*((tau[j]-tau[4])/(tau[3]-tau[4]))*((tau[j]-tau
[5])/(tau[3]-tau[5]))
phi4[j]=((tau[j]-tau0)/(tau[4]-tau0))*((tau[j]-tau
[1])/(tau[4]-tau[1]))*((tau[j]-tau[2])/(tau[4]-tau
[2]))*((tau[j]-tau[3])/(tau[4]-tau[3]))*((tau[j]-tau
[5])/(tau[4]-tau[5]))
phi5[j]=((tau[j]-tau0)/(tau[5]-tau0))*((tau[j]-tau
[1])/(tau[5]-tau[1]))*((tau[j]-tau[2])/(tau[5]-tau
[2]))*((tau[j]-tau[3])/(tau[5]-tau[3]))*((tau[j]-tau
[4])/(tau[5]-tau[4]))

# element i 1
# collocation pt 1 till 5
0=ya[1][0]*phi0[j]+ya[1][1]*phi1[j]+ya[1][2]*phi2[j]
+ya[1][3]*phi3[j]+ya[1][4]*phi4[j]+ya[1][5]*phi5[j]
+(k1*ya[1][j]-k2*yb[1][j])*dalpha[1]
```

```
0=yb[1][0]*phi0[j]+ya[1][1]*phi1[j]+yb[1][2]*phi2[j]
+yb[1][3]*phi3[j]+yb[1][4]*phi4[j]+yb[1][5]*phi5[j]
+(-k1*ya[1][j]+(k2+k3)*yb[1][j]-k4*yc[1][j])*dalpha
[1]
0=yc[1][0]*phi0[j]+yc[1][1]*phi1[j]+yc[1][2]*phi2[j]
+yc[1][3]*phi3[j]+yc[1][4]*phi4[j]+yc[1][5]*phi5[j]
+(-k3*yb[1][j]+k4*yc[1][j])*dalpha[1]

# element i 2
# collocation pt 1 till 5
0=ya[2][0]*phi0[j]+ya[2][1]*phi1[j]+ya[2][2]*phi2[j]
+ya[2][3]*phi3[j]+ya[2][4]*phi4[j]+ya[2][5]*phi5[j]
+(k1*ya[2][j]-k2*yb[2][j])*dalpha[2]
0=yb[2][0]*phi0[j]+yb[2][1]*phi1[j]+yb[2][2]*phi2[j]
+yb[2][3]*phi3[j]+yb[2][4]*phi4[j]+yb[2][5]*phi5[j]
+(-k1*ya[2][j]+(k2+k3)*yb[2][j]-k4*yc[2][j])*dalpha[2]
0=yc[2][0]*phi0[j]+yc[2][1]*phi1[j]+yc[2][2]*phi2[j]
+yc[2][3]*phi3[j]+yc[2][4]*phi4[j]+yc[2][5]*phi5[j]
+(-k3*yb[2][j]+k4*yc[2][j])*dalpha[2]

# element i 3
# collocation pt 1 till 5
0=ya[3][0]*phi0[j]+ya[3][1]*phi1[j]+ya[3][2]*phi2[j]
+ya[3][3]*phi3[j]+ya[3][4]*phi4[j]+ya[3][5]*phi5[j]
+(k1*ya[3][j]-k2*yb[3][j])*dalpha[3]
0=yb[3][0]*phi0[j]+yb[3][1]*phi1[j]+yb[3][2]*phi2[j]
+yb[3][3]*phi3[j]+yb[3][4]*phi4[j]+yb[3][5]*phi5[j]
+(-k1*ya[3][j]+(k2+k3)*yb[3][j]-k4*yc[3][j])*dalpha[3]
0=yc[3][0]*phi0[j]+yc[3][1]*phi1[j]+yc[3][2]*phi2[j]
+yc[3][3]*phi3[j]+yc[3][4]*phi4[j]+yc[3][5]*phi5[j]
+(-k3*yb[3][j]+k4*yc[3][j])*dalpha[3]


# element i 4
# collocation pt 1 till 2, here only 2 colloc points
phi_end0[1]=((tau_end[1]-tau_end[1])/(tau_end0-
tau_end[1]))*((tau_end[1]-tau_end[2])/(tau_end0-
tau_end[2]))
phi_end1[1]=((tau_end[1]-tau_end0)/(tau_end[1]-
tau_end0))*((tau_end[1]-tau_end[2])/(tau_end[1]-
tau_end[2]))
phi_end2[1]=((tau_end[1]-tau_end0)/(tau_end[2]-
tau_end0))*((tau_end[1]-tau_end[1])/(tau_end[2]-
tau_end[1]))
0=ya_end[0]*phi_end0[1]+ya_end[1]*phi_end1[1]+ya_end
[2]*phi_end2[1]+(k1*ya_end[1]-k2*yb_end[1])*dalpha[4]
```

```
0=yb_end[0]*phi_end0[1]+yb_end[1]*phi_end1[1]+yb_end
[2]*phi_end2[1]+(-k1*ya_end[1]+(k2+k3)*yb_end[1]-
k4*yc_end[1])*dalpha[4]
0=yc_end[0]*phi_end0[1]+yc_end[1]*phi_end1[1]+yc_end
[2]*phi_end2[1]+(-k3*yb_end[1]+k4*yc_end[1])*dalpha[4]

# connectivity condition
# defines state variable y at the beginning of the element i
# for each element

# first calc phi at tau equal 1
phi0_tau1=((1-tau[1])/(tau0-tau[1]))*((1-tau[2])/
(tau0-tau[2]))*((1-tau[3])/(tau0-tau[3]))*((1-tau
[4])/(tau0-tau[4]))*((1-tau[5])/(tau0-tau[5]))
phi1_tau1=((1-tau0)/(tau[1]-tau0))*((1-tau[2])/(tau
[1]-tau[2]))*((1-tau[3])/(tau[1]-tau[3]))*((1-tau
[4])/(tau[1]-tau[4]))*((1-tau[5])/(tau[1]-tau[5]))
phi2_tau1=((1-tau0)/(tau[2]-tau0))*((1-tau[1])/(tau
[2]-tau[1]))*((1-tau[3])/(tau[2]-tau[3]))*((1-tau
[4])/(tau[2]-tau[4]))*((1-tau[5])/(tau[2]-tau[5]))
phi3_tau1=((1-tau0)/(tau[3]-tau0))*((1-tau[1])/(tau
[3]-tau[1]))*((1-tau[2])/(tau[3]-tau[2]))*((1-tau
[4])/(tau[3]-tau[4]))*((1-tau[5])/(tau[3]-tau[5]))
phi4_tau1=((1-tau0)/(tau[4]-tau0))*((1-tau[1])/(tau
[4]-tau[1]))*((1-tau[2])/(tau[4]-tau[2]))*((1-tau
[3])/(tau[4]-tau[3]))*((1-tau[5])/(tau[4]-tau[5]))
phi5_tau1=((1-tau0)/(tau[5]-tau0))*((1-tau[1])/(tau
[5]-tau[1]))*((1-tau[2])/(tau[5]-tau[2]))*((1-tau
[3])/(tau[5]-tau[3]))*((1-tau[4])/(tau[5]-tau[4]))
# last element (contains only 2 colloc points)
#phi_end0_tau1=((1-tau_end[1])/(tau_end0-tau_end
[1]))*((1-tau_end[2])/(tau_end0-tau_end[2]))
#phi_end1_tau1=((1-tau_end0)/(tau_end[1]-tau_end0))*
((1-tau_end[2])/(tau_end[1]-tau_end[2]))
#phi_end2_tau1=((1-tau_end0)/(tau_end[2]-tau_end0))*
((1-tau_end[1])/(tau_end[2]-tau_end[1]))

# element i 1
#0=ya[0][0]*phi0_tau1+ya[0][1]*phi1_tau1+ya[0][2]
*phi2_tau1+ya[0][3]*phi3_tau1+ya[0][4]*phi4_tau1+ya
[0][5]*phi5_tau1-ya[1][0]
#0=yb[0][0]*phi0_tau1+yb[0][1]*phi1_tau1+yb[0][2]
*phi2_tau1+yb[0][3]*phi3_tau1+yb[0][4]*phi4_tau1+yb
[0][5]*phi5_tau1-yb[1][0]
#0=yc[0][0]*phi0_tau1+yc[0][1]*phi1_tau1+yc[0][2]
*phi2_tau1+yc[0][3]*phi3_tau1+yc[0][4]*phi4_tau1+yc
[0][5]*phi5_tau1-yc[1][0]
```

```
# element i 2
0=ya[1][0]*phi0_tau1+ya[1][1]*phi1_tau1+ya[1][2]
*phi2_tau1+ya[1][3]*phi3_tau1+ya[1][4]*phi4_tau1+ya
[1][5]*phi5_tau1-ya[2][0]
0=yb[1][0]*phi0_tau1+yb[1][1]*phi1_tau1+yb[1][2]
*phi2_tau1+yb[1][3]*phi3_tau1+yb[1][4]*phi4_tau1+yb
[1][5]*phi5_tau1-yb[2][0]
0=yc[1][0]*phi0_tau1+yc[1][1]*phi1_tau1+yc[1][2]
*phi2_tau1+yc[1][3]*phi3_tau1+yc[1][4]*phi4_tau1+yc
[1][5]*phi5_tau1-yc[2][0]
# element i 3
0=ya[2][0]*phi0_tau1+ya[2][1]*phi1_tau1+ya[2][2]
*phi2_tau1+ya[2][3]*phi3_tau1+ya[2][4]*phi4_tau1+ya
[2][5]*phi5_tau1-ya[3][0]
0=yb[2][0]*phi0_tau1+yb[2][1]*phi1_tau1+yb[2][2]
*phi2_tau1+yb[2][3]*phi3_tau1+yb[2][4]*phi4_tau1+yb
[2][5]*phi5_tau1-yb[3][0]
0=yc[2][0]*phi0_tau1+yc[2][1]*phi1_tau1+yc[2][2]
*phi2_tau1+yc[2][3]*phi3_tau1+yc[2][4]*phi4_tau1+yc
[2][5]*phi5_tau1-yc[3][0]
# element i 4
0=ya[3][0]*phi0_tau1+ya[3][1]*phi1_tau1+ya[3][2]
*phi2_tau1+ya[3][3]*phi3_tau1+ya[3][4]*phi4_tau1+ya
[3][5]*phi5_tau1-ya_end[0]
0=yb[3][0]*phi0_tau1+yb[3][1]*phi1_tau1+yb[3][2]
*phi2_tau1+yb[3][3]*phi3_tau1+yb[3][4]*phi4_tau1+yb
[3][5]*phi5_tau1-yb_end[0]
0=yc[3][0]*phi0_tau1+yc[3][1]*phi1_tau1+yc[3][2]
*phi2_tau1+yc[3][3]*phi3_tau1+yc[3][4]*phi4_tau1+yc
[3][5]*phi5_tau1-yc_end[0]


# calculation of objective function
# elements are between
# 0and0.25, 0.228775 and 0.478775, 0.45755 and
0.70755, 0.686325 and 0.986325

#element i 1, 0and0.25, ycalc
phi0_ex[1]=((tex[1]-t[1][1])/(t[1][0]-t[1][1]))*
((tex[1]-t[1][2])/(t[1][0]-t[1][2]))*((tex[1]-t[1]
[3])/(t[1][0]-t[1][3]))*((tex[1]-t[1][4])/(t[1][0]-t
[1][4]))*((tex[1]-t[1][5])/(t[1][0]-t[1][5]))
phi1_ex[1]=((tex[1]-t[1][0])/(t[1][1]-t[1][0]))*
((tex[1]-t[1][2])/(t[1][1]-t[1][2]))*((tex[1]-t[1]
[3])/(t[1][1]-t[1][3]))*((tex[1]-t[1][4])/(t[1][1]-t
[1][4]))*((tex[1]-t[1][5])/(t[1][1]-t[1][5]))
phi2_ex[1]=((tex[1]-t[1][0])/(t[1][2]-t[1][0]))*
((tex[1]-t[1][1])/(t[1][2]-t[1][1]))*((tex[1]-t[1]
```

```
[3])/(t[1][2]-t[1][3]))*((tex[1]-t[1][4])/(t[1][2]-t
[1][4]))*((tex[1]-t[1][5])/(t[1][2]-t[1][5]))
phi3_ex[1]=((tex[1]-t[1][0])/(t[1][3]-t[1][0]))*
((tex[1]-t[1][1])/(t[1][3]-t[1][1]))*((tex[1]-t[1]
[2])/(t[1][3]-t[1][2]))*((tex[1]-t[1][4])/(t[1][3]-t
[1][4]))*((tex[1]-t[1][5])/(t[1][3]-t[1][5]))
phi4_ex[1]=((tex[1]-t[1][0])/(t[1][4]-t[1][0]))*
((tex[1]-t[1][1])/(t[1][4]-t[1][1]))*((tex[1]-t[1]
[2])/(t[1][4]-t[1][2]))*((tex[1]-t[1][3])/(t[1][4]-t
[1][3]))*((tex[1]-t[1][5])/(t[1][4]-t[1][5]))
phi5_ex[1]=((tex[1]-t[1][0])/(t[1][5]-t[1][0]))*
((tex[1]-t[1][1])/(t[1][5]-t[1][1]))*((tex[1]-t[1]
[2])/(t[1][5]-t[1][2]))*((tex[1]-t[1][3])/(t[1][5]-t
[1][3]))*((tex[1]-t[1][4])/(t[1][5]-t[1][4]))

phi0_ex[2]=((tex[2]-t[1][1])/(t[1][0]-t[1][1]))*
((tex[2]-t[1][2])/(t[1][0]-t[1][2]))*((tex[2]-t[1]
[3])/(t[1][0]-t[1][3]))*((tex[2]-t[1][4])/(t[1][0]-t
[1][4]))*((tex[2]-t[1][5])/(t[1][0]-t[1][5]))
phi1_ex[2]=((tex[2]-t[1][0])/(t[1][1]-t[1][0]))*
((tex[2]-t[1][2])/(t[1][1]-t[1][2]))*((tex[2]-t[1]
[3])/(t[1][1]-t[1][3]))*((tex[2]-t[1][4])/(t[1][1]-t
[1][4]))*((tex[2]-t[1][5])/(t[1][1]-t[1][5]))
phi2_ex[2]=((tex[2]-t[1][0])/(t[1][2]-t[1][0]))*
((tex[2]-t[1][1])/(t[1][2]-t[1][1]))*((tex[2]-t[1]
[3])/(t[1][2]-t[1][3]))*((tex[2]-t[1][4])/(t[1][2]-t
[1][4]))*((tex[2]-t[1][5])/(t[1][2]-t[1][5]))
phi3_ex[2]=((tex[2]-t[1][0])/(t[1][3]-t[1][0]))*
((tex[2]-t[1][1])/(t[1][3]-t[1][1]))*((tex[2]-t[1]
[2])/(t[1][3]-t[1][2]))*((tex[2]-t[1][4])/(t[1][3]-t
[1][4]))*((tex[2]-t[1][5])/(t[1][3]-t[1][5]))
phi4_ex[2]=((tex[2]-t[1][0])/(t[1][4]-t[1][0]))*
((tex[2]-t[1][1])/(t[1][4]-t[1][1]))*((tex[2]-t[1]
[2])/(t[1][4]-t[1][2]))*((tex[2]-t[1][3])/(t[1][4]-t
[1][3]))*((tex[2]-t[1][5])/(t[1][4]-t[1][5]))
phi5_ex[2]=((tex[2]-t[1][0])/(t[1][5]-t[1][0]))*
((tex[2]-t[1][1])/(t[1][5]-t[1][1]))*((tex[2]-t[1]
[2])/(t[1][5]-t[1][2]))*((tex[2]-t[1][3])/(t[1][5]-t
[1][3]))*((tex[2]-t[1][4])/(t[1][5]-t[1][4]))

phi0_ex[3]=((tex[3]-t[1][1])/(t[1][0]-t[1][1]))*
((tex[3]-t[1][2])/(t[1][0]-t[1][2]))*((tex[3]-t[1]
[3])/(t[1][0]-t[1][3]))*((tex[3]-t[1][4])/(t[1][0]-t
[1][4]))*((tex[3]-t[1][5])/(t[1][0]-t[1][5]))
phi1_ex[3]=((tex[3]-t[1][0])/(t[1][1]-t[1][0]))*
((tex[3]-t[1][2])/(t[1][1]-t[1][2]))*((tex[3]-t[1]
[3])/(t[1][1]-t[1][3]))*((tex[3]-t[1][4])/(t[1][1]-t
[1][4]))*((tex[3]-t[1][5])/(t[1][1]-t[1][5]))
```

```
phi2_ex[3]=((tex[3]-t[1][0])/(t[1][2]-t[1][0]))*
((tex[3]-t[1][1])/(t[1][2]-t[1][1]))*((tex[3]-t[1]
[3])/(t[1][2]-t[1][3]))*((tex[3]-t[1][4])/(t[1][2]-t
[1][4]))*((tex[3]-t[1][5])/(t[1][2]-t[1][5]))
phi3_ex[3]=((tex[3]-t[1][0])/(t[1][3]-t[1][0]))*
((tex[3]-t[1][1])/(t[1][3]-t[1][1]))*((tex[3]-t[1]
[2])/(t[1][3]-t[1][2]))*((tex[3]-t[1][4])/(t[1][3]-t
[1][4]))*((tex[3]-t[1][5])/(t[1][3]-t[1][5]))
phi4_ex[3]=((tex[3]-t[1][0])/(t[1][4]-t[1][0]))*
((tex[3]-t[1][1])/(t[1][4]-t[1][1]))*((tex[3]-t[1]
[2])/(t[1][4]-t[1][2]))*((tex[3]-t[1][3])/(t[1][4]-t
[1][3]))*((tex[3]-t[1][5])/(t[1][4]-t[1][5]))
phi5_ex[3]=((tex[3]-t[1][0])/(t[1][5]-t[1][0]))*
((tex[3]-t[1][1])/(t[1][5]-t[1][1]))*((tex[3]-t[1]
[2])/(t[1][5]-t[1][2]))*((tex[3]-t[1][3])/(t[1][5]-t
[1][3]))*((tex[3]-t[1][4])/(t[1][5]-t[1][4]))

phi0_ex[4]=((tex[4]-t[1][1])/(t[1][0]-t[1][1]))*
((tex[4]-t[1][2])/(t[1][0]-t[1][2]))*((tex[4]-t[1]
[3])/(t[1][0]-t[1][3]))*((tex[4]-t[1][4])/(t[1][0]-t
[1][4]))*((tex[4]-t[1][5])/(t[1][0]-t[1][5]))
phi1_ex[4]=((tex[4]-t[1][0])/(t[1][1]-t[1][0]))*
((tex[4]-t[1][2])/(t[1][1]-t[1][2]))*((tex[4]-t[1]
[3])/(t[1][1]-t[1][3]))*((tex[4]-t[1][4])/(t[1][1]-t
[1][4]))*((tex[4]-t[1][5])/(t[1][1]-t[1][5]))
phi2_ex[4]=((tex[4]-t[1][0])/(t[1][2]-t[1][0]))*
((tex[4]-t[1][1])/(t[1][2]-t[1][1]))*((tex[4]-t[1]
[3])/(t[1][2]-t[1][3]))*((tex[4]-t[1][4])/(t[1][2]-t
[1][4]))*((tex[4]-t[1][5])/(t[1][2]-t[1][5]))
phi3_ex[4]=((tex[4]-t[1][0])/(t[1][3]-t[1][0]))*
((tex[4]-t[1][1])/(t[1][3]-t[1][1]))*((tex[4]-t[1]
[2])/(t[1][3]-t[1][2]))*((tex[4]-t[1][4])/(t[1][3]-t
[1][4]))*((tex[4]-t[1][5])/(t[1][3]-t[1][5]))
phi4_ex[4]=((tex[4]-t[1][0])/(t[1][4]-t[1][0]))*
((tex[4]-t[1][1])/(t[1][4]-t[1][1]))*((tex[4]-t[1]
[2])/(t[1][4]-t[1][2]))*((tex[4]-t[1][3])/(t[1][4]-t
[1][3]))*((tex[4]-t[1][5])/(t[1][4]-t[1][5]))
phi5_ex[4]=((tex[4]-t[1][0])/(t[1][5]-t[1][0]))*
((tex[4]-t[1][1])/(t[1][5]-t[1][1]))*((tex[4]-t[1]
[2])/(t[1][5]-t[1][2]))*((tex[4]-t[1][3])/(t[1][5]-t
[1][3]))*((tex[4]-t[1][4])/(t[1][5]-t[1][4]))

phi0_ex[5]=((tex[5]-t[1][1])/(t[1][0]-t[1][1]))*
((tex[5]-t[1][2])/(t[1][0]-t[1][2]))*((tex[5]-t[1]
[3])/(t[1][0]-t[1][3]))*((tex[5]-t[1][4])/(t[1][0]-t
[1][4]))*((tex[5]-t[1][5])/(t[1][0]-t[1][5]))
```

```
phi1_ex[5]=((tex[5]-t[1][0])/(t[1][1]-t[1][0]))*
((tex[5]-t[1][2])/(t[1][1]-t[1][2]))*((tex[5]-t[1]
[3])/(t[1][1]-t[1][3]))*((tex[5]-t[1][4])/(t[1][1]-t
[1][4]))*((tex[5]-t[1][5])/(t[1][1]-t[1][5]))
phi2_ex[5]=((tex[5]-t[1][0])/(t[1][2]-t[1][0]))*
((tex[5]-t[1][1])/(t[1][2]-t[1][1]))*((tex[5]-t[1]
[3])/(t[1][2]-t[1][3]))*((tex[5]-t[1][4])/(t[1][2]-t
[1][4]))*((tex[5]-t[1][5])/(t[1][2]-t[1][5]))
phi3_ex[5]=((tex[5]-t[1][0])/(t[1][3]-t[1][0]))*
((tex[5]-t[1][1])/(t[1][3]-t[1][1]))*((tex[5]-t[1]
[2])/(t[1][3]-t[1][2]))*((tex[5]-t[1][4])/(t[1][3]-t
[1][4]))*((tex[5]-t[1][5])/(t[1][3]-t[1][5]))
phi4_ex[5]=((tex[5]-t[1][0])/(t[1][4]-t[1][0]))*
((tex[5]-t[1][1])/(t[1][4]-t[1][1]))*((tex[5]-t[1]
[2])/(t[1][4]-t[1][2]))*((tex[5]-t[1][3])/(t[1][4]-t
[1][3]))*((tex[5]-t[1][5])/(t[1][4]-t[1][5]))
phi5_ex[5]=((tex[5]-t[1][0])/(t[1][5]-t[1][0]))*
((tex[5]-t[1][1])/(t[1][5]-t[1][1]))*((tex[5]-t[1]
[2])/(t[1][5]-t[1][2]))*((tex[5]-t[1][3])/(t[1][5]-t
[1][3]))*((tex[5]-t[1][4])/(t[1][5]-t[1][4]))

phi0_ex[6]=((tex[6]-t[1][1])/(t[1][0]-t[1][1]))*
((tex[6]-t[1][2])/(t[1][0]-t[1][2]))*((tex[6]-t[1]
[3])/(t[1][0]-t[1][3]))*((tex[6]-t[1][4])/(t[1][0]-t
[1][4]))*((tex[6]-t[1][5])/(t[1][0]-t[1][5]))
phi1_ex[6]=((tex[6]-t[1][0])/(t[1][1]-t[1][0]))*
((tex[6]-t[1][2])/(t[1][1]-t[1][2]))*((tex[6]-t[1]
[3])/(t[1][1]-t[1][3]))*((tex[6]-t[1][4])/(t[1][1]-t
[1][4]))*((tex[6]-t[1][5])/(t[1][1]-t[1][5]))
phi2_ex[6]=((tex[6]-t[1][0])/(t[1][2]-t[1][0]))*
((tex[6]-t[1][1])/(t[1][2]-t[1][1]))*((tex[6]-t[1]
[3])/(t[1][2]-t[1][3]))*((tex[6]-t[1][4])/(t[1][2]-t
[1][4]))*((tex[6]-t[1][5])/(t[1][2]-t[1][5]))
phi3_ex[6]=((tex[6]-t[1][0])/(t[1][3]-t[1][0]))*
((tex[6]-t[1][1])/(t[1][3]-t[1][1]))*((tex[6]-t[1]
[2])/(t[1][3]-t[1][2]))*((tex[6]-t[1][4])/(t[1][3]-t
[1][4]))*((tex[6]-t[1][5])/(t[1][3]-t[1][5]))
phi4_ex[6]=((tex[6]-t[1][0])/(t[1][4]-t[1][0]))*
((tex[6]-t[1][1])/(t[1][4]-t[1][1]))*((tex[6]-t[1]
[2])/(t[1][4]-t[1][2]))*((tex[6]-t[1][3])/(t[1][4]-t
[1][3]))*((tex[6]-t[1][5])/(t[1][4]-t[1][5]))
phi5_ex[6]=((tex[6]-t[1][0])/(t[1][5]-t[1][0]))*
((tex[6]-t[1][1])/(t[1][5]-t[1][1]))*((tex[6]-t[1]
[2])/(t[1][5]-t[1][2]))*((tex[6]-t[1][3])/(t[1][5]-t
[1][3]))*((tex[6]-t[1][4])/(t[1][5]-t[1][4]))
```

```
ya_calc[1]=ya[1][0]*phi0_ex[1]+ya[1][1]*phi1_ex[1]
+ya[1][2]*phi2_ex[1]+ya[1][3]*phi3_ex[1]+ya[1][4]
*phi4_ex[1]+ya[1][5]*phi5_ex[1]
yb_calc[1]=yb[1][0]*phi0_ex[1]+yb[1][1]*phi1_ex[1]
+yb[1][2]*phi2_ex[1]+yb[1][3]*phi3_ex[1]+yb[1][4]
*phi4_ex[1]+yb[1][5]*phi5_ex[1]
yc_calc[1]=yc[1][0]*phi0_ex[1]+yc[1][1]*phi1_ex[1]
+yc[1][2]*phi2_ex[1]+yc[1][3]*phi3_ex[1]+yc[1][4]
*phi4_ex[1]+yc[1][5]*phi5_ex[1]

ya_calc[2]=ya[1][0]*phi0_ex[2]+ya[1][1]*phi1_ex[2]
+ya[1][2]*phi2_ex[2]+ya[1][3]*phi3_ex[2]+ya[1][4]
*phi4_ex[2]+ya[1][5]*phi5_ex[2]
yb_calc[2]=yb[1][0]*phi0_ex[2]+yb[1][1]*phi1_ex[2]
+yb[1][2]*phi2_ex[2]+yb[1][3]*phi3_ex[2]+yb[1][4]
*phi4_ex[2]+yb[1][5]*phi5_ex[2]
yc_calc[2]=yc[1][0]*phi0_ex[2]+yc[1][1]*phi1_ex[2]
+yc[1][2]*phi2_ex[2]+yc[1][3]*phi3_ex[2]+yc[1][4]
*phi4_ex[2]+yc[1][5]*phi5_ex[2]

ya_calc[3]=ya[1][0]*phi0_ex[3]+ya[1][1]*phi1_ex[3]
+ya[1][2]*phi2_ex[3]+ya[1][3]*phi3_ex[3]+ya[1][4]
*phi4_ex[3]+ya[1][5]*phi5_ex[3]
yb_calc[3]=yb[1][0]*phi0_ex[3]+yb[1][1]*phi1_ex[3]
+yb[1][2]*phi2_ex[3]+yb[1][3]*phi3_ex[3]+yb[1][4]
*phi4_ex[3]+yb[1][5]*phi5_ex[3]
yc_calc[3]=yc[1][0]*phi0_ex[3]+yc[1][1]*phi1_ex[3]
+yc[1][2]*phi2_ex[3]+yc[1][3]*phi3_ex[3]+yc[1][4]
*phi4_ex[3]+yc[1][5]*phi5_ex[3]

ya_calc[4]=ya[1][0]*phi0_ex[4]+ya[1][1]*phi1_ex[4]
+ya[1][2]*phi2_ex[4]+ya[1][3]*phi3_ex[4]+ya[1][4]
*phi4_ex[4]+ya[1][5]*phi5_ex[4]
yb_calc[4]=yb[1][0]*phi0_ex[4]+yb[1][1]*phi1_ex[4]
+yb[1][2]*phi2_ex[4]+yb[1][3]*phi3_ex[4]+yb[1][4]
*phi4_ex[4]+yb[1][5]*phi5_ex[4]
yc_calc[4]=yc[1][0]*phi0_ex[4]+yc[1][1]*phi1_ex[4]
+yc[1][2]*phi2_ex[4]+yc[1][3]*phi3_ex[4]+yc[1][4]
*phi4_ex[4]+yc[1][5]*phi5_ex[4]

ya_calc[5]=ya[1][0]*phi0_ex[5]+ya[1][1]*phi1_ex[5]
+ya[1][2]*phi2_ex[5]+ya[1][3]*phi3_ex[5]+ya[1][4]
*phi4_ex[5]+ya[1][5]*phi5_ex[5]
yb_calc[5]=yb[1][0]*phi0_ex[5]+yb[1][1]*phi1_ex[5]
+yb[1][2]*phi2_ex[5]+yb[1][3]*phi3_ex[5]+yb[1][4]
*phi4_ex[5]+yb[1][5]*phi5_ex[5]
```

```
yc_calc[5]=yc[1][0]*phi0_ex[5]+yc[1][1]*phi1_ex[5]
+yc[1][2]*phi2_ex[5]+yc[1][3]*phi3_ex[5]+yc[1][4]
*phi4_ex[5]+yc[1][5]*phi5_ex[5]

ya_calc[6]=ya[1][0]*phi0_ex[6]+ya[1][1]*phi1_ex[6]
+ya[1][2]*phi2_ex[6]+ya[1][3]*phi3_ex[6]+ya[1][4]
*phi4_ex[6]+ya[1][5]*phi5_ex[6]
yb_calc[6]=yb[1][0]*phi0_ex[6]+yb[1][1]*phi1_ex[6]
+yb[1][2]*phi2_ex[6]+yb[1][3]*phi3_ex[6]+yb[1][4]
*phi4_ex[6]+yb[1][5]*phi5_ex[6]
yc_calc[6]=yc[1][0]*phi0_ex[6]+yc[1][1]*phi1_ex[6]
+yc[1][2]*phi2_ex[6]+yc[1][3]*phi3_ex[6]+yc[1][4]
*phi4_ex[6]+yc[1][5]*phi5_ex[6]


#element i 2, 0.228775 and 0.478775, ycalc
phi0_ex[7]=((tex[7]-t[2][1])/(t[2][0]-t[2][1]))*
((tex[7]-t[2][2])/(t[2][0]-t[2][2]))*((tex[7]-t[2]
[3])/(t[2][0]-t[2][3]))*((tex[7]-t[2][4])/(t[2][0]-t
[2][4]))*((tex[7]-t[2][5])/(t[2][0]-t[2][5]))
phi1_ex[7]=((tex[7]-t[2][0])/(t[2][1]-t[2][0]))*
((tex[7]-t[2][2])/(t[2][1]-t[2][2]))*((tex[7]-t[2]
[3])/(t[2][1]-t[2][3]))*((tex[7]-t[2][4])/(t[2][1]-t
[2][4]))*((tex[7]-t[2][5])/(t[2][1]-t[2][5]))
phi2_ex[7]=((tex[7]-t[2][0])/(t[2][2]-t[2][0]))*
((tex[7]-t[2][1])/(t[2][2]-t[2][1]))*((tex[7]-t[2]
[3])/(t[2][2]-t[2][3]))*((tex[7]-t[2][4])/(t[2][2]-t
[2][4]))*((tex[7]-t[2][5])/(t[2][2]-t[2][5]))
phi3_ex[7]=((tex[7]-t[2][0])/(t[2][3]-t[2][0]))*
((tex[7]-t[2][1])/(t[2][3]-t[2][1]))*((tex[7]-t[2]
[2])/(t[2][3]-t[2][2]))*((tex[7]-t[2][4])/(t[2][3]-t
[2][4]))*((tex[7]-t[2][5])/(t[2][3]-t[2][5]))
phi4_ex[7]=((tex[7]-t[2][0])/(t[2][4]-t[2][0]))*
((tex[7]-t[2][1])/(t[2][4]-t[2][1]))*((tex[7]-t[2]
[2])/(t[2][4]-t[2][2]))*((tex[7]-t[2][3])/(t[2][4]-t
[2][3]))*((tex[7]-t[2][5])/(t[2][4]-t[2][5]))
phi5_ex[7]=((tex[7]-t[2][0])/(t[2][5]-t[2][0]))*
((tex[7]-t[2][1])/(t[2][5]-t[2][1]))*((tex[7]-t[2]
[2])/(t[2][5]-t[2][2]))*((tex[7]-t[2][3])/(t[2][5]-t
[2][3]))*((tex[7]-t[2][4])/(t[2][5]-t[2][4]))

phi0_ex[8]=((tex[8]-t[2][1])/(t[2][0]-t[2][1]))*
((tex[8]-t[2][2])/(t[2][0]-t[2][2]))*((tex[8]-t[2]
[3])/(t[2][0]-t[2][3]))*((tex[8]-t[2][4])/(t[2][0]-t
[2][4]))*((tex[8]-t[2][5])/(t[2][0]-t[2][5]))
phi1_ex[8]=((tex[8]-t[2][0])/(t[2][1]-t[2][0]))*
((tex[8]-t[2][2])/(t[2][1]-t[2][2]))*((tex[8]-t[2]
[3])/(t[2][1]-t[2][3]))*((tex[8]-t[2][4])/(t[2][1]-t
[2][4]))*((tex[8]-t[2][5])/(t[2][1]-t[2][5]))
```

```
phi2_ex[8]=((tex[8]-t[2][0])/(t[2][2]-t[2][0]))*
((tex[8]-t[2][1])/(t[2][2]-t[2][1]))*((tex[8]-t[2]
[3])/(t[2][2]-t[2][3]))*((tex[8]-t[2][4])/(t[2][2]-t
[2][4]))*((tex[8]-t[2][5])/(t[2][2]-t[2][5]))
phi3_ex[8]=((tex[8]-t[2][0])/(t[2][3]-t[2][0]))*
((tex[8]-t[2][1])/(t[2][3]-t[2][1]))*((tex[8]-t[2]
[2])/(t[2][3]-t[2][2]))*((tex[8]-t[2][4])/(t[2][3]-t
[2][4]))*((tex[8]-t[2][5])/(t[2][3]-t[2][5]))
phi4_ex[8]=((tex[8]-t[2][0])/(t[2][4]-t[2][0]))*
((tex[8]-t[2][1])/(t[2][4]-t[2][1]))*((tex[8]-t[2]
[2])/(t[2][4]-t[2][2]))*((tex[8]-t[2][3])/(t[2][4]-t
[2][3]))*((tex[8]-t[2][5])/(t[2][4]-t[2][5]))
phi5_ex[8]=((tex[8]-t[2][0])/(t[2][5]-t[2][0]))*
((tex[8]-t[2][1])/(t[2][5]-t[2][1]))*((tex[8]-t[2]
[2])/(t[2][5]-t[2][2]))*((tex[8]-t[2][3])/(t[2][5]-t
[2][3]))*((tex[8]-t[2][4])/(t[2][5]-t[2][4]))

phi0_ex[9]=((tex[9]-t[2][1])/(t[2][0]-t[2][1]))*
((tex[9]-t[2][2])/(t[2][0]-t[2][2]))*((tex[9]-t[2]
[3])/(t[2][0]-t[2][3]))*((tex[9]-t[2][4])/(t[2][0]-t
[2][4]))*((tex[9]-t[2][5])/(t[2][0]-t[2][5]))
phi1_ex[9]=((tex[9]-t[2][0])/(t[2][1]-t[2][0]))*
((tex[9]-t[2][2])/(t[2][1]-t[2][2]))*((tex[9]-t[2]
[3])/(t[2][1]-t[2][3]))*((tex[9]-t[2][4])/(t[2][1]-t
[2][4]))*((tex[9]-t[2][5])/(t[2][1]-t[2][5]))
phi2_ex[9]=((tex[9]-t[2][0])/(t[2][2]-t[2][0]))*
((tex[9]-t[2][1])/(t[2][2]-t[2][1]))*((tex[9]-t[2]
[3])/(t[2][2]-t[2][3]))*((tex[9]-t[2][4])/(t[2][2]-t
[2][4]))*((tex[9]-t[2][5])/(t[2][2]-t[2][5]))
phi3_ex[9]=((tex[9]-t[2][0])/(t[2][3]-t[2][0]))*
((tex[9]-t[2][1])/(t[2][3]-t[2][1]))*((tex[9]-t[2]
[2])/(t[2][3]-t[2][2]))*((tex[9]-t[2][4])/(t[2][3]-t
[2][4]))*((tex[9]-t[2][5])/(t[2][3]-t[2][5]))
phi4_ex[9]=((tex[9]-t[2][0])/(t[2][4]-t[2][0]))*
((tex[9]-t[2][1])/(t[2][4]-t[2][1]))*((tex[9]-t[2]
[2])/(t[2][4]-t[2][2]))*((tex[9]-t[2][3])/(t[2][4]-t
[2][3]))*((tex[9]-t[2][5])/(t[2][4]-t[2][5]))
phi5_ex[9]=((tex[9]-t[2][0])/(t[2][5]-t[2][0]))*
((tex[9]-t[2][1])/(t[2][5]-t[2][1]))*((tex[9]-t[2]
[2])/(t[2][5]-t[2][2]))*((tex[9]-t[2][3])/(t[2][5]-t
[2][3]))*((tex[9]-t[2][4])/(t[2][5]-t[2][4]))

phi0_ex[10]=((tex[10]-t[2][1])/(t[2][0]-t[2][1]))*
((tex[10]-t[2][2])/(t[2][0]-t[2][2]))*((tex[10]-t[2]
[3])/(t[2][0]-t[2][3]))*((tex[10]-t[2][4])/(t[2][0]-
t[2][4]))*((tex[10]-t[2][5])/(t[2][0]-t[2][5]))
```

```
phi1_ex[10]=((tex[10]-t[2][0])/(t[2][1]-t[2][0]))*
((tex[10]-t[2][2])/(t[2][1]-t[2][2]))*((tex[10]-t[2]
[3])/(t[2][1]-t[2][3]))*((tex[10]-t[2][4])/(t[2][1]-
t[2][4]))*((tex[10]-t[2][5])/(t[2][1]-t[2][5]))
phi2_ex[10]=((tex[10]-t[2][0])/(t[2][2]-t[2][0]))*
((tex[10]-t[2][1])/(t[2][2]-t[2][1]))*((tex[10]-t[2]
[3])/(t[2][2]-t[2][3]))*((tex[10]-t[2][4])/(t[2][2]-
t[2][4]))*((tex[10]-t[2][5])/(t[2][2]-t[2][5]))
phi3_ex[10]=((tex[10]-t[2][0])/(t[2][3]-t[2][0]))*
((tex[10]-t[2][1])/(t[2][3]-t[2][1]))*((tex[10]-t[2]
[2])/(t[2][3]-t[2][2]))*((tex[10]-t[2][4])/(t[2][3]-
t[2][4]))*((tex[10]-t[2][5])/(t[2][3]-t[2][5]))
phi4_ex[10]=((tex[10]-t[2][0])/(t[2][4]-t[2][0]))*
((tex[10]-t[2][1])/(t[2][4]-t[2][1]))*((tex[10]-t[2]
[2])/(t[2][4]-t[2][2]))*((tex[10]-t[2][3])/(t[2][4]-
t[2][3]))*((tex[10]-t[2][5])/(t[2][4]-t[2][5]))
phi5_ex[10]=((tex[10]-t[2][0])/(t[2][5]-t[2][0]))*
((tex[10]-t[2][1])/(t[2][5]-t[2][1]))*((tex[10]-t[2]
[2])/(t[2][5]-t[2][2]))*((tex[10]-t[2][3])/(t[2][5]-
t[2][3]))*((tex[10]-t[2][4])/(t[2][5]-t[2][4]))

ya_calc[7]=ya[2][0]*phi0_ex[7]+ya[2][1]*phi1_ex[7]
+ya[2][2]*phi2_ex[7]+ya[2][3]*phi3_ex[7]+ya[2][4]
*phi4_ex[7]+ya[2][5]*phi5_ex[7]
yb_calc[7]=yb[2][0]*phi0_ex[7]+yb[2][1]*phi1_ex[7]
+yb[2][2]*phi2_ex[7]+yb[2][3]*phi3_ex[7]+yb[2][4]
*phi4_ex[7]+yb[2][5]*phi5_ex[7]
yc_calc[7]=yc[2][0]*phi0_ex[7]+yc[2][1]*phi1_ex[7]
+yc[2][2]*phi2_ex[7]+yc[2][3]*phi3_ex[7]+yc[2][4]
*phi4_ex[7]+yc[2][5]*phi5_ex[7]

ya_calc[8]=ya[2][0]*phi0_ex[8]+ya[2][1]*phi1_ex[8]
+ya[2][2]*phi2_ex[8]+ya[2][3]*phi3_ex[8]+ya[2][4]
*phi4_ex[8]+ya[2][5]*phi5_ex[8]
yb_calc[8]=yb[2][0]*phi0_ex[8]+yb[2][1]*phi1_ex[8]
+yb[2][2]*phi2_ex[8]+yb[2][3]*phi3_ex[8]+yb[2][4]
*phi4_ex[8]+yb[2][5]*phi5_ex[8]
yc_calc[8]=yc[2][0]*phi0_ex[8]+yc[2][1]*phi1_ex[8]
+yc[2][2]*phi2_ex[8]+yc[2][3]*phi3_ex[8]+yc[2][4]
*phi4_ex[8]+yc[2][5]*phi5_ex[8]

ya_calc[9]=ya[2][0]*phi0_ex[9]+ya[2][1]*phi1_ex[9]
+ya[2][2]*phi2_ex[9]+ya[2][3]*phi3_ex[9]+ya[2][4]
*phi4_ex[9]+ya[2][5]*phi5_ex[9]
yb_calc[9]=yb[2][0]*phi0_ex[9]+yb[2][1]*phi1_ex[9]
+yb[2][2]*phi2_ex[9]+yb[2][3]*phi3_ex[9]+yb[2][4]
*phi4_ex[9]+yb[2][5]*phi5_ex[9]
```

```
yc_calc[9]=yc[2][0]*phi0_ex[9]+yc[2][1]*phi1_ex[9]
+yc[2][2]*phi2_ex[9]+yc[2][3]*phi3_ex[9]+yc[2][4]
*phi4_ex[9]+yc[2][5]*phi5_ex[9]

ya_calc[10]=ya[2][0]*phi0_ex[10]+ya[2][1]*phi1_ex
[10]+ya[2][2]*phi2_ex[10]+ya[2][3]*phi3_ex[10]+ya[2]
[4]*phi4_ex[10]+ya[2][5]*phi5_ex[10]
yb_calc[10]=yb[2][0]*phi0_ex[10]+yb[2][1]*phi1_ex
[10]+yb[2][2]*phi2_ex[10]+yb[2][3]*phi3_ex[10]+yb[2]
[4]*phi4_ex[10]+yb[2][5]*phi5_ex[10]
yc_calc[10]=yc[2][0]*phi0_ex[10]+yc[2][1]*phi1_ex
[10]+yc[2][2]*phi2_ex[10]+yc[2][3]*phi3_ex[10]+yc[2]
[4]*phi4_ex[10]+yc[2][5]*phi5_ex[10]

#element i 3, 0.45755 and 0.70755, ycalc
phi0_ex[11]=((tex[11]-t[3][1])/(t[3][0]-t[3][1]))*
((tex[11]-t[3][2])/(t[3][0]-t[3][2]))*((tex[11]-t[3]
[3])/(t[3][0]-t[3][3]))*((tex[11]-t[3][4])/(t[3][0]-
t[3][4]))*((tex[11]-t[3][5])/(t[3][0]-t[3][5]))
phi1_ex[11]=((tex[11]-t[3][0])/(t[3][1]-t[3][0]))*
((tex[11]-t[3][2])/(t[3][1]-t[3][2]))*((tex[11]-t[3]
[3])/(t[3][1]-t[3][3]))*((tex[11]-t[3][4])/(t[3][1]-
t[3][4]))*((tex[11]-t[3][5])/(t[3][1]-t[3][5]))
phi2_ex[11]=((tex[11]-t[3][0])/(t[3][2]-t[3][0]))*
((tex[11]-t[3][1])/(t[3][2]-t[3][1]))*((tex[11]-t[3]
[3])/(t[3][2]-t[3][3]))*((tex[11]-t[3][4])/(t[3][2]-
t[3][4]))*((tex[11]-t[3][5])/(t[3][2]-t[3][5]))
phi3_ex[11]=((tex[11]-t[3][0])/(t[3][3]-t[3][0]))*
((tex[11]-t[3][1])/(t[3][3]-t[3][1]))*((tex[11]-t[3]
[2])/(t[3][3]-t[3][2]))*((tex[11]-t[3][4])/(t[3][3]-
t[3][4]))*((tex[11]-t[3][5])/(t[3][3]-t[3][5]))
phi4_ex[11]=((tex[11]-t[3][0])/(t[3][4]-t[3][0]))*
((tex[11]-t[3][1])/(t[3][4]-t[3][1]))*((tex[11]-t[3]
[2])/(t[3][4]-t[3][2]))*((tex[11]-t[3][3])/(t[3][4]-
t[3][3]))*((tex[11]-t[3][5])/(t[3][4]-t[3][5]))
phi5_ex[11]=((tex[11]-t[3][0])/(t[3][5]-t[3][0]))*
((tex[11]-t[3][1])/(t[3][5]-t[3][1]))*((tex[11]-t[3]
[2])/(t[3][5]-t[3][2]))*((tex[11]-t[3][3])/(t[3][5]-
t[3][3]))*((tex[11]-t[3][4])/(t[3][5]-t[3][4]))

phi0_ex[12]=((tex[12]-t[3][1])/(t[3][0]-t[3][1]))*
((tex[12]-t[3][2])/(t[3][0]-t[3][2]))*((tex[12]-t[3]
[3])/(t[3][0]-t[3][3]))*((tex[12]-t[3][4])/(t[3][0]-
t[3][4]))*((tex[12]-t[3][5])/(t[3][0]-t[3][5]))
phi1_ex[12]=((tex[12]-t[3][0])/(t[3][1]-t[3][0]))*
((tex[12]-t[3][2])/(t[3][1]-t[3][2]))*((tex[12]-t[3]
```

```
[3])/(t[3][1]-t[3][3]))*((tex[12]-t[3][4])/(t[3][1]-
t[3][4]))*((tex[12]-t[3][5])/(t[3][1]-t[3][5]))
phi2_ex[12]=((tex[12]-t[3][0])/(t[3][2]-t[3][0]))*
((tex[12]-t[3][1])/(t[3][2]-t[3][1]))*((tex[12]-t[3]
[3])/(t[3][2]-t[3][3]))*((tex[12]-t[3][4])/(t[3][2]-
t[3][4]))*((tex[12]-t[3][5])/(t[3][2]-t[3][5]))
phi3_ex[12]=((tex[12]-t[3][0])/(t[3][3]-t[3][0]))*
((tex[12]-t[3][1])/(t[3][3]-t[3][1]))*((tex[12]-t[3]
[2])/(t[3][3]-t[3][2]))*((tex[12]-t[3][4])/(t[3][3]-
t[3][4]))*((tex[12]-t[3][5])/(t[3][3]-t[3][5]))
phi4_ex[12]=((tex[12]-t[3][0])/(t[3][4]-t[3][0]))*
((tex[12]-t[3][1])/(t[3][4]-t[3][1]))*((tex[12]-t[3]
[2])/(t[3][4]-t[3][2]))*((tex[12]-t[3][3])/(t[3][4]-
t[3][3]))*((tex[12]-t[3][5])/(t[3][4]-t[3][5]))
phi5_ex[12]=((tex[12]-t[3][0])/(t[3][5]-t[3][0]))*
((tex[12]-t[3][1])/(t[3][5]-t[3][1]))*((tex[12]-t[3]
[2])/(t[3][5]-t[3][2]))*((tex[12]-t[3][3])/(t[3][5]-
t[3][3]))*((tex[12]-t[3][4])/(t[3][5]-t[3][4]))

phi0_ex[13]=((tex[13]-t[3][1])/(t[3][0]-t[3][1]))*
((tex[13]-t[3][2])/(t[3][0]-t[3][2]))*((tex[13]-t[3]
[3])/(t[3][0]-t[3][3]))*((tex[13]-t[3][4])/(t[3][0]-
t[3][4]))*((tex[13]-t[3][5])/(t[3][0]-t[3][5]))
phi1_ex[13]=((tex[13]-t[3][0])/(t[3][1]-t[3][0]))*
((tex[13]-t[3][2])/(t[3][1]-t[3][2]))*((tex[13]-t[3]
[3])/(t[3][1]-t[3][3]))*((tex[13]-t[3][4])/(t[3][1]-
t[3][4]))*((tex[13]-t[3][5])/(t[3][1]-t[3][5]))
phi2_ex[13]=((tex[13]-t[3][0])/(t[3][2]-t[3][0]))*
((tex[13]-t[3][1])/(t[3][2]-t[3][1]))*((tex[13]-t[3]
[3])/(t[3][2]-t[3][3]))*((tex[13]-t[3][4])/(t[3][2]-
t[3][4]))*((tex[13]-t[3][5])/(t[3][2]-t[3][5]))
phi3_ex[13]=((tex[13]-t[3][0])/(t[3][3]-t[3][0]))*
((tex[13]-t[3][1])/(t[3][3]-t[3][1]))*((tex[13]-t[3]
[2])/(t[3][3]-t[3][2]))*((tex[13]-t[3][4])/(t[3][3]-
t[3][4]))*((tex[13]-t[3][5])/(t[3][3]-t[3][5]))
phi4_ex[13]=((tex[13]-t[3][0])/(t[3][4]-t[3][0]))*
((tex[13]-t[3][1])/(t[3][4]-t[3][1]))*((tex[13]-t[3]
[2])/(t[3][4]-t[3][2]))*((tex[13]-t[3][3])/(t[3][4]-
t[3][3]))*((tex[13]-t[3][5])/(t[3][4]-t[3][5]))
phi5_ex[13]=((tex[13]-t[3][0])/(t[3][5]-t[3][0]))*
((tex[13]-t[3][1])/(t[3][5]-t[3][1]))*((tex[13]-t[3]
[2])/(t[3][5]-t[3][2]))*((tex[13]-t[3][3])/(t[3][5]-
t[3][3]))*((tex[13]-t[3][4])/(t[3][5]-t[3][4]))

phi0_ex[14]=((tex[14]-t[3][1])/(t[3][0]-t[3][1]))*
((tex[14]-t[3][2])/(t[3][0]-t[3][2]))*((tex[14]-t[3]
```

```
[3])/(t[3][0]-t[3][3]))*((tex[14]-t[3][4])/(t[3][0]-
t[3][4]))*((tex[14]-t[3][5])/(t[3][0]-t[3][5]))
phi1_ex[14]=((tex[14]-t[3][0])/(t[3][1]-t[3][0]))*
((tex[14]-t[3][2])/(t[3][1]-t[3][2]))*((tex[14]-t[3]
[3])/(t[3][1]-t[3][3]))*((tex[14]-t[3][4])/(t[3][1]-
t[3][4]))*((tex[14]-t[3][5])/(t[3][1]-t[3][5]))
phi2_ex[14]=((tex[14]-t[3][0])/(t[3][2]-t[3][0]))*
((tex[14]-t[3][1])/(t[3][2]-t[3][1]))*((tex[14]-t[3]
[3])/(t[3][2]-t[3][3]))*((tex[14]-t[3][4])/(t[3][2]-
t[3][4]))*((tex[14]-t[3][5])/(t[3][2]-t[3][5]))
phi3_ex[14]=((tex[14]-t[3][0])/(t[3][3]-t[3][0]))*
((tex[14]-t[3][1])/(t[3][3]-t[3][1]))*((tex[14]-t[3]
[2])/(t[3][3]-t[3][2]))*((tex[14]-t[3][4])/(t[3][3]-
t[3][4]))*((tex[14]-t[3][5])/(t[3][3]-t[3][5]))
phi4_ex[14]=((tex[14]-t[3][0])/(t[3][4]-t[3][0]))*
((tex[14]-t[3][1])/(t[3][4]-t[3][1]))*((tex[14]-t[3]
[2])/(t[3][4]-t[3][2]))*((tex[14]-t[3][3])/(t[3][4]-
t[3][3]))*((tex[14]-t[3][5])/(t[3][4]-t[3][5]))
phi5_ex[14]=((tex[14]-t[3][0])/(t[3][5]-t[3][0]))*
((tex[14]-t[3][1])/(t[3][5]-t[3][1]))*((tex[14]-t[3]
[2])/(t[3][5]-t[3][2]))*((tex[14]-t[3][3])/(t[3][5]-
t[3][3]))*((tex[14]-t[3][4])/(t[3][5]-t[3][4]))

phi0_ex[15]=((tex[15]-t[3][1])/(t[3][0]-t[3][1]))*
((tex[15]-t[3][2])/(t[3][0]-t[3][2]))*((tex[15]-t[3]
[3])/(t[3][0]-t[3][3]))*((tex[15]-t[3][4])/(t[3][0]-
t[3][4]))*((tex[15]-t[3][5])/(t[3][0]-t[3][5]))
phi1_ex[15]=((tex[15]-t[3][0])/(t[3][1]-t[3][0]))*
((tex[15]-t[3][2])/(t[3][1]-t[3][2]))*((tex[15]-t[3]
[3])/(t[3][1]-t[3][3]))*((tex[15]-t[3][4])/(t[3][1]-
t[3][4]))*((tex[15]-t[3][5])/(t[3][1]-t[3][5]))
phi2_ex[15]=((tex[15]-t[3][0])/(t[3][2]-t[3][0]))*
((tex[15]-t[3][1])/(t[3][2]-t[3][1]))*((tex[15]-t[3]
[3])/(t[3][2]-t[3][3]))*((tex[15]-t[3][4])/(t[3][2]-
t[3][4]))*((tex[15]-t[3][5])/(t[3][2]-t[3][5]))
phi3_ex[15]=((tex[15]-t[3][0])/(t[3][3]-t[3][0]))*
((tex[15]-t[3][1])/(t[3][3]-t[3][1]))*((tex[15]-t[3]
[2])/(t[3][3]-t[3][2]))*((tex[15]-t[3][4])/(t[3][3]-
t[3][4]))*((tex[15]-t[3][5])/(t[3][3]-t[3][5]))
phi4_ex[15]=((tex[15]-t[3][0])/(t[3][4]-t[3][0]))*
((tex[15]-t[3][1])/(t[3][4]-t[3][1]))*((tex[15]-t[3]
[2])/(t[3][4]-t[3][2]))*((tex[15]-t[3][3])/(t[3][4]-
t[3][3]))*((tex[15]-t[3][5])/(t[3][4]-t[3][5]))
phi5_ex[15]=((tex[15]-t[3][0])/(t[3][5]-t[3][0]))*
((tex[15]-t[3][1])/(t[3][5]-t[3][1]))*((tex[15]-t[3]
[2])/(t[3][5]-t[3][2]))*((tex[15]-t[3][3])/(t[3][5]-
t[3][3]))*((tex[15]-t[3][4])/(t[3][5]-t[3][4]))
```

```
ya_calc[11]=ya[3][0]*phi0_ex[11]+ya[3][1]*phi1_ex
[11]+ya[3][2]*phi2_ex[11]+ya[3][3]*phi3_ex[11]+ya[3]
[4]*phi4_ex[11]+ya[3][5]*phi5_ex[11]
yb_calc[11]=yb[3][0]*phi0_ex[11]+yb[3][1]*phi1_ex
[11]+yb[3][2]*phi2_ex[11]+yb[3][3]*phi3_ex[11]+yb[3]
[4]*phi4_ex[11]+yb[3][5]*phi5_ex[11]
yc_calc[11]=yc[3][0]*phi0_ex[11]+yc[3][1]*phi1_ex
[11]+yc[3][2]*phi2_ex[11]+yc[3][3]*phi3_ex[11]+yc[3]
[4]*phi4_ex[11]+yc[3][5]*phi5_ex[11]

ya_calc[12]=ya[3][0]*phi0_ex[12]+ya[3][1]*phi1_ex
[12]+ya[3][2]*phi2_ex[12]+ya[3][3]*phi3_ex[12]+ya[3]
[4]*phi4_ex[12]+ya[3][5]*phi5_ex[12]
yb_calc[12]=yb[3][0]*phi0_ex[12]+yb[3][1]*phi1_ex
[12]+yb[3][2]*phi2_ex[12]+yb[3][3]*phi3_ex[12]+yb[3]
[4]*phi4_ex[12]+yb[3][5]*phi5_ex[12]
yc_calc[12]=yc[3][0]*phi0_ex[12]+yc[3][1]*phi1_ex
[12]+yc[3][2]*phi2_ex[12]+yc[3][3]*phi3_ex[12]+yc[3]
[4]*phi4_ex[12]+yc[3][5]*phi5_ex[12]

ya_calc[13]=ya[3][0]*phi0_ex[13]+ya[3][1]*phi1_ex
[13]+ya[3][2]*phi2_ex[13]+ya[3][3]*phi3_ex[13]+ya[3]
[4]*phi4_ex[13]+ya[3][5]*phi5_ex[13]
yb_calc[13]=yb[3][0]*phi0_ex[13]+yb[3][1]*phi1_ex
[13]+yb[3][2]*phi2_ex[13]+yb[3][3]*phi3_ex[13]+yb[3]
[4]*phi4_ex[13]+yb[3][5]*phi5_ex[13]
yc_calc[13]=yc[3][0]*phi0_ex[13]+yc[3][1]*phi1_ex
[13]+yc[3][2]*phi2_ex[13]+yc[3][3]*phi3_ex[13]+yc[3]
[4]*phi4_ex[13]+yc[3][5]*phi5_ex[13]

ya_calc[14]=ya[3][0]*phi0_ex[14]+ya[3][1]*phi1_ex
[14]+ya[3][2]*phi2_ex[14]+ya[3][3]*phi3_ex[14]+ya[3]
[4]*phi4_ex[14]+ya[3][5]*phi5_ex[14]
yb_calc[14]=yb[3][0]*phi0_ex[14]+yb[3][1]*phi1_ex
[14]+yb[3][2]*phi2_ex[14]+yb[3][3]*phi3_ex[14]+yb[3]
[4]*phi4_ex[14]+yb[3][5]*phi5_ex[14]
yc_calc[14]=yc[3][0]*phi0_ex[14]+yc[3][1]*phi1_ex
[14]+yc[3][2]*phi2_ex[14]+yc[3][3]*phi3_ex[14]+yc[3]
[4]*phi4_ex[14]+yc[3][5]*phi5_ex[14]

ya_calc[15]=ya[3][0]*phi0_ex[15]+ya[3][1]*phi1_ex
[15]+ya[3][2]*phi2_ex[15]+ya[3][3]*phi3_ex[15]+ya[3]
[4]*phi4_ex[15]+ya[3][5]*phi5_ex[15]
yb_calc[15]=yb[3][0]*phi0_ex[15]+yb[3][1]*phi1_ex
[15]+yb[3][2]*phi2_ex[15]+yb[3][3]*phi3_ex[15]+yb[3]
[4]*phi4_ex[15]+yb[3][5]*phi5_ex[15]
yc_calc[15]=yc[3][0]*phi0_ex[15]+yc[3][1]*phi1_ex
[15]+yc[3][2]*phi2_ex[15]+yc[3][3]*phi3_ex[15]+yc[3]
[4]*phi4_ex[15]+yc[3][5]*phi5_ex[15]
```

```
#element i 4, 0.686325 and 0.986325, ycalc
phi0_ex[16]=((tex[16]-t_end[1])/(t_end[0]-t_end[1]))
*((tex[16]-t_end[2])/(t_end[0]-t_end[2]))
phi1_ex[16]=((tex[16]-t_end[0])/(t_end[1]-t_end[0]))
*((tex[16]-t_end[2])/(t_end[1]-t_end[2]))
phi2_ex[16]=((tex[16]-t_end[0])/(t_end[2]-t_end[0]))
*((tex[16]-t_end[1])/(t_end[2]-t_end[1]))

phi0_ex[17]=((tex[17]-t_end[1])/(t_end[0]-t_end[1]))
*((tex[17]-t_end[2])/(t_end[0]-t_end[2]))
phi1_ex[17]=((tex[17]-t_end[0])/(t_end[1]-t_end[0]))
*((tex[17]-t_end[2])/(t_end[1]-t_end[2]))
phi2_ex[17]=((tex[17]-t_end[0])/(t_end[2]-t_end[0]))
*((tex[17]-t_end[1])/(t_end[2]-t_end[1]))

ya_calc[16]=ya_end[0]*phi0_ex[16]+ya_end[1]*phi1_ex
[16]+ya_end[2]*phi2_ex[16]
yb_calc[16]=yb_end[0]*phi0_ex[16]+yb_end[1]*phi1_ex
[16]+yb_end[2]*phi2_ex[16]
yc_calc[16]=yc_end[0]*phi0_ex[16]+yc_end[1]*phi1_ex
[16]+yc_end[2]*phi2_ex[16]

ya_calc[17]=ya_end[0]*phi0_ex[17]+ya_end[1]*phi1_ex
[17]+ya_end[2]*phi2_ex[17]
yb_calc[17]=yb_end[0]*phi0_ex[17]+yb_end[1]*phi1_ex
[17]+yb_end[2]*phi2_ex[17]
yc_calc[17]=yc_end[0]*phi0_ex[17]+yc_end[1]*phi1_ex
[17]+yc_end[2]*phi2_ex[17]

# objective function

helpa=sum_d((ya_calc[d]-ya_exp[d])^2)
helpb=sum_d((yb_calc[d]-yb_exp[d])^2)
helpc=sum_d((yc_calc[d]-yc_exp[d])^2)

Obj=(helpa+helpb+helpc)/(3*NDAT)
```

A1.12.1 Controlled release model

This problem was not solved with ICAS-MoT

A1.12.2 Fermentation model: ch-12-2-fermentation.mot

```
# Chapter 12.2
# Batch fermenttion model
#
```

```
#O_set=if((t<5) then (z1) else (z2))

#z2=if ((t<10) then (z3) else (4))

#z1=if((t<15) then (0.006)else (0.004))

e_O=O-O_set

delta=Kc_DO*e_O+Ki_DO*I

V_air=if((V_air_0<-delta) then (0.00001) else (V_air_0
+delta))

U_O=V_air/(A_air*3600)

#N= if((t<5) then (8.0) else (10.0))

Pow=Pn*N^3*di^5*rho

kla1=0.00495*(Pow/V)^0.593*U_O^0.94

kla=3600*kla1

kla_CO2=kla*(CO2/O)^0.5

e_pH=pH-pH_set

delta_NH3=Kc_pH*e_pH+Ki_pH*I_NH3

V_NH3=if((V_NH3_0<-delta_NH3) then (0.00001) else
(V_NH3_0+delta_NH3))


U_NH3=V_NH3/(A*3600)

kla_NH31=0.00495*(Pow/V)^0.593*U_NH3^0.94

kla_NH3=kla_NH31*3600

e_T=T-T_set

delta_T=Kc_T*e_T+Ki_T*I_T

F_w=if((F_w_0<-delta_T) then (0.00001) else (F_w_0
+delta_T))

mue=mue_max*S*O*NH3*H3PO4/((Ks+S)*(Ko+O)*(Kn+NH3)*
(Kp+H3PO4))
```

```
ms=ms1*S*O/((Ks+S)*(Ko+O))

bx=bx1*O/(Ko+O)

#O_sat=(14.161-0.3943*(T-273.15)+0.007714*(T-273.15)
^2-0.0000646*(T-273.15)^3)*0.001

M_x=V*x

qo=mue/Yxo

Qo=qo*x

qco2=mue/Yxco2

Q_co2=qco2*x

P=1

yy=(0.006*(3600*N)*(di/D_T)^2.17*(D_T/HT)^0.5)

HO=1-exp(-yy*t)

K_d=K_do*exp(-E/(R*T));

H_gr=154370*x

r_CO2=kf1*(CO2-H2CO3/K1)

r_H2CO3=kf2*(H2CO3-HCO3*H/K2)

r_HCO3=kf3*(HCO3-CO3*H/K3)

r_H3PO4=kfp1*(H3PO4-H2PO4*H/KP1)*k

r_H2PO4=kfp2*(H2PO4-HPO4*H/KP2)

r_HPO4=kfp3*(HPO4-PO4*H/KP3)

r_NH4=kf_NH4*(NH4-NH3*H/K_NH4)

r_H2SO4=kfs1*(H2SO4-HSO4*H/KS1)

r_HSO4=kfs2*(HSO4-SO4*H/KS2)

r_H2O=kfw*(1-H*OH/KW)
```

```
NH3_T=kla_NH3*(NH3_sat-NH3)

dT=(-U1*A1*(T-T_w)+V*(H_gr+H_ag-H_surr-H_evp-
H_sen))/(rho*V*C_p)

dT_w=(rho_w*F_w*C_pw*(T_w_in-T_w)+U1*A1*(T-T_w)
+U2*A2*(T_surr-T_w))/(rho_w*C_pw*V_w)

dH2CO3=r_CO2-r_H2CO3

dHCO3=r_H2CO3-r_HCO3

dCO3=r_HCO3

dH2PO4=r_H3PO4-r_H2PO4

dHPO4=r_H2PO4-r_HPO4

dPO4=r_HPO4

dNH4=-r_NH4

dHSO4=r_H2SO4-r_HSO4

dSO4=r_HSO4

dx=(mue-K_d)*x

dS=-(mue/Yxs+ms)*x

dO=kla*(O_sat-O)-Qo-ms*x*0.1-bx*x*0.1

dCO2=kla_CO2*(CO2_sat-CO2)+Q_co2-ms*x+bx*x-r_CO2

dH3PO4=(bx-mue)*x/Yxpi-r_H3PO4

dNH3=(bx-mue)*x/Yxn+kla_NH3*(NH3_sat-NH3)+r_NH4

dH2SO4=(bx-mue)*x/Yxsu-r_H2SO4

dOH=r_H2O

dH=(bx-mue)*i_px*x+r_H3PO4+r_H2PO4+r_HPO4+r_H2SO4
+r_HSO4+r_H2CO3+r_HCO3+r_H2O+r_NH4

dI=e_O
```

```
dI_NH3=e_pH

dI_T=e_T

pH=-log(H)
```

A1.12.3 Milk pasteurisation model: ch-12-3-pasteurization-ch.mot

```
# Chapter 12.3
# Pasteurization process model

# Cross sectional area of inner pipe in preheating heat
exchanger

A_pre=pi*r_pre^2

# Residence time in the preheating heat exchanger

t_pre=A_pre*l_pre/F

# Cross sectional area of inner pipe in heating heat
exchanger

A_h=pi*r_h^2

# Residence time in the heating heat exchanger

t_h=A_h*l_h/F

# Cross sectional area of pipe in holding section

A_hold=pi*r_hold^2

# Holding time

t_hold=A_hold*l_hold/F

# Flow area in shell side of preheating heat exchanger

A_c1=pi*(r_c1^2-r_pre^2)

# Residence time

t_c1=A_c1*l_pre/F

# Cross sectional area of inner pipe in cooling heat
exchanger
```

```
A_c=pi*r_c^2

# Residence time in the cooling heat exchanger

t_c=A_c*l_c/F

# For heating section.........................
b_pre=2*U_pre/(rho_milk*r_pre*Cp_milk)

a_h=F*rho_milk*Cp_milk/(F_hw*rho_w*Cp_w)

b_h=2*U_h/(rho_milk*r_h*Cp_milk)

beta_h=1-exp((a_h-1)*b_h*t_h)

c_h=a_h*beta_h/(a_h-1)

# For cooling section.........................

a_c=F*rho_milk*Cp_milk/(F_cw*rho_w*Cp_w)

b_c=2*U_c/(rho_milk*r_c*Cp_milk)

beta_c=1-exp((a_c-1)*b_c*t_c)

c_c=a_c*beta_c/(a_c-1)

# Implicit equations.........................

0=-Tpre_f+T0+b_pre*(T1-T0)*t_pre

0=-T1+Tp+b_pre*(Tpre_f-Tp)*t_c1

0=-Tp+Tpre_f+(Tpre_f-Thw_out)*beta_h/(a_h-1)

0=-Thw_out+(c_h*Tpre_f-Thw_in)/(c_h-1)

0=-Tf+T1+(T1-Tcw_out)*beta_c/(a_c-1)

0=-Tcw_out+(c_c*T1-Tcw_in)/(c_c-1)

# Explicit equationa..........................

gama_h=1-exp((a_h-1)*b_h*(t-t_pre))

gama_c=1-exp((a_c-1)*b_c*(t-t_pre-t_h-t_hold-t_c1))
```

```
Tpre=if((t<t_pre) then (T0+b_pre*(T1-T0)*t) else
(Tpre_f))

Th=if((t<t_pre) then (Tpre_f) else (W0))

W0=if((t<(t_pre+t_h)) then (Tpre_f+(Tpre_f-Thw_out)
*gama_h/(a_h-1)) else (Tp))

Tc1=if((t<(t_pre+t_h+t_hold)) then (Tp) else (W1))

W1=if((t<(t_pre+t_h+t_hold+t_c1)) then (Tp+b_pre*
(Tpre_f-Tp)*(t-t_pre-t_h-t_hold)) else (T1))

Tc=if((t<(t_pre+t_h+t_hold+t_c1)) then (T1) else (W2))

W2=if((t<(t_pre+t_h+t_hold+t_c1+t_c)) then (T1+(T1-
Tcw_out)*gama_c/(a_c-1)) else (Tf))

#Main stream (milk) temperature

T=if((t<t_pre) then (Tpre) else (Z0))

Z0=if((t<(t_pre+t_h)) then (Th) else (Z1))

Z1=if((t<(t_pre+t_h+t_hold)) then (Tp) else (Z2))

Z2=if((t<(t_pre+t_h+t_hold+t_c1)) then (Tc1) else (Z3))

Z3=if((t<(t_pre+t_h+t_hold+t_c1+t_c)) then (Tc) else
(Tf))

#Death kinetic constant of the microorganism

K_d=K_do*exp(-E/(R*(T+273.15)));

#Number of alived microorganism

N= N_0*exp(x)

dx=if((T<T_ster) then (0) else (-K_d+x-x;));

#Sterility ratio

r=log(N_0/N)
```

A1.12.4 Grinding model: ch-12-4-granulation.mot

```
# Chapter 12-4
# Granulation process model

# Number of initial particles
N_0=6*M_0/(pi*d_0^3*rho_s)

# Coagulation half time
tau=2/(K*N_0)

#Number of particles at time t
r=1/(1+t_gr/tau)

N=N_0*r

#Moisture control
error_X_w=X_w_set-X_w

delta_X_w=Kc_X_w*error_X_w+Ki_X_w*I_X_w

#F_b=if((F_b_0<-delta_X_w) then (0.00001) else (F_b_0
+delta_X_w))

# Weight of water in binder
F_bw=X_bw*F_b

#Amount of binder added in time t
M_b=(1-X_bw)*F_b*t_gr

# Total weight of bed
M_T=M_b+M_w+M_0

#Temperature in bed is defined as follows
T=H/(M_T*Cp_T)

#Temperature control

error_T=T_set-T

delta_T=Kc_T*error_T+Ki_T*I_T

fa_in=if((fa_in_0<-delta_T) then (0.00001) else
(fa_in_0+delta_T))

#Inlet mass flow rate of air
F_a_in=fa_in*rho_g
```

```
#Outlet mass flow rate of air
F_a_out=fa_out*rho_g

#Heat lose to surrounding
Q_en=U*Ar*(T-T_en)

#Enthalpies
h_b=F_b*Cp_b*(T_b-T)

h_a_out=F_a_out*Cp_a_out*(T_a_out-T)

h_a_in=F_a_in*Cp_a_in*(T_a_in-T)

#Saturation vapor pressure
P_sat=exp(A-B/(T+273.15+C))

#Moisture content
X_w=M_w/M_T
#0=-X_w+0.04*10*fi*(1-9*fi^8+8*fi^9)/((1-fi)*(1
+9*fi-10*fi^9))

#Equilibrium pressure
P_eq=P_sat*fi

#Molar mass of gas
M_g_mol=M_g/MW_g

#Molar mass of moisture
M_w_mol=M_gw/MW_w

#Equilibrium moisture content
Y_eq=MW_w*P_eq/(MW_g*(P-P_eq))

# Weight fraction of binder
X_b=M_b/M_T

#Granules density
rho_p=rho_s*rho_b*rho_w/(rho_b*rho_w*(1-X_w-X_b)
+rho_s*rho_w*X_b+rho_s*rho_b*X_w)

#Granules volume
V_p=M_T/(N*rho_p)

#Granules diameter
d_p=6*V_p^(1/3)/pi

#Granule size control
error_d_p=d_p-d_p_set
```

```
xx=if((F_b_0<-delta_X_w) then (0.00001) else (F_b_0
+delta_X_w))

F_b=if((error_d_p<0) then (xx) else (xx*(1-K_switch)))

#Total granules surface area
A_bed=pi*d_p^2*N

#Moisture content in air
#Y_s=F_ev/(F_ev+F_a_in-F_a_out)
Y_s=M_gw/M_g

# Rate of evaporation
F_ev=if(((Y_eq-Y_s)<0) then (0) else
(rho_g*beta_pg*A_bed*(Y_eq-Y_s)*nue_dot))

#Differential equations
dM_w=F_bw-F_ev

dM_gw=F_ev-F_a_out*Y_s+F_a_in*Y_w_in

dM_g=F_a_in-F_a_out+F_ev

dH=h_a_in-h_a_out+h_b+F_ev*deltaH-Q_en

dt_gr=if((error_d_p<0) then (1) else (1-K_switch))
+t_gr-t_gr

dI_X_w=error_X_w
dI_T=error_T
```

A1.12.5 Milling model: ch-12-5-milling.mot

```
# Chapter 12-5
#Milling machine model

#Weight of solid in feed

W_s=y_s*W

# Mass of one particle in size interval K

m_p[k]=rho*Kv*(x[k])^3

m_p_fine=rho*Kv*(x_fine)^3

# Mass of particles in size interval K
```

```
m[k]=W[k]*W_s

# Weight fraction of fine particles

W_fine=1-sum_k(W[k])

# Weight of fine particles

m_fine=W_fine*W_s

# Number of particles in size interval k

N[k]=m[k]/m_p[k]

# Number of fine particles

N_fine=m_fine/m_p_fine

#Total number of particles

N_total=sum_k(N[k])+N_fine

#Average size of particles

L_avg=(sum_k(N[k]*x[k])+N_fine*x_fine)/N_total

# On-off controller

Error_L_avg=L_avg-L_avg_set

N=if((Error_L_avg>0) then (N_0) else (xx))
xx=(1-K_switch)*N_0

# Critical rotational speed

N_c=54.19/(R^0.5)

#Critical velocity fraction

fi_c=N/N_c

# Correction factor

a=k*fi_c/(1+exp(15.7*(fi_c-0.95)))

#Specific rate of breakage
```

```
S[k]=a*S*(x[k]/x[1])^alpha/(1+x[k]/x_m)^beta

# Weight fraction of particles in size interval k

Q[10]=sum_k(S[k]*W[k])-S[10]*W[10]

Q[9]=Q[10]-S[9]*W[9]

Q[8]=Q[9]-S[8]*W[8]

Q[7]=Q[8]-S[7]*W[7]

Q[6]=Q[7]-S[6]*W[6]

Q[5]=Q[6]-S[5]*W[5]

Q[4]=Q[5]-S[4]*W[4]

Q[3]=Q[4]-S[3]*W[3]

Q[2]=Q[3]-S[2]*W[2]

Q[1]=Q[2]-S[1]*W[1]

dW[k]=-S[k]*W[k]+a*b*Q[k]
```

A1.12.6 Tablet pressing model: ch-12-6-tablet-press.mot

```
# Chapter 12.6
# Tablet press
# Volume of tablet

Ar=(3.14*d^2)/4

V_pre=Ar*L_pre

#Random Noise terms

n1=0.0009*(t+t^2-2.5*t^3+t^4-sin(t*10))

#Deviation from set point (compression pressure)
#Tablet weight control through the controlling of pre-
compression pressure

a=a_0+n1

V_0_set_M=M_set/((1-a)*rho)
```

```
x_set_M=b*(V_0_set_M*(a-1)+V_pre)

P_set_M=(V_0_set_M-V_pre)/x_set_M

F_set_M=P_set_M*Ar*10^6

error_P_M=P_pre-P_set_M

#Tablet hardness control through the controlling of main
compression pressure

error_P_H=H-H_set

# Actuator for controlling the weight

delta=Kc_P*error_P_M+Ki_P*I

V_m=if((V_m_0<-delta+V_pre) then (V_pre) else (V_m_0
+delta))

# Actuator for controlling the hardness

delta_H=Kc_P_H*error_P_H+Ki_P_H*I_H

#V_m_H=if((V_m_0<-delta_H+V) then (V) else (V_m_0
+delta_H))

L=L_0+delta_H

V=Ar*L

#Random Noise terms

n=0.000000003*(t-t^2+2*t^3-t^4+sin(t*10))

#Initial volume of powder with noise

V_0=if((V_m<V/(1-a)) then (V_m+n) else (0.98*V/(1-a)))

# Depth of fill

L_depth=V_0/Ar

#punch displacement

L_punch_disp=L_depth-L
```

```
#Dual time
t_dual=L_punch_disp/u
# Pre-compression pressure
x_pre=b*(V_0*(a-1)+V_pre)
P_pre=(V_0-V_pre)/x_pre
F_pre=P_pre*Ar*10^6
# Main compression pressure
a_main=1-(V_0/V_pre)*(1-a)
x_main=b*(V_pre*(a_main-1)+V)
P_main=(V_pre-V)/x_main
F_main= P_main*Ar*10^6
# Weight of tablet
M=(1-a)*V_0*rho
#Tablet hardness
V_s=(1-a)*V_0
rho_r=V_s/V
x=ln((1-rho_r)/(1-rho_r_c))
H=H_max*(1-exp(rho_r-rho_r_c+x))
# Total number of tablet produced in time t
dTab_no=180000+Tab_no-Tab_no
#Integral error
dI=error_P_M
dI_H=error_P_H
```

Model Library

In this appendix, a list of models (see Table A2-1) that can be downloaded from the book website (http://www.elsevierdirect.com/companion.jsp?ISBN=9780444531612) is given.

Model	Type	Short Description	Reference	File Name
CPA equation of state	AE-implicit	The CPA *equation of state* is a *constitutive property* model relating pressure–volume–temperature of the pure compound and/or mixture. Fugacity coefficients can be derived from this equation of state for use in VLE calculations involving non-ideal mixtures. Version developed by Kontogeorgis et al. (1999).	Gani et al. (2006)	cpa-eos.mot
PC-SAFT equation of state	AE-implicit	The PC-SAFT equation of state is a constitutive property model relating pressure–volume–temperature of the pure compound and/or mixture. Fugacity coefficients can be derived from this equation of state for use in VLE calculations involving non-ideal mixtures. Version developed by Gross and Sadowski (2002).	Gani et al. (2006)	pc-saft-eos.mot
Original UNIFAC VLE model	AE-explicit	This is a constitutive property model based on the group contribution concept. The "original" and "VLE" terms in the model name refer to the parameter set corresponding to this model. The model provides activity coefficients of compounds in the liquid phase. Version developed by Fredenslund et al. (1977).	Gani et al. (2006)	unifac-mot file
NRTL model	AE explicit	This is a constitutive property model based on the group contribution concept. The model provides activity coefficients of compounds in the liquid phase. Version developed by Renon and Prausnitz (1968).	CAPEC MoT model library	nrtl.mot

Elec-UNIQUAC model	AE-explicit	This is a constitutive property model. It is applicable to aqueous electrolyte systems and provides the activity coefficients of the ions in the liquid phase. Version developed by Thomsen et al. (1996)	Gani et al. (2006)	elec-UNIQUAC.mot
Two-phase flash	AE-implicit	This is a process (steady-state) model for a 2-phase VLE flash operation. The VLE is assumed to be ideal. Mass and energy balance are considered.	ICAS-MoT model library	ss-flash-ideal.mot
Two-phase flash	DAE	This is a process (dynamic) model for a two-phase VLE flash operation. The VLE is assumed to be ideal. Mass and energy balance are considered.	ICAS-MoT model library	dynamic-flasf-ideal.mot
Bioreactor (anaerobic oxidation)	DAE	This is a model for the anaerobic oxidation of bimass to gases (for example methane). The kinetic model parameters are given. Variations of this model can also be used to regress the kinetic model parameters.	M. Sales-Cruz (2006)	bioreactor-model.mot
Two-phase reactor model	DAE	This is a process model for a two-phase (vapour-liquid) reactor. The reaction takes place in the reactor. The reaction rate model needs to be specified (or changed from the one given). Physical equilibrium is considered for the vapour phase that is in equilibrium with the reacting liquid.	ICAS-MoT model library	dynamic-2phase-react.mot
Emulsion polymerisation reactor	DAE	Dynamic model for emulsion coplymerisation reactor model of Dimitratos et al. (1989). Test system. Styrene/methyl methacrylate.	M. Sales-Cruz (2006)	dynamic-cstr-poly.mot

(continued)

(continued)

Model	Type	Short Description	Reference	File Name
Solution copolymerisation reactor Dyn_Polym_State1.mot	AE - implicit	Solution copolymerisation model of Richards and Congalides (1989).	M. Sales-Cruz (2006)	
Bifurcation analysis in polymerisation reactor	DAE	Bifuraction analysis on a polymerisation reactor	M. Sales-Cruz (2006)	f.poly-bifurcation.mot
Membrane distillation	AE implicit	This is a membrane distillation model for the separation of a binary mixture. The specified data is for ethanol-water separation.		mbd-ideal.mot
TE-control	DAE	This model allows the simulation of the Tennessee Eastman process in closed loop (with controllers and a specified control scheme). Ricker model	M. Sales-Cruz (2006)	TE_CONTROL.mot
TE-steady state	AE-implicit	This model allows the steady-state simulation of the Tennessee Eastman process. Ricker model (Ricker and Lee, 1995)	M. Sales-Cruz (2006)	TE_SS_SOL.mot
Crsytallisation	PDAE	1-D method of class solution technique 1d_classes_linearcool_independent.mot	Samad et al. (2011)	
Crystallisation	PDAE	2-D method of class solution technique 2d_classes_linearcool_independent.mot	Samad et al. (2011)	
Crystallisation	PDAE	2-D method of method of moments solution technique 2d_moment_linearcool_independent.mot	Samad et al. (2011)	

REFERENCES

Dimitratos, J., Georgiakis, C., El-Asser, S., Klein, A., 1989. Computers and Chemical Engineering. 13, 21–33.

Fredenslund, Aa., Rasmussen, P., Gmehling, J., 1977. Vapor-Liquid Equilibria Using UNIFAC. Elsevier Scientific Publishing Company.

Gani, R., Muro-Suñé, N., Sales-Cruz, M., Leibovici, C., O'Connell, J.P., 2006. Fluid Phase Equilibria. 250, 1–32.

Gross, J., Sadowski, G., 2002. Industrial and Engineering Chemistry Research. 41, 1084–1093.

Kontogeorgis, G.M., Yakoumis, I.V., Meijer, H., Hendriks, E.M., 1999. Fluid Phase Equilibria. 158-160, 201–211.

Renon, H., Prausnitz, J.M., 1968. AIChE J. 14, 135–144.

Richards, J.R., Congalides, J.P., 1989. J Applied Polymer Science. 27, 2727–2756.

Ricker, N.L., Lee, J.H., 1995. Computers and Chemical Engineering. 19, 983–1005.

M. Sales-Cruz, "Development of a computer aided modelling system for bio and chemical process and product design, PhD-thesis, Technical University of Denmark, 2006, lyngby, Denmark.

Thomsen, K., Rasmussen, P., Gani, R., 1996. Chemical Engineering Science. 51, 3675–3683.

Index

E

F

N

O

P

R

Acronyms

Product and Process Modelling: A Case Study Approach

Acronyms

AEs	Algebraic equations
AI	Active ingredient
AIBN	Azo-iso-butyronitrile
ANN	Artificial neural network
API	Active pharmaceutical ingredient
AR	Auto-regressive
ARMAX	Auto-regressive, moving average exogenous
ATP	Alberta-Taciuk Processor
B	Benzene
BDF	Backward differentiation formula
BJ	Box-Jenkins
BUD	Balance volume diagram
CAPEC	Computer Aided Product-Process Engineering Center
CFD	Computational fluid dynamics
COM	Component object model
CPA	Cubic plus association
CPM	Critical path method
CSD	Crystal size distribution
CSTR	Continuous stirred tank reactor
DAEs	Differential algebraic equations
DAP	Diammonium phosphate
DCB	Dichlorobenzene
DCS	Distributed control system
DEE	Diethyl ether
DEM	Discrete element method
DIRK	Diagonally implicit Runge-Kutta
DMFC	Direct methanol fuel cell
DTU	Danmarks Tekniske Universitet
EA	Ethanol
EOS	Equations of state
FEM	Finite element method
FSA	Finite state automata
FSM	Finite state machine

ICAS	Integrated computer aided system
IPODAEs	Integral partial ordinary differential algebraic equations
LHS	Left hand side
LLE	Liquid liquid equilibrium
LP	Linear programming
MAP	Monoammonium phosphate
MCB	Monochlorobenzene
MILP	Mixed integer linear programming
MINLP	Mixed integer non-linear programming
MMA	Methyl methacrylate
ModDev	Model development software
MoT	Modelling tool-box
MS	Multi-scale
MS	Microsoft
NLP	Nonlinear programming
NRTL	Non-random two-liquid
ODEs	Ordinary differential equations
PAT	Process analytical technology
PC-SAFT	Perturbed chain statistical associating fluid theory
PDAEs	Partial differential algebraic equations
PDEs	Partial differential equations
PEM	Proton electrolyte membrane
PFR	Plug flow reactor
PLC	Programmed logic controllers
PODAEs	Partial ordinary differential algebraic equations
PODEs	Partial ordinary differential equations
PSD	Particle size distribution
PSE	Process System Enterprise
PSE	Process system engineering
PVT	Pressue, volume, temperature
RBM	Risk based maintenance
RHS	Right hand side
RM	Risk management
RPN	Reverse polish notation
SCOR	Supply chain operations reference
SLE	Solid liquid equilibrium
SRK	Soave-Redlich-Kwong
SS	Single-scale
SSCF	Simultaneous saccharification and co-fermentation
UNIFAC	Universal functional activity coefficient
UNIQUAC	Universal quasi-chemical
VLE	Vapour liquid equilibrium
VPPDL	Virtual product process design lab
W	Water
XML	Extensible markup language

Printed and bound by CPI Group (UK) Ltd, Croydon, CR0 4YY
08/05/2025
01864812-0001